Failure Analysis of Microbiologically Influenced Corrosion

Microbes, Materials, and the Engineered Environment

Series Editors

Torben Lund Skovhus and Richard B. Eckert

Failure Analysis of Microbiologically Influenced Corrosion

Edited by Richard B. Eckert and Torben Lund Skovhus

For more information about this series, please visit: www.routledge.com

Failure Analysis of Microbiologically Influenced Corrosion

Edited by

Richard B. Eckert and Torben Lund Skovhus

CRC Press is an imprint of the
Taylor & Francis Group, an **informa** business

First edition published 2022
by CRC Press
6000 Broken Sound Parkway NW, Suite 300, Boca Raton, FL 33487-2742

and by CRC Press
2 Park Square, Milton Park, Abingdon, Oxon, OX14 4RN

CRC Press is an imprint of Taylor & Francis Group, LLC

ISBN: 978-0-367-35680-4 (hbk)
ISBN: 978-1-032-12232-8 (pbk)
ISBN: 978-0-429-35547-9 (ebk)

DOI: 10.1201/9780429355479

Typeset in Times
by MPS Limited, Dehradun

Contents

PART II MIC Failure Analysis Case Studies

Industry Foreword

During my 33-year professional career as a petroleum microbiologist for ConocoPhillips and now as an industry consultant, I have had the opportunity to be involved in some very high-profile MIC failures. These included the first reported MIC failure in the Norwegian sector of the North Sea of a production line in 1997, resulting in the shut in of nearly 200,000 BOPD, and the MIC failure of the BP Alaska oil transit line at Prudhoe Bay in 2006 resulting in the leakage of over 4700 bbl. of oil to the tundra. I have seen significant progress in the general awareness of MIC in the oil and gas industry during this time, but slower progress in terms of adoption of new microbiological technologies and practices when it comes to failure analyses. Early in my career, a senior corrosion engineer with Phillips Petroleum Company commented to me that he never really believed microorganisms could cause corrosion but admitted there were failures he could not explain by other causes or known mechanisms. Unfortunately, this thinking has permeated throughout the oil and gas industry and led to the practice of implicating microorganisms as the cause of failure only when abiotic mechanisms have been eliminated.

It is encouraging, however, that over the past 15 years significant support from industry has been garnered by several MIC research programs. These include the NSERC-sponsored Petroleum Microbiology Group at the University of Calgary and the Biocorrosion Center activities at the University of Oklahoma, as well as an ongoing Genome Canada–sponsored large-scale applied research project on "Managing Microbial Corrosion in Canadian Offshore and Onshore Production Operations" spearheaded by several Canadian universities. It is a goal of the latter program to develop and disseminate new guidelines for MIC failure analyses and it is notable that many of the authors of this book have been associated with at least one of the MIC programs mentioned. I have also known the editors of this book for many years and know them to be knowledgeable and passionate about improving the way industry conducts MIC failure analyses, and know they have been intimately involved in updating and improving industry standards in the areas of testing, detection, diagnosis, and evaluation of MIC for the oil and gas industries.

Too many times in my career I was presented with bead blasted pipe, months-old pipe wrapped in filthy towels, or dirty containers filled with brine and asked to determine if MIC was involved. Current practitioners of MIC failure analyses, however, should be aware of the importance of good specimen preservation and obtaining information about the chemical, metallurgical, operational, and microbiology of the failure. For those wanting more detailed information, Part 1 of this book covers the general approach to conducting an MIC failure analyses as well as a review of MIC mechanisms. The recent revelation that some microorganisms, including sulfate reducing bacteria and methanogens, can directly uptake electrons from Fe^0 has transformed our thinking about MIC mechanisms. In addition, some of the best evidence for MIC can be obtained by analyzing mineralogical phases of corrosion products and determining whether they could have formed biotically which necessitates proper sample preservation by limiting exposure of pipe to oxygen and desiccation. The same can be said with water chemistry, as properly preserved and analyzed samples are critical to confirming the mineralogical phases and identifying key nutrients that could provide clues to microbial activity. Included in Part 4 are chapters on sampling and preservation of metallurgical samples as well as processes and protocols for collecting and analyzing chemical samples that should be useful.

An important aspect, however, of any failure analysis involving microorganisms is the confirmation of their presence, physiological types, and activity. It is simply not possible to conclude the involvement of microorganisms based solely on pit morphologies and metallurgical features. Unfortunately, it is my experience that detection and identification of microorganisms are often the most neglected or inadequately performed analyses by many commercial labs. Too much attention is given to involvement of sulfate reducing bacteria and acid producing bacteria as detected by traditional culture-based methods. Reasons for this include inadequate microbiological expertise and perceived expense. Since the presence of microorganisms identified in the brine or aqueous phase cannot be necessarily correlated with the corrosive biofilm, it is imperative that the corrosion products, pits, and areas surrounding pits be interrogated for microbial types. Even if samples received by a lab are days to weeks old and dried out, analysis of DNA to identify microbial types can still be successful, whereas culture-based testing becomes a futile exercise. Although molecular-based tools have limitations and biases, they still provide the best current means to assess involvement of microorganisms, especially if fresh samples are unavailable. Additionally, the cost of these analyses continues to decline. Chapters in Part 4 of this book should be especially useful in explaining how to take MIC samples in the field, preserve them, and analyze them to get useful information on microorganisms.

It is also useful to discern during a failure analysis if there are similar case studies where microorganisms have been implicated in the failure of a similar material or process. This information is often used to recommend new materials or alloys as replacements or new construction. Therefore, knowing the failed material was within the chemical and metallurgical specifications when it was installed is valuable information. Microorganisms can create environmental conditions under which microstructural anomalies formed during the manufacturing process become preferred areas of localized corrosion. The history of one such case occurs in a high-pressure seawater line in Part 2 of this book. Other case studies provide useful analogues for conducting MIC failure analyses, while Part 3 contains useful accounts of MIC failures in engineered systems found in some oil and gas facilities.

Failures due to MIC can be consequential and very costly to industry. The total cost of corrosion alone for the oil and gas industry has been estimated at over $1 billion USD annually of which MIC is estimated to account for up to 20% of all corrosion. A recent catastrophic MIC failure of well casings in Aliso Canyon, CA, was attributed to attack by methanogens and resulted in release of over 100,000 tons of methane at a total cost of over $1 billion USD to the operator in terms of repairs, litigation, and fines. I am confident this book will be of value to metallurgists, corrosion engineers, chemists, and microbiologists working in any industry to assess the involvement of microorganisms in corrosion and in the selection of materials for replacement of old construction or installation of new materials.

Gary Jenneman

Preface

This book contains the latest technical and scientific contributions to the field of failure analysis with the inclusion of microbiologically influenced corrosion (MIC) for engineered systems. In 26 chapters, over 60 authors have contributed by sharing their industry and academic perspectives on failure analysis and MIC through extensive investigations of insightful case studies for others to learn from. As editors we are grateful for all of these high-quality contributions and the hard work provided by all the authors.

Investigating the role of microorganisms in corrosion damage and failures is a challenging and interesting topic, and there has been no shortage of controversy over the years as to the best way to diagnose MIC. However, nearly all can agree that corrosion is caused by a combination of factors, including physical, chemical, and microbiological conditions, design and metallurgy, and operational factors; thus, corrosion needs to be diagnosed based upon integration and understanding of all these factors. A further challenge to MIC diagnosis is the transdisciplinary nature of the topic, encompassing chemical engineering, physics, microbiology and bioanalytics, metallurgy, asset operation, maintenance and inspection, and other technical fields that each have their own history, ontology, and worldviews. It has become clear in recent years that in order to substantially advance MIC diagnosis and management capabilities, focused and combined efforts are needed to bridge the barriers to transdisciplinary collaboration. The preparation of this book is seen by the editors as at least a small step in that direction, bringing together a variety of experts from different fields to share their experience with MIC diagnosis.

This book is broken down into four main parts. Part 1 helps to set the stage for the case studies to follow in subsequent sections by reviewing the overall approach to MIC failure analysis and applicable analytical methods based on current understanding, a review of MIC corrosion mechanisms, an excellent review of the state of MIC diagnosis in Alberta's oil and gas sector, and new insights on the important role of methanogens in MIC.

Part 2 provides a deep dive into nine unique case studies of MIC failure analysis in different parts of the oil ang gas industry where MIC is a ubiquitous problem leading to premature failures and reductions in the useable life of assets, such as tanks, pipelines, and offshore oil production equipment. Each case study integrates information about the chemical and microbiological conditions present, materials properties, and operational conditions to help determine the extent to which the failure mechanism resulted from the presence and activities of microorganisms. One case describes an abiotic corrosion mechanism that occurred in a pipeline with abundant and diverse microorganisms; a reminder that the mere presence of bacteria, archaea, or fungi cannot be the sole the basis for identifying MIC since microorganisms can be found nearly everywhere.

Part 3 of the book provides MIC failure analysis case studies from other types of assets, such as mooring chains for large offshore vessels and fire protection system piping in buildings; both being environments that are abundant in both microorganisms and MIC.

Part 4, MIC Failure Analysis Processes and Protocols, contains a wealth of practical information in a concise format for those seeking to improve their processes for MIC assessments and MIC failure investigations. In this section, readers will find guidance for root cause analysis (RCA) of corrosion failures and several chapters on different aspects of sample collection and preservation, including insights on best practices for sample handling, the types of samples to collect for chemical and microbiological analysis, and strategies for performing assessments of MIC in pipeline systems. Part 4 also includes forms developed to help collect vital sample background data and checklists for collecting different types of liquid and solid samples that were developed as part of the geno-MIC project (Managing Microbial Corrosion in Canadian Offshore and Onshore Oil Production Operations), a Large Scale Applied Research Project funded by Genome Canada from 2017 to 2021, and a number of academic and industry partners. This part also includes a chapter on the role of standards for MIC management in engineered systems that identifies various standards that exist today or are currently under development and will soon be published.

Readers may apply the information and protocols in this text in a number of different ways, including referring to individual case studies that are similar to an existing issue they are facing, developing their own company guidelines for sampling, preservation, testing and failure analysis, or finding a broader understanding of MIC mechanisms and how they can be characterized. The application of molecular microbiological methods (MMM) is discussed in many of the case studies in this book, along with the ways to integrate data from MMM with other information about chemistry, electrochemistry, and metallurgy when determining the role of MIC. Although the case studies may cover technical concepts or methods that are new to the reader, the cases are above all meant to be practical to help readers apply and interpret new methods to promote more reliable diagnosis of MIC.

The editors wish to thank all of the authors and reviewers for their valuable time and commitment to preparing, editing, and finalizing their chapters to bring this unique edition together. Many of them are regarded as experts in their field; engineers, scientists, researchers, and educators whose schedules are always beyond full, but who were kind enough to make time for their important contributions. Finally, while the planning of this book by the editors in 2019 did not include a global pandemic, our authors and co-authors persevered and delivered regardless of the unprecedented stresses on families and society.

We hope this book helps improve the state-of-the-art in failure analysis of MIC, and further opens up avenues for transdisciplinary communication and collaboration to advance our understanding of MIC and the means to manage this threat in the years to come.

Richard B. Eckert, DNV GL USA, Inc
Torben Lund Skovhus, VIA University College

Editors

Richard B. Eckerthas been involved in pipeline corrosion/failure investigation and forensic corrosion engineering for over 40 years. He is a recognized expert in the field of microbiologically influenced corrosion (MIC), having published more than 50 symposium and peer-reviewed papers and three books on internal corrosion.

Mr. Eckert earned a BS in engineering metallurgy at Western Michigan University. His career has included 22 years in corrosion and failure analysis roles for a natural gas gathering, transmission and storage company, and for an oil major in upstream crude production and transportation in Alaska. The balance of his career has been in technical advisory roles for internal corrosion mitigation and management, corrosion failure investigation, and particularly, on microbiologically influenced corrosion (MIC).

His present work at DNV focuses on MIC diagnosis and mitigation using state-of-the-art molecular methods and applied research. Mr. Eckert provides corrosion and materials support for projects in close collaboration with DNV Laboratories, Integrity, Research, Forensic, Risk Management, Litigation, and other groups. He also provides litigation support and expert opinions in the area of MIC.

Mr. Eckert chaired the development of NACE Standard TM0212 on internal MIC management for pipelines and chaired/contributed to many other industry consensus standards. He served on the NACE Board of Directors for three years as a publications director and chaired an ISO Corrosion Management committee.

Mr. Eckert has received numerous awards including most notably the NACE Presidential Achievement Award, NACE Distinguished Service Award, NACE T.J. Hull Award, and the AUSC Colonel George C. Cox Award.

Mr. Eckert has been an internal corrosion instructor and MIC workshop speaker for SPE, NACE, Energy Institute, government agencies, and major oil and gas companies. He is a NACE-certified Internal Corrosion Specialist.

Mr. Eckert contributed to and managed corrosion research projects for PRCI, GRI, SwRI, and others, including development of a method for microscopic characterization of corrosion initiation by microorganisms that is widely used today and included in industry standards. Most recently, Mr. Eckert was a contributor to the development of new recommended guidelines, models, and tools for MIC control and failure analysis as part of the Genome Canada Large Scale Applied Research Project (LSARP), Managing Microbial Corrosion in Canadian Offshore and Onshore Oil Production.

Torben Lund Skovhus, PhD, is a docent and project manager at VIA University College in the Centre of Applied Research and Development in Building, Energy and Environment (Horsens, Denmark). He earned a master's degree (cand.scient.) in biology and a PhD in microbiology at Aarhus University, Denmark. In 2005, Dr. Torben was employed at the Danish Technological Institute (DTI) in the Centre for Chemistry and Water Technology, where he was responsible for the consultancy activities for the oil and gas industry around the North Sea. Dr. Torben headed the

DTI Microbiology Laboratory while developing several consultancy and business activities with the oil and gas industry. He founded DTI Oil and Gas in both Denmark and Norway, where he was team and business development leader for five years. Thereafter Dr. Torben worked as a project manager at DNV GL (Det Norske Veritas) in the field of corrosion management in both Bergen and Esbjerg. He is currently chair of NACE/AMPP TEG286X, NACE/AMPP SC-22 on Biodeterioration and ISMOS TSC, an organization he cofounded in 2006. He is an international scientific reviewer and the author of 80+ technical and scientific papers and book chapters related to industrial microbiology, applied biotechnology, corrosion management, oilfield microbiology, water treatment and safety, reservoir souring, and biocorrosion. He is a scientific/technical reviewer with over 20 international journals in the same fields. He is coeditor of *Applied Microbiology and Molecular Biology in Oilfield Systems* (Springer, 2011); *3rd International Symposium on Applied Microbiology and Molecular Biology in Oil Systems* (Elsevier, 2013); *Applications of Molecular Microbiological Methods* (Caister Academic Press, 2014); *Microbiologically Influenced Corrosion in the Upstream Oil and Gas Industry* (CRC Press, 2017); and *Oilfield Microbiology* (CRC Press, 2019). Dr. Torben was honored with the NACE Technical Achievement Award in 2020 for outstanding research on MIC in the energy sector.

Contributors

A.A. Abilio
University of Alberta
Edmonton, Alberta, Canada

Farah Al-Tabbakh
Kuwait Oil Company
Ahmadi, Kuwait

Annie Biwen An
Bundesanstalt für Materialforschung und -prüfung (BAM)
Berlin, Germany

Recep Avci
Department of Physics
Montana State University
Bozeman, Montana, USA
and
The Biocorrosion Center
University of Oklahoma
Norman, Oklahoma, USA

Mahsan Basafa
Memorial University of Newfoundland
St. John's, Newfoundland, Canada

Øystein Bjaanes
DNV GL
Høvik, Norway

Caitlin L. Bojanowski
Soft Matter Materials Branch
Materials and Manufacturing Directorate
Air Force Research Laboratory
Wright-Patterson Air Force Base
Dayton, Ohio, USA

Christina S. Bottaro
Memorial University of Newfoundland
St. John's, Newfoundland, Canada

Treva T. Brown
Naval Research Laboratory
NASA Stennis Space Center
Bay Saint Louis, Mississippi, USA

Katherine M. Buckingham
DNV GL
Columbus, Ohio, USA

K. Crippen
GTI Energy
Des Plaines, Illinois, USA

A. Darwin
Woodside Energy Ltd.
Perth, Australia

Kathleen Duncan
Department of Botany and Microbiology
Institute for Energy and the Environment
University of Oklahoma
Norman, Oklahoma, USA

Richard B. Eckert
DNV GL
Commerce Township, Michigan, USA

James G. Floyd
Department of Microbiology and Plant Biology
University of Oklahoma
Norman, Oklahoma, USA

Nuno Fragoso
Petroleum Microbiology Research Group
University of Calgary
Calgary, Alberta, Canada

Lisa M. Gieg
Petroleum Microbiology Research Group
University of Calgary
Calgary, Alberta, Canada

Wendy J. Goodson
Soft Matter Materials Branch
Materials and Manufacturing Directorate
Air Force Research Laboratory
Wright-Patterson Air Force Base
Dayton, Ohio, USA

Tesfa Haile
InnoTech Alberta
Edmonton, Alberta, Canada

David Hampton
ConocoPhillips Company
Bartlesville, Oklahoma, USA

A. Harmon
GTI Energy
Des Plaines, Illinois, USA

Kelly A. Hawboldt
Memorial University of Newfoundland
St. John's, Newfoundland, Canada

Abdulhaqq Ibrahim
Memorial University of Newfoundland
St. John's, Newfoundland, Canada

Thomas R. Jack
Department of Biological Science
University of Calgary
Calgary, Alberta, Canada

Akhil Jaithlya
Kuwait Oil Company
Ahmadi, Kuwait

Amer Jarragh
Kuwait Oil Company
Ahmadi, Kuwait

Gary Jenneman
GJ Microbial Consulting, LLC
Bartlesville Oklahoma, USA
and
The Biocorrosion Center
University of Oklahoma
Norman, Oklahoma, USA

R. Johnsen
Norwegian University of Science and Technology
Trondheim, Norway

B. Kinsella
Curtin Corrosion Centre
Western Australia School of Mines
Minerals, Energy and Chemical Engineering
Curtin University
Bentley, Australia

Sherin Kleinbub
Bundesanstalt für Materialforschung und -prüfung (BAM)
Berlin, Germany

J.W. Klijnstra
Endures B.V.
Den Helder, The Netherlands

Andrea Koerdt
Bundesanstalt für Materialforschung und -prüfung (BAM)
Berlin, Germany

Susmitha Purnima Kotu
DNV GL
Columbus, Ohio, USA

Sandip Anantrao Kuthe
Kuwait Oil Company
Ahmadi, Kuwait

Jason S. Lee
Naval Research Laboratory
USA

Tim Lee
AMOG Pty. Ltd.
Melbourne, Australia

S. Leleika
GTI Energy
Des Plaines, Illinois, USA

Vitor Liduino
Federal University of Rio de Janeiro
Rio de Janeiro, Brazil

Bjarte Lillebø
DNV GL
Høvik, Norway

Jo-Inge Lilleengen
DNV GL
Høvik, Norway

Brenda J. Little
B.J. Little Corrosion Consulting, LLC
Diamondhead, Mississippi, USA

Márcia Lutterbach
National Institute of Technology
Rio de Janeiro, Brazil

L.L. Machuca
Curtin Corrosion Centre
Western Australia School of Mines
Minerals, Energy and Chemical Engineering
Curtin University
Bentley, Australia

Robert E. Melchers
Centre for Infrastructure Performance and Reliability
University of Newcastle
Callaghan, Australia

Ali Modir
Department of Chemistry
Memorial University of Newfoundland
St. John's, Newfoundland, Canada

Israa Mohammad
Kuwait Oil Company
Ahmadi, Kuwait

N. Noël-Hermes
Endures B.V.
Den Helder, The Netherlands

H. Parow
Norwegian University of Science and Technology
Trondheim, Norway

João Payão-Filho
Federal University of Rio de Janeiro
Rio de Janeiro, Brazil

Trevor Place
Enbridge
Edmonton, Alberta, Canada

T. Pojtanabuntoeng
Curtin Corrosion Centre
Western Australia School of Mines
Minerals, Energy and Chemical Engineering
Curtin University
Bentley, Australia

Natalie M. Rachel
Petroleum Microbiology Research Group
University of Calgary
Calgary, Alberta, Canada

Angham Saeed
Memorial University of Newfoundland
St. John's, Newfoundland, Canada

S. Salgar-Chaparro
Curtin Corrosion Centre
Western Australia School of Mines
Minerals, Energy and Chemical Engineering
Curtin University
Bentley, Australia

Jennifer Sargent
Suez – Water Technologies and Solutions
Edmonton, Alberta, Canada

Eliana Sérvulo
Federal University of Rio de Janeiro
Rio de Janeiro, Brazil

Mohita Sharma
Petroleum Microbiology Research Group
University of Calgary
Calgary, Alberta, Canada

Torben Lund Skovhus
Centre of Applied Research and Development
VIA University College
Aarhus, Denmark

Blake W. Stamps
Airman Systems Directorate
Air Force Research Laboratory
Wright-Patterson Air Force Base
and
Integrative Health and Performance Sciences Division
UES Inc.
Dayton, Ohio, United States

Bradley S. Stevenson
Department of Microbiology and Plant Biology
University of Oklahoma
Norman, Oklahoma, USA

E. Suarez
Curtin Corrosion Centre
Western Australia School of Mines
Minerals, Energy and Chemical Engineering
Curtin University
Bentley, Australia

Joseph M. Suflita
Department of Microbiology and Plant Biology
and
The Biocorrosion Center
University of Oklahoma
Norman, Oklahoma, USA

Nicolas Tsesmetzis
Shell International Exploration and Production, Inc.
Houston, Texas, USA

J. Wolodko
University of Alberta
Edmonton, Alberta, Canada

T. Zintel
TC Energy Pipelines
Troy, Michigan, USA

Part I

Introduction

1 History of Failure Analysis for Microbiologically Influenced Corrosion

Thomas R. Jack
University of Calgary

CONTENTS

1.1 THE ROLE OF FAILURE ANALYSIS

Failure analysis is a form of triage in which all the possible causes of a failure are assessed with the aim of discovering the critical processes and factors that caused the problem. It forms the basis for identifying timely and cost effective ways to restore system performance and to avoid future failures occurring by the same mechanism. For facility operators, public or private, the business goal is to reduce the risk posed by possible future failures to tolerable levels in a swift, affordable, and certain manner. In the case of corrosion this is accomplished through the various activities that make up an overall Integrity Management Plan (Figure 1.1). The enabling factors cited in Figure 1.1 are conditions or processes that can foster, support and accelerate a given corrosion mechanism. Identification of these factors is often helpful in locating or identifying MIC and in finding targets for control and mitigation procedures.

Failure analysis plays a seminal role in the development and implementation of the Integrity Management Plan. Once the original failure has been repaired, Condition Assessment is necessary to show that the affected facility is fit to return to service.

DOI: 10.1201/9780429355479-2

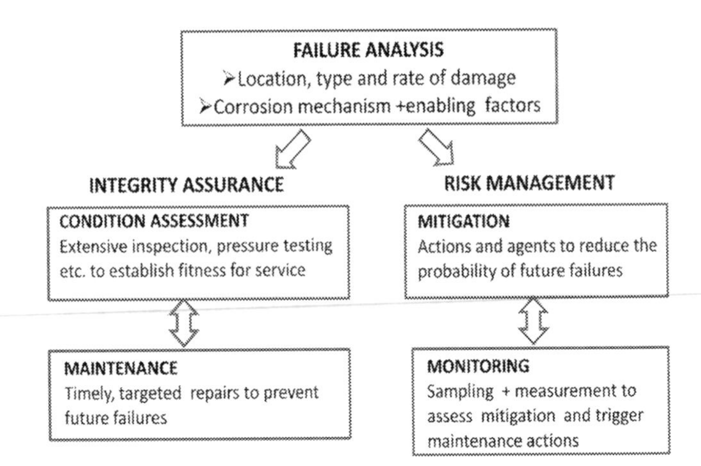

FIGURE 1.1 Failure analysis in the context of an overall Integrity Management Plan.

Failure analysis informs the design of the Condition Assessment Program. The kind of defect responsible for the failure and where it was located as well as the enabling factors identified in the failure analysis process provide the basis for selecting appropriate inspection techniques and targeting locations where similar defects are most likely to be found. Once the system is free of imminent threats a Mitigation Program can be put in place to reduce the probability of future failures. This can include actions such as changing operating conditions, mechanical cleaning of accessible surfaces (by pigging for example) or the use of agents such as corrosion inhibitors and biocides. The choice of actions and agents and how they are implemented is based on the results of failure analysis. A Monitoring Program is needed to assess the efficacy of the Mitigation Program on an ongoing basis. Selection of the analytical methods, targets and timing for the Monitoring Program again depends on the findings of the failure analysis as does the development of Key Performance Indicators that can be used to decide whether the Mitigation Program is satisfactory or whether improvements are needed. Finally a proactive Maintenance Program is required to prevent further failures through timely intervention at corrosion sites that continue to develop despite best efforts at mitigation. The value proposition for failure analysis is based on its ability to influence the design and implementation of an efficient, cost effective Integrity Management Plan for existing facilities and to improve the design and construction of new ones.

The amount of effort put into failure analysis will depend on the business context. Failure analysis is an exercise in risk reduction. Where risks are relatively small, evidence of high levels of a microbial activity associated with past MIC failures is often taken as sufficient rationale to implement a simple Mitigation Program. This might be the case for small diameter water lines in an upstream oil and gas operation for example, where biocide injection can be readily put in place to reduce the probability of a future MIC failure. In high risk situations, a very serious effort in failure analysis may be warranted and may require significant changes to operations and facilities in the implementation of an effective Mitigation Program.

1.2 ORIGINS OF FAILURE ANALYSIS FOR MIC

For more than a century the development of failure analysis for MIC has been intertwined with the discovery of sulfate reducing bacteria (SRB) and investigation of the interaction of these microorganisms with iron and its alloys. This history focuses specifically on the involvement of SRB in corrosion failures seen on the outside of buried pipelines. This type of failure was chosen as the theme for this history because it has been a focal point for research and development from the first suggestion that sulfur cycle microorganisms were somehow promoting the corrosion process (Gaines 1910). Over the decades failure analysis has been extended as required to include other types of microorganisms and other mechanisms and has stimulated the development of a wide range of analytical, monitoring, inspection and other supporting technologies needed to implement an effective Integrity Management Plan.

1.3 SULFATE REDUCING BACTERIA

Beijerinck (1895) isolated the first SRB from anaerobic environments blackened by iron sulfide precipitation using novel enrichment procedures and the optical microscope as his principal tools. He assigned the name *Spirillum desulfuricans* (now *Desulfovibrio desulfuricans*) to a newly discovered microorganism characterized by its ability to use sulfate as an electron acceptor for anaerobic metabolism, cell morphology (curved rods), motility (polar flagellum) and a negative Gram stain.

Microbial taxonomy is a work in progress. Over the last century SRB of interest have been classified and reclassified numerous times. Microorganisms now grouped with the *Desulfovibrio* have at various times in the past been classified as *Spirillum, Vibrio and Microspira* (Davis 1967; Miller 1971). Such name changes are not trivial to those seeking to find and interpret the early literature. Prior to the discovery of the structure of DNA by Watson and Crick (1953) the identification and taxonomic classification of a microorganism was based on the physical characteristics of its cells and colonies, its metabolic capabilities, and its tolerance to various physical and chemical conditions – in other words its phenotype. This descriptive approach to taxonomy dates back to the time of Aristotle (Leroi 2014) but a shift has taken place. The context in which microorganisms are named and classified has been fundamentally altered by genomics. Very limited genetic information appeared in the 8th edition of *Bergey's Manual of Determinative Bacteriology* (1974). The content of the nucleic acid bases guanosine and cytosine (G+C content, %) was included as a factor that could be used to discriminate different species of *Desulfovibrio* for example. Woese and Fox (1977) proposed a fundamental change based on research into DNA sequences for the 16S rRNA gene which indicated that the SRB in the original taxonomic scheme actually consisted of two genetically distinct kinds of microorganism. Accordingly the sulfate reducing archaea (SRA) were split from the SRB with both groups making up the sulfate reducing prokaryotes (SRP). Today genomics is used to build family trees (phylogeny), to fingerprint (identify) new organisms and to show the relatedness of various microorganisms (classify) through the wide spread use of readily available DNA sequencing technologies. This shift represents a profound change and is a fine example of how a disruptive new technology can impact an established area. The full impact of this technology on the practice of failure analysis is yet to be seen.

1.4 MIC INVOLVING SRB

In 1910, Gaines suggested that iron and sulphur bacteria might be responsible for the corrosion of ferrous metals. Von Wohlzogen Kuhr (1923) refined this suggestion by proposing that SRB were responsible for severe corrosion (graphitization) observed on iron pipelines in black anaerobic soil that was rich in organic matter and often waterlogged (Miller 1971). In a widely cited failure analysis published in 1934 Von Wohlzogen Kuhr and Van Der Vlugt (1934) proposed a corrosion mechanism that coupled the oxidation of iron to microbial sulfate reduction. In the corrosion cell, the oxidation of iron at the anode (Eq. 1.1) is coupled to the reduction of protons derived from water to yield atomic hydrogen at the cathode (Eq. 1.2). In abiotic circumstances a build-up of atomic hydrogen on the metal surface is assumed to inhibit the overall corrosion process. A consequent buildup of electrons at the metal surface gives the cathode a negative charge, a phenomenon referred to as cathodic polarization. The microbial uptake of hydrogen was seen as a way to depolarize the cathode and accelerate corrosion. The net reaction for iron oxidation coupled to sulfate reduction was summarized by (Eq. 1.3).

$$\text{Fe} \rightarrow \text{Fe}^{2+} + 2\ \text{electrons} \quad \text{(Anodic Reaction)} \tag{1.1}$$

$$2\text{H}^{+} + 2\ \text{electrons} \rightarrow 2\text{H}^{\cdot} \rightarrow \text{H}_2 \text{(Cathodic Reaction)} \tag{1.2}$$

$$4\text{Fe} + \text{SO}_4^{2-} + 8\text{H}^{+} \rightarrow \text{FeS} + 3\text{Fe}^{2+} + 4\text{H}_2\text{O} \tag{1.3}$$

This mechanism was widely accepted and continues to appear in presentations, publications, and textbooks to this day despite enduring dispute and criticism.

In 1939, Richard Hadley working for the Susquehanna Pipe Line Co. in the USA brought this mechanism to the attention of the corrosion community (Hadley 1939; Hadley 1940a; Hadley 1940b; Davis 1967). The American company had encountered severe external corrosion on oil pipelines it operated in Ohio, Pennsylvania, and New York notably in low-lying wet areas. Failure of pipe with a 0.71 cm wall thickness took less than 10 years. These failures triggered a Condition Assessment Program that involved an extensive campaign of excavations to expose sections of the buried pipe for inspection. The program involved the excavation of hundreds of locations along the pipeline system (Hadley 1939; Davis 1967). Anaerobic microbial corrosion (Eq. 1.3) was identified by comparing analytical results for deposits from corroded and un-corroded areas (Hadley 1940). Chemical analysis using classic "wet chemistry" methods showed that deposits from corroded areas were enriched in organic matter (loss on ignition), ferrous ions (and to a lesser extent ferric ions), sulfur species including sulfate, sulfite and sulfide and carbonate, but were depleted in silica. Corroded areas showed a lower pH (5.8 -7.0) than uncorroded areas (7.2). Based on these results ferrous sulfide and ferrous carbonate were identified as primary corrosion products. The increase in ferric ion concentration and drop in pH in samples from corroded areas suggested that secondary oxidation due to air exposure had occurred in the collection and shipment of field samples to the analytical lab. Soil adjacent to corroded pipe had a blackish grey color. The burial of organic matter (such as unwanted wooden skids and vegetation) in the backfill around the pipe was associated with fostering detrimental local microbial activity. Identification of this mechanism displaced an earlier assumption that the failures had been caused by long-circuit differential aeration corrosion cells between exposed pipe surfaces in the wet anaerobic soils and nearby exposed pipe surfaces in aerobic zones. Hadley (1940) proposed that bacterial analysis showing the presence of SRB and chemical analysis of corrosion products showing sulfide could be taken as sufficient hallmarks for this scenario.

The maximum pit penetration rate observed in the field, 0.5 mm/year, compared favourably with the maximum pitting rate obtained from laboratory studies reported by Bunker (1939) in which mild steel corrosion coupons were exposed to a semi-continuous culture of *Sporovibrio desulfuricans* growing in a lactate-sulfate medium for 156 days.

The presence of viable SRB in corrosion products was detected by Hadley presumably using the sort of enrichment, isolation and microscopic inspection techniques pioneered by Beijerinck in the late 1800's. No microbial numbers were reported in his failure analysis work but Bunker had noted in 1939 that SRB were more abundant close to the corroding metal and in corrosion products than in adjacent soil (Bunker 1939; Starkey 1958). Viable cell numbers may have been assessed by the Most Probable Number (MPN) method using a selective growth medium based on the enrichment media being used by microbiologists to isolate new species of SRB at that time. Standard probability tables to support this technique were widely available by 1939 (Ministry of Health 1939). The MPN method in use today is based on selective growth media specified in various industry standards but it is now recognized that use of these standard media may miss many sulfate reducing prokaryotes (SRP) of potential interest. By 1987 at least nine genera of sulfate reducers had been identified, many of which are unable to grow in the standard media used for MPN determinations (Hamilton 1987). Today all these microorganisms can be identified in a single assay using DNA sequencing technology.

The results obtained by Hadley (1940) fulfil the requirements of failure analysis shown in Figure 1.1;

- location for failures
    - on external pipe surfaces
    - buried in wet and swampy areas
- type of failure
    - localized pitting corrosion (areas of overlapping pitting)
- rate of damage
    - corrosion rate from time to failure, 0.7 mm/year.

  - maximum pit penetration rate measured in field, 0.5 mm/year.
- corrosion mechanism
  - cathodic depolarization by SRB (based on identification of bacteria in corrosion products and the presence of sulfide)
- enabling factors
  - bare pipe exposed directly to soil environment
  - wet anaerobic soil at near neutral pH in contact with unprotected steel, temperature and water chemistry suitable for microbial growth, organic matter buried next to the pipe

Actions for Mitigation of this threat suggested by Hadley (1948) included application of cathodic protection (CP) and application of concrete, asphalt or bituminous enamel coatings on the external surface of the pipe. Use of CP for this purpose was an innovation at the time. Interestingly, Hadley specified that the last two coating options should be supplemented by CP. While CP could have been put in place on the existing pipeline without excessive exposure of buried pipe, the field application of coatings would have required complete excavation of buried pipe in susceptible areas. There was no practical way to implement an effective on-going Monitoring Program for external corrosion at the time. An on-going Condition Assessment Program based on focused excavation of susceptible sites may have been the only option. The urgency of interventions in the Maintenance Program would have been informed by the corrosion rates noted above in terms of the urgency of intervention. The design and construction of future facilities may have been improved by selecting low risk pipeline routes that avoided areas of wet anaerobic soil in the planning stage and by avoiding the burial of organic matter in backfill along the right of way and applying coatings and CP during construction as recommended by Hadley (1948).

Between 1941 and 1945 the American Gas Association funded an extensive field program to determine better ways to locate probable corrosion sites (Davis 1967). Using soil probes to measure the redox potential, E_h, at pipe depth, Starkey and Wight (1945) found that severe corrosion occurred when the E_h at pipe depth was <100 mV (NHE). This is consistent with the development of an SRB-based mechanism. The *Desulfovibrio* described in the 8th edition of Bergey's Manual of Determinative Bacteriology (1974) all require an E_h < 80mV (NHE) for sustained growth in near neutral pH medium. Soil probes that measure conditions including E_h at pipe depth have seen limited use in the intervening years. For example such probes were used in selecting excavation sites for assessing the presence of stress corrosion cracking on buried pipelines by Wilmott et al. (1995).

1.5 MECHANISTIC STUDIES

Despite the observation by Stephenson & Strickland (1931) that SRB could take up molecular hydrogen while reducing sulfate (Eq. 1.4), the scheme presented by Von Wohlzogen Kuhr and Van Der Vlugt (1934) showed that cathodic depolarization by bacteria occurred through the consumption of atomic hydrogen (Eq. 1.5).

$$4H_2 + SO_4^{2-} \rightarrow S^{2-} + 4H_2O \tag{1.4}$$

$$8H\cdot + SO_4^{2-} \rightarrow S^{2-} + 4H_2O \tag{1.5}$$

In a critical review H.K. Hadley (1948) concluded that "*… conflicting reports in the literature regarding bacterial utilization of hydrogen are such as to warrant additional research on the subject.*"

A surge of intensive and extensive laboratory research followed in the next two decades, largely focused on proving or disproving the hypothesis that microorganisms involved in cathodic depolarization had to have the hydrogenase enzyme to take up molecular hydrogen (Eq. 1.4). This period was characterized by conflicting results and new discoveries. Electrochemical techniques employing corrosion test cells which allowed processes occurring on the anode or cathode to be investigated separately confirmed cathodic depolarization in the presence of growing hydrogenase-positive SRB (in certain cases); however, the overall corrosion scenario proved to be far more complex. What emerged was not a single process that could account for the damage seen but a corrosion scenario in which a number of complementary processes could be active. It also became apparent that corrosion rates measured for suspensions of hydrogenase-positive microorganisms in contact with steel in simple short term laboratory experiments were generally too low to account for rapid failures seen in the field. After a critical review of available material, Miller (1971) concluded that the anaerobic microbiological scenario could involve;

A. The utilization of hydrogen by SRB (and possibly other microorganisms that possess a suitable enzyme system)
B. Cathodic depolarization by precipitated ferrous sulfide
C. The prevention of the formation of protective sulfide films in the presence of excess ferrous ion
D. Anodic stimulation by the sulfide ions
E. The formation of local concentration cells

Enning and Garrelfs (2014) provided an updated view of the various processes that can be involved in the interaction of SRB with ferrous metals and assessed their relative importance in MIC.

Although statement A above does encompass the cathodic depolarization process by SRB originally suggested by Von Wohlzogen Kuhr and Van Der Vlugt (1934) it leaves the door open to other possibilities. The possibility that the microbiological reduction of nitrate or carbon dioxide in place of sulfate might be coupled to iron oxidation in anaerobic MIC had been suggested at that time (Von Wohlzogen Kuhr 1937) but in his 1948 review, Hadley deemed the involvement of nitrate reducing bacteria (NRB) or methanogenic microorganisms able to reduce carbon dioxide to be "speculative" in the absence of any laboratory or field evidence (Hadley 1948).

Evans (1960) had noted in a review that "*no cases of corrosion caused by the activity of methane bacteria appear to be on record*" but in 1961, Kuznetsov and

Pantskava (Kuznetsov and Pantskava 1961; Davis 1967) isolated a pure culture of the methanogen, *Methanobacterium formicicum,* from corroding oilfield equipment through anaerobic enrichment in a formate-containing medium that was maintained without formate under an atmosphere of hydrogen and carbon dioxide. Corrosion was demonstrated for iron plates exposed to an active culture of this microorganism for 25 days, but only in those cases where growth was seen and a significant decline in redox potential, E_h, occurred. The overall reaction was summarized as shown in (Eq. 1.6); however, the results were not clear cut in terms of its importance as a primary cause of corrosion failures.

$$8H^{\cdot} + CO_2 \rightarrow CH_4 + 2H_2O \quad (1.6)$$

Analysis of corrosion products from the field showed the presence of a mixed microbial population with 10^6 and 10^7 cells/g of methane-forming microorganisms and SRB, respectively. Further, Hadley had noted in 1948 that methanogens were often associated with SRB in soils and that corrosion failures where methanogens were present usually showed the presence of iron sulfide as a corrosion product. These observations combined with the general difficulty of growing methanogens reliably in short term lab tests may have discouraged the incorporation of assays for these organisms as a routine practice in failure analysis. Daniels et al. (1987) confirmed that use of cathodic hydrogen from the corrosion of steel (Eq. 1.6) can support microbial growth of certain methanogens; however, corrosion rates observed in laboratory tests have been modest (Park et al. 2014). Use of MMM has provided more insight into mixed communities involving acetogens, methanogens and SRP in MIC (Mand et al. 2014a).

Caldwell and Ackerman (1946) identified NRB as agents of internal corrosion in an eight inch (20.3 cm) steel waterline at a U. S. navy station in California. Nitrate initially present at 4 – 7 ppm (as nitrogen) in water from source wells was reduced to nitrite and ammonia on passage through a 3.8 km pipeline. The concomitant release of iron turned the water orange. Nitrate reduction was reduced or eliminated and water quality restored by increasing the pH of the input water to 8 and by adding chlorine. In his 1961 Ph.D. thesis at the University of Edinburgh, F. D. Cook showed that steel wool and nails could be corroded by hydrogen utilizing strains of denitrifying bacteria including soil isolates growing in batch culture; however, subsequent attempts to establish cathodic depolarization as the corrosion mechanism were frustrated by apparently contradictory results. Corrosion rates for steel coupons exposed to semi-continuous cultures of a hydrogenase-positive *Escherichia coli* NCIB 8666 and hydrogenase-negative *Pseudomonas stutzeri* NCIB 9040 in a nitrate medium for 35 days under anaerobic conditions were determined by Mara and Williams in 1971 (Mara and Williams 1971). The hydrogenase-positive microorganism showed a corrosion rate of about 0.07 mm/year, five times that seen with the hydrogenase-negative culture and six times that seen for a sterile control. The authors considered these results confirmation that hydrogenase-positive microorganisms could accelerate corrosion through cathodic depolarization with the reduction of nitrate to nitrite taking the place of sulfate reduction to sulfide; however, an attempt to replicate the experiments with *E. coli* NCIB 8666 led to very different conclusions (Ashton et al. 1973a, 1973b). No correlation was found between nitrate reduction and iron oxidation. Further the nitrite formed by nitrate reduction reacted directly with the steel surface to form a passivating film of ferric oxide, γ-Fe_2O_3. The corrosion observed after five weeks exposure took the form of corrosion pits under adherent clumps of microbial cells on the metal surface and was attributed to organic acid formation by the sessile microorganisms. In 1948, Hadley had expressed doubt that MIC by NRB was likely to be of any great economic importance. This opinion may have reflected the generally low concentration of nitrate commonly seen in failure analyses in his day (Hadley 1948). Real interest in NRB as potentially important agents of MIC emerged decades later with the use of nitrate and nitrite at relatively high concentrations to control souring due to the production of H_2S by SRP in oilfield applications (Hubert et al. 2005).

King and Miller deemed processes A and B to be the most important drivers for MIC but went on to suggest that these two processes could be combined (King and Miller 1971; King et al. 1976a). In the mechanism shown in Figure 1.2A, solid iron sulfide in contact with steel forms a galvanic couple in which the iron sulfide acts as the cathode (process B above). The importance of this coupling in the corrosion of steel was discovered in the 1920's in laboratory tests by Stumper (Stumper 1923; Starkey 1958) and in failure analyses done in flow tanks in sour oilfield operations by Ginter (Ginter 1927; Starkey 1958). Cathodic depolarization of the iron sulfide by SRB (process A) accelerates the corrosion and sustains the corrosion current while producing more iron sulfide (Eq. 1.3). This activity in turn increases the size of the corrosion cell and the rate of corrosion by depositing more iron sulfide. The very large surface area of finely divided iron sulfide deposits can support abiotic forms of cathodic depolarization in addition to the microbiological process shown in Figure 1.2A. This mechanism is consistent with observations made for external corrosion on buried pipelines and for internal corrosion in many operating facilities

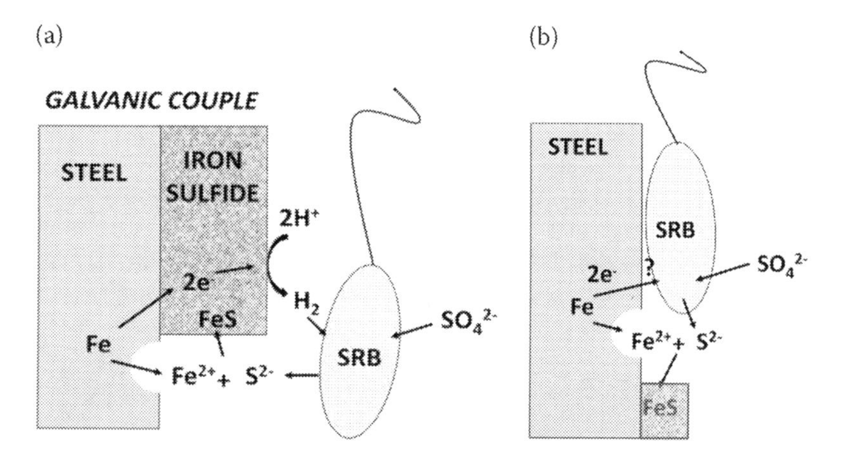

FIGURE 1.2 A. Mechanism proposed by King and Miller 1971. B. The electrical microbiologically influenced corrosion (EMIC) mechanism proposed by Enning et al. 2012.

notably in offshore oil and gas production (Jack 2002; Hamilton 1987). Korea Gas and Tokyo Gas developed correlations and predictive models for the probability of this form of external MIC based on extensive field investigations carried out on buried pipelines in the late 1990's (Jack 2002).

More recently Enning et al. (2012) have reported the isolation of SRP able to take up electrons from the metal surface without hydrogen acting as an intermediate, Figure 1.2B (Enning and Garrelfs 2014). These microorganisms may also be able to take up electrons from the surface of iron sulfide deposits in a galvanic couple with steel. Direct electron uptake bypasses the kinetically slow cathodic reduction reactions involving hydrogen that constrain the rate of abiotic anaerobic corrosion (Eq. 1.2). The mechanism was dubbed electrical MIC (EMIC). The proposed mechanism, Figure 1.2B, immediately raised the question of how electrons can be transferred from a metal surface into the microbial cell. The question is of key importance in a number of application areas beyond MIC including biofuel cells, microbiologically catalyzed electrosynthesis, bioremediation of groundwaters using buried "curtains" of iron nanoparticles and so on (Kato 2016; Blackwood 2018; Jack 2020).

Both mechanisms in Figure 1.2 have been shown to match very high field corrosion rates in laboratory studies (Jack 2002; Enning and Garrelfs 2014). Either can be regarded as a primary corrosion mechanism in failure analysis and where found should be considered in the development of an appropriate Integrity Management Plan.

1.6 IMPLICATIONS FOR FAILURE ANALYSIS

The impact of these mechanistic details on the applicability of the original approach to failure analysis described by Hadley in 1948 is surprisingly small. Observation of enhanced levels of SRB in corrosion deposits rich in iron sulfide is consistent with the operation of some or all the processes identified (A to E above). There is however a shift in focus from seeking to identify a unique corrosion mechanism dependent on one kind of microbial activity to the identification of a corrosion scenario in which a number of processes may be active.

The additional information provided by the mechanism shown in Figure 1.2A allows the use of field kits to verify this mechanism on-site during a failure Investigation. Commercial kits are available for the on-site assay of hydrogenase enzyme activity (Boivin et al. 1990) as is a quick test for iron sulfide. These tests can be completed in a matter of a few hours. The test for sulfide is based on the addition of a drop of acid to a corrosion product to liberate H_2S. Field assays for the EMIC mechanism, Figure 1.2B, have yet to appear.

1.7 IRON SULFIDES – A CONTROLLING FACTOR

Research through the 1970's and 1980's showed that the form of iron sulfide and the manner of its deposition are critical parameters in the anaerobic corrosion scenario (Hamilton and Lee 1995). Information on various forms of iron sulfide and their electrical properties provides the basis for interpretation of complex behaviour

(Smith and Miller 1975; Vaughan and Craig, 1978). A film of mackinawite (FeS_{1-x}) deposited in a uniform integral layer will passivate a metal surface and prevent further corrosion. This has been the bane of short term laboratory tests in which a metal coupon is simply immersed in an active culture of SRB. The protection offered by deposition of iron sulfide is dependent on a continuous integral layer being sustained that can act as a barrier to prevent contact between electrolyte from the operating or soil environment and the underlying metal surface. Where high concentrations of soluble ferrous ion are present, iron sulfide tends to precipitate in suspension rather than on the metal surface (Adams and Farrar 1953). This precludes the formation of a passivating layer (noted as Process C above). Where this occurs the ferrous ion concentration can be classed as an enabling factor in failure analysis. In any case the passivating layer will tend to rupture over time (a period of months) as further sulfidization converts the mackinawite to greigite Fe_3S_4 (cubic crystals) (Mara and Williams 1972; King et al. 1976b). Where the concentration of soluble Fe^2 exceeds 125 μM smythite Fe_3S_4 (hexagonal crystals) or pyrrohotite, $Fe_{1-x}S$ can also be formed (King et al. 1976b). Once the layer ruptures aggressive pitting can ensue with small areas of exposed steel acting as the anode for a larger area of undisrupted cathodic iron sulfide on the surrounding metal surface. The corrosion rate depends on the ratio of surface areas for the cathode and anode and can be very high.

Deposition of a mixture of corrosion products on the metal surface is unlikely to form an effective passivating layer. Where the electrolyte contains sufficient carbonate, the excess Fe^{2+} released by the corrosion process will precipitate as $FeCO_3$ (siderite) theoretically in a 3:1 ratio with FeS (Eq. 1.7). Despite the fact that siderite is a white solid, the presence of black iron sulfide is sufficient to give the mixture a black appearance.

$$4Fe + SO_4^{2-} + 5H^+ + 3HCO_3^- \rightarrow FeS + 3FeCO_3 + 4H_2O \qquad (1.7)$$

1.8 SECONDARY OXIDATION

A factor not included in Miller's summary (Miller 1971) was the effect of adventitious oxygen exposure on the anaerobic corrosion scenario (Hamilton and Lee 1995). There were early indications that this factor could play an important role in accelerating the anaerobic corrosion process. Hadley (1948) noted that soil drainage had been tried as a mitigation measure for pipelines in wet anaerobic areas but risked severe corrosion through the oxidation of ferrous sulfate to ferric sulfate as the soil environment became more aerobic. In 1953, Kulman attempted to correlate seasonal parameters such as temperature, rainfall, pH and so on with the occurrence of leaks due to external corrosion on a pipeline (Kulman 1953). Most leaks occurred in early summer as increased soil aeration shifted conditions *in situ* from anaerobic to aerobic. In 1958, Meyer et al. (1958) found that surface deposits rich in iron sulfide taken from corroded field equipment caused severe corrosion when placed in contact with steel. A corrosion rate of 22 mm/year was seen over 25 days of exposure. This greatly exceeded the corrosion rate seen for coupons exposed to samples of the iron sulfide, FeS (0.076 mm/year) and FeS_2 (0.66 mm/year), but was similar to the corrosion rate seen for exposure of steel to elemental sulfur (28 mm/year) (Meyer et al. 1958). It was suggested that oxidation of initially deposited anaerobic corrosion products by intermittent or low level exposure to oxygen had generated more corrosive chemical species such as sulfur through oxidation of the iron sulfides present. Von Wohlzogen Kuhr had discussed the ability of both iron sulfides and elemental sulfur to act as an electron acceptor in the anaerobic corrosion of steel in his 1937 review (Von Wohlzogen Kuhr 1937). Farrer and Wormwell (1953) demonstrated extremely high corrosion rates for steel exposed to suspensions of elemental sulfur in controlled laboratory experiments.

Corrosion products can be used as indicators of the past history of a failure site (Jack et al., 1995). They can be analyzed by Energy Dispersive X-ray (EDX) analysis for elemental composition and by X-ray Diffraction (XRD) for identification of crystalline forms. Analysis of corrosion products from external corrosion sites on an extensive large diameter gas pipeline network distinguished a number of different corrosion scenarios. Field measurement of corrosion damage seen in each case enabled the calculation of a corrosion rate for each scenario (Table 1.1) (Jack 2002).

The primary anaerobic MIC (SRB) scenario, Table 1.1, was characterized by a mixture of mackinawite and siderite as corrosion products in accord with (Eq. 1.7). This scenario has been replicated in the laboratory by exposing corrosion coupons made of pipeline steel to microbial activity in experimental boxes filled with soil from excavation sites and housed in an anaerobic chamber. The sustained corrosion rate and damage seen in the laboratory matched field observations.

TABLE 1.1
Indicator Minerals Found as Corrosion Products in External Corrosion Scenarios (Based on Jack et al. 1995; Jack 2002)

Corrosion Scenario	Corrosion Products	Corrosion Rate (mm/year)
Primary Corrosion		
Aerobic corrosion	Yellow/orange/brownish- black iron (III) oxides (goethite, magnetite, maghemite, hematite)	0.04 – 0.2 (pitting)
Anaerobic corrosion	White iron (II) carbonate (siderite)	0.002 – 0.01 (general corrosion)
Anaerobic MIC (SRB)	Black mixture of iron (II) sulfides (macknawite, greigite) plus iron (II) carbonate (siderite) where carbonate is sufficient concentration	0.7 (pitting)
Secondary Oxidation		
Anaerobic MIC (SRB) to Aerobic	Elemental sulfur, iron (III) oxides over residual anaerobic corrosion products	2–5 (pitting)

Field sites with ferric oxides and possibly elemental sulfur along with the original anaerobic corrosion products show a higher corrosion rate due to secondary oxidation (Table 1.1). This too could be replicated in the laboratory by aggressively sparging the soil boxes described above with air (Jack et al. 1998).

In failure analysis secondary oxidation of the anaerobic microbial corrosion scenario through intermittent or low level oxygen exposure can be treated as an important or even crucial enabling factor for very rapid failure.

Building on all of these observations and field results for the basic anaerobic corrosion scenario a failure analysis done in the late 1990's might have looked like this:

- location for failures
    - damage seen under disbonded field applied polyolefin tape wrap coatings in wet anaerobic high clay soils regardless of the application of cathodic protection (CP)
- type of failure
    - localized corrosion featuring serious metal loss in areas of overlapping pitting
- rate of damage
    - estimated penetration rate based on pit depth measurements, 0.7 mm/year
- corrosion mechanism
    - MIC initiated and sustained by SRB according to (Eq. 1.7) based on enhanced numbers of SRB (MPN viable cell count) being found in black deposits of $FeS/FeCO_3$ overlying the corrosion damage (XRD/EDX)
    - the absence of key indicators for secondary oxidation (XRD/EDX)
- enabling factors
    - shielding coating disbondments that allow groundwater access to the underlying pipe surface but block CP.
    - wet anaerobic high clay content soil, groundwater at near neutral pH in contact with exposed steel under disbonded coating, temperature, and water chemistry suitable for microbial growth.

The outcome mirrors the first failure analysis described above using Hadley's results from 1940 but the newer version has a much higher level of certainty and precision enabled by a better understanding of relevant mechanisms and the use of much more sophisticated analytical techniques (Little et al., 2006). The improvement in outcome is a consequence of advances in the scientific and analytical context for failure analysis rather than a change in the aim or role of failure analysis. Options for the design and implementation of an Integrity Management Plan in the 1990's were also enriched by advances in supporting technologies. In particular, the development of in-line inspection (ILI) tools that could be pushed by the flow of gas or fluid through large diameter pipelines to map metal loss due to external and internal pitting made Condition Assessment more comprehensive and an on-going Monitoring Program practical. Application of coatings and CP as Mitigative measures as recommended by Hadley (1948) had become normal practice in the construction of buried pipeline systems by this time. A crucial enabling factor was the disbondment of a shielding coating that allowed groundwater access to the underlying pipe surface while blocking effective CP potentials. Resistance to cathodic disbondment and shielding is a significant factor in the design and selection of modern pipeline coatings.

In failure analysis, the observation of corrosion on a metal surface free of calcite scale in contact with an aqueous phase at the ambient pH for the surrounding environment provides evidence that CP had not reached a failure location. In contrast metal surfaces under effective CP show no corrosion, an elevated pH and hard white deposits of calcite ($CaCO_3$) where groundwater with sufficient carbonate content is present.

Although a history of the development of analytical techniques to support failure analysis lies beyond the scope of this chapter, some sense of the progress made in this period can be gained by comparing summaries made in 1982 and 2006. Pope et al. (1982) reviewed the assays being used at that time for failure analysis to detect, enumerate and determine the viability of various kinds of organisms based on project work done for the U.S. DOE. Discussion was limited to the use of optical microscopy, electron microscopy and culture methods (e.g. MPN). In 2006, Little et al. (2006) covered the use of a much wider range of analytical methods able to provide the metallurgical, chemical and biological evidence required to establish multiple lines of evidence for MIC. An overview of the expanding scope of analytical methods captured in this later review indicates a trend towards to use of growth-independent methods that do not suffer the constraints of the selective growth media as well as a trend towards using assays with a rapid turnaround time that can be performed in the field. A number of these new methods are focused on measuring metabolites or potential metabolic activity rather than identifying microorganisms directly.

1.9 A WIDER VIEW

Although an MIC failure involving SRP and FeS can occur very rapidly and is easily identified; most MIC related failures are not due to this scenario. This was demonstrated in a major project funded by the Gas Research Institute in the late 1980's which surveyed a significant number of internal and external corrosion sites on natural gas pipelines and other facilities (Pope et al. 1988; Pope et al. 1990; Pope et al. 1992).

The possibility that other kinds of microorganisms could be involved in MIC was recognized by Gaines in 1910 (Gaines 1910). Von Wohlzogen Kuhr and Van der Vlugt (1934) had suggested that cathodic depolarization might occur by the reduction of carbon dioxide or nitrate in place of sulfate (Eq. 1.4). In 1948, Hadley (1948) considered whether methanogens, nitrate reducers, sulfur oxidizers (e.g. *Thiobacilli*), sulfur bacteria, iron bacteria (e.g. *Leptothrix, Gallionella, Siderocapsa, Sideromonas, Crenothrix*) and miscellaneous microorganisms including slime formers, fungi, algae, protozoa, diatoms and bryozoa, important in biofouling and biofilm development, could be involved in MIC. In 1982, Pope et al. (1982) listed organic acid producers, aerobes, anaerobes, sulfuric acid producers, slime producers, iron and manganese oxidizing microorganisms as targets for investigation in a review of methods for detecting and enumerating microorganisms in failure analysis. Other field guides for failure analysis at the time focused on the identification and enumeration of those microorganisms able to account for the observation of thick biofilms, tubercles, acids, manganic or ferric oxide deposits and so on as well as the damage seen under the growth conditions found at the failure site (Ostroff 1979; Herro and Port, 1993).

In the course of extensive field work, Pope et al. (1988) discovered the wide occurrence of surprisingly high concentrations (thousands of ppm) of the conjugate bases of organic acids (e.g. acetate for acetic acid) in produced waters from oilfields and in corrosion pits (Jack 2002; Little et al. 2006). This was taken as a sign of microbial activity by organic acid producing bacteria (APB) and a commercially available MPN medium was duly developed based on the fermentation of sugar. This medium supports the growth of a relatively wide range of microorganisms and has been widely adopted for use in failure analysis despite lingering uncertainty about the relevance of sugar fermentation in corrosion scenarios. More recent work using MMM has established that acetogenic microorganisms found on metal surfaces can form acetate by coupling the uptake of cathodic hydrogen to reduction of carbon dioxide (Mand et al. 2014a). Although corrosion rates observed in simple laboratory tests with these microorganisms are low, this activity could account at least in part for the observation of acetate in failure analyses.

Enhanced levels of APB and organic acids came to be associated with a distinct corrosion morphology that featured tunnelling extending into the walls of deep corrosion pits in the direction of rolling. Pope et al. (1988) were able to reproduce this morphology in the laboratory in pipeline steel exposed to 20000 ppm acetic acid for 7 days at pH 2 but the pattern was not observed at pH 4 or at pH 2 when sulfuric acid was used in place of acetic acid. Because organic acids could not be expected to create a pH 2 environment, a complex MIC mechanism was suggested in which pits initiated under biologically active deposits producing organic acids develop into deep pits with very acidic conditions in the pit bottom where chloride has taken the place of acetate as the counter ion (Pope et al. 1990, Pope 1993).

An overview of MIC scenarios that reflects the diversity noted above is presented in Table 1.2. The table shows how different microorganisms are connected to corrosion consequences through their metabolic activity. The examples given are grouped into five categories. The first two categories are sometimes referred to as active MIC because the corrosion process is sustained by the metabolic activity indicated. These MIC modes can be classified as primary corrosion mechanisms in the scheme shown in Figure 1.1. Alteration of chemical conditions at the metal surface is sometimes referred to as a passive form of MIC because the role of the microbes is to set the stage for an abiotic corrosion process, not support corrosion directly on an on-going basis. The last two categories involve the loss of protection for a metal surface through the degradation of mitigative measures and conditions. Processes in the last three categories can be classed as enabling factors in the scheme for failure analysis shown in Figure 1.1. Indicators that can be used to identify various mechanisms are given in (Jack 2002).

TABLE 1.2
Five Basic Types of MIC with Examples (Based on Jack 2002)

Type of Microorganism	Activity	Consequence	Enabling Condition
Direct involvement in the electrochemical corrosion cell			
Sulfate Reducers	Uptake of cathodic hydrogen Production of FeS	Cathodic depolarization Galvanic corrosion	Anaerobic
Generation of corrosive metabolites			
Sulfidogenic	Generation of H_2S	Depolarization of corrosion cells	Anaerobic
Thiobacilli	Generation of sulfuric acid	Acid attack at low pH	Aerobic
Manganese Oxidizers	Oxidation of Mn^{2+} to MnO_2	MnO_2 oxidizes metal	Aerobic
Acid Producing Bacteria (APB)	Generation of organic acids Generation of acetate	Cathodic depolarization Enhanced CO_2 corrosion	Anaerobic
S-cycle Microorganisms	Generation of corrosive species	Sulfur, polysulfide etc. oxidize metal	Various scenarios
Alteration of chemical conditions at the metal surface			
Iron Oxidizers	Production of ferric oxides as tubercles on metal surface	Differential aeration cells Concentration cells Anaerobic conditions at metal surface promote SRP	Aerobic
Slime Formers	Generation of biofilms	Differential aeration cells Anaerobic conditions at metal surface promote SRP	Various scenarios
Disruption of passivating layers			
Iron Reducers	Ferric oxides reduced to soluble ferrous species	Dissolution of passivating layer on metal surface	Anaerobic
APB	Generation of organic acid	Dissolution of passivating layer on metal surface Under deposit corrosion	Anaerobic
Degradation of coatings, chemicals and cathodic protection			
Various	Biodegradation of coatings and treatment chemicals	Corrosion proceeds	Aerobic
Sulfidogenic	Formation of FeS deposits	Increased CP current demand	Anaerobic

The wider scope of MIC was recognized in GRI field guides published in 1991 and 1992 for the identification of MIC in external and internal failure investigations (GRI-88/0113 1991; GRI-92/0005 1992). Unlike other guides for failure analysis available at the time, the GRI guides were published in a small format that could be easily carried in the field and were laminated to protect them from damage by rain, snow, or mud. They were intended to be used by a person carrying out the investigation of a failure in the field. The aim was to improve the cost, speed, and reliability of failure analysis by avoiding the expense, delay, and risk of exposure to temperature changes, secondary oxidation or even loss in transit for field samples being sent to a laboratory. All the tools and chemicals needed were specified with instructions on how to use them as was a list of information required to describe the affected facility and failure site. A training module in VHS videotape format was available. The objective of these field guides was not to identify MIC scenarios but rather to estimate the probability that a failure had involved MIC. This was done using a score card approach in which points were assigned to various biological factors (MPN results using SRB, APB and MC media), chemical descriptions and qualitative analyses (presence of deposits containing $CaCO_3$, $FeCO_3$, sulfide) and metallurgical features (color of contents in pits, pit shape, presence and orientation of striations and tunneling). The sum of the points obtained determined whether MIC was very unlikely, likely, or very likely to have been involved in the failure. Expressing the outcome as a probability was very timely. It coincided with the increasing use of probability-based Risk Assessment in Integrity Management. Hard copy reporting forms were available that automatically generated a duplicate copy that could be sent by mail to a pan-industry database. The aim was to leverage findings from single company investigations by considering them in a wider industry context. Unfortunately the scheme was not widely embraced by facility operators.

The scope of tests in the GRI scorecard covered active SRP and APB scenarios individually but MIC could still be deemed “likely” without seeing high viable cell counts for either of these types of microorganisms based on metallurgical results alone. This became a contentious issue. The current view is that the pit morphology described above may be indicative of MIC but is neither unique to nor an essential feature of MIC failures (Little et al. 2006). Despite the obvious appeal of being able to identify MIC by visual examination of corrosion damage at a failure site, the current consensus is that multiple lines of evidence based on complementary biological, chemical and metallurgical analyses are needed to identify an MIC site with any degree of certainty (Little et al. 2006).

1.10 SUMMARY

Failure analysis plays a seminal role in crafting a practical, cost effective Integrity Management Plan that can reduce the risk posed by the possibility of future failures to acceptable levels. The role of failure analysis has remained unchanged for decades. The improved output in failure investigations seen over time reflects advances in the science and technology platforms that support the scheme and has been influenced by the broad implementation of probability-based Risk Management.

Failure analysis for MIC was inspired by the proposal in 1934 of a single corrosion mechanism based on the action of one kind of microorganism, the SRB. The original objective was to identify the presence of that microorganism and its iconic iron sulfide corrosion product as the hallmarks of this mechanism. In 1940 this was sufficient to recommend the application of protective coatings and CP to the outside of pipelines passing through wet organic rich anaerobic soils as a Mitigative strategy. By 1971 however, it had become apparent that SRB might be better considered an indicator of a wider anaerobic scenario in which a number of consequent and complementary processes might be occurring. A state-of-the- art review by Hamilton and Lee in 1995 included the statement, "… *the microbiological component is seldom simple or easily identified as a single microorganism or a unique mechanism*" (Hamilton and Lee 1995). The advent of readily available, cost effective MMM able to characterize whole microbial communities in about 2006 was timely. Today failure analysis is expected to identify the influence of a range of microorganisms in an overall corrosion scenario through the development of multiple lines of evidence based on complementary metallurgical, chemical and biological assays interpreted within the context of the operating history of the affected facility.

As was seen by its impact on Bergey's Manual of Determinative Bacteriology in the late 1970's MMM can be considered a disruptive technology. It should not be a surprise that MMM was initially used as a better way to provide the usual information needed to support a traditional approach to failure analysis, i.e. to provide cell numbers for specific kinds of microorganisms; however, the broader scope of information made available by these new methods is having a wider impact. By providing information on the nature and metabolic activity for the whole microbial community MMM can enable unique insights into the chemical and operational environment in which a failure occurred and can suggest synergistic relationships between microorganisms (Mand et al. 2014b). Early examples of this sort of broader contribution include the observation of aerobic communities in a supposedly deoxygenated seawater injection stream (Mand et al. 2014b) and the conversion of excess sulfite injected as an oxygen scavenger into elemental sulfur rather than sulfate in process water piping (Park et al. 2011; Mand et al. 2014b). These insights upset assumptions that had been made in the management of the operating facilities affected. Use of MMM to analyze field samples has also enabled a new approach to predictive modelling for MIC failures in offshore oil and gas production facilities (Larsen et al. 2013). In this new approach, estimates of remaining life are made using corrosion rates determined through laboratory tests under appropriate conditions for selected members of the active microbial community identified by MMM. No doubt there is more to come.

REFERENCES

Adams, M. E. and T. W. Farrar. 1953. The influence of ferrous iron on bacterial corrosion. *J. Appl. Chem.* 3: 117–120.

Ashton, S. A., King, R. A. and J. D. A. Miller. 1973a. Protective film formation on ferrous metals in semi-continuous cultures of nitrate-reducing bacteria. *Br. Corros. J.* 8: 132–136.

Ashton, S. A., King, R. A. and J. D. A. Miller. 1973b. Corrosion of ferrous metals in batch cultures of nitrate-reducing bacteria. *Br. Corros. J.* 8: 185–189.

Beijerinck, M. W.1895. Uber *Spirillum desulfuricans* als Ursuche van Sulfate Reduktion, *Zentr. Bakteriol., Parasitenk., Abt II*, 1: 1–9, 49-59, 104-114.

Bergey's Manual of Determinative Bacteriology, 8th edition. 1974. Ed. R. E. Buchanan and N. E. Gibbons. Baltimore, MD: The Williams & Wilkins Company.

Blackwood, D. J.2018. An electrochemist perspective of microbiologically influenced corrosion. *Corros. Mater. Degrad.*, 1: 59–76.

Boivin, J., Laishley, E. J., Bryant, R. D. and J. W. Costerton. 1990. The influence of enzyme systems on MIC, Paper 128, CORROSION/90, Houston, TX: NACE.

Bunker, H. J. 1939. Microbiological experiments in anaerobic conditions. *J. Soc. Chem. Ind. (London)* 58: 93–100.

Caldwell, D. H. and J. B. Ackerman. 1946. Anaerobic corrosion of steel pipe due to nitrate. *J. A. W. W. A.* 38: 61–64.

Cook, F. D.1961. Denitrifying Bacteria of Soil, PhD thesis, University of Edinburgh, pp. 97–130.

Daniels, L., Belay, B. N., Rajagopal, S., and P. Weimer. 1987. Bacterial methanogenesis and growth from CO_2 with elemental iron as the sole source of electrons. *Science* 237: 509–511.

Davis, J. B. 1967. *Petroleum Microbiology*, Amsterdam, NL: Elsevier Publishing Company.

Enning, D., Venzlaff, H., Garrelfs, J., Dinh, H. T., Meyer, V., Mayrhofer, K., Hassel, A. W., Stratmann, M., and F. Widdel. 2012. Marine sulphate-reducing bacteria cause serious corrosion of iron under electroconductive biogenic mineral crust. *Environ. Microbiol.* 14: 1772–1787.

Enning, D. and J. Garrelfs. 2014. Corrosion of iron by sulfate-reducing bacteria: New views of an old problem. *Applied and Environmental Microbiology* 80 (4): 1226–1236.

Farrer, T. W. and F. Wormwell. 1953. Corrosion of iron and steel by aqueous suspensions of sulphur. *Chemistry and Industry* (January), 106–107.

Evans, U. R. 1960. *The Corrosion and Oxidation of Metals: Scientific Principles and Practical Applications*. London: Edward Arnold (Publishers) Ltd.

Gaines, R.1910. Bacterial activity as a corrosive influence in the soil. *J. Ind. Eng. Chem.* **2**: 128–130.

Ginter, R.1927. Interior corrosion of oil flow tanks in field where the sulfur conditions are bad. *Proc. Am. Petrol. Inst.* 8: 400–410.

D. Pope, et al. 1991. *GRI-88/0113 Microbiologically Influenced Corrosion (MIC): Methods of Detection in the Field*. Chicago, IL: Gas Research Institute.

D. Pope, et al. 1992. *GRI-92/0005 Microbiologically Influenced Corrosion (MIC) II: Internal MIC and Testing of Mitigation Measures*. Chicago, IL: Gas Research Institute.

Hadley, R. F.1939. Microbiological anaerobic corrosion of steel pipelines. *Oil and Gas J.* 38 (19): 92–98.

Hadley, R. F. 1940a. Methods of studying microbiological anaerobic corrosion of pipe lines. Part 1. *Petrol. Engr.* 11 (March): 171–177.

Hadley, R. F. 1940b. Methods of studying microbiological anaerobic corrosion of pipe lines. Part 2. *Petrol. Engr.* 11 (April): 112–116.

Hadley, R. F. 1948. Corrosion by micro-organisms in aqueous and soil environments. In *The Corrosion Handbook*, ed. H. H. Uhlig. 406–481. London, UK: John Wiley & Sons, Inc.

Hamilton, W. A. 1987. Mechanisms of microbial corrosion. In *Microbial Problems in the Offshore Oil Industry, Proceedings of an International Conference organized by The Institute of Petroleum Microbiology, Aberdeen*, April 1986, ed. E. C. Hill, J. L. Shennan and R. J. Watkinson. 1–11. NY, NY: John Wiley & Sons.

Hamilton, W. A. and W. Lee. 1995. Biocorrosion. In *Sulfate-Reducing Bacteria*, ed. L. L. Barton. 243 to 264. NY, NY: Plenum Press.

Herro, H. M. and Port, R. D. 1993. *The Nalco Guide to Cooling Water System Failure Analysis*, NY, NY: McGraw-Hill Inc.

Hubert, C., Nemati, M., Jenneman, G. and G. Voordouw. 2005. Corrosion risk associated with microbial souring control using nitrate and nitrite. *Applied and Environmental Microbiology* 68: 272–282. 10.1007/s00253-005-1897-2

Jack, T. R., Wilmott, M. J. and R. L. Sutherby. 1995. Indicator minerals formed during external corrosion of line pipe. *Materials Performance* 34(11): 19–22.

Jack, T. R., Wilmott, M., Stockdale, J., Van Boven, G., Worthingham, R. G. and R. L. Sutherby. 1998. Corrosion consequences of secondary oxidation of microbial corrosion. *Corrosion* 54(3): 246–252. 10.5006/1.3284850

Jack, T. R. Biological Corrosion Failures. 2002. In *ASM Handbook, Failure Analysis and Prevention*, Volume 11, ed. W. T. Becker and R. J. Shipley. 881 to 898. Materials Park, OH: ASM International.

Jack, T. R.Biological Corrosion failures. In *Failure Analysis and Prevention*, Volume 11, Materials Park, OH: ASM International, in preparation for release in 2020.

Kato, S. 2016. Microbial extracellular electron transfer and its relevance to iron corrosion. *Microbial Biotechnology* 9(2): 141–148.

King, R. A. and J. D. A. Miller. 1971. Corrosion by sulphate-reducing bacteria, *Nature* 233: 491–492.

King, R. A., Dittmer, C. K. and J. D. A. Miller. 1976a. The mechanism of sulphide corrosion by sulphate-reducing bacteria. In *the Proceedings of the Second International Biodeterioration Symposium*. 103 to113. Lunteren, Netherlands.

King, R. A., Dittmer, C. K. and J. D. A. Miller. 1976b. Effect of ferrous ion concentration on the corrosion of iron in semi-continuous cultures of sulphate-reducing bacteria, *Brit. Corr. J.* 11: 105–107.

Kulman, F. E. 1953. Microbiological corrosion of buried steel pipe. *Corrosion* 9: 11–18.

Kuznetsov, S. I. and E. S. Pantskava. 1961. Effect of methane–producing bacteria on the intensification of electrochemical corrosion of metal. *Dokl. Akad. Nauk. S. S. S. R.* 139: 478–480.

Larsen, J., Juhler, S., Sorensen, K. B. and D. S. Pedersen. 2013. *The application of molecular microbiological methods for early warning of MIC in pipelines. Paper 2029. Corrosion 2013*. Houston, NACE International.

Leroi, A. M. 2014. *The Lagoon, How Aristotle Invented Science*. NY, NY: Viking.

Little, B. J., Lee, J. S. and R. I. Ray. 2006. Diagnosing microbiologically influenced corrosion: A state-of-the-art review. *Corrosion* 62: 1006–1017.

Mand, J., Park, H. S., Jack, T. R. and G. Voordouw. 2014a. The role of acetogens in microbially influenced corrosion of steel. *Frontiers in Microbiology*, *5*(JUN), 1–1. 10.3389/fmicb.2014.00268

Mand, J., Park, H. S., Jack, T. R., Voordouw, G. and H. Hoffmann, H. 2014b. Use of molecular methods (pyrosequencing) for evaluating MIC potential in water systems for oil production in the North Sea, SPE-169638-MS. *Society of Petroleum Engineers International Oilfield Corrosion Conference and Exhibition*, (15), SPE-169638-MS. 10.2118/169638-MS

Mara, D. D. and D. J. A. Williams. 1971. Corrosion of mild steel by nitrate reducing bacteria. *Chemistry and Industry* May 22: 566–567.

Mara, D. D. and D. J. A. Williams. 1972. The mechanism of sulphide corrosion by sulphate-reducing bacteria. In *Biodeterioration of Materials*, Volume 2, ed. A. M. Walters and E. H. Hueck van der Plas. 103–113. London: Applied Science Publishers.

Meyer, F. H., Riggs, O. L., McGlasson, R. L. and J. D. Sudbury. 1958. Corrosion products of mild steel in hydrogen sulfide environments. *Corrosion* 14: 109t–115t.

Miller, J. D. A. 1971. Microbial corrosion of buried and immersed metal. In *Microbial Aspects of Metallurgy*, ed. J. D. A. Miller. 63–105. Chiltern House Aylesbury UK: MTP Medical and Technical Publishing Co. Ltd.

Ministry of Health. 1939. *The Bacteriological Examination of Water Supplies*, 2nd edition. (Reports on Public Health and Medical Subjects, No. 71). London: His Majesty's Stationary Office.

Ostroff, A. G. 1979. *Introduction to Oilfield Water Technology*, 2nd edition, Houston TX: NACE. (a reissue of the first edition, Prentice-Hall, Inc. 1965)

Park, H. S., Chatterjee, I., Dong, X., Wang, S.-H., Sensen, C. W., Caffrey, S. M., Jack, T. R., Boivin, J. and G. Voordouw. 2011. Effect of sodium bisulfite injection on the microbial community composition in a brackish-water-transporting pipeline. *Applied and Environmental Microbiology*, *77*(19): 6908–6917. 10.1128/AEM.05891-11

Park, H. S., Mand, J., Jack, T. R. and G. Voordouw. 2014. Next-generation sequencing approach to understand pipeline biocorrosion. In *Molecular Methods and Applications in Microbiology*, ed. T. L. Skovhus, C. Hubert and S. Caffrey. 35–41. Norfolk, UK: Horizon Press, Caister Academic Press.

Pope, D. H., Soracco, R. J. and E. W. Wilde. 1982. Methods of detecting, enumerating and determining viability of microorganisms involved in biologically induced corrosion. Paper 23. Corrosion/82, Houston: NACE.

Pope, D. H., Zintel, T. P., Kuruvilla, A. K. and O. W. Siebert. 1988. Organic acid corrosion of carbon steel: A mechanism of microbiologically influenced corrosion. Paper 79. Corrosion/88. Houston, NACE.

Pope, D. H., Dziewulski, D. and J. R. Frank. 1990. *Recent Advances in Understanding Microbiologically Influenced Corrosion in the Gas Industry and New Approaches to Mitigation, GRI 90-DT-62*. Chicago, IL, Gas Research Institute.

Pope, D. H., Dziewulski, Lockwood, S. F., Werner, D. P. and J. R. Frank. 1992. *Microbiologically Influenced Corrosion of Gas Transmission Lines, GRI 92-DT-66*, Chicago, IL, Gas Research Institute.

Pope, D. H. 1993. Microbiologically influenced corrosion of internal aspects o*f natural gas industry pipelines and associated equipment: mechanisms, diagnos* is and mitigation. In *Microbiologically Influenced Corrosion*, Volume 2, ed. J. G. Stoecker. 6.13–6.25. Houston TX: NACE International.

Smith, J. S. and J. D. A. Miller. 1975. Nature of sulphides and their corrosive effect on ferrous metals: A review. *British Corros. J.* 10: 136–143.

Starkey, R. L. and K. M. Wight. 1945. *Anaerobic Corrosion of Iron in Soil*. NY, NY, Am. Gas Assoc.

Starkey, R. L. 1958. The general physiology of the sulfate reducing bacteria in relation to corrosion. *Producers Monthly* (June): 13–30.

Stephenson, M. and L. H. Strickland. 1931. Hydrogenase II. The reduction of sulphate to sulphide by molecular hydrogen. Biochem. J. 25: 215–220.

Stumper, R. 1923. La corrosion du fer en presence du sulfure de fer. *Compt. Rend.* 176: 1316–1317.

Vaughan, D. J. and J. R. Craig. 1978. *Mineral Chemistry of Metal Sulfides*. Appendix A. pp. 391. Cambridge UK: Cambridge University Press.

C. A. H. Von Wohlzogen Kuhr. 1923. Die sulfaatreductie als corzaak der nantasting vani jzeren buisleddingen (Sulfate reduction as the cause of the corrosion of cast iron pipelines) *Water and Gas VII* (no. 26): 277.

Von Wohlzogen Kuhr, C. A. H. and L. S. Van der Vlugt. 1934. The graphitization of cast iron as an electrochemical process in anaerobic soil. *Water (The Hague)* 18: 147–165.

Von Wohlzogen Kuhr, C. A. H. 1937. Unity of anaerobic and aerobic iron corrosion process in soil. In *Proceedings of the 4th National Bureau of Standards Soil Corrosion Conference*. Reprinted as Corrosion 17 (June): 293t – 299t, 1961 by NACE.

Watson J. D. and F. H. Crick. 1953. "Molecular structure of nucleic acids; a structure for deoxyribose nucleic acid. Nature. 171(4356): 737–738.

Wilmott, M. J., Jack, T. R., Geerligs, J., Sutherby, R. L., Diakow, D. and B. Dupuis. 1995. Soil probe measures several properties to predict corrosion. *Oil & Gas Journal* 93 (14): 54–58.

Woese, C. R. and G. E. Fox. 1977. Phylogenetic structure of the prokaryotic domain: The primary kingdoms. *Proc. Natl. Acad. Sci. USA*. 74 (11): 5088–5090.

2 Review and Gap Analysis of MIC Failure Investigation Methods in Alberta's Oil and Gas Sector

A.A. Abilio and J. Wolodko
University of Alberta

Richard B. Eckert
DNV GL USA

Torben Lund Skovhus
VIA University College

CONTENTS

DOI: 10.1201/9780429355479-3

2.1 INTRODUCTION

Microbiologically influenced corrosion (MIC) is a form of corrosion that is caused by the presence and activity of microorganisms such as bacteria, archaea, and fungi. It is routinely found in oil and gas production facilities and pipelines and in other sectors where a water environment is present (e.g., utilities, wastewater processing, marine). It commonly occurs as a result of the formation of biofilms or deposits on the internal surfaces of pipelines, tanks or vessels, in particularly where there is low flow or stagnant conditions. MIC can also occur on external pipe surfaces due to the presence of microorganisms in soils.

While MIC has been known as a corrosion mechanism for over a century (von Wolzogen Kühr 1923, von Wolzogen Kühr and van der Vlugt 1934), its assessment and diagnosis is still an area of ongoing research and development (Skovhus et al. 2017b, Hashemi et al. 2018, Little et al. 2020). Part of the issue is the complexity and challenge in characterizing MIC indicators conclusively. As a result, diagnosing MIC is difficult due to the synergistic interaction between a large number of biotic (microorganisms) and abiotic (chemical and physical environment) factors. From a practical point of view, this also necessitates expertise in a number of technical areas and disciplines (e.g., microbiology, chemistry, materials science, and engineering), which is often challenging to find from one source.

Although MIC has been recognized as a problem in the oil and gas industry for some time, its occurrence relative to other corrosion threats has not been well documented or quantified. Many studies in the literature regularly cite that MIC is responsible for anywhere between 10% and 40% of all corrosion issues in the sector (De Romero et al. 2000, Beavers and Thompson 2006, Rajasekar et al. 2007, Little and Lee 2009, Liengen et al. 2014, Kaduková et al. 2014, Koch et al. 2016, Hashemi et al. 2018). This range of estimated numbers; however, is often quoted without citing the original source or providing the methodology used to justify these numbers. In fact, the original sources of these prevalence numbers are difficult to find. For example, one of the most commonly cited references in the MIC-related literature is Graves and Sullivan (Graves and Sullivan 1966). In reviewing the original paper (which was published in 1966 not 1996 as commonly cited), it was determined that there, in fact, was no reference to MIC prevalence. It is unclear when the first instance of this erroneous citation occurred, but it has since been copied in subsequent MIC-related papers.

The economic impact due to MIC has also been estimated in a few studies (Maxwell et al. 2014, Rajasekar et al. 2007). However, the determination of MIC-related economic impact is often calculated simply by taking the total cost of corrosion and multiplying it by the estimated prevalence, i.e., 10% to 40% (Liengen et al. 2014). As such, in order to better inform both industry and the research community, there is a need to better quantify and document the prevalence of MIC relative to other corrosion threats.

In theory, the prevalence of MIC can be determined using a variety of methodologies: 1) the occurrence of MIC-related corrosion defects in a system based on maintenance or inspection data/reports, 2) the total cost of maintenance/repair due to MIC-related corrosion based on accounting records, or 3) the number of failure incidents due to MIC. Each of these methods will likely produce a different value or percentage relative to other corrosion threats. Of the three methods, quantifying failure incidents is likely the easiest method to estimate MIC due to the availability of public pipeline failure statistics and reports in many jurisdictions. Many of these reports include third-party failure assessments, which provide details on the likely root cause of failure including MIC.

A bigger question, however, is whether the failure analysis methodologies used in both current and historical assessments are accurate, particularly with regard to MIC, which has seen significant growth over the past few decades in terms of new knowledge and technological developments (Whitby and Skovhus 2009, Skovhus et al. 2017b, Hashemi et al. 2018, Little et al. 2020). The key technological development has been the slow adoption of DNA-based Microbiological Molecular Methods (MMM) for identifying and characterizing microorganisms in systems and continued improvement in diagnostic and integrative methodologies for assessing MIC threats and failures (Little et al. 2006, Lee and Little 2017, Eckert and Buckingham 2017, Eckert and Skovhus 2019, Kotu and Eckert 2019). In addition to microbiology, a multilayered integration between chemical, metallurgical, and operating-related evidence is required for a conclusive MIC diagnosis.

As a result, the objective of this chapter is two-fold: 1) to quantify the prevalence and most common conditions for MIC failures in the Alberta oil and gas sector over a three-year period (2017–2019), and 2) to assess the current state of the art with respect to MIC failure investigation methods and techniques, and compare these to available best practices. The intent is to highlight key areas necessary to improve diagnose of MIC failures, and to provide guidance on the development of future recommended practices.

2.2 REVIEW OF MIC FAILURE INVESTIGATION METHODS IN THE LITERATURE

There have been numerous studies published in the literature highlighting MIC failure cases and their approaches. Historically, there are two high-profile failures that are associated with MIC in the oil and gas industry: 1) Carlsbad, New Mexico and 2) Prudhoe Bay, Alaska (Kotu and Eckert 2019).

The Carlsbad incident occurred in 2000 with the explosion of a 75 cm (30 in) diameter natural gas transmission line. An ignition resulted in a rupture that killed 12 people nearby and resulted in over $1 million in damages. Investigations confirmed the presence of blackish oily solids blocking 70% of the cross section of the pipe (National Transportation Safety Board 2003, Petrović 2016). The features associated with MIC included: interconnecting pits with undercutting features; chloride concentration increasing from top to bottom inside the pits; the presence of sulfate-reducing bacteria (SRB); acid-producing bacteria (APB); and general aerobic and anaerobic bacteria at corrosion pits 630 m away from the rupture; and pH from fluid and solid debris of between 6.2 and 6.8.

The Prudhoe Bay incident occurred in 2006 on an aboveground oil production pipeline, releasing over 200,000 gallons of fluid (United States Department of

Transportation Pipeline and Hazardous Materials Safety Administration 2006, Jacobson 2007). A 6.4 mm hole at the bottom of the pipe resulted in an unnoticed leak that lasted for several days; which led to the largest oil spill on Alaska's north slope to date (Alaska Department of Environmental Conservation 2010, 2020). Microorganisms (e.g., SRB) were found to be present in the system and were associated to the internal pitting that eventually led to the through-wall perforation. As the system was working at low capacity, the stagnant flow conditions resulted in accumulation of both water and solids at the bottom of the pipe.

Both the Carlsbad and Prudhoe Bay cases, although being high-profile MIC incidents, were diagnosed without the integration of all biological, chemical, metallurgical, and operating layers of evidence together. While technology (e.g., MMM) and the understanding of MIC has significantly progressed since these incidents (in the past 15 to 20 years), the uptake and use of novel methods and techniques has still been limited in terms of analyzing suspected MIC failures.

In terms of microbiological evidence, culture-based analysis methods (as opposed to more modern MMM techniques) are still primarily used by industry due to their availability and convenience. Some failure studies have highlighted the use of DNA based tests; however, their significance and potential has not yet been fully appreciated or utilized (Sreekumari et al. 2004, San et al. 2012, Al-Nabulsi et al. 2015, Rizk et al. 2017). In particular, quantification and identification of microorganisms are not necessarily correlated with available chemical, operating, and/or metallurgical data.

Another common indicator that has been historically used to identify MIC failures is pitting morphology. A number of studies have suggested that pitting morphology alone can be used to identify the activities of specific microorganisms or microbial functional groups (Sreekumari et al. 2004). For example, terraced pits were commonly associated with SRB, while serrations and tunneling were associated with APB (Al-Sulaiman et al. 2010, Islam et al. 2016). The evidence for these types of relationships, however, is weak and does not account for the fact that these specific pitting morphologies could also be caused by abiotic conditions that often accompany MIC-related mechanisms. While MIC generally results in localized corrosion damage (pitting), this idea of MIC-specific morphological fingerprints has since been discounted (Eckert 2003, Little and Lee 2009, Lee and Little 2017).

The presence of specific chemical signatures (i.e., elements or microbial nutrients) is also common in historical studies. For example, high amounts of both carbon and oxygen (between 20% to 30% and above) from chemical analyses of solid samples were used as indicators for MIC based on the assumption that the presence of biomass (biofilm) would be an additional source of such elements (Liu 2014b) in seawater systems. In oil-related systems, both oxygen and carbon can be present from sources other than biofilms. For instance, organic acids (i.e., carbon sources) are abundantly present in oil and gas systems while oxygen may be related to CO_2 conditions and not necessarily indicative of biofilms.

Finally, the integration of data during MIC failure analyses has also evolved over the past number of years. For example, many studies have integrated nutrient availability in MIC failure analyses (Rao and Nair 1998, Renner 1998, Samant et al. 1999, Borenstein and Lindsay 2002, Garber and Chakrapani 2004, Starosvetsky et al. 2007, Al-Sulaiman et al. 2010, Xu et al. 2013, Liu 2014a, Al-Nabulsi et al.

2015, Prithiraj et al. 2019), while only fewer carried out DNA testing (Sreekumari et al. 2004, San et al. 2012, Al-Nabulsi et al. 2015, Rizk et al. 2017). The lack of failure investigations that use both chemical and microbiological evidence over the years illustrate that procedures and integration steps to diagnose MIC are still inconsistent.

In summary, MIC failure case studies in the literature over the past two decades (1998 to 2019) have shown a lack of procedural consistency with respect to the diagnosis of MIC. As the understanding of MIC has evolved, testing and integration methodologies have improved over time. However, even newer case studies often do not fully adopt the current best practices and instead rely on older approaches to determine the root cause of corrosion which, ultimately, may not be accurate or conclusive.

2.3 OVERVIEW OF ALBERTA'S OIL AND GAS SECTOR AND REGULATORY STRUCTURE

The Province of Alberta is the center of the oil and gas industry in Canada with the third largest proven oil reserves in the world (only behind Venezuela and Saudi Arabia). The sector accounts for 97% of Canada's petroleum inventory (Alberta Government 2020), and is responsible for approximately 82% of Canada's oil production (Government of Canada 2020). The vast majority of Alberta's reserves (over 90%) are related to unconventional production, which includes in-situ heavy oil and mined oilsands. Since the first discovery of oil in the province over 70 years ago, the oil and gas sector has grown substantially in Alberta to create Canada's largest petrochemical supply chain that includes upstream production (conventional and unconventional), pipeline transport, and downstream processing (upgrading, refining, and associated petrochemical processing).

Regulation of these pipelines mainly fall under the jurisdiction of two government bodies: a) the Alberta Energy Regulator (AER) which regulates all pipelines that do not cross provincial or international borders, and b) the Canadian Energy Regulator (CER) that regulates all pipelines (mainly large transmission lines) that cross provincial or international borders. Within Alberta, the AER regulates over 433,000 km of pipelines that are used for the transportation of both produced crude and refined products, and includes operating, discontinued, abandoned, and permitted pipelines (Alberta Energy Regulator 2020). These pipelines consist of a broad range of pipe sizes (upstream, midstream, and transmission pipelines) and products types (e.g., crude oil, water, natural gas, sour gas, natural gas liquids, refined products). The AER enforces compliance with various Canadian legislative and technical standards and requirements (e.g., CSA Z662:19, Pipeline Act, Pipeline Rules) throughout the operating life of pipelines and facilities in order to ensure public safety and environmental protection (Alberta Energy Regulator Report 2013). The regulatory framework starts with the application process (before construction begins), and extends to inspection of the construction, operation, and maintenance (integrity management), monitoring, surveillance, leak detection, emergency response, discontinuation and abandonment.

In the event of an incident, the responsible party (e.g., operating company) must report it immediately to the regulatory bodies, and is responsible for all levels of incident response (e.g., isolation, containment, recovery of lost product, remediation, and reclamation). In the case of a pipeline damage incident, the operating companies are required to investigate and implement measures to prevent reoccurrence (CSA Z662:19 2019). The AER conducts the incident review by assessing the information the companies provide. Companies often use third-party labs and consultants to aid in their assessment (e.g., materials failure analysis labs). The regulatory bodies provide oversight to this process in order to understand the causes of the incident and to ensure that compliance is met and deficiencies are addressed. As failure assessments are conducted, various pieces of evidence are gathered, including failure reports carried out by third-party consultants to identify the degradation mechanism that led to the failure. Information of pipeline incidents are made publicly available by both federal and provincial regulators to provide transparency to both the public and industry, and to better track and classify these incidents in order to ultimately identify trends, improve safety, and ensure compliance.

In Alberta, a dedicated database managed by AER is made publicly available for incident classification and inventory. Pipeline-related incidents in the province are classified under a number of different damage types (Alberta Energy Regulator Report 2019), as shown in Table 2.1. These include important damage modes that commonly occur during normal operation of pipelines such as leaks, ruptures, and mechanical damage; and also include damage modes that occur during pipeline construction, commissioning and maintenance, such as integrity tests and pressure tests (specific definitions of each damage type are provided as a footnote in Table 2.1). Leaks are further classified into a number of descriptors including "leaks" (which typically encompass failures in main pipe bodies due, for example, to corrosion), "GSPT Releases" (which include any releases due failure of gaskets, seals, packing glands, or threaded fittings), and "installation leaks" (which encompass releases specifically at auxiliary sites along the pipeline such as compressor, pumping, or metering stations).

Table 2.1 also lists the number of incidents in the province of Alberta based on damage type between January 1, 2017, and December 31, 2019. Of the 1,479 total incidents during the three-year period, "leaks" account for the majority (62%) of pipeline-related incidents in the province, while "GSPT releases" (i.e., leaks in gaskets or fittings) represent 16% of total incidents. Releases resulting from pipeline "integrity tests" and "pressure tests" only represent 8% and 4% of all incidents, respectively. It should be noted that failures from these tests are somewhat beneficial in that they successfully locate weaknesses in the system and do not take place under normal operating conditions (Alberta Energy Regulator Report 2013). While pipeline "ruptures" (bursts) have the potential for most harm due to an abrupt release of fluid content, it can be seen that these are relatively infrequent (only 3% of all incidents over the three-year period).

In addition to the main damage types outlined in Table 2.1, the AER database also categorizes specific failure modes within these broad damage types. For pipeline corrosion failures, for example, two categories have been historically used to track incidents since 1975: "internal corrosion" and "external corrosion." While

TABLE 2.1

Number of Pipeline Incidents by Pipe Damage Type in Alberta, Canada, from January 1, 2017, to December 31, 2019, as per the AER Database Classification (Alberta Energy Regulator Report 2019)[a,b]

Pipe Damage Type Category	2017	2018	2019	Total Number of Incidents per Category	Total Percentage per Category
Leak	309	324	291	924	62%
GSPT Release	63	79	80	222	16%
Integrity Test Failure	44	46	30	120	8%
Hit	41	28	21	90	6%
Pressure Test Failure	22	18	20	60	4%
Rupture	21	12	14	47	3%
Installation Leak	2	5	9	16	1%
Total per Year	502	512	465	1,479	100%

Notes

a In the table, "Leak" refers to incidents where the substance released does not immediately stop the operation. "GSPT Release" refers to leaks specifically due to failure (or lack of sealing) of gaskets, seals, packing glands, or threaded fittings (GSPT). "Integrity Test Failure" are leak incidents originating from failures occurring during an integrity test. "Pressure Test Failure" refers to a failure during qualification of new pipe construction. A "Hit" is an incident that occurs due to contact damage to a pipe or its coating due to a ground disturbance event that do not result in substance release (e.g., backhoe impact during construction). "Rupture" means a burst where the pipeline immediate ceases to operate. "Installation Leak" relates to releases which take place at auxiliary sites along the pipeline (e.g., compressor, pumping, or metering stations).

b Note that these statistics are dynamic in nature, and may not represent the AER database at any specific point in time due to possible changes during ongoing reviews.

these two general categories are useful for classifying and tracking corrosion failures, they do not identify and differentiate specific mechanisms involved in the incident. Subcategories have historically been added to the database to better identify these various mechanisms, and are based on the failure assessments by the operator or third-party consultants. This extended classification allows for improved tracking and classification of common corrosion issues.

2.4 PREVALENCE OF MIC-RELATED FAILURES IN ALBERTA

This section aims to quantify the prevalence of MIC-related failure incidents in Alberta. The intent is to provide a better understanding of failure statistics associated with MIC in comparison with other common failure modes. The data was further analyzed based on operating factors such as pipe diameter/length and fluid type to identify where MIC is most commonly found.

This analysis was conducted by reviewing the AER's internal database and files associated with all corrosion incidents classified in the "leak" damage category over a three-year period from January 1, 2017, to December 31, 2019. For this period, there was a total of 573 corrosion-related incidents with 447 classified as internal corrosion (78% of all corrosion-related incidents) and 126 classified as external corrosion failures (22%). Each of these 573 incident files were manually reviewed to flag those cases where MIC was identified as either the main corrosion mechanism or a contributing cause. This was determined based on third-party failure reports and/or by information provided by the pipeline operator.

Overall, there was a total of 67 MIC-related failures (or 11.7%) out of the 573 corrosion-related leak incidents reported over the three-year period. These MIC-related incidents were then subdivided into internal corrosion failures or external corrosion failures based on categorizations in the database and files. As shown in Tables 2.2 and 2.3, MIC was involved in 13.6% of all internal corrosion incidents (61 out of 447 cases) and 4.8% of all external corrosion incidents (6 out of 126 cases), respectively. For the internal corrosion cases, MIC was found to be the main cause or a contributing factor under a number of corrosion subcategories in the AER database including microbiologically influenced corrosion (as expected), multi-mechanism corrosion, under deposit corrosion, CO_2 corrosion, and others. For the external corrosion cases, MIC occurred in two subcategories: equipment failure (external) and coating disbondment. This broad range of categories where MIC was found highlights the breadth and complexity of potential MIC impacts, and the associated challenges faced by the industry and regulators in assessing and classifying MIC-related failures.

It should be noted that these numbers reflect failures caused by MIC and not the prevalence of MIC found from inspection and integrity management programs (i.e., non-failure events). To the authors' knowledge, these MIC failure statistics (13.6% of all internal corrosion incidents and 4.8% of all external corrosion incidents) represent the first time that the prevalence of MIC has been accurately quantified in the literature based on a single, reliable data set.

To further understand where MIC failures most commonly occur, the 67 MIC-related cases were further categorized based on the pipeline contents and size of pipe. Table 2.4 highlights the number of MIC incidents and their frequency (per km per year) for various pipeline fluid types. It can be seen that MIC is most prevalent in water and multiphase pipelines at 150×10^{-5} and 45×10^{-5} MIC incidents per km per year, respectively. This translates to one MIC failure (on average) every 667 km in water pipelines and every 2,222 km in multiphase pipelines, each year.

This same analysis was also performed based on pipeline outside diameter (O.D.), as shown in Table 2.5. This table shows that all recorded MIC incidents occurred in relatively small-diameter pipelines (i.e., 220.3 mm O.D./8 in. nominal or less), which are more representative of upstream production flowlines and collection systems. Furthermore, a majority of MIC failures occurred in pipelines with outside diameters less than 101.6 mm (3½ in. nominal pipe sizes or less). The highest MIC failure frequency was 20×10^{-5} MIC incidents per km per year which was found in pipelines with an O.D. between 82.6 mm and 101.6 mm (3 in. and 3½ in. nominal pipe diameters). This translates to one MIC failure (on average) every 5,000 km per year. The next-highest MIC failure frequency was 18×10^{-5} MIC

TABLE 2.2
Breakdown of All Internal Corrosion-Related "Leak" Incidents in Alberta Based on Reported Damage Mechanism (Including MIC) over a Three-Year Period from January 1, 2017, through December 31, 2019

Subcategories for Internal Corrosion	Total Incidents per Subcategory	Number of MIC-Related Failures	MIC-Related Failures as a Percentage of Total Internal Corrosion Incidents
Equipment Failure—Internal corrosion	140	19	4.3%
Multi Mechanism	124	13	2.9%
Under Deposit Corrosion	60	7	1.6%
Microbiologically Influenced Corrosion	27	19	4.3%
CO_2 Corrosion	24	2	0.4%
Corrosion Under Internal Coating	18	1	0.2%
Interference Corrosion	13	–	–
Corrosion at Valve or Fitting	7	–	–
H_2S Corrosion	6	–	–
Corrosion at Internally Coated Riser	5	–	–
Preferential Weld Corrosion	5	–	–
Corrosion Behind Plastic Liner	4	–	–
Other	4	–	–
Corrosion Under Cement Lining	3	–	–
Oxygen Induced Corrosion	3	–	–
Corrosion at Inline Coupler	2	–	–
Erosion Corrosion	2	–	–
Total Internal Corrosion Incidents	447	61	13.6%

incidents per km per year which was found in pipelines with an O.D. less than 80.9 mm (2½ in. nominal pipe sizes or less). This translates to one MIC failure (on average) every 5,556 km each year. The main reason for the high MIC failure frequencies in small diameter pipelines is likely due to the fact that many of these

TABLE 2.3
Breakdown of All External Corrosion "Leak" Incidents in Alberta Based on Reported Damage Mechanism (Including MIC) over a Three-Year Period from January 1, 2017, through December 31, 2019

Subcategories for External Corrosion	Total Incidents per Subcategory	Number of MIC-Related Failures	MIC-Related Failures as a Percentage of Total External Corrosion Incidents
Equipment Failure—External Corrosion	43	4	3.2%
Coating Disbonded or Shielding	34	2	1.6%
Missing or Damaged Coating	30	–	–
Corrosion under Insulation	11	–	–
Atmospheric	5	–	–
Soil-to-Air Interface	2	–	–
Corrosion at Valve or Fitting	1	–	–
Total External Corrosion Incidents	126	6	4.8%

systems have significant water content in the fluid (i.e., produced water or unprocessed crude emulsions), which is a necessary precursor for MIC to occur. Additionally, smaller diameter lines are often more difficult to inspect and maintain (e.g., pig) relative to larger-diameter pipelines that may also contribute to the frequency of MIC incidents.

Finally, the fluid type and outer diameter data sets were combined to identify specific field conditions where MIC was found to be most prevalent. As shown in Table 2.6, water pipelines with outer diameters smaller than 80.9 mm (2-1/2 in. nominal pipe size) and 114–116.8 mm (4 in. nominal pipe size) had the highest MIC failure frequencies at 470×10^{-5} and 247×10^{-5} incidents per km per year, respectively. This translates to one MIC failure (on average) every 213 and 405 km, respectively, each year. The remainder of fluid-diameter combinations in Table 2.6 have much lower failure frequencies, and consist of a mixture of multiphase and water pipelines of varying diameters. Natural gas pipelines had the lowest MIC failure frequencies between 3×10^{-5} and 10×10^{-5} incidents per km per year, suggesting these lines were typically carrying dehydrated natural gas.

TABLE 2.4
Frequency of MIC Incidents by Total Length of Steel Pipelines for Various Carrier Fluid Types over a Three-Year Period from January 1, 2017, through December 31, 2019

Fluid Type	Number of MIC Incidents over 3-Year Period	Total Steel Pipeline Length (km)[a]	MIC Incidents per km per Year
Water	11	2,438	150×10^{-5}
Multiphase	31	22,735	45×10^{-5}
Natural Gas	23	152,913	5×10^{-5}
Sour Gas	1	13,952	2×10^{-5}
Other	1	31,937	1×10^{-5}
Crude Oil	0	13,809	0
Total	67	237,784	N/A

Note

a Note that these lengths only represent steel pipelines (excludes non-metallics) that are either in operation or discontinued (excludes abandoned and permitted lines).

2.5 REVIEW AND GAP ANALYSIS OF MIC FAILURE INVESTIGATION METHODS

In this section, MIC failure investigation and analysis methods were reviewed from assessments of recent pipeline failure incidents in Alberta over a three-year period from January 1, 2017, to December 31, 2019 (see previous section for more details). Fifty failure assessments were sampled for the period of interest and each was reviewed to identify the information that was typically collected during these assessments and how this information was used to confirm MIC. These assessments consisted of reports from third-party consultants or from summaries prepared by AER inspectors based on available information/data from the incident and operations. All incident information regarding specific operators, asset name and location, and the third-party consultant has been blinded to ensure confidentiality.

As shown in Table 2.7, a broad range of assessments was used in this review, covering cases where MIC was either identified as the primary failure mechanism or as a contributing (secondary) factor in conjunction with some other possible failure mechanism (e.g., CO_2 or H_2S corrosion). Out of the 50 incidents that were reviewed, 19 were categorized as "MIC," 10 as "multi-mechanism", 6 as "under deposit corrosion", 1 as "under internal coating" while 14 were categorized simply as internal corrosion (subcategory not discriminated). Although there is a dedicated category for MIC, it can be seen that MIC is often reported in addition with several other failure causes.

The information compiled from these assessments was then compared to the latest best practises used to diagnose MIC failures as determined through expert elicitation and from methodologies published in the open literature (Eckert and

TABLE 2.5
Frequency of MIC Incidents by Total Length of Steel Pipelines for Various Outer Diameters of Pipe over a Three-Year Period from January 1, 2017, through December 31, 2019

Pipe Outer Diameter	Number of MIC Incidents over 3-Year Period	Total Steel Pipeline Length (km)[a]	MIC Incidents per km per Year
82.6 mm–101.6 mm (3 in. and 3½ in. nominal)	29	49,205	20×10^{-5}
≤80.9 mm (≤2½ in. nominal)	11	20,698	18×10^{-5}
114.0 mm–116.8 mm (4 in. nominal)	14	63,405	7×10^{-5}
139.7 mm–177.8 mm (5 in. and 6 in. nominal)	9	47,107	6×10^{-5}
219.0 mm–220.3 mm (8 in. nominal)	4	20,782	6×10^{-65}
≥255.3 mm (≥10 in. nominal)	0	36,587	0
Total	67	237,784	N/A

Note
a Note that these lengths only represent steel pipelines (excludes non-metallics) that are either in operation or discontinued (excludes abandoned and permitted lines).

Buckingham 2017, Eckert and Skovhus 2018, Kotu and Eckert 2019). The intent of this exercise was to identify gaps in current MIC failure assessment methodologies, and suggest improvements on how industry and regulators can better identify potential MIC incidents. To accomplish this task, a checklist of information and methods was created based on the current best practises, and was used to evaluate each of the 50 MIC assessments from the AER database to identify potential gaps in both the data and/or analysis methods used. The checklist was grouped into five major information and analysis groups, as shown in Figure 2.1. These categories outline the main layers of evidence required to conduct a successful assessment for MIC. These include four key information groups (microbiological, chemical, metallurgical, and operating data) and a data integration step, which is required to properly identify MIC as a possible failure mechanism. In addition to these five main categories, this gap analysis also examined a number of other important factors that may affect a successful MIC diagnosis, including proper sampling and the availability/use of standards.

Details for each of these categories and results from the gap analysis are provided in the following sections. The results are presented in comprehensive figures

TABLE 2.6

Frequency of MIC Incidents by Total Length of Steel Pipelines for Various Outer Diameters and Fluid Types of Pipe over a Three-Year Period from January 1, 2017, through December 31, 2019

Fluid Type	Pipe Outer Diameter	Number of MIC Incidents over 3-Year Period	Total Steel Pipeline Length (km)[a]	MIC Incidents per km per Year
Water	≤80.9 mm (≤2½ in. nominal)	7	497	470×10^{-5}
Water	114.0 mm–116.8 mm (4 in. nominal)	3	405	247×10^{-5}
Multiphase	82.6 mm–101.6 mm (3 in. and 3½ in. nom.)	18	9,297	65×10^{-5}
Water	82.6 mm– 101.6 mm (3 in. and 3½ in. nom.)	1	594	56×10^{-5}
Multiphase	114.0 mm–116.8 mm (4 in. nominal)	7	5,155	45×10^{-5}
Multiphase	219.0 mm–220.3 mm (8 in. nominal)	2	1,526	44×10^{-5}
Multiphase	≤80.9 mm (≤2½ in. nominal)	2	2,037	33×10^{-5}
Multiphase	139.7 mm–177.8 mm (5 in. and 6 in. nominal)	2	3,692	18×10^{-5}
Other	219.0 mm–220.3 mm (8 in. nominal)	1	3,043	11×10^{-5}
Natural Gas	82.6 mm–101.6 mm (3 in. and 3½ in. nom.)	10	34,323	10×10^{-5}
Sour Gas	139.7 mm–177.8 mm (5 in. and 6 in. nominal)	1	4,594	7×10^{-5}
Natural Gas	139.7 mm–177.8 mm (5 in. and 6 in. nominal)	6	34,309	6×10^{-5}
Natural Gas	≤80.9 mm (≤2½ in. nominal)	2	12,301	5×10^{-5}
Natural Gas	219.0 mm–220.3 mm (8 in. nominal)	1	11,521	3×10^{-5}
Natural Gas	114.0 mm–116.8 mm (4 in. nominal)	4	50,005	3×10^{-5}
Remaining Combinations of Fluid Type and Pipe Outer Diameter		0	64,486	0
Total		67	237,784	N/A

Note

a Note that these lengths only represent steel pipelines (excludes non-metallics) that are either in operation or discontinued (excludes abandoned and permitted lines).

TABLE 2.7
Breakdown of the 50 MIC Failure Assessments Considered in This Study Based on Corrosion Subcategory and Primary/Secondary Contribution

Corrosion Subcategory Classification Database	Number of Reviewed Assessments Where MIC Was the Primary Mechanism	Number of Reviewed Assessments Where MIC Was a Secondary Mechanism	Total Number of Reviewed Assessments
Microbiologically Influenced Corrosion	*17*	2	*19*
Multi-Mechanism	4	6	10
Under Deposit Corrosion	2	4	6
Under Internal Coating	1	–	1
<Subcategory not discriminated>	8	6	14
Total	32	18	50

that identify whether a particular MIC failure assessment (numbered 1 thru 50) has included a specific data set/analysis method or not. This allows quantification of select layers of evidence, and an overall indicator of whether certain analytical practises are being followed. In addition, the same 50 MIC failure cases are used throughout the remainder of the chapter to ensure consistency in the gap analysis.

2.5.1 Overall Summary

One goal of a failure assessment is to understand why the failure took place at a specific location and the underlying cause. For assessing suspected MIC-related corrosion failures, there is a need to determine whether the failure mechanism was due to abiotic factors (e.g., CO_2, H_2S, O_2, Cl^- in the system), biotic factors (due to microbiological processes), or a combination of both. It is the contrast and integration between the multiple layers of evidence, as shown in Figure 2.1 (microbiological, chemical, metallurgical and operating data, and its integration), that can lead to the most conclusive and reliable diagnosis given the limits of current technology and knowledge.

In general, all five of these layers of evidence were to some extent fulfilled in a majority of the 50 MIC assessments reviewed, as listed in Table 2.8. Over 90% of the 50 MIC assessments reviewed took chemical data, operating data, metallurgical data, and integration steps into consideration. However, the greatest gap is observed for microbiological data, where only 70% of the MIC assessments reviewed conducted some sort of microbiological analysis. While this shows that most of the MIC assessments incorporated the five general categories of information in their reports (other than microbiological), a larger discrepancy can be seen when exploring specific details within each category as outlined in the following sections.

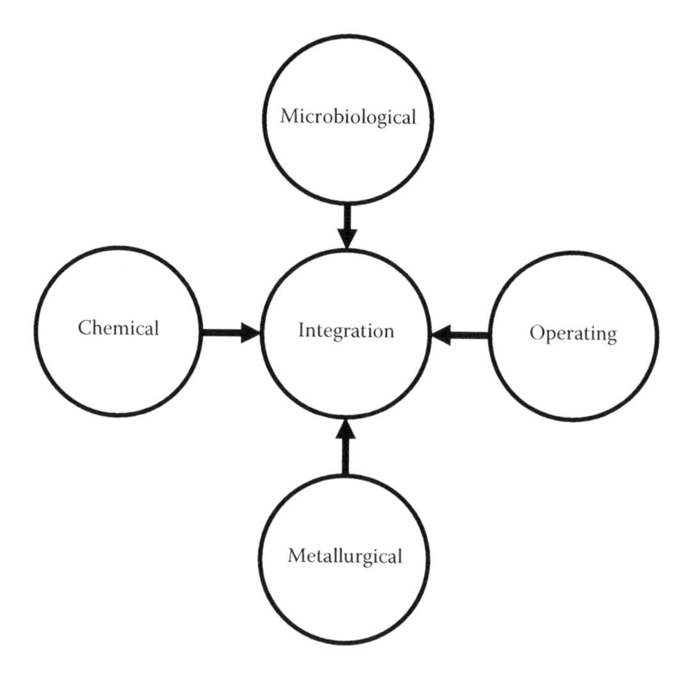

FIGURE 2.1 Major information and analysis groups required for conclusive and reliable MIC diagnosis.

2.5.2 Reported Microbiological Data

Microbiological information is a key component to identify MIC. MIC can only occur if specific types of microbial functional groups (MFG) are present at the surface of the pipe as part of a biofilm (sessile microorganisms). Furthermore, these microorganisms must be active for MIC to occur. It should be noted that many species of microorganisms commonly exist in oil and gas systems; however, not all are responsible for MIC. For this reason, it is important to assess three key microbiological factors in the investigation: microbial diversity, activity, and abundance. Diversity characterizes the community of microorganisms present in a system (both MIC and non-MIC-related functional groups). Activity indicates whether the microorganisms present are actively involved in metabolic functions (versus those that are either dormant or deceased). Finally, abundance quantifies the numbers of specific microorganisms present in the system.

A detailed listing of the microbiological information used in the 50 MIC failure assessments is shown in Figure 2.2, including sample type and whether microbiological analysis was performed using modern MMM or older culture-based techniques (e.g., Most Probable Number [MPN] from serial dilution or Biological Activity Reaction Tests, [BART]). The figure represents a matrix highlighting the pieces of information or analyses that were performed (rows) for each MIC failure assessment 1 through 50 (columns). Those intersections that are marked in "black" indicate that a given piece of information or method was used in that particular failure assessment/report. The failure assessment numbers are consistent throughout

TABLE 2.8
Inclusion of Key Information and Analysis Groups in the Reviewed MIC Failure Assessments

Information and Analysis Group	Percentages of Failure Assessments That Included Specific Information and Analysis Groups
Microbiological	70%
Chemical	92%
Metallurgical	98%
Operating	92%
Integration	90%

each figure in the remainder of the chapter, which allows the reader to cross-reference different data sets based upon the same set of data.

2.5.2.1 Microbiological Sample Type

As shown in Figure 2.2, a majority of the 50 MIC failure assessments performed microbiological testing using liquid samples (52%), followed by solids (10%) and surface swabs (16%). In terms of best practices, swabbing at or near the failure location is considered the most representative microbiological sample since MIC is highly dependent on activity happening at the pipeline surface. Microbiological sampling of solids is also considered a reasonable indicator if the sample is related to the corrosion location. Both swab and solid samples represent potential sessile microorganisms (those found to congregate at surfaces in the form of biofilms), while fluid samples can only identify planktonic (free-floating) microorganisms in the system. While fluid samples are relatively easy to obtain, it has been shown that planktonic microorganisms do not necessarily represent sessile (surface) populations and should only be used as supplementary information (Jack 2002, Sooknah et al. 2007, Larsen et al. 2008, Wrangham and Summer 2013, Eckert and Skovhus 2018).

It should be noted that a number of failure reports commented that no microbiological samples were collected due to the absence of liquid material on the cut-out pipe samples. In such cases, surface solids collection via swabbing is still a viable option for microbiological assessment. However, in some cases, swab or solid sampling may not be possible due to poor preservation of the failure location, or possible disruption of the surface during the investigation (e.g., inadvertent cleaning to examine the corrosion defect). In these cases, fluid sampling may be the only option.

2.5.2.2 Microbiological Analysis Methods

In terms of microbiological testing methodologies, 34% of assessments used some aspect of newer MMM while 44% of the assessments used culture-based techniques including MPN or BART, as shown in Figure 2.2.

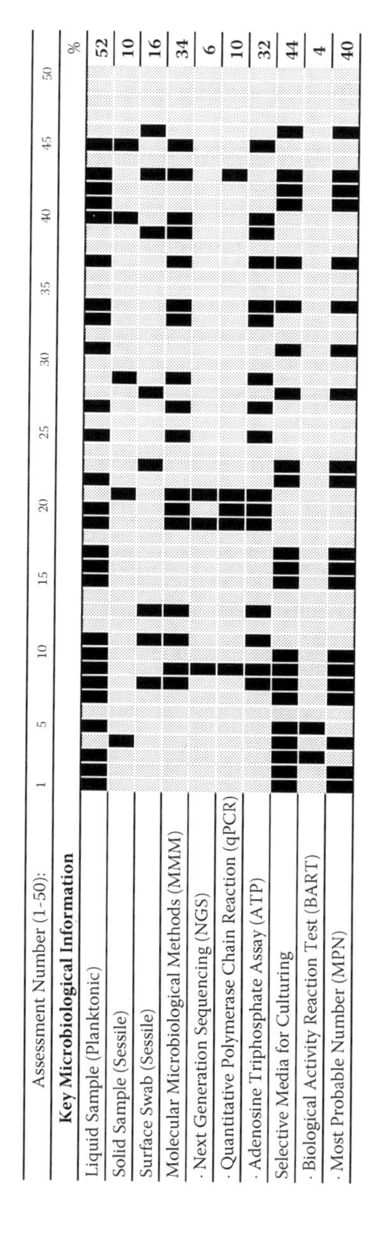

FIGURE 2.2 Inclusion of specific microbiological information and analysis methods for each of the failure assessments reviewed (1 through 50).

Of the MMM tests conducted, the Adenosine Triphosphate assay (ATP) was most commonly used (32%) followed by Quantitative Polymerase Chain Reaction (qPCR) at 10% and Next Generation Sequencing (NGS) at 6%. ATP measures microbiological activity in the system which is critical for MIC to occur. NGS measures microbiological diversity, which can identify a range of MIC-related MFG present in any given sample, and is a key factor for identifying MIC as the links between microbiology and chemistry (both from potential corrosion products to energy sources and electron acceptors in the environment measured in other samples). qPCR measures the abundance of a particular microorganism or functional gene present in a sample and it may be a good indicator of the MIC-related microorganisms that are most likely responsible for the observed corrosion. In terms of portability, ATP assays can be performed in the field (Fichter and Summer 2017), while qPCR and NGS testing are still moving to be more portable and are mostly conducted in a laboratory setting. However, results from ATP tests only represent microbiological activity at one specific point in time (i.e., one value does not indicate historical activity). In all three cases (ATP, qPCR, and NGS), tests should be conducted on actual or preserved samples as soon as possible to ensure representative results.

DNA sequencing of 16S ribosomal RNA (rRNA) genes via NGS allows for the identification of all bacteria and archaea present in the system, while qPCR of functional genes uses primers targeted to specific MFG based on their metabolic pathways (Lee and Little 2017, Eckert and Skovhus 2019). When it comes to assessing microbial diversity, activity and abundance to diagnose MIC, MMM outputs reliable results that can be linked to other layers of evidence, such as corrosion products and environment chemical composition (Larsen et al. 2008, Whitby and Skovhus, 2009, Larsen et al. 2013, Skovhus and Eckert 2014, Fichter and Summer 2017, Skovhus et al. 2017a, De Paula et al. 2018, Eckert and Skovhus 2018, Kotu and Eckert 2019, Salgar-Chaparro and Machuca 2019, Gieg et al. 2020).

Of the culture-based methods conducted, the determination of MPN via serial dilution was most common (40%) with only a few of the assessments using BART (4%). MPN and BART both identify the presence and quantity of specific microorganism groups or traits (e.g., slime-forming bacteria) based on their response to conditions in the culture medium. Many third-party consultants and test labs are familiar with these methods, and prefer them due to availability, low cost and ease of use (e.g., they do not require sophisticated equipment for analysis). Culture-based methods can also provide some clues regarding the operating environment inside the pipeline if properly linked to other layers of evidence (such as chemical, corrosion, and/or operating data). However, the main challenge with culture-based methods is the fact they can only identify a limited number of MIC related microorganisms based on selective growth media available. As a result, they may not be able to characterize the full diversity of the actual microbiological community present in the sample (Lee and Little 2017). By contrast, MMM-based NGS has a distinct advantage in that it can provide a full analysis of the microbial community present in any given sample. This does not preclude the use of culture-based methods; however, there is a risk that some MIC-related microorganism present in the system are not being properly identified (i.e., false negatives due to a lack of selective media for a particular microbial function group). Another challenge with culture-based methods is that they do not distinguish between active and dormant microorganisms in the actual environment since all living microorganisms (whether active or dormant) can grow during lab-based culturing. Conversely, commercially available ATP assays currently only identify those microorganisms that are active (and not dormant) at the time of testing.

The difference between MMM and culture-based methods in their ability to measure microbiological diversity can be clearly seen in Figure 2.3 for each of the MIC failure incidents studied. Note that the figure only represents the types of tests conducted, not necessarily the microorganisms that were found. Those assessments that used MMM (NGS and qPCR) were able to distinguish a wide range of possible MFG, including sulfate-reducing bacteria, sulfate-reducing archaea, thiosulfate-reducing bacteria, methanogenic archaea, acid-producing bacteria, nitrate-reducing bacteria, sulfate-oxidizing bacteria, iron-reducing bacteria, iron-oxidizing bacteria, manganese-reducing bacteria, and manganese-oxidizing bacteria. Conversely, those failure assessments that used culture-based techniques were only able to test for a limited number of MIC-related microbial functional groups (e.g., APB, SRB). The list of culture-based tests; however, also includes assays for a number of broad groups of microorganisms, such as aerobic (AERO), anaerobic (ANA), iron-related bacteria (IReB), low nutrient bacteria (LNB), slime-forming bacteria (SLYM), and heterotrophic anaerobic bacteria (HAB). While these media may provide some additional information on the pipeline environment, they have limited use in assessing MIC directly as they are not specific to MIC-related microorganisms. Historically, these general groups were often characterized to compensate for the limited ability to identify the microbiological diversity of MIC species. The intent was to look for other traits that could suggest the presence of microorganisms in the system. With the increasing availability (and declining cost) of more comprehensive Microbiological Molecular Methods, these assays are not as relevant today.

In terms of microorganisms linked to the failure assessments reviewed, 60% of all failure incidents examined were attributed to either SRB and/or APB, while the remaining 40% of assessments diagnosed MIC without identification of any specific MFG. Instead of using available microbiological tests, such as serial dilution, qPCR or NGS, some of these latter assessments inferred MIC-related MFG or mechanisms based on either chemical analysis of the corrosion products (26%) or visual analysis of the pitting morphology (12%). While analysis of corrosion products is an important step in confirming specific MIC mechanisms, it alone cannot give a conclusive diagnosis of MIC without other layers of evidence including microbiological diversity data. Conversely, correlating corrosion morphology with specific MFGs has been shown to be inconclusive since common MIC morphologies can also be associated with abiotic corrosion mechanisms (Eckert 2003, Little et al. 2006, Little and Lee 2009, Lee and Little 2017). It should be pointed out that a number of these MIC assessments based solely on corrosion products or morphologies, also conducted ATP measurements to confirm the presence of microbiological activity in the system. However, ATP results are unable to confirm whether any of the "active" microorganisms detected are, in fact, MIC related.

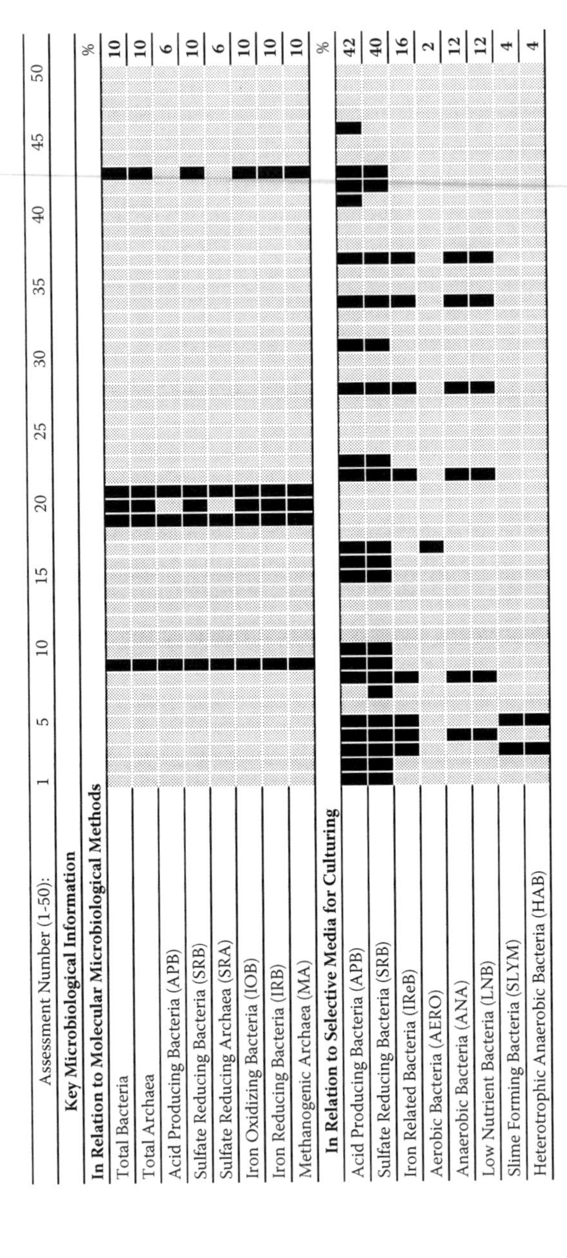

FIGURE 2.3 Contrast between the microbial functional groups (MFG) identified by molecular microbiological methods and Culture-based methods for each of the failure assessments reviewed (1 through 50).

2.5.3 Reported Chemical Analysis Data

Chemical analysis data can validate or invalidate MIC as a viable failure mechanism. The hypothesis is based on the contrast between corrosion products (e.g., solids) and the corrodents present in the water and in the gas. Therefore, the chemical composition of solids, liquids, and gases should be known in order to answer to what extent the environment supported either microbiological processes or abiotic mechanisms (e.g., CO_2, H_2S).

Figure 2.4 lists the chemical compounds tested in the different MIC failure assessments for solids, liquids, and gases. As Table 2.8 displays, 92% of the failure assessments gathered some sort of chemical data; supporting the importance of chemical considerations for a conclusive failure diagnosis and also demonstrating a better industry familiarity with abiotic drivers.

The importance of chemical evidence to the MIC investigation framework is twofold: 1) it indicates the abiotic and biotic potential for corrosion by comparing dissolved chemical species in the water and in the gas with those found in the corrosion products; and 2) it provides evidence for linking microbiological diversity with corrosion products and dissolved nutrients (i.e., energy sources and electron acceptors) in the water.

2.5.3.1 Reported Solids Data

Identification of corrosion products is used as evidence to show that a corrosion process took place. In order to track whether the source of corrosion products is abiotic or biotic, their composition is contrasted to the chemistry of the water and the gas in the system, along with microbiological community characterization. As Figure 2.4 shows, 82% of incidents provided characterization of solids, either by qualitative chemical spot testing (66%), elemental analysis (46%, Energy Dispersive X-ray Spectroscopy [EDS]), and/or mineral analysis (42%, X-Ray Diffraction [XRD]). Spot testing was most routinely performed due to its simplicity and low cost. Qualitative spot testing is commonly used as a high-level screening step to identify expected corrosion products and was used to check for sulfides (62%), carbonates (54%), and chlorides (44%).

EDS and XRD analyses of solids provide more specific evidence than spot testing as EDS identifies the chemical species at an elemental level (e.g., sulfur, carbon, oxygen, iron, calcium, etc.), while XRD identifies crystalline (e.g., oxides, carbonates, sulfides, etc.) compounds (Craig 2002, Waseda et al. 2011, Goldstein et al. 2018). Iron carbonates (40%) and iron sulfides (34%) were the compounds most found by XRD. Mackinawite, an iron sulfide compound, was found in 26% of the failure assessments, accounting for 13 of the reviewed incidents. In 11 of these incidents, SRB were diagnosed as a contributing cause, showcasing how strongly SRB-driven MIC is currently attributed to the presence of mackinawite. The correlation between mackinawite as a corrosion product and SRB is believed to be more significant when the system is known to have no obvious sources of hydrogen sulfide (the main abiotic cause of iron sulfide formation).

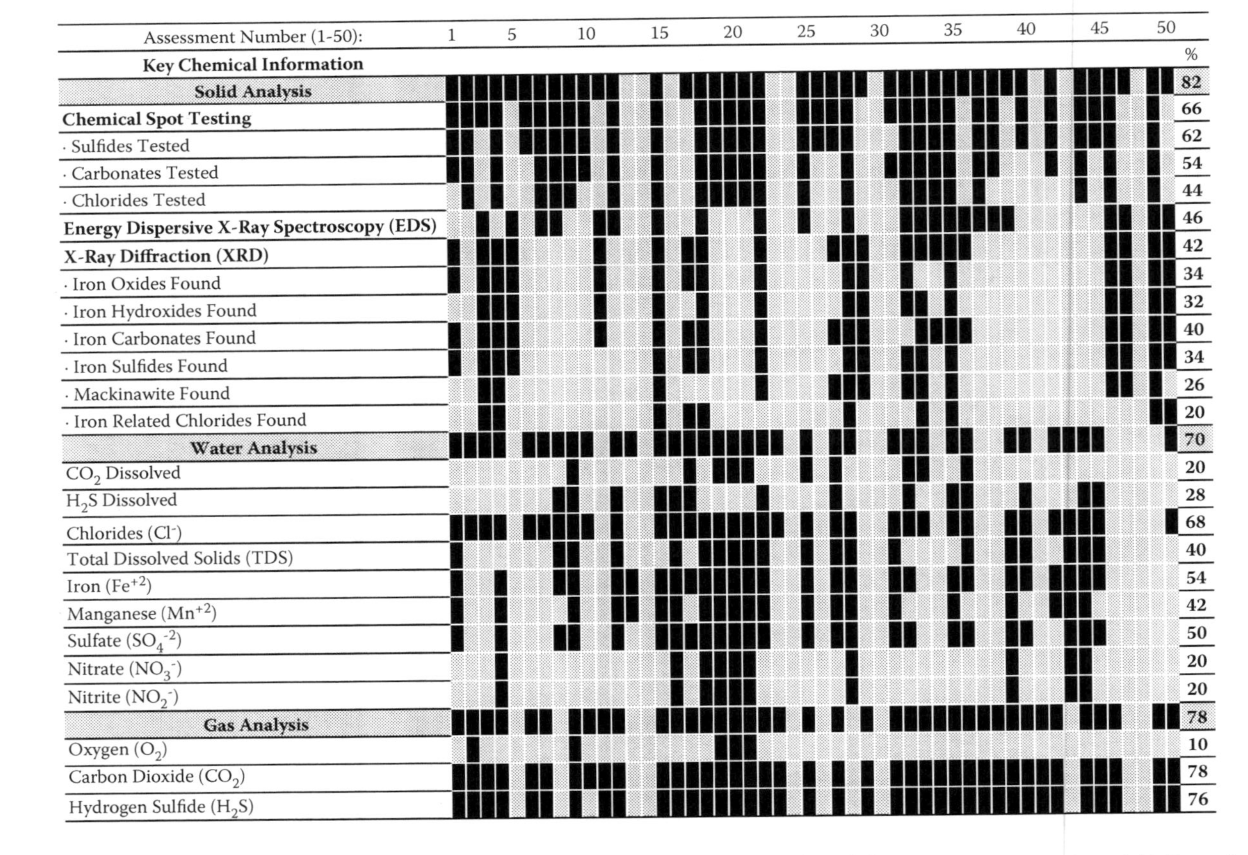

FIGURE 2.4 Inclusion of specific chemical information and analysis methods for each of the failure assessments reviewed (1 through 50).

2.5.3.2 Reported Water Data

Water-soluble chemical species are both the source of nutrients for microbiological activity and potential abiotic corrosion drivers. The evaluation of the water content and composition provides a means to track the source of corrosive constituents and corrosion products and, therefore, to add discrimination as to whether the corrosion is driven abiotically or biotically. Chemical analysis of water samples, present in 70% of the incidents, was used less as evidence of corrosion as compared to solids analyses, 82% (Figure 2.4). Although 68% of failure assessments tested for chlorides, only 20% tested for dissolved CO_2 and 28% for dissolved H_2S in the water.

None of the failure assessments tested for O_2 dissolved in the water, most likely because production pipelines are anaerobic environments where CO_2 and H_2S, along with chlorides, are the most significant drivers for abiotic corrosion and oxygen is rarely present. Yet, the presence of oxygen should be taken into consideration, as it can be more corrosive than both CO_2 and H_2S (Ibrahim et al. 2018) and its concentrations may lead to shifts in MIC potential and microbiological community populations (Dolla et al. 2006, Reddy and DeLaune 2008).

Chloride concentration can also impact MIC in terms of both limitations on growth and the diversity of microorganisms present (Oren 2006, Oren 2011, Ebrahimi et al. 2015). Therefore, the presence of dissolved chlorides, which contributes to the salinity and the total dissolved solids (TDS) of the water, may help indicate to what extent MIC is a viable failure cause.

In regard to both CO_2 and H_2S corrosion, the levels of these species in a flowing system can give some initial indication of whether there is an abiotic contribution to the corrosion. The presence of dissolved CO_2 and H_2S in water is a function of their concentration (partial pressure) in the gas and system pressure (Henry's law), along with water chemistry. Hence, based on corrosion rate modeling, the CO_2 and H_2S levels provide abiotic corrosion rates that can be used as a starting point for assessing a possible abiotic origin of the corrosion. In contrast, very low gas flow rates or stagnant systems may experience local accumulation of high concentrations of CO_2 and H_2S that are biotically produced. The relation between CO_2 and H_2S gases and solids sample evidence considered by the reviewed failure assessments is discussed later in this chapter (Sections 2.5.3.3 and 2.5.6).

The concentration of both CO_2 and H_2S in the water indicate to some extent the potential of these corrodents to drive the corrosion process (abiotic) as compared to MIC (biotic). Carbon dioxide is required for the activity of some microorganisms, such as methanogenic archaea and acetogenic bacteria (Suflita et al. 2008, Mand et al. 2014). In flowing systems where CO_2 is abundant, iron carbonates will tend to form abiotically. In stagnant systems, consumption of carbon dioxide by microorganisms can result in an increase in both bicarbonate and carbonate alkalinity, along with pH, which can cause precipitation of carbonate minerals (e.g., iron carbonates). Enning and Garrelfs 2014 mention a case where electrical MIC (EMIC) by SRB also can result in iron carbonate precipitation.

On the other hand, the presence of sulfidic corrosion products can be more meaningful evidence to differentiate biotic from abiotic corrosion. Mackinawite formation is associated with the same environmental conditions in which SRB and

SRA can be present (McNeil and Little 1990, Craig et al. 1991, McNeil et al. 1991). Therefore, the presence of mackinawite may be considered a potential indicator of the contribution of SRB and/or SRA to the corrosion process when they are found to be present in the system; while other iron sulfide forms (e.g., troilite, pyrrhotite, pyrite, marcasite, greigite, etc.) indicate either abiotic H_2S corrosion or potential decomposition of mackinawite (Eckert 2003) or other iron sulfides. Hence, when H_2S is not present in the gas (and consequently not dissolved in the water) but iron sulfides are present, that can be a significant indication that SRB and/or SRA may be contributing to the degradation process, since there is no apparent source for sulfide formation except for microbial activity. However, when H_2S is present, this comparison becomes more complex, as there are now two potential sources for iron sulfide formation. One way to discriminate between biotic and abiotic sulfidic corrosion products is by comparison of stable S isotope ratios (Chambers and Trudinger 1979, Krouse and Coplen, 1997, Hubert et al. 2009); however, this was not observed to be common practice.

Regarding dissolved chemical species that influence MIC, electron acceptors/donors (e.g., Fe, Mn, S, SO_4^{-2}, NO_3^-, NO_2^-, etc.) are of particular importance (Sooknah et al. 2007, Ibrahim et al. 2018) as they may represent a limiting factor for microbial growth in oil and gas systems. Therefore, their presence indicates to what extent the operating environment could support the microorganisms identified (diversity) and their potential activity. However, while electron donors/acceptors were used as indication for abiotic potential, the reviewed failure assessments overlooked the ability to use this evidence to support MIC as a credible failure cause.

In terms of carbon that can be used as energy source by microorganisms, as the reviewed pipeline environments handle hydrocarbons, some form of carbon will nearly always be present for microorganisms to metabolize (Davidova and Suflita 2005, Suflita and Duncan 2009, Ibrahim et al. 2018). Additionally, oilfield chemicals related to production, drilling, completion, and hydraulic fracturing can also work as sources of carbon and nutrients.

While 52% of the failure assessments concluded the failure to be SRB driven (either due to microbiological, chemical and/or corrosion related evidence), and 50% tested for SO_4^{-2} (Figure 2.4), only 24% of the cases both tested for SO_4^{-2} and concluded the failure to be SRB driven. Therefore, as sulfate was not considered in many cases as an essential electron acceptor for SRB activity, this identifies a current shortcoming in the approach to water composition evidence to support a MIC diagnosis.

Hence, for the chemical species listed in Figure 2.4, it is possible that relevant chemical data were assessed in some cases; however, some failure assessments missed the opportunity to use the information to support the MIC diagnosis.

2.5.3.3 Reported Gas Data

Gas concentration is used with operating pressure to determine the partial pressure of each gas present. The partial pressure of some gases allows one to assess to some extent whether abiotic drivers are reasonable contributors to the corrosion process or not (SP0106 2018).

In respect to gas analysis, 78% of failure assessments determined gas compositions (Figure 2.4). CO_2 and H_2S were tested in 78% and 76% of the time, respectively, as they are the main drivers for abiotic corrosion in anaerobic oil and gas systems (Smith 2015). As these compounds were tested in the gas, that may explain why they were not as extensively tested in water samples.

2.5.4 Reported Metallurgical Data

Evidence of metal loss as supporting proof of corrosion was gathered in 98% of failure assessments reviewed (Table 2.8). Visual features related to pipe wall loss due to chemical and/or microbiological reactions are the first indicators of corrosion degradation. It is the contrast of microbiological (diversity, activity, abundance), chemical, and operating factors between the corroded and non-corroded areas that best helps determine the failure mechanism.

Figure 2.5 displays the metallurgical data gathered by the failure assessments reviewed. Only 42% of failure assessments had samples analyzed from the failure/corrosion location. In respect to MIC, the comparison between the evidence at the failure/corroded area and the non-corroded area is essential. MIC is biofilm dependent, therefore, it is highly localized (Campbell and Maxwell 2006, Maxwell 2006, Pots et al. 2002, Sooknah et al. 2008, Wrangham and Summer 2013, Sørensen et al. 2012, Eckert 2015, Kotu and Eckert 2019). Consequently, diversity, activity, and abundance results may vary considerably between the corroded area and the non-corroded area when MIC is the cause of failure.

Additionally, although ruptures account for only 3% of the incidents in Alberta between 2017 and 2019 (Table 2.1) and no rupture was caused by MIC between this time period, the effect of ruptures can affect the validity of MIC failure samples. Sometimes damage by post rupture events (e.g., leak, fire, ignitions, groundwater intrusion in underground pipes) affect the substance at the through-wall perforation where the leak originates, which causes the evidence no longer to be representative. Hence, when evidence at the leak site is compromised by post-rupture events, sampling from a nearby corroded area (where corrosion products are still present in the pits) is a valid strategy to obtain undisturbed representative evidence.

In terms of corrosion, pitting was the dominant feature, present in 96% of incidents, while general corrosion was present in only 10% (and always in conjunction with pitting). The vast majority (92%) of the failure assessments identified that the failure occurred at the bottom half of the pipes, which may be interpreted as demonstrating the contribution of water accumulation on the degradation process.

Pitting morphology was described in 82% of the failure assessments, showcasing how strongly diagnoses take this layer of evidence into consideration. MIC was reported to be associated with tunneling, undercutting, pits within pits ("bulls' eye" appearance) and faceted pitting morphologies, even though pitting alone is not diagnostic for MIC as it cannot be related to a unique type of MIC morphology (Eckert 2003, Little and Lee 2007, Lee and Little 2017).

Solid deposition (e.g., deposits, scales, corrosion products, biomass) on the pipe surface was identified by 86% of failure assessments. However, no clear distinction

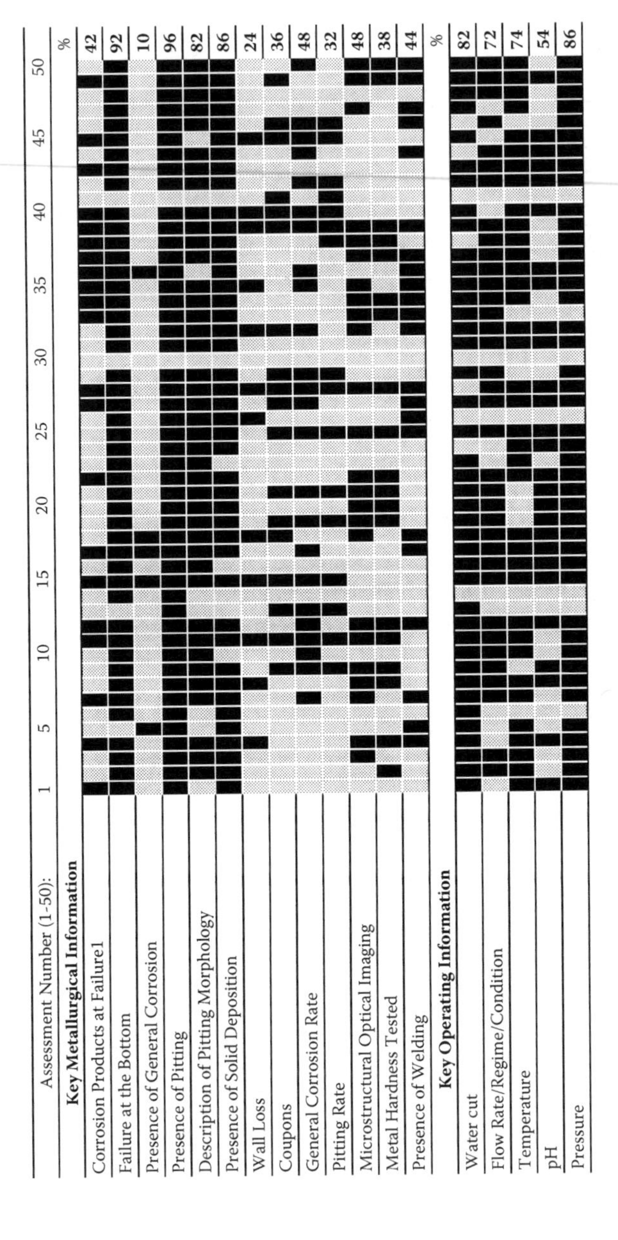

FIGURE 2.5 Inclusion of specific metallurgical and operating information and analysis methods for each of the failure assessments reviewed (1 through 50).

between deposits, scales, and corrosion products was made and such terms were used interchangeably.

General corrosion rates were estimated in 48% of failure assessments, either as part of the third-party failure report (where the wall loss was measured and divided by the length of time the pipe was operating) or as provided by coupon monitoring results. Averaging corrosion rates over the entire life of an asset may be non-conservative, as higher corrosion rates may have initiated later in the asset life when conditions changed.

Coupon monitoring, corrosion pitting rates, and general wall loss, either from previous inspection or from examination of the pipe sample, were pieces of information included in 36%, 32%, and 24% of the failure assessments, respectively. MIC is known to occur at unpredictably high rates, being normally associated with "premature" failures. However, when a leak occurs on a pipeline that is 30 years old, there is no way to determine when the corrosion began and therefore a linkage between corrosion rate and MIC is unfeasible.

Metallographic examination of the failure cross-section surface and metal hardness testing were assessed in 48% and 38% of the failure assessments, respectively. Based on these two factors, none of the cut-outs were deemed to be out of specification. However, neither spectrographic analysis for chemical composition nor mechanical strength testing (e.g., tensile strength, yield strength, percent of elongation) were carried out to augment material compatibility verification for comparison with pipe manufacturing specifications.

In terms of welding, electric resistance welding (ERW) seams and/or girth welds were present in 44% of the failure assessments reviewed. Aside from one specific case where MIC was directly related to the welding position due to the presence of morphological features (e.g., tunneling and undercutting) over the weld line, MIC was not associated with welding features in the reviewed failure assessments. Although four cases tested for the hardness of the weld heat affected zone (HAZ), no failure assessment included a cross-section of the weld to evaluate the potential influence of weld microstructure on the failure.

2.5.5 Reported Operating Data

Operating parameters serve to validate (or invalidate) either MIC and abiotic failure mechanisms. The corrosion products and microorganisms (e.g., metabolic functional groups) identified are correlated with operating conditions in order to determine which corrosion mechanisms (biotic, abiotic, or both combined) could be supported.

Operating parameters (e.g., temperature, pH, flow velocity, oil/water wetting) not only are some of the most influential parameters towards pipeline internal corrosion (Sooknah et al. 2007) in general, but also play a major role in MIC as they influence the activity and abundance of the MFG (diversity) present in the system. Consequently, operating data may rule out MIC if historical temperature, pH, and water accumulation do not support MIC as a credible failure cause.

Figure 2.5 displays the operating data gathered by the failure assessments tabulated. Water cut and flow conditions were gathered by 82% and 72% of failure assessments, respectively. Although MIC is directly related to the presence of water (biofilm formation and corrosion requires water wetting on the metal surface), water data was not significantly correlated to the potential for MIC.

Temperature and pH were reported in 74% and 54% of the failure assessments, respectively. pH was either calculated or measured as a part of the water analyses. The fact that pH is a parameter not as readily available as temperature explains its lower percentage. Even though specific ranges of temperature and pH have an effect on microbiological proliferation (Dexter 2003, Eckert and Skovhus 2018), no failure assessment took these two parameters into account as supporting evidence to diagnose MIC.

Similarly, pressure was not explicitly linked to either biotic or abiotic conditions, although 86% of failure assessments gathered such information. Pressure is relevant when combined with gas composition. The partial pressure of H_2S and CO_2 provides some degree of aid to determine if these abiotic corrosion drivers are expected to contribute to the failure.

2.5.6 Integration of Data

Although not an information group per se, data integration is a key element for effective MIC failure analysis, as no information group is diagnostic on its own; a consequence of the multidisciplinary nature of MIC. Therefore, only by adequately integrating multiple layers of evidence (microbiological, chemical, corrosion, and operating factors), is it possible to conclusively diagnose MIC as the failure cause with the technologies currently available.

Figure 2.6 shows the integration steps taken into consideration by the reviewed MIC failure assessments. Microbiological data was considered at the highest level for the diagnosis. Subsequently, at the next level, chemical considerations to either validate or invalidate MIC (diversity and abiotic drivers contrasted to corrosion products, water, and gas) were considered. At the lowest level (bottom of Figure 2.6), corrosion and operating-related considerations were considered.

Although microbiological activity was tested for 32% of failure assessments (Figure 2.2), only 10% of assessments considered that activity in their MIC diagnoses. Similarly, 10% of failure assessments tested for microbial abundance (Figure 2.2), but only 4% used abundance as an integration piece. It has been stated that no layer of evidence (microbiologically related or not) is diagnostic on its own. However, ATP activity was used as the sole layer of microbiological evidence for 18% of the failure assessments. Therefore, as microbiological evidence is not as frequently used for integration (Figure 2.6) as it is gathered (Figure 2.2), it is reasonable to say that the use and integration of microbiological evidence to draw a conclusive diagnosis is yet not fully understood. Therefore, the significance of microbiological evidence in the MIC diagnosis is under appreciated. This lack of appreciation results in the statistical mismatch observed between Figures 2.6 and 2.2.

Microbiological diversity was gathered either by culture-based techniques or MMM (Figures 2.2 and 2.3). However, only 18% of failure assessments compared microbial diversity to corrosion product chemistry. More than twice as many failure

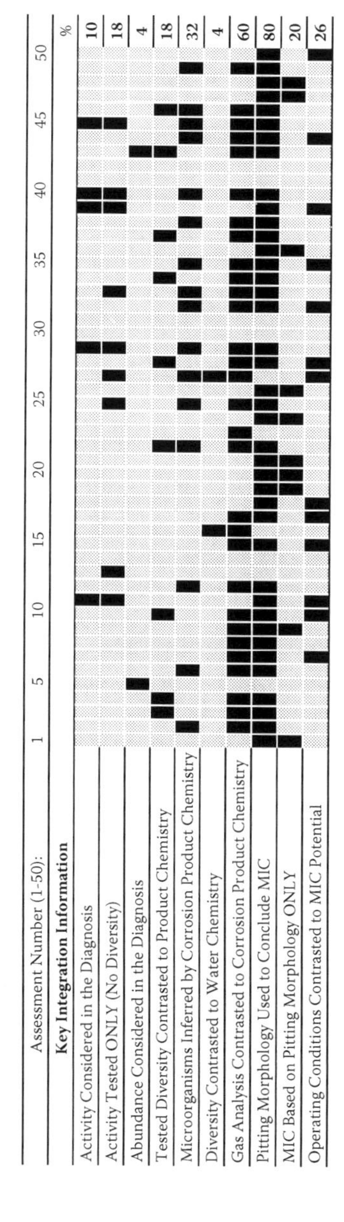

FIGURE 2.6 Inclusion of key integration steps for each of the failure assessments reviewed (1 through 50).

assessments, 38%, inferred a contribution by APB and/or SRB either from carbonate and sulfide corrosion products, respectively, or due to corrosion morphology; without any evidence of microbiological diversity to support the conclusions.

In terms of solids, chemistry, and microbial diversity integration, the reviewed failure assessments strongly related mackinawite to SRB, particularly when the gas composition analysis found no H_2S to be present. As a result, the identified sulfides were assumed to be of biotic origin. A similar rationale was used for APB when CO_2 was absent: the presence of iron carbonate (siderite) was the key evidence for assuming the participation of acid producing bacteria in the failure. However, the relationship between APB and siderite finds no support in the literature as it currently stands, as siderite formation is a function of abiotic CO_2 corrosion (Kermani and Morshed 2003), rather than APB-driven MIC. APB participation in MIC is usually associated with production of organic acids, not CO_2 or siderite.

The correlation between the composition of the gas and the composition of the solids was the strongest piece for integration used in the reviewed failure assessments and it was used to either conclude or invalidate the role of MIC. When corrosion products were present in the absence of their abiotic drivers, MIC was concluded as the failure cause. Additionally, the extent of contribution by specific damage mechanisms in the degradation process was directly related to the amounts of their respective mineral corrosion products (XRD) in the solids.

The significance of water chemistry and operating conditions to the MIC diagnosis were in general overlooked in the reviewed failure assessments. In regard to water chemistry and microbial diversity evaluation, only 4% of failure incidents linked nutrient availability to microbiological potential; despite the fact that 70% of failure assessments had water composition data in their records. Operating conditions (temperature, pH, and the potential for water accumulation) were taken into consideration in 26% of failure assessments for the MIC diagnosis. The majority of the cases related to water settlement and how this factor is conducive to MIC. The low percentage of nutrient availability and operating conditions being considered in the failure assessments indicates that the importance of integrating supporting conditions to the potential for MIC is yet being overlooked in the MIC diagnosis.

Metal loss features were frequently associated with MIC diagnosis. In fact, out of the 45 cases (90% of the total 50 assessments reviewed) that made some sort of integration between the different information groups included in the diagnosis, 30 used pitting morphology as supporting evidence to conclude MIC as the mechanism, along with other complementary layers of evidence. In fact, 10 (20% of the 50 total) of the cases reviewed used pitting morphology as the *sole* diagnostic evidence to conclude MIC, even though MIC has no unique morphological fingerprint.

As per Figure 2.6, pit morphology (80%) and the comparison between gas composition and corrosion product chemistry (60%) were the two layers of evidence most often considered in the reviewed failure assessments to diagnose MIC. Consequently, it can be concluded that although the reviewed failure assessments did a fair job in terms of integrating the data gathered, there are still gaps related to integration that must be addressed to increase the reliability of MIC diagnoses.

2.5.7 Sampling and Preservation

Sampling and preservation have a marked importance in the validity of the analytical data used to diagnose MIC. Microbiological and chemical changes to samples take place over time and as soon as the samples are exposed to air. Since the chemical and microbiological conditions must be integrated to identify MIC, significant misconclusions can occur if potential shifts in microbiological community and/or degradation of corrosion products after sampling are not taken into consideration (De Paula et al. 2018, Eckert and Skovhus 2018).

Figure 2.7 displays factors that influence validity of microbiological and chemical samples in terms of time of sampling and analysis. Assessing the time between the failure, sample collection and sample analysis can provide aid to determine the degree of degradation that a sample may have undergone. Only 12% of failure assessments explicitly mention cut-out dates and even fewer, 8%, mention when the cut-out sections were received for analysis. Consequently, only the 4% of failure assessments (which recorded the dates of the incident, the cut-out, and the lab analysis) would be able to evaluate to what degree the effect of time may have shifted sample characteristics.

Based on the failure assessments that recorded the date when pipe sample cut-outs were received for analysis, it was observed that cut-outs take from up to 3 days to be received by the third-party failure consultants after being cut, and may take between 10 and 18 days to be received by the third-party failure consultants after the failure date. Despite these delays in sampling and analysis, which could have been enough for significant shifts in both chemical and biological related evidence, no considerations to sample validity were mentioned in the reviewed failure assessments.

In respect to microbiological sampling, only 28% of failure assessments reported the date of sample collection and only 32% indicated the date of testing. Additionally, 18% of failure assessments considered historical microbiological data (as tests were run prior to the failure as part of monitoring programs) while 20% of the cases gathered microbiological data after the failure dates.

Additionally, out of the 16% of failure assessments that had samples taken near the failure location and 22% that had samples taken distal from it, only 8% had samples taken from both locations. Therefore, only 8% of the failure assessments would be able to make a comparison between microbiological diversity, activity, and abundance at the location of the corrosion failure and distal from it (where corrosion was not present); voiding the ability to more accurately determine to what extent MIC was a driver of the corrosion.

In terms of sample preservation, none (0%) of the incidents indicated they preserved the samples and no preservation-related comments were made in the cases reviewed. Consequently, the lack of information on preservation decreases the confidence in the final diagnosis, as considerations about sample validity are precluded.

In respect to chemical samples, only one incident mentions the date for solid sample collection, which occurred 33 days after the incident. In terms of water samples, out of the 48% of failure assessments that recorded sampling dates, 10%

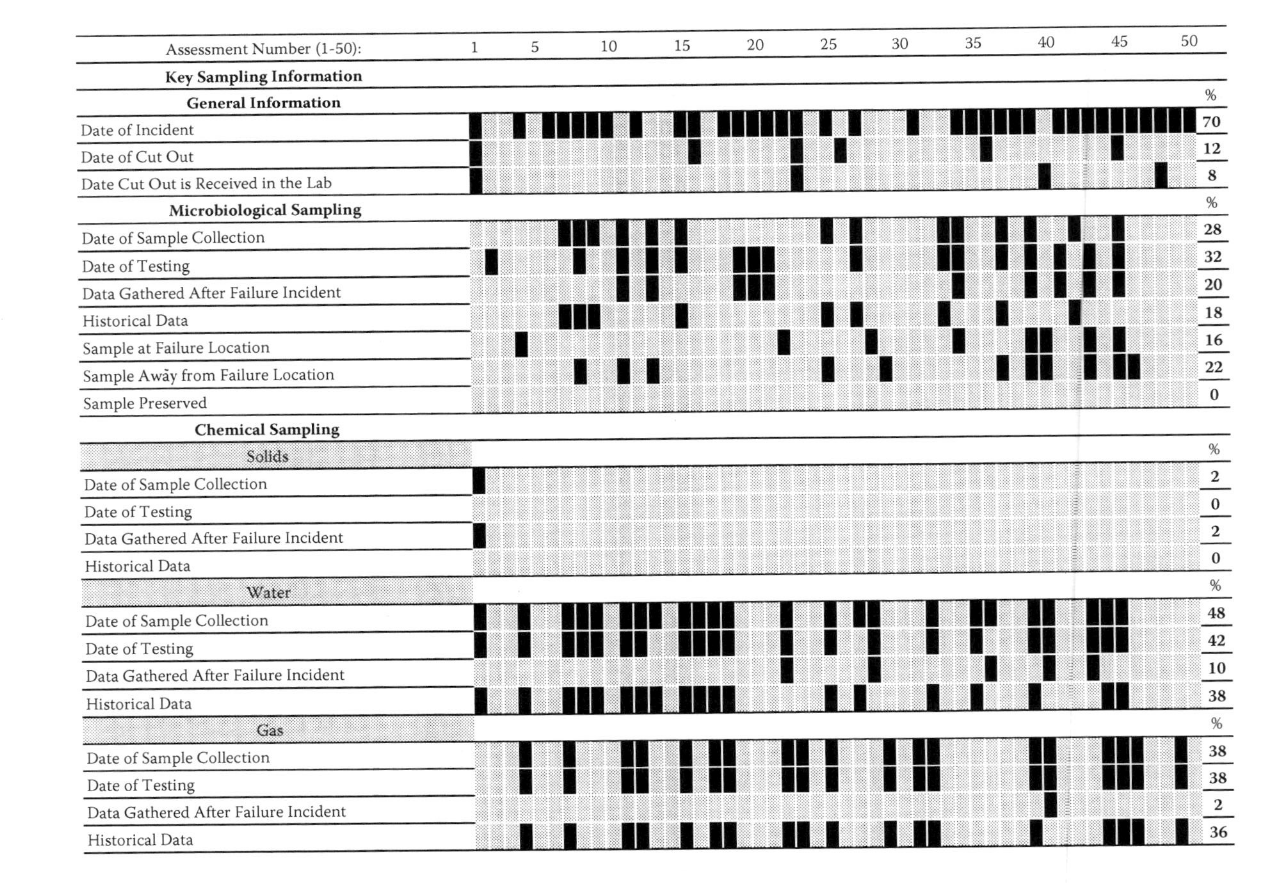

FIGURE 2.7 Inclusion of specific sampling considerations reported for each of the failure assessments reviewed (1 through 50).

included dates for samples collected after the incident, which varied between 0 and 36 days. The other 38% of failure assessments took into consideration the results of water samples taken over a variety of time periods prior to the incident. Those taken closest to the time of the incident varied from three to seven months after, while others varied from one to three years and the most extreme were taken five to nine years prior to the leak.

Regarding gas sampling, out of the 38% of failure assessments that mention sampling dates, only 2% were taken after the incident (32 days after). The other 36% of failure assessments were taken at a variety of time periods prior to the incident. Those closest to the incidents varied from two to ten months, while others varied from one to three years and the most extreme were taken five to eight years prior to the leak.

Therefore, in most cases both water (38%) and gas (36%) chemical considerations are based on historical evidence, as the information taken into account was collected prior to the incident. In cases where it was unfeasible to collect a reliable sample at the time of failure due to post rupture events, it is reasonable to utilize chemical samples gathered prior to the incident, as they will tend to more accurately represent the conditions at the time of the failure.

Craig (2002), Eckert (2003), Larsen et al. (2008), Kilbane (2014), De Paula et al. (2018), and Gieg et al. (2020) discuss variations that both microbiological and chemical samples may undergo due to changes caused by sampling and preservation methods. Additionally, Price (2012) and Eroini et al. (2015, 2017) provided suggestions in terms of sampling and preservation methods that could be used in order to avoid changes in sample composition during collection and transport.

2.5.8 Standards Referenced in the Failure Assessments

There are few standards that provide guidance on MIC sampling, testing, and interpretation that have been published and updated over the years. NACE TM0194 (2014) discuss MPN best practices and provides a high-level introduction to MMM. Conversely, TM0106 (2016) and NACE TM0212 (2018) are best practices dedicated to the use of MMMs to address both external and internal MIC, respectively.

However, only 22% of the reviewed failure assessments explicitly refer to the use of standards in their failure assessments. This number is even lower when it comes to microbiological standards. Only 4% of the failure assessments reviewed mentioned a standard dedicated to microbiological testing: NACE TM0194. No mention was either made to NACE TM0106 or NACE TM0212. These low percentages may be one more indicator of the lack of familiarity of the sector with microbiological considerations, which are even more pronounced in relation to applying and interpreting MMM data. Table 2.9 lists all the standards mentioned in the tabulated failure assessments and how often they were referenced.

2.6 CONCLUSIONS

This chapter highlights a review and analysis of MIC-related pipeline incidents in the province of Alberta, Canada, over a three-year period (2017–2019). The intent

TABLE 2.9
List of Standards Mentioned in the Reviewed MIC Failure Assessments

Standard ID	Standard Title	Number of Mentions
General		
ASTM E3	Standard Guide for Preparation of Metallographic Specimens	7
ASTM A370	Standard Test Methods and Definitions for Mechanical Testing of Steel Products	2
ASTM D-93	Standard Test Methods for Flash Point by Pensky-Martens Closed Cup Tester	1
ASTM D422-63	Standard Test Method for Particle-Size Analysis of Soils	1
NACE SP0775 (2018)	Preparation, Installation, Analysis, and Interpretation of Corrosion Coupons in Oilfield Operations	3
ASTM E1404	Standard Specification for Laboratory Glass Conical Flasks	1
Microbiological		
NACE TM0194	Field Monitoring of Bacterial Growth in Oil and Gas Systems	2

was to present an analysis of the occurrence of MIC failures relative to other corrosion mechanisms, and to conduct a gap analysis of MIC failure investigation techniques being used relative to the current state of the art.

Over this three-year period, MIC was found to be involved in 13.6% of all internal corrosion incidents (61 out of 447 cases) and 4.8% of all external corrosion incidents (6 out of 126 cases), either as the main failure mechanism or as a contributing factor. Furthermore, all of these failures occurred in small diameter upstream pipelines (with less than or equal to 220.3 mm outside diameter) mainly carrying produced water or multiphase fluids (oil-water emulsions). The highest MIC failure frequencies occurred in water pipelines with outer diameters smaller than 80.9 mm (2-1/2 in. nominal pipe size) and 114–116.8 mm (4 in. nominal pipe size) at 470×10^{-5} and 247×10^{-5} incidents per km per year, respectively. This translates to one MIC failure (on average) every 213 and 405 km, respectively, each year. To the authors' knowledge, this is one of the few well documented instances documenting MIC prevalence in the open literature with supporting evidence and data.

A gap analysis was also performed in this study to assess MIC failure investigation and analysis methods compared to current state-of-the-art and best practices as determined by expert elicitation and recent literature. A total of 50 failure assessments/third-party failure reports from MIC-related pipeline failures in Alberta that occurred over a three-year period (2017–2019) were reviewed in detail to identify the information that was typically collected during these assessments and how this information was used to confirm MIC as a cause. Specific analysis methods, collected data, and integration steps of key information groups (microbiological, chemical, metallurgical, and operating data) were evaluated.

Overall, over 90% of the 50 MIC assessments reviewed included chemical data, operating data, metallurgical data, and integration steps, however, only 70% of the MIC assessments reviewed conducted some sort of microbiological testing. In addition, only half of all assessments reviewed included tests to identify specific MFG, either using culturing methods or more advanced Microbiological Molecular Methods (MMM). This lack of identification of both diversity and abundance of specific microbial groups significantly reduces the confidence in a MIC diagnosis.

Furthermore, it was found that current failure assessments are not utilizing more modern microbiological (MMM) testing approaches, such as qPCR and Next Generation Sequencing. These advanced methods provide more robust data sets as compared to culture-based methods that can only identify a fraction of the possible MIC-related microorganisms in a system. While MMM approaches are currently more expensive than MPN (but decreasing over time) their value can be justified by a more accurate diagnosis of MIC (and specific MIC mechanisms), which allows for improved selection of tailored mitigation approaches to prevent future incidents.

The lack of microbiological testing suggests that industry and the failure analysis community are most familiar with methods used for assessing abiotic corrosion mechanisms (e.g., CO_2 and H_2S corrosion in oil and gas systems). This is not surprising since most failure analyses are performed by materials/metallurgical engineers/technicians who are not specifically trained in microbiology, which is a significant challenge in the multidisciplinary field of MIC.

In terms of metallurgical analysis, a significant number of failure assessments diagnosed MIC solely based on corrosion products and/or pit morphology without identification of specific microbial functional groups (a requirement to confirm MIC). Those diagnoses based on corrosion products used chemical analysis to infer the presence of specific microorganisms, which cannot be conclusive without microbiological testing. While MIC most commonly manifests itself as localized corrosion (pitting), the use of morphological features alone (such as tunnelling and undercutting) has been refuted in current best practices and standards since these morphological features can also be created by abiotic (non-MIC) corrosion mechanisms. Again, microbiological testing is an indispensable step to help confirm MIC mechanisms.

In terms of integration of data, the biggest gap was found in the correlation of microbiological data with chemical, metallurgical, and operating data to conclusively confirm specific MIC mechanisms, and to identify MIC as a primary or secondary contributor to the failure. In general, the integration of microbiological data (activity, diversity, and abundance) along with the chemical composition of corrosion products, gas analysis and elimination of abiotic corrosion mechanisms, are the key pieces of information required to conclusively confirm MIC specific mechanisms.

Overall, this gap analysis highlights the need for further education and training with respect to diagnosis of MIC particularly with respect to microbiological methods, and integration of biotic and abiotic data to confidently confirm MIC.

Standardized guidelines, protocols and analysis tools (e.g., online checklists) should also be developed to allow both specialists and non-specialists to better assess and diagnose MIC for both failure analysis and maintenance/integrity management activities. Finally, further work is also required to develop low-cost, rapid and easy-to-use microbiological diagnostic methods that can be handled in the field by non-specialists. These recommendations will go a long way to improve how the industry diagnoses and prevents MIC failures in the future.

ACKNOWLEDGMENTS

This work is part of a project entitled "Managing Microbial Corrosion in Canadian Offshore and Onshore Oil Production Operations." The authors would like to thank the Alberta Energy Regulator for kindly hosting the researchers and for providing access to available data, files, and personnel support. In addition, the authors would like to acknowledge the funding contributions for this current study from Genome Canada, the Government of Alberta, Alberta Innovates, and DNV GL Corrosion Management Group, along with in-kind support from a number of industry partners. This work is also supported by the Alberta Innovates Graduate Student Scholarship awarded to Andre Abilio.

REFERENCES

Alaska Department of Environmental Conservation (North Slope Spills Analysis Report). "Final report on North Slope spills analysis and expert panel recommendations on mitigation measures." Seldovia: Alaska Department of Environmental Conservation, NSSA, 2010.

Alaska Department of Environmental Conservation (Division of Spill Prevention and Response, Situation Report). "GC-2 oil transit line release." Seldovia: Alaska Department of Environmental Conservation, 2020. https://dec.alaska.gov/spar/ppr/spill-information/response/2006/07-gc2/

Alberta Energy Regulator. "What we do." Alberta: Alberta Energy Regulator, 2020. https://www.aer.ca/providing-information/about-the-aer/what-we-do.html

Alberta Energy Regulator Report. "Report 2013-B: Pipeline performance in Alberta, 1990-2012." Alberta: Alberta Energy Regulator, 2013.

Alberta Energy Regulator Report. "Release report – Definitions for pipeline details section." Alberta: Alberta Energy Regulator, 2019.

Alberta Government. "Let's talk royalties: Alberta's oil reserves compared to other countries." Alberta: Alberta Government, 2020. https://open.alberta.ca/publications/alberta-s-oil-reserves-compared-to-other-countries

Al-Nabulsi, Khalid M., Faisal M. Al-Abbas, Tony Y. Rizk, and Ala'a Edin M. Salameh. "Microbiologically assisted stress corrosion cracking in the presence of nitrate reducing bacteria." *Engineering Failure Analysis* 58, no. 1 (2015): 165–172. 10.1016/j.engfailanal.2015.08.003

Al-Sulaiman, S., A. Al-Shamari, A. Al-Mithin, M. Islam, and S. S. Prakash. "Microbiologically influenced corrosion of a crude oil pipeline." *Proceedings of the CORROSION 2010*, no. 10210. NACE International (2010): 1–17.

ASTM A370-2020, ASTM Standard. "Standard test methods and definitions for mechanical testing of steel products." West Conshohocken: ASTM International, 2020. 10.1520/A0370-20

ASTM D422-63-2007, ASTM Standard. "Standard test method for particle-size analysis of soils (withdrawn 2016)." West Conshohocken: ASTM International, 2007. 10.1520/D0422-63R07E02

ASTM D-93-2020, ASTM Standard. "Standard test methods for flash point by Pensky-Martens closed cup tester." West Conshohocken: ASTM International, 2020. 10.1520/D0093-20

ASTM E1404-2019, ASTM Standard. "Standard specification for laboratory glass conical flasks." West Conshohocken: ASTM International, 2019. 10.1520/E1404-94R19

ASTM E3-2017, ASTM Standard. "Standard guide for preparation of metallographic specimens." West Conshohocken: ASTM International, 2017. 10.1520/E0003-11R17

Beavers, John A., and Neil G. Thompson. "External corrosion of oil and natural gas pipelines." In *Corrosion: Environments and Industries*, Vol 13C, ASM Handbook, edited by Stephen D. Cramer and Bernard S. Covino Jr., 1015–1025. Ohio:ASM International, 2006. 10.31399/asm.hb.v13c.a0004213

Borenstein, Susan W., and Philip B. Lindsay. "Microbiologically influenced corrosion failure analysis of 304L stainless steel piping system left stagnant after hydrotesting with city water." *Proceedings of the CORROSION 2002*, no. 02446. NACE International (2002): 1–10.

Campbell, Scott, and Stephen Maxwell. "Monitoring the mitigation of MIC risk in pipelines." *Proceedings of the CORROSION 2006*, no. 06662. NACE International (2006): 1–16.

Chambers, L. A., and P. A. Trudinger. "Microbiological fractionation of stable sulfur isotopes: A review and critique." *Geomicrobiology Journal* 1, no. 3 (1979): 249–293. 10.1080/01490457909377735

Craig, Bruce. "Corrosion product analysis—a road map to corrosion in oil and gas production." *Materials Performance* 41, no. 8 (2002): 2–4.

Craig, B. D., M. B. McNeil, and B. J. Little. "Discussion of "Mackinawite formation during microbial corrosion"." *Corrosion* 47, no. 5 (1991): 329–329. 10.5006/1.3585260

CSA Z662:19, CSA Group National Standard of Canada. "Oil and gas pipeline systems." Toronto: CSA Group, 2019. ISBN 978-1-4883-1771-2

Davidova, Irene A., and Joseph M. Suflita. "Enrichment and isolation of anaerobic hydrocarbon-degrading bacteria." *Methods in Enzymology* 397 (2005): 17–34. 10.1016/S0076-6879(05)97002-X

De Paula, Renato, Cruz St Peter, Ian Alex Richardson, Jep Bracey, Ed Heaver, Kathleen Duncan, Mary Eid, and Ralph Tanner. "DNA sequencing of oilfield samples: Impact of protocol choices on the microbiological conclusions." *Proceedings of the CORROSION 2018*, no. 11662. NACE International (2018): 1–11.

De Romero, M., Z. Duque, O. de Rincón, O. Pérez, I. Araujo, and A. Martinez. "Online monitoring systems of microbiologically influenced corrosion on Cu-10% Ni alloy in chlorinated, brackish water." *Corrosion* 56, no. 8 (2000): 867–876. 10.5006/1.3280590

Dexter, Stephen C. "Microbiologically influenced corrosion." In *Corrosion: Fundamentals, Testing,* and Protection, Vol 13A, ASM Handbook, edited by Stephen D. Cramer and Bernard S. Covino Jr., 398–416. Ohio:ASM International, 2003. 10.31399/asm.hb.v13a.a0003637

Dolla, Alain, Marjorie Fournier, and Zorah Dermoun. "Oxygen defense in sulfate-reducing bacteria." *Journal of Biotechnology* 126, no. 1 (2006): 87–100. 10.1016/j.jbiotec.2006.03.041

Ebrahimi, Shelir, Thi Hau Nguyen, and Deborah J. Roberts. "Effect of temperature and salt concentration on salt tolerant nitrate-perchlorate reducing bacteria: Nitrate degradation kinetics." *Water Research* 83 (2015): 345–353. 10.1016/j.watres.2015.07.006

Eckert, Richard B. *Field guide for investigating internal corrosion of pipelines*. Houston: NACE International, 2003. ISBN: 1575901714

Eckert, Richard B. "Emphasis on biofilms can improve mitigation of microbiologically influenced corrosion in oil and gas industry." *Corrosion Engineering, Science and Technology* 50, no. 3 (2015): 163–168. 10.1179/1743278214Y.0000000248

Eckert, Richard B., H. C. Aldrich, C. A. Edwards, and B. A. Cookingham. "Microscopic differentiation of internal corrosion initiation mechanisms in a natural gas gathering systems." *Proceedings of the CORROSION 2003*, no. 03544. NACE International (2003): 1–13.

Eckert, Richard B., and Kathy Buckingham. "Investigating pipeline corrosion failures." *Inspectioneering Journal* 23, no. 4 (2017): 2–9.

Eckert, Richard B., and Torben L. Skovhus. "Advances in the application of molecular microbiological methods in the oil and gas industry and links to microbiologically influenced corrosion." *International Biodeterioration and Biodegradation* 126 (2018): 169–176. 10.1016/j.ibiod.2016.11.019

Eckert, Richard B., and Torben L. Skovhus. "Pipeline failure investigation: Is it MIC?." *Materials Performance* 58, no. 2 (2019): 40–43.

Enning, Dennis, and Julia Garrelfs. "Corrosion of iron by sulfate-reducing bacteria: New views of an old problem." *Applied and Environmental Microbiology* 80, no. 4 (2014): 1226–1236. 10.1128/AEM.02848-13

Eroini, Violette, Hilde Anfindsen, and Anthony F. Mitchell. "Investigation, classification and remediation of amorphous deposits in oilfield systems." *Proceedings of the SPE International Symposium on Oilfield Chemistry*, no. 173719. Society of Petroleum Engineers (2015): 1–13. 10.2118/173719-MS

Eroini, Violette, Mike Christian Oehler, Brit Kathrine Graver, Anthony Mitchell, Kari Lønvik, and Torben Lund Skovhus. "Investigation of amorphous deposits and potential corrosion mechanisms in offshore water injection systems." *Proceedings of the CORROSION 2017*, no. 9433. NACE International (2017): 1–11.

Fichter, Jennifer, and Elizabeth J. Summer. "The use of multiple microbial population analysis techniques to diagnose microbially influenced corrosion potential in oil and gas systems." In *Microbiologically Influenced Corrosion in the Upstream Oil and Gas Industry*, edited by Torben L. Skovhus, Dennis Enning, and Jason S. Lee, 435–462. Boca Raton: CRC Press, 2017. 10.1201/9781315157818

Garber, Richard I., and Durgam G. Chakrapani. "Some recent failures of fire sprinkler system components: Corrosion case histories." *Proceedings of the CORROSION 2004*, no. 04511. NACE International (2004): 1–29.

Gieg, Lisa, Jennifer Sargent, Hitesh Bagaria, Trevor Place, Mohita Sharma, Yin Shen, and Danielle Kiesman. "Synergistic effect of biocide and biodispersant to mitigate microbiologically influenced corrosion in crude oil transmission pipelines." *Proceedings of the CORROSION 2020*, no. 15090. NACE International (2020): 1–13.

Goldstein, Joseph I., Dale E. Newbury, Joseph R. Michael, Nicholas WM Ritchie, John Henry J. Scott, and David C. Joy. *Scanning electron microscopy and X-ray microanalysis*. New York: Springer, 2018. ISBN 978-1-4939-6676-9.

Government of Canada, Natural Resources Canada. "Energy fact book 2020 – 2021." Canada: Government of Canada, 2020. ISSN 2370-3105.

Graves, John W., and E. H. Sullivan. "Internal corrosion in gas gathering systems and transmission lines." *Materials Protection* 5, no. 6 (1966): 33–37.

Hashemi, Seyed Javad, Nicholas Bak, Faisal Khan, Kelly Hawboldt, Lianne Lefsrud, and John Wolodko. "Bibliometric analysis of microbiologically influenced corrosion (MIC) of oil and gas engineering systems." *Corrosion* 74, no. 4 (2018): 468–486. 10.5006/2620

Hubert, Casey, Gerrit Voordouw, and Bernhard Mayer. "Elucidating microbial processes in nitrate- and sulfate-reducing systems using sulfur and oxygen isotope ratios: The example of oil reservoir souring control." *Geochimica et Cosmochimica Acta* 73, no. 13 (2009): 3864–3879. 10.1016/j.gca.2009.03.025

Ibrahim, Abdulhaqq, Kelly Hawboldt, Christina Bottaro, and Faisal Khan. "Review and analysis of microbiologically influenced corrosion: The chemical environment in oil and gas facilities." *Corrosion Engineering, Science and Technology* 53, no. 8 (2018): 549–563. 10.1080/1478422X.2018.1511326

Islam, Moavin, Saleh Al-Sulaiman, Abdul Razzaq Al-Shamari, Surya Prakash, and Allen Biedermann. "Premature failure of access fittings installed on high pressure effluent water lines due to microbiologically induced corrosion." *Proceedings of the CORROSION 2016*, no. 7367. NACE International (2016): 1–15.

Jack, Thomas R. "Biological corrosion failures." In *Failure Analysis and Prevention*, Vol 11, ASM Handbook, edited by W. T. Becker and R. J. Shipley, 881–898. Ohio:ASM International, 2002. 10.31399/asm.hb.v11.a0003556

Jacobson, Gretchen A. "Corrosion at Prudhoe Bay: A lesson on the line." *Materials Performance* 46, no. 8 (2007): 26–34.

Kaduková, Jana, Erika Škvareková, Vojtech Mikloš, and Renáta Marcinčáková. "Assessment of microbially influenced corrosion risk in Slovak pipeline transmission network." *Journal of Failure Analysis and Prevention* 14, no. 2 (2014): 191–196. 10.1007/s11668-014-9782-x

Kermani, M. B., and A. Morshed. "Carbon dioxide corrosion in oil and gas production—a compendium." *Corrosion* 59, no. 8 (2003): 659–683. 10.5006/1.3277596

Kilbane, John. "Effect of sample storage conditions on oilfield microbiological samples." *Proceedings of the CORROSION 2014*, no. 3788. NACE International (2014): 1–6.

Koch, Gerhardus, Jeff Varney, Neil Thompson, Oliver Moghissi, Melissa Gould, and Joe Payer. "International measures of prevention, application, and economics of corrosion technologies study." NACE International (2016): 1–216.

Kotu, Susmitha Purnima, and Richard B. Eckert. "A framework for conducting analysis of microbiologically influenced corrosion failures." *Inspectioneering Journal* 25, no. 4 (2019): 1–10.

Krouse, H. R., and Tyler B. Coplen. "Reporting of relative sulfur isotope-ratio data (technical report)." *Pure and Applied Chemistry* 69, no. 2 (1997): 293–295.

Larsen, Jan, Torben Lund Skovhus, Aaron Marc Saunders, Bo Højris, and Mikkel Agerbæk. "Molecular identification of MIC Bacteria from scale and produced water: Similarities and differences." *Proceedings of the CORROSION 2008*, no. 08652. NACE International (2008): 1–21.

Larsen, Jan, Ketil Bernt Sørense, Susanne Juhler, and Dorthe Skou Pedersen. "The application of molecular microbiological methods for early warning of MIC in pipelines." *Proceedings of the CORROSION 2013*, no. 2029. NACE International (2013): 1–9.

Lee, Jason S., and Brenda J. Little. "Diagnosing microbiologically influenced corrosion." In *Microbiologically Influenced Corrosion in the Upstream Oil and Gas Industry*, edited byTorben L. Skovhus, Dennis Enning, and Jason S. Lee, 3–34. Boca Raton: CRC Press, 2017. 10.1201/9781315157818

Liengen, Turid, R. Basseguy, Damien Feron, I. Beech, and V. Birrien. *Understanding biocorrosion: Fundamentals and applications*. Cambridge: Elsevier, 2014. ISBN 978-1-78242-125-2

Little, B. J., D. J. Blackwood, J. Hinks, F. M. Lauro, E. Marsili, A. Okamoto, S. A. Rice, S. A. Wade, and H.-C. Flemming. "Microbially influenced corrosion–any progress?." *Corrosion Science* 170 (2020): 108641. 10.1016/j.corsci.2020.108641

Little, Brenda J., and Jason S. Lee. *Microbiologically influenced corrosion*. New Jersey: John Wiley and Sons, 2007. ISBN 978-0-471-77276-7

Little, B. J., and Jason. S. Lee. "Microbiologically influenced corrosion." *Kirk-Othmer Encyclopedia of Chemical Technology* (2009): 1–42. 10.1002/0471238961.micrlitt.a01

Little, B. J., Jason. S. Lee, and R. I. Ray. "Diagnosing microbiologically influenced corrosion: A state-of-the-art review." *Corrosion* 62, no. 11 (2006): 1006–1017. 10.5006/1.3278228

Liu, William. "High temperature MIC on an offshore pipeline and the strong arsenate-reduction function in the hyperthermophiles." *Engineering Failure Analysis* 45 (2014a): 376–386. 10.1016/j.engfailanal.2014.07.001

Liu, William. "Rapid MIC attack on 2205 duplex stainless steel pipe in a yacht." *Engineering Failure Analysis* 42 (2014b): 109–120. 10.1016/j.engfailanal.2014.04.001

Mand, Jaspreet, Hyung Soo Park, Thomas R. Jack, and Gerrit Voordouw. "The role of acetogens in microbially influenced corrosion of steel." *Frontiers in Microbiology* 5, no. 268 (2014): 1–14. 10.3389/fmicb.2014.00268

Maxwell, Stephen. "Predicting microbially influenced corrosion in seawater injection systems." *Proceedings of the SPE International Symposium on Oilfield Chemistry*, no. 100519. Society of Petroleum Engineers (2006): 1–8. 10.2118/100519-MS

Maxwell, Stephen, Carol Devine, Fiona Rooney, Iain Spark. "Monitoring and control of bacterial biofilms in oilfield water handling systems." *Proceedings of the CORROSION 2004*, no. 04752. NACE International (2014): 1–16.

McNeil, M. B., J. M. Jones, and B. J. Little. "Mineralogical fingerprints for corrosion processes induced by sulfate reducing bacteria." *Proceedings of the CORROSION 1991*, no. 580. NACE International (1991): 1–26.

McNeil, M. B., and B. J. Little. "Technical note. Mackinawite formation during microbial corrosion." *Corrosion* 46, no. 7 (1990): 599–600. 10.5006/1.3585154

National Transportation Safety Board, Pipeline Accident Report NTSB/PAR-03/01. "Natural gas pipeline rupture and fire near Carlsbad, New Mexico, August 19, 2000." Washington: National Transportation Safety Board, 2003.

Oren, Aharon. "Life at high salt concentrations." In *Prokaryotes*, Vol. 2,edited by Martin Dworkin, Stanley Falkow, Eugene Rosenberg, Karl-Heinz, and Erko Stackebrandt, 263–282. New York: Springer, 2006. 10.1007/0-387-30742-7_9

Oren, Aharon. "Thermodynamic limits to microbial life at high salt concentrations." *Environmental Microbiology* 13, no. 8 (2011): 1908–1923. 10.1111/j.1462-2920.2010.02365.x

Petrović, Zoran C. "Catastrophes caused by corrosion." *Vojnotehnički glasnik* 64, no. 4 (2016): 1048–1064. 10.5937/vojtehg64-10388

Pots, Bernardus F. M., Sergio D. Kapusta, Randy C. John, M. J. J. Thomas, Ian J. Rippon, T. S. Whitham, and Magdy Girgis. "Improvements on de Waard-Milliams corrosion prediction and applications to corrosion management." *Proceedings of the CORROSION 2002*, no. 02235. NACE International (2002): 1–19.

Price, Andy. "Novel DNA extraction and preservation for identification of micro-organisms on-site using novel nucleic acid extraction cards for the oil and gas industry." *Proceedings of the CORROSION 2012*, no. 1758. NACE International (2012): 1–15.

Prithiraj, Alicia, Iyiola Olatunji Otunniyi, Peter Osifo, and Josias van Der Merwe. "Corrosion behaviour of stainless and carbon steels exposed to sulphate–reducing bacteria from industrial heat exchangers." *Engineering Failure Analysis* 104 (2019): 977–986. 10.1016/j.engfailanal.2019.06.042

Province of Alberta, Pipeline Act. "Revised statutes of Alberta 2000 Chapter P-15." Edmonton: Alberta Queens's Printer, 2020.

Province of Alberta, Pipeline Act, Pipeline Rules. "Alberta regulation 91/2005." Edmonton: Alberta Queens's Printer, 2020.

Rajasekar, A., T. Ganesh Babu, S. Karutha Pandian, S. Maruthamuthu, N. Palaniswamy, and A. Rajendran. "Biodegradation and corrosion behavior of manganese oxidizer Bacillus cereus ACE4 in diesel transporting pipeline." *Corrosion Science* 49, no. 6 (2007): 2694–2710. 10.1016/j.corsci.2006.12.004

Rao, T. S., and K. V. K. Nair. "Microbiologically influenced stress corrosion cracking failure of admiralty brass condenser tubes in a nuclear power plant cooled by freshwater." *Corrosion Science* 40, no. 11 (1998): 1821–1836. 10.1016/S0010-938X(98)00079-1

Reddy, K. Ramesh, and Ronald D. DeLaune. *Biogeochemistry of wetlands: Science and applications*. Boca Raton: CRC Press, 2008. 10.1201/9780203491454

Renner, Michael H. W. "Corrosion engineering aspects regarding MIC related failures on stainless steels." *Proceedings of the CORROSION 1998*, no. 98285. NACE International (1998): 1–27.

Rizk, Tony Y., Khalid M. Al-Nabulsi, and Min Hyum Cho. "Microbially induced rupture of a heat exchanger shell." *Engineering Failure Analysis* 76 (2017): 1–9. 10.1016/j.engfailanal.2016.11.004

Salgar-Chaparro, Silvia J., and Laura L. Machuca. "Complementary DNA/RNA-based profiling: Characterization of corrosive microbial communities and their functional profiles in an oil production facility." *Frontiers in Microbiology* 10 (2019): 2587–2605. 10.3389/fmicb.2019.02587

Samant, A. K., V. K. Sharma, S. Thomas, P. F. Anto, and S. K. Singh. "Investigation of premature failure of a well fluid pipeline in an Indian Offshore installation." In *Advances in Corrosion Control and Materials in Oil and Gas Production*, edited byP. S. Jackman and L. M. Smith, 180–187. London: IOM Communications Ltd, 1999. ISBN 1-86125-092-4

San, Nalan Oya, Hasan Nazır, and Gönül Dönmez. "Microbiologically influenced corrosion failure analysis of nickel–copper alloy coatings by Aeromonas salmonicida and Delftia acidovorans bacterium isolated from pipe system." *Engineering Failure Analysis* 25 (2012): 63–70. 10.1016/j.engfailanal.2012.04.007

Skovhus, Torben Lund, and Richard Bruce Eckert. "Practical aspects of MIC detection, monitoring and management in the oil and gas industry." *Proceedings of the CORROSION 2014*, no. 3920. NACE International (2014): 1–13.

Skovhus, Torben Lund, Richard B. Eckert, and Edgar Rodrigues. "Management and control of microbiologically influenced corrosion (MIC) in the oil and gas industry—Overview and a North Sea case study." *Journal of biotechnology* 256 (2017a): 31–45. 10.1016/j.jbiotec.2017.07.003

Skovhus, Torben Lund, Dennis Enning, and Jason S. Lee. *Microbiologically Influenced Corrosion in the Upstream Oil and Gas Industry*. Boca Raton: CRC Press, 2017b. 10.1201/9781315157818

Smith, Stephen N. "The carbon dioxide/hydrogen sulfide ratio-use and relevance." *Materials Performance* 54, no. 5 (2015): 64–67.

Sooknah, Reeta, Sankara Papavinasa, and R. Winston Revie. "Modelling the occurrence of microbiologically influenced corrosion." *Proceedings of the CORROSION 2007*, no. 07515. NACE International (2007): 1–12.

Sooknah, Reeta, Sankara Papavinasam, and R. Winston Revie. "Validation of a predictive model for microbiologically influenced corrosion." *Proceedings of the CORROSION 2008*, no. 08503. NACE International (2008): 1–17.

Sørensen, Ketil Bernt, Uffe Sognstrup Thomsen, Susanne Juhler, and Jan Larsen. "Cost efficient MIC management system based on molecular microbiological methods." *Proceedings of the CORROSION 2012*, no. 1111. NACE International (2012): 1–15.

SP0106-2018, NACE International Standard Practice. "Control of internal corrosion in steel pipelines and piping systems." Houston: NACE International, 2018.

SP0775-2018, NACE International Standard Practice. "Preparation, installation, analysis, and interpretation of corrosion coupons in oilfield operations." Houston: NACE International, 2018.

Sreekumari, Kurissery R., Kyozo Hirotani, Yasushi Kikuchi, Katsuya Akamatsu, and Takashi Imamichi. "Microbiologically influenced corrosion failure of AISI type 304 stainless steel in a wastewater treatment system."*Proceedings of the CORROSION 2004*, no. 04600. NACE International (2004): 1–8.

Starosvetsky, J., D. Starosvetsky, and R. Armon. "Identification of microbiologically influenced corrosion (MIC) in industrial equipment failures." *Engineering Failure Analysis* 14, no. 8 (2007): 1500–1511. 10.1016/j.engfailanal.2007.01.020

Suflita, Joseph M., and Kathleen E. Duncan. "Problems caused by microbes and treatment strategies – Anaerobic hydrocarbon biodegradation and biocorrosion: A case study." In *Applied Microbiology and Molecular Biology in Oilfield Systems*, edited by Corinne Whitby and Torben L. Skovhus, 141–149. New York: Springer, 2009. 10.1007/978-90-481-9252-6_17

Suflita, Joseph M., Tommy J. Phelps, and B. Little. "Carbon dioxide corrosion and acetate: A hypothesis on the influence of microorganisms." *Corrosion* 64, no. 11 (2008): 854–859. 10.5006/1.3279919

TM0106-2016, NACE International Test Method. "Detection, testing, and evaluation of microbiologically influenced corrosion (MIC) on external surfaces of buried pipelines." Houston: NACE International, 2016.

TM0194-2014, NACE International Test Method. "Field monitoring of bacterial growth in oil and gas systems." Houston: NACE International, 2014.

TM0212-2018, NACE International Test Method. "Detection, testing, and evaluation of microbiologically influenced corrosion on internal surfaces of pipelines." Houston: NACE International, 2018.

United States Department of Transportation Pipeline and Hazardous Materials Safety Administration. "BP's pipeline spills at prudhoe bay: What went wrong?." *Hearing Before the Committee on Energy and Commerce Subcommittee on Oversight and Investigations United States House of Representatives*, no. 109–135 (2006).

von Wolzogen Kühr, C. A. H. "Sulfate reduction as cause of corrosion of cast iron pipeline." *Water Gas* 7, no. 26 (1923): 277.

von Wolzogen Kühr, C. A. H., and L. S. Van Der Vlugt. "The graphitization of cast iron as an electrochemical process in anaerobic soils." *Water* 18 (1934): 147–165.

Waseda, Yoshio, Eiichiro Matsubara, and Kozo Shinoda. *X-ray diffraction crystallography: Introduction, examples and solved problems*. New York: Springer, 2011. 10.1007/978-3-642-16635-8

Whitby, Corinne, and Torben Lund Skovhus. *Applied microbiology and molecular biology in oilfield systems*. New York: Springer, 2009. 10.1007/978-90-481-9252-6

Wrangham, Jodi B., and Elizabeth J. Summer. "Planktonic microbial population profiles do not accurately represent same location sessile population profiles." *Proceedings of the CORROSION 2013*, no.2780. NACE International (2013): 1–7.

Xu, Dake, Yingchao Li, Fengmei Song, and Tingyue Gu. "Laboratory investigation of microbiologically influenced corrosion of C1018 carbon steel by nitrate reducing bacterium Bacillus licheniformis." *Corrosion Science* 77 (2013): 385–390. 10.1016/j.corsci.2013.07.044

3 A Practical Approach to Corrosion Failure Analysis

Richard B. Eckert
DNV GL USA

CONTENTS

3.1 INTRODUCTION: BACKGROUND AND DRIVING FORCES

An essential part of all asset integrity and risk management programs is learning from incidents or near misses, yet all too often, a rash of similar incidents will occur within an organization or across an entire industry sector over a relatively short period of time. Corrosion incidents are no different. Recently, a number of leaks associated with internal corrosion in crude oil facility piping dead legs (where there

DOI: 10.1201/9780429355479-4

is normally no flow), have been observed. This is not a surprising new phenomenon; however, it is also one for which better remedial solutions should already be in place, as the threat of internal corrosion in dead legs due to microbiologically influenced corrosion (MIC) and under deposit corrosion (UDC) is broadly known. Industry-accepted mitigation and prevention measures have simply not been established based on lessons learned from these dead leg failures and leaks in many cases, so the problem continues. Learning from thorough investigation of corrosion-related incidents is one way to extend the life cycle of assets, based on a clear understanding of the mechanisms, and contributing causes that led to the incident.

Cases for corrosion investigation are not exclusively provided by leaks or ruptures. Corrosion may be discovered, e.g., on a pipeline during inline inspection (ILI) or direct assessment (DA) or during routine maintenance. In the United States, buried, regulated, hazardous liquid and natural gas pipelines must be examined for external corrosion whenever they are exposed, and for internal corrosion whenever the internal surface is made visible, e.g., during a pipeline repair. If corrosion is found, this provides the operator an opportunity to collect samples and other information to help determine why the corrosion occurred and to determine the appropriate mitigation measures. In plant piping, corrosion may be discovered during routine maintenance or equipment replacement. For example, cleaning and inspection of separator vessels during a plant turn-around may lead to the discovery of internal corrosion under accumulated deposits. Being prepared to take advantage of these "learning opportunities" where corrosion is discovered is essential so that evidence of the corrosion mechanism is not lost. In this case, being prepared means having trained personnel available who know how to preserve and sample corrosion evidence, who have access to the necessary tools, containers and materials needed to collect the samples, and information needed to perform the work. This chapter will focus on the corrosion investigation process, types of samples and strategies for sample collection, field and laboratory tests that can provide useful data for corrosion investigation and integration of various groups of data to determine the cause of corrosion.

3.2 OVERVIEW OF CORROSION ANALYSIS PROCESS

All corrosion is caused by anodic and cathodic electrochemical reactions that occur concurrently on a metal surface that is in contact with an electrolyte (e.g., water) or a material that can hold water (e.g., soil). The electrolyte and any other solid material or biofilm present on the metal surface provide chemical species that may contribute to, or facilitate, anodic and cathodic reactions. Surface conditions, such as the flow rate of electrolyte across the metal surface, contribute to the distribution and rate of the corrosion damage by removing reaction products that would otherwise accumulate around the cathodic and anodic sites and slow the corrosion reactions. Increasing temperature also increases the rate of aqueous corrosion to a point. Therefore, if the goal of a corrosion investigation is to determine the corrosion mechanism, the chemical and physical environment to which a corroded sample was exposed must be understood. Typically, this understanding comes from analysis of the deposits associated with the corrosion damage, examining the form in which corrosion is manifested, and analyzing historical information about the environment that was in contact with the metal.

Reviewing the historical operating conditions to which the component was exposed and relating the metallurgical, chemical, and microbiological conditions on the component surface to the operating environment provides insight into the types of corrosion mechanisms that are likely to occur in that environment. Historical operating conditions and supporting data include physical parameters (phases present, temperature, pressure, flow rate and flow regime, cathodic protection potentials, etc.), fluid chemistry, soil type/drainage, mitigation history (inhibitors, biocides, scale inhibitor, coating system, CP application, AC mitigation, etc.), inspection history, and system design. Having this information available helps the investigator connect the analytical findings to the environment or conditions that led to the corrosion.

Ideally, a corrosion failure analysis should be based upon proper characterization of (1) the conditions that are present in the "bulk" environment (e.g., soil, water, process fluid); (2) the "layer" between the metal/environment interface (e.g., in deposits or biofilms) and the (3) metal surface itself, both where corrosion has formed and where it has not. What is not commonly considered is that these three environments can be quite different from one another, even though they are present at the same time in the same place. For example, Wrangham and Summer (2013) showed that the types and numbers of microorganisms present in the bulk fluid phase are known to be quite different than those located on a surface within a biofilm. Deposits on a metal surface can also vary in composition and physical properties throughout their thickness as demonstrated by Larsen et al. (2008) who showed significant differences in both corrosion product composition and microbiology in thick deposits from offshore oil and gas production.

Background information about a corroded component is used to corroborate the data being collected through field and laboratory testing and to more accurately determine the corrosion mechanism that led to the damage. In many cases, obtaining reliable historical operating data about a corrosion failure is essential to finding a solution.

ASTM Standard G 161 provides a useful overview of the corrosion analysis process and a checklist that can be used when collecting information that will be used to support the corrosion failure analysis (ASTM G 161 2018).

Table 3.1 provides examples of the types of background and analytical information that are generally collected for corrosion investigation and some potential sources of that information.

3.3 IMPORTANCE OF SAMPLE SELECTION AND PRESERVATION

In many cases a corroded sample is not handled or preserved properly and the opportunity to diagnose the corrosion mechanism using reliable, accurate analytical results is lost. Uninformed observers may remove and discard any deposits associated with corrosion to assess the severity of the damage; thereby ruining valuable evidence.

TABLE 3.1
Types and Sources of Information Typically Used for Corrosion Failure Analysis

Type of Information to Collect	Sources of Information
Failure incident site conditions	Images and recordings of the site, 3D laser mapping of the incident site showing position of failed components, process and design drawings with annotations from field inspection, eye witness reports, visual inspection, damage caused by the leak or failure
Operating conditions	Operating temperature, pressure, flow rate recorded by instrumentation, process variables, weather, normal or upset conditions, recent equipment or process flow changes, proximity to mixing points, accumulation of water or solids, operator interviews and operating logs, alarm conditions
Operating history	Initial in-service date, list of changes or repairs, corrosion or failures in similar equipment, history and changes in corrosion mitigation, maintenance practices, inspection and test records, operating issues, e.g., fouling, plugging; hydrostatic test records, corrosion monitoring records from coupons, probes, etc. Previous root-cause analysis reports.
Corrosion products	Collection and analysis of deposits directly within or associated with corrosion damage and those adjacent to the corrosion in undamaged areas
Operating environment sampling	Gas samples, liquid or water samples, solids e.g., wax, black powder, scale, sludge, etc. for chemical and microbiological analysis; samples of known process materials and chemicals for comparison with unknowns. Samples of process fluids (water, oil, coolant, heat transfer fluid, etc.), soil, coating, etc. as relevant.
Metallurgical materials	Samples of the corroded component, material test reports from original construction, comparative samples from unaffected components. Results from material qualification and corrosion testing. Material specifications and procurement record; fabrication and welding records.
Corrosion damage morphology	Nature and severity of the corrosion damage; general, localized (pitting), cracking; clock position in pipe, proximity to interfaces (gas-liquid, liquid-solid, etc.), relationship to process conditions, association with heat transfer and velocity/flow regime; relationship to process or phase changes or mixing points; distribution of corrosion within the asset.

In some cases alteration or contamination of the sample is unavoidable, such as when a rupture occurs, a failure results in a fire or the failed component is exposed to fire suppressant. When a corroded sample has been altered, a representative sample may still be available if corrosion is present at another location on the same asset that was not affected by post-failure events or contamination. To obtain reliable samples for analysis of corrosion, it is essential that the samples be:

- Collected and preserved immediately upon exposing the corroded metal to atmospheric conditions, in order to prevent oxidation of corrosion products and changes to microbiology

- Collected using appropriate, sterile tools, placed into the recommended types of containers to avoid contamination and preserved until they can be analyzed
- Stored/shipped under recommended conditions to avoid damage and degradation of the samples

As discussed previously, sampling should ideally be performed to provide material to characterize the bulk phase, the layer between the bulk phase and metal (if present), and the material directly in contact with the metal surface itself, where applicable. Samples should also be collected from corroded and uncorroded areas for comparison. ASTM Standard E 14925 provides details on practices for receiving, documenting, storing, and retrieving samples for laboratory analysis and proper sampling and preservation techniques (ASTM E 14925 2017). It is typical for the laboratory performing a given analysis to provide guidance on sample collection and preservation to obtain the information desired from the test. Advance planning is required to ensure that the right materials are available to personnel performing the sampling.

3.4 SAMPLING STRATEGIES

3.4.1 Bulk Environment

In a leak or rupture situation due to internal corrosion, a bulk fluid sample may be obtainable at a nearby location away from the point where the fluid was released. Fluids collected once the failed component is removed may no longer be representative of actual conditions due to exposure to air and atmospheric pressure. A fluid sample should be collected under the same conditions at which the component was operating. For example, if there was no flow and water collected along the bottom of a pipe, it would be important to collect the accumulated water from a nearby similar source for analysis.

Fluid samples for dissolved gas analysis should generally be collected at the operating pressure of the asset. For crude oil or natural gas containing hydrogen sulfide (H_2S), special sample cylinders with a non-reactive lining should be used. Bulk liquid samples may be collected in glass or Nalgene bottles, depending on the requirements of the laboratory where the samples will be analyzed. Since dissolved gas will evolve from liquid samples collected from pressurized systems, field tests of the water phase for pH, alkalinity, and dissolved gasses are ideally performed in the field when the samples are taken since these parameters will change quickly after the sample is exposed to atmosphere.

When investigating external corrosion on buried pipelines it is important to collect various soil samples and ground water samples as the pipeline is being excavated and exposed (e.g., at various elevations relative to the pipe, at different distances laterally away from the pipe, soil directly in contact with the pipe, etc.). These samples must be taken before the corrosion/coating damage is fully exposed, particularly since the precise location of the damage may not be proven until the pipeline area of interest is actually exposed. During this time, on and off pipe-to-soil potential and soil resistivity measurements may be taken to accurately characterize the external environment. Note that the backfill used in a pipeline ditch may be a different composition and texture in comparison to the soil outside the ditch. Once the soils begin to dry out and are exposed to oxygen, their composition and texture will change. Therefore, it is best to collect samples while the excavation is taking place. Soil samples are typically collected in 2 L size quick-seal bags that are filled and then double bagged. Depending on the analysis to be performed the soil samples may need to be kept cool. Prior to shipping soil samples for analysis, contact the analytical laboratory for special shipping instructions as special labeling and permits are often required for shipping soils between regions or countries. Field pH measurements should be taken and recorded of any ground water collected. If external corrosion on aboveground piping has occurred under insulation, samples of the insulation near the corrosion damage and away should be collected. Note the location and orientation of the insulation sample relative to the corrosion to guide subsequent analysis. If moisture is present within the insulation and/or on the surface of the pipe beneath the insulation, pH measurements should be taken in the field and recorded.

Bulk samples are used to provide general chemical and microbiological information about the environment in which the corrosion occurred; however, Zintel et al. (2003) showed that characterization of planktonic microorganisms in the bulk phase alone does not provide sufficient information to identify MIC. Thin film samples on a surface may be collected using sterile fiber swabs or sterile fiber pads from which the sample can be later extracted in a lab setting.

3.4.2 Surface-Associated Samples

Since corrosion is an interfacial phenomenon, samples collected from the corroded surface are of particular significance in the corrosion investigation as discussed by Kagarise et al. (2017); however, it is also important to consider the potential origin of the surface materials. In crude pipelines for example, solid particulates of upstream origin, e.g., mineral scales, corrosion products, wax, formation solids, etc., can settle out of the fluid under low velocity or stagnant conditions. In some cases these solids from upstream sources may create conditions conducive to under deposit corrosion and facilitate the development of biofilms, yet their composition does not represent the corrosion reactions occurring beneath them. For example, iron carbonate (Siderite) may be detected in a pipeline where no carbon dioxide (CO_2) is present, because its origin was far upstream where CO_2 was present during production and later removed during processing. In some cases, thick layers of sludge and solids (Figure 3.1) or tightly adherent, dried deposits, such as salts (Figure 3.2), actually prevent visual examination of the metal surface for corrosion until they can be removed. Thick layers of material may need to be gradually removed and sampled during removal, until the steel surface is visible and corrosion is observed. Samples where different layers are visually noted should ideally have samples collected from each layer, as the differences between the layers may help in understanding the events that occurred prior to the corrosion damage.

When selecting surface-associated samples and interpreting solids analysis data, it is important to consider that the corrosion damage of interest happened over some

FIGURE 3.1 Thick sludge found when investigating internal corrosion in a pipeline.

FIGURE 3.2 A pipe cross section showing the accumulation of salt layers that occurred over years of service in a natural gas pipeline that experienced occasional upsets of high salinity brine.

period of time, either recently or perhaps further in the past when operating conditions were different that those in the current operation. Therefore the solids present during examination may not be representative of conditions that caused the corrosion in the past.

Samples are often collected from corroded and uncorroded areas on the affected component to look for differences in chemistry and microbiology that might help identify the corrosion mechanism. Collecting these comparative samples is easily achievable when there are multiple corroded areas; however, during field sampling it is also desirable to leave some areas undisturbed for subsequent laboratory analysis including surface composition techniques if the component will be examined in a lab. Obtaining comparative samples may be challenging when there are few (or only one) corrosion features. If there has been a leak associated with corrosion, often the surface materials collected at the point of the leak are contaminated or altered by the leaking fluid or even groundwater that enters back into the pipeline if it is not under pressure.

Corrosion may also be associated with coatings failures, and Cavallo (2017) showed that the coating system itself may be an integral part of the failure investigation, thus various types of coating samples may need to be collected to determine the root cause of a corrosion failure. The type of coating (or coatings) used, their condition, thickness, adherence to the pipe or to each other, presence of gaps or disbondment, fluid under blisters, presence of calcareous deposits, presence of mechanical damage to the coating, and the relationship of these features with the corrosion damage are important data to collect for the failure investigation.

One of the more challenging parts of a corrosion investigation is determining whether the corrosion was influenced by microbiological conditions (biotic) or formed strictly through electrochemical conditions with no microbiological input (abiotic). Microorganisms exist under a wide range of temperature, salinity, pH, and oxygen conditions, and can use many substrates for growth. Buried and submerged pipelines and metal structures are potentially susceptible to external or internal microbiologically influenced corrosion (MIC). MIC is a primary concern for pipelines carrying produced water or seawater, or hydrocarbon pipelines with any water accumulation. Cooling water, fire water, and waste water systems are all normally susceptible to MIC. Even deposits on a metal surface that adsorb and retain water can provide a suitable environment for microbial growth.

Samples for microbiological analysis must be collected using sterile tools and containers, and should ideally be collected as soon as the corrosion is exposed. Typically, samples for microbiological analysis in the lab are kept near zero C (not frozen) in a portable cooler with ice packs, and delivered to the lab within 24 hours. Other chemical preservation methods are available from different providers of microbiological analysis when samples cannot be shipped overnight on ice, such as from a remote offshore platform. In preserving corrosion deposits collected from offshore pipelines, a 100 mL volume screw top glass bottle and a lid with two hose connections was used to purge the bottle with inert gas (nitrogen) after sampling. This approach described by Eroini et al. (2015) also helped keep oxygen from changing the corrosion products and reduced desiccation.

Since most crude oil, gas, and products pipelines operate with little or no oxygen in the fluid, exposing the corrosion products (particularly iron sulfides) to air during sampling can change the mineral composition of the deposits. Packing corrosion deposits into a small glass bottle immediately after collection can also help reduce the effects of oxidation on the iron corrosion products. Surface samples that are covered in oil are somewhat more protected from immediate oxidation; however, these must also be collected immediately and stored in sealed glass vials to reduce oxidation. Corrosion products are particularly useful in determining the mechanism of corrosion when they are not altered after removal from the component. External corrosion deposits on buried pipelines can form under anaerobic conditions and require immediate collection upon exposure and proper preservation to avoid oxidizing the iron sulfides that could be present due to the activities of sulfate reducing bacteria (SRB).

3.4.3 Component Samples

A sample of the corroded component itself is highly desirable if any laboratory/metallurgical investigation will be performed. Ideally, the sample should be removed in a way that does not introduce contaminants to the corroded area; this is sometimes achieved using cold cutting methods and cutting some distance away from the corroded area. Extra material may be needed by the laboratory if mechanical property testing will be performed. It is good practice to either break the edges of flame cut metal with a grinder and/or wrap the edges of pipe sections with heavy cardboard and tape to prevent injury of those handling the sections.

For shipping, often the entire component is wrapped in heavy polyethylene film, sealed shut with tape and secured in a box or on a pallet. Care should be taken when strapping samples to a pallet to ensure that the strapping is not in direct contact with the area of interest. Where internal corrosion is of concern, such as in a pipe, the use of plastic end caps can help to seal the pipe to protect it from contamination. The corroded surface should not be cleaned or disturbed in the field, other than to collect surface samples as described earlier. If the component has experienced a leak or rupture and the corroded area has been grossly affected, it may be possible to collect another nearby sample that has corrosion but which has not been damaged during post-failure events. Once samples have been collected and preserved, field and subsequent laboratory testing are performed.

3.5 FIELD PROCEDURES

An overview of the typical field procedures associated with corrosion analysis is shown in Figure 3.3.

As mentioned previously, field testing is performed to measure sample compositional parameters that can change when a sample is exposed to atmospheric conditions. Chemical and microbiological changes can occur due to dissolved gas evolution, the introduction of oxygen, temperature changes, and oxidation of solids.

Chemical field test methods are often used for measuring dissolved gases, e.g., carbon dioxide, hydrogen sulfide, and oxygen, in aqueous samples. The pH and alkalinity of an aqueous sample are also typically measured immediately after a sample is collected, as these parameters will change due to gas evolution and entrainment. Temperature and pressure are parameters that should be recorded when samples are collected, but from a failure analysis viewpoint, the temperature and pressure trends experienced at the point of failure over time are more meaningful and can sometimes be obtained from historical operating records. Chemical field test methods are also often used for measuring the presence of calcium, sulfide,

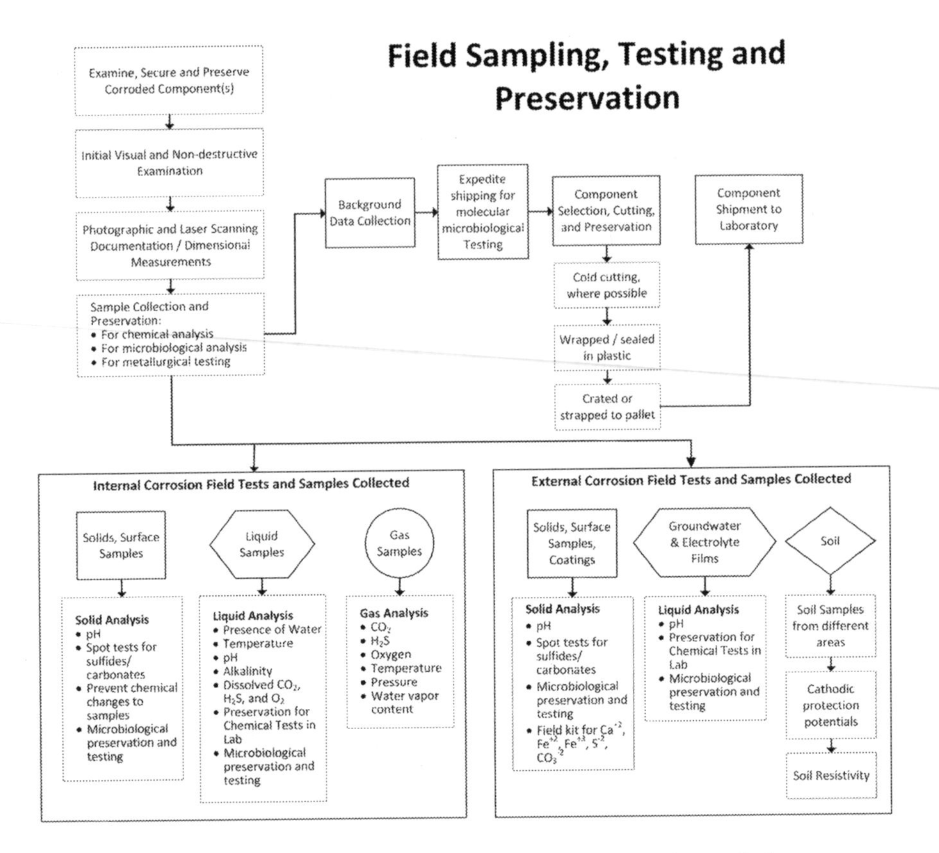

FIGURE 3.3 Field procedures for samples associated with corrosion analysis.

bicarbonate, ferrous, and ferric iron solid deposit samples. The color, odor, and texture of all collected samples, including the presence of distinct phases, should be documented.

One microbiological test that produces immediate results in the field is adenosine triphosphate (ATP) measurement. ATP is a measure of microbiological activity and the test value is often converted into a microbial equivalent (ME) number of active bacteria in a sample. Liquid or gel culture test media have historically been inoculated in the field as soon as samples are collected; however, the results are not available until after days to weeks of incubation and the resolution of the test method is low. Samples for molecular microbiological analysis (based on DNA) are also collected and preserved in the field and results are produced after laboratory analysis. One molecular method based on quantitative polymerase chain reaction (qPCR) is currently being developed in a portable field test unit.

Field procedures may also include nondestructive testing (NDT) if environmental cracking mechanisms, e.g., stress corrosion cracking (SCC) and sulfide stress cracking (SSC) are possible. If the cracking phenomenon is limited to a small area, care must be taken during NDT to avoid contaminating at least some of the surface exhibiting cracking for subsequent lab analysis. Other mechanisms including corrosion fatigue, erosion, and erosion-corrosion may be better characterized by inspecting larger areas in-situ in the field.

3.6 LABORATORY PROCEDURES

Testing in a laboratory may involve samples of the corroded component, solids, liquids, and gas samples in some cases. All of these samples help provide information about the potential mechanism of corrosion. Some typical laboratory tests and procedures are illustrated in Figure 3.4.

The following steps are often performed as part of the laboratory analysis of corroded samples.

3.6.1 Examination and Documentation

Visual examination and dimensional inspection are often performed before any cleaning or sectioning is initiated. It is important to compare the surface appearance of the corrosion as-received at the lab to that from the field photographs, as often there are significant changes that occur such that the surface no longer accurately represents the as-found condition because of oxidation and dehydration. Cleaning should not occur until all surface deposit samples are collected. Cleaning all or

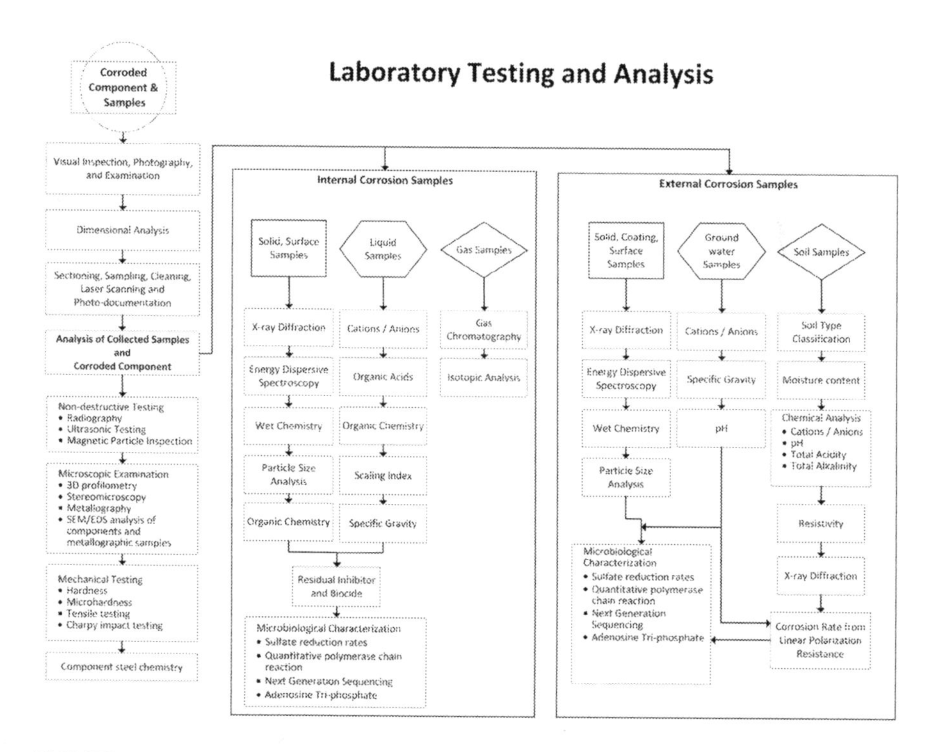

FIGURE 3.4 Overview of the typical laboratory testing and procedures used for internal and external corrosion analysis.

some of the corroded surface is performed to ensure no corrosion is overlooked and the morphology is captured correctly. Typically, inhibited acids are used to remove surface deposits using a soft plastic brush; ISO Standard 8407 (2009) provides information about a range of chemical cleaning processes that are intended to preserve the metal surface while removing deposits.

Three-dimensional (3D) laser scanning is usually performed after cleaning to document the locations and dimensions of the corrosion damage. In some cases both the internal and external surface of a component can be scanned and the two images combined to quantify the metal loss.

3.6.2 Sample Preparation and Sectioning

Additional samples for chemical analysis may be collected prior to any sectioning of the sample. In some cases where corrosion deposits are present, a small coupon (e.g., 4 cm x 4 cm square) may be removed for subsequent analysis, for example using SEM/EDS. Other samples are removed to prepare metallographic mounts of cross sections through the corrosion. Careful handling is required to prevent the pit deposits from being removed (or contaminated) during sample preparation. When the deposits are friable, sometimes the sample is vacuum impregnated with a two part epoxy before sectioning, or other non-viscous, fast-setting adhesives are used.

3.6.3 Stereo Microscopy and Surface Examination

Once samples of the corroded material are removed and can be easily handled, stereo microscopy and digital profilometry may be performed. Low magnification examination is used to document the morphology of the corrosion damage, e.g., general or localized distribution, physical features of isolated corrosion pits, relationship to welds, relationship to deposits, evidence of previous liquid levels or phase interfaces, etc. Careful visual and optical examination of the manner in which corrosion damage has manifested before and after cleaning the sample surface can provide information and clues about the conditions that were present. For example, Figure 3.5 shows an corrosion pit on a surface that is free of any general corrosion damage and in comparison, Figure 3.6 shows a number of isolated pits that have occurred in an area where general corrosion damage is also evident. These two corroded samples represent the different environmental conditions that were present in each case, even though both samples exhibited surface deposits when first examined.

The pit in Figure 3.5 was covered with a deposit or occlusion and the remainder of the surface had no deposits. The corrosion occurred only under the deposits. In Figure 3.6, a layer of non-protective scale was present over the entire surface and pit coincided with breaks in the scale layer.

3.6.4 Qualitative Chemical Tests

When very little surface deposits are present, qualitative chemical spot tests may be performed directly on the component surface, e.g., using hydrochloric acid and lead

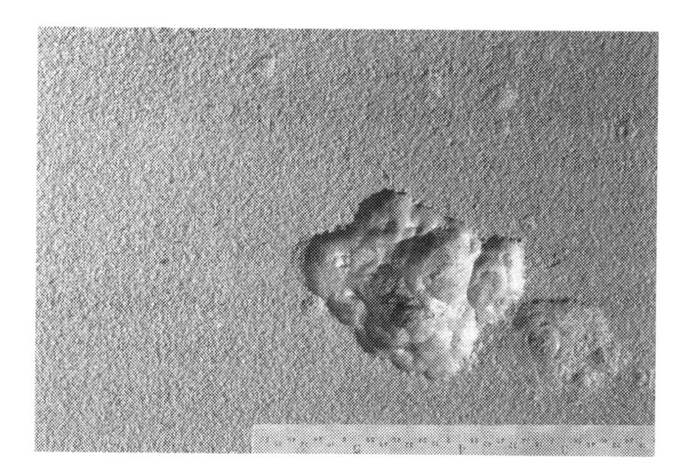

FIGURE 3.5 A large isolated corrosion pit on a component from a hydrocarbon system that exhibits no general corrosion on the remaining surface.

FIGURE 3.6 Numerous isolated pits and general corrosion are both present on a steel sample exposed to a brine solution.

acetate test paper for sulfide. Any collected solids may also have been tested this way in the field.

3.6.5 METALLOGRAPHY

Cross sections through corrosion damage are often prepared as metallographic specimens so that the microstructure of the component, any attached deposits and the corrosion morphology, can be examined at magnifications to 1,000x.

Metallurgical examination is used to determine whether the corrosion was selective to a particular constituent of the material, e.g., grain boundaries, a specific phase, or non-metallic inclusions. Metallography is also used to determine whether cracks are associated with the corrosion, such as sulfide stress cracks or stress corrosion cracks. If the deposits are well-preserved within the mounted pit cross section, the section can be examined using SEM and EDS to determine the different elements that are present.

3.6.6 Scanning Electron Microscopy (SEM)

SEM is used to characterize the condition of the component surface, the corrosion damage, cross sections and deposits at magnifications to over 5,000x. High resolution imaging can be used to document intergranular attack, for example. Deposit samples can be characterized as to the size and shape of individual particles. SEM is often used with energy dispersive x-ray spectroscopy (EDS) to determine the elemental chemical composition of corrosion features and deposits.

Characteristic corrosion products produced by iron-oxidizing bacteria (IOB) may be found (using SEM or environmental SEM [ESEM]) in the form of twisted stalks or sheaths as reported by Chan et al. (2011); however, these microscopic features are fragile and can be easily destroyed by rough handling.

In terms of visualizing microorganisms associated with corrosion, drying and oxidation of the sample makes it difficult to identify cells and additionally, the cells are often buried in inorganic deposits and corrosion products and thereby obscured. Environmental SEM (ESEM) allows imaging of "less dry" samples, but in some cases the exopolymers surrounding the biofilm cover over the cells.

3.6.7 Energy Dispersive X-Ray Spectroscopy (EDS)

Used in combination with SEM, EDS provides semiquantitative elemental composition of deposits. It does not provide the composition of compounds or minerals, for example, a deposit of iron sulfide would show only iron and sulfur using EDS. The method is particularly helpful for mapping the distribution of elements in corrosion deposits and pits, which in turn helps determine the corrosion mechanism.

3.6.8 X-Ray Diffraction (XRD)

XRD is an important tool for corrosion investigation as it provides the mineral composition of solid materials, such as pit deposits. The method can identify the mineral forms of iron oxides and oxyhydroxides, iron sulfides, iron carbonate; scales, e.g., calcium carbonate; sodium chloride, elemental sulfur (S_8), sand, and other materials in mineral form. XRD does not measure materials that are present as amorphous (non-crystalline) solids as is sometimes the case with sulfur and carbon.

Corrosion of steel under abiotic conditions produces iron oxides and oxyhydroxides, such as magnetite (Fe_3O_4), goethite (FeOOH) hematite (Fe_2O_3), maghemite (γ-Fe_2O_3), and lepidocrocite (FeOOH). In fresh and marine waters, two-line ferrihydrite is produced by iron-oxidizing bacteria (IOB), e.g., Leptothrix and

Gallionella; however, it is unstable and transforms into goethite and/or hematite upon air exposure. As described by McNeil and Little (1990), Mackinawite is a form of iron sulfide (FeS_{1-x}) characteristically produced by SRB; however, it can be produced in some cases by abiotic conditions as well. Manganese oxide (MnO_2) is a mineral characteristic of manganese oxidizing bacteria (MOB) that is more likely to be found under more aerobic conditions.

3.6.9 Other Chemical Analyses

The composition of aqueous samples may be determined e.g., using ion chromatography, ion-coupled plasma spectroscopy (ICP), liquid chromatography, and colorimetric titration methods, depending upon the analyte of interest. Organic liquids may be analyzed e.g., using Fourier-transform infrared (FTIR) methods, high performance liquid chromatography (HPLC), and mass spectrometry (MS) or gas chromatography-mass spectrometry (GC-MS).

The composition and properties of soil samples may be characterized e.g., by particle size distribution, moisture content, resistivity, cations and anions, alkalinity, etc. Since soil can dry out after collection, it is important to not only consider the soil properties in the as-received condition, but also when water-saturated. Ideally, resistivity and saturation measurements should be made in the field when samples are being collected.

3.6.10 Microbiological Analysis

Microbiological analyses most often seek to answer three questions; who is there, how many are there, and what are they doing? There currently is not a single test that can answer all three of these questions at the same time, but rather, different types of microbiological analyses are employed to provide specific answers. Skovhus and Eckert (2014) discussed the application of molecular microbiological methods (MMM) to MIC investigation, and these methods based on DNA have become widely utilized in the diagnosis of MIC in industry. The remainder of this book will go into detail about many of these methods, so they will not be repeated here.

A recent and interesting advancement in the use of molecular methods to diagnose MIC was reported by Lahme et al. (2020) where a novel [NiFe] hydrogenase from a specific methanogen was demonstrated to accelerate corrosion. A qPCR assay was developed to detect the gene (labelled micH) in biofilms that were highly corrosive (>0.15 mm/yr). The micH gene was found to be absent in non-corrosive biofilms, despite an abundance of methanogens. This proposed MIC biomarker was detected in multiple oil fields, indicating widespread involvement of this [NiFe] hydrogenase in MIC. New developments in molecular microbiology may eventually produce tools that are diagnostic for MIC.

3.6.11 Mechanical/Chemical Properties

Although not typically related to corrosion, mechanical property testing may be performed on component samples to characterize the hardness, yield and tensile strength, percent elongation, etc. The chemical composition of the component material may also be analyzed to compare with the standards to which the component was manufactured.

3.7 DATA INTEGRATION PROCESS

Once all of the data is collected and analyses are completed, the mechanistic cause of the corrosion is evaluated by examining and integrating the information available. The first step is to consider all of the relevant corrosion mechanisms for the alloy of interest based on the operating environment. Papavinasam (2014) and Eckert (2016) describe the common corrosion mechanisms affecting oil and gas industry assets and numerous other texts discuss corrosion mechanisms plaguing various engineered systems. The general approach to identifying the corrosion mechanism in a failure investigation is to integrate the chemical, metallurgical, and microbiological testing results with the design and operating condition history of the asset to gain an understanding of the corrosive environment that could have been present. This integrated approach as applied to MIC has been described by Kotu and Eckert (2019), Starosvetsky et al. (2007) and Stoecker (1984) and a generalized version of this concept is shown in Figure 3.7.

For corrosion failure analysis, the main data categories can be used to help address a series of questions that the investigation is attempting to answer, through data and sample collection and analysis; see Table 3.2.

The historical review also needs to consider any unusual or upset conditions, mitigation methods, cleaning, and previous repairs. Corrosion products, in particular, provide important clues about the conditions that were present when the corrosion damage occurred. Certain mineral forms of iron oxide, for example, occur only in the presence of chlorides, carbon dioxide, or hydrogen sulfide as shown by Burns (2016). The importance of data integration relative to the threat of internal MIC is discussed in NACE Standard Test Method TM0212 (2018).

Metallographic examination is often relied upon to confirm environmental cracking mechanisms, such as near neutral stress corrosion cracking. Chemical and physical conditions can have an effect on both biotic and abiotic corrosion. Determining the extent to which microbiological conditions caused the corrosion requires sufficient background and laboratory test data. The mere presence of microorganisms does not confirm that MIC has occurred.

Eckert and Skovhus (2018) discussed a general approach for linking microbiological activity to corrosion by considering the metabolism of the predominant, active groups of microorganisms that are present in biofilms on the material. One must then link the metabolic function of the microorganisms either directly or indirectly to a probable anodic and/or cathodic corrosion reaction. For example, an indirect link is the production of organic acids lowering the pH within the biofilm, removing passive layers, increasing the solubility of iron in solution and thereby promoting concurrent corrosion reactions. A direct link between metabolic function and corrosion could be cathodic hydrogen uptake by microorganisms in the biofilm, or production of sulfide ion leading to iron dissolution. While this approach linking metabolism to corrosion reactions is desirable, it is partially limited by a lack of

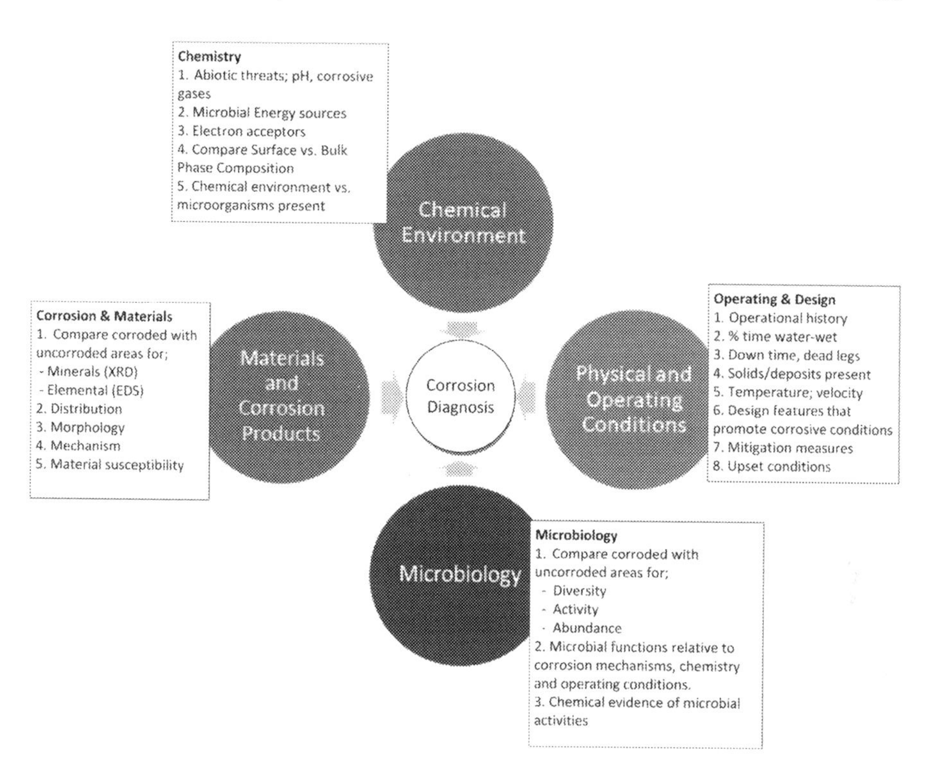

FIGURE 3.7 Overview of the data integration scheme for corrosion mechanism determination in failure analysis showing the type of data used from the four groups of information.

understanding of how microorganisms behave in mixed communities. For forensic analysis of corrosion damage, Lee and Little (2017) identified that the following information are generally accepted as being suitable evidence for identifying MIC:

1. A sample of the corrosion product from the affected surface that has not been altered by collection or storage (i.e., contamination, oxidation, etc.),
2. Identification of a complete corrosion mechanism (or a combination of mechanisms) including anodic and cathodic reactions,
3. Identification of microorganisms capable of growth in the particular environment, and,
4. Demonstration of an association of those microorganisms with the observed corrosion, both spatially and chemically/electrochemically.

The use of probabilistic, mechanistic, and empirical MIC, CO_2 or other models is another approach for integrating data from a corrosion investigation. An overview of the state of the art in MIC models is presented in Wolodko et al. (2018).

Regardless of the cause of corrosion, appropriate and effective future prevention methods can only be identified if the corrosion mechanism is identified correctly in the first place. This chapter has sought to provide some practical guidelines for

TABLE 3.2
Main Data Categories for Corrosion Failure Analysis and Questions to Be Considered Based upon the Data and Information Collected

Main Data Category	Questions to Be Answered in the Investigation
Design and Operation	• When did the corrosion occur relative to changes in the operational history of the component/asset? • How long was the component exposed to water? • Are conditions different in down time or during maintenance and could they be contributing to the corrosion? • Are surface deposits present and for how long? What are the possible origins? • How do temperature and velocity contribute to the corrosive conditions? • Are there design features that contribute to the corrosive environment? • What mitigation measures have been used and how have they been applied? How were they monitored? • Could process upsets contribute to the corrosive conditions?
Chemistry	• Based on the chemical environment what abiotic corrosion mechanisms are viable? (e.g., low pH? corrosive gas?) • What types of microbial nutrients are present? • How does the composition at the corroded location differ from areas where there is no corrosion or from the bulk phase? • How does the composition change over time or during upsets or maintenance? • How does the chemical environment compare with the microbial functional groups or types of microorganisms that were detected? Are there bioproducts of microbial activity present? Does the environment support their growth?
Microbiology	• How do the numbers, types, or activities of microorganisms at the corroded location differ from areas where there is no corrosion or from the bulk phase? • Which microorganisms could thrive under the chemical and physical conditions that are present? Which would not? • What differences are observed in microbiology at different locations within the asset
Corrosion and Metallurgy	How does the elemental/mineralogical composition differ between corroded and uncorroded areas or from the bulk phase? Are there corrosion products present that can only be produced by microbial activity? How does the morphology of the corrosion relate to the chemical, microbiological and physical conditions present? Are both general corrosion and pitting corrosion observed? Is the damage observed characteristic for this alloy in this environment? Does the metallurgy of the component conform with the standards to which is was manufactured? Did manufacturing or fabrication methods or errors influence the corrosion or potential for corrosion? Was there a repair? Are there identical components in similar service that did not corrode?

conducting analysis of any form of corrosion, with some examples of experiences from the oil and gas world. Clearly, advance preparation is needed in terms of training, supplies, and staffing to ensure the correct steps are taken when a corrosion failure occurs or severe corrosion damage is identified.

3.8 CONCLUSIONS

Corrosion failure analysis that supports incident investigation and root cause analysis can lead to risk reduction, as discussed by Pitblado et al. (2011).

Corrosion damage morphology is not a sufficient basis determine its cause; analysis and verification are essential steps if an accurate diagnosis is desired. Field and laboratory test data from samples collected and handled properly and sufficient background information are the most important factors for reliably determining the mechanism of corrosion damage.

To prevent or mitigate future corrosion, understanding the corrosion mechanism is an essential first step; although it is not enough. RCA is a tool that operators can use to increase organizational learning from incidents and to reduce the likelihood of repeat incidents. Understanding the immediate causes and root causes of corrosion will help asset owners identify and implement appropriate measures to reduce the likelihood of corrosion and help ensure the reliability and integrity of their operations. Although there are still many questions about MIC to be answered, as recently pointed out by Little et al. (2020), approaching MIC (or any) corrosion threat using an integrated, multidisciplinary, and scientific approach can provide a rational basis for selection of mitigation and monitoring activities, leading to improved asset integrity, extension of asset life and a step in the direction of sustainability.

REFERENCES

ASTM E 1492. 2017. *Practice for Receiving, Documenting, Storing, and Retrieving Evidence in a Forensic Science Laboratory*. ASTM International, West Conshohocken, PA.

ASTM G 161 – 00. 2018. *Standard Guide for Corrosion-Related Failure Analysis*. ASTM International, West Conshohocken, PA.

Burns, M. 2016. The Tales That Rust Can Tell: The Use of Corrosion Product Analysis in Corrosion Failure Analysis. SPE-179927-MS, presented at SPE International Oilfield Conference in Aberdeen, Scotland, May 2016. Society of Petroleum Engineers.

Cavallo, J. 2017. *CorrCompilation: Coating Failure Analysis*. NACE International, Houston, TX. ISBN 978-1-57590-343-9

Chan, C. S., Fakra, S. C., Emerson, D., Fleming, E. J., and Edwards, K. J. 2011. Lithotrophic iron -oxidizing bacteria produce organic stalks to control mineral growth: implications for biosignature formation. *The ISME Journal*, 5, 717–727. 10.1038/ismej.2010.173

Eckert, R. 2016. *Field Guide for Internal Corrosion Monitoring and Mitigation of Pipelines*. NACE Press, Houston, TX. ISBN 978-1-57590-328-6

Eckert, R., and Skovhus, T. L. 2018. Advances in the application of molecular microbiological methods in the oil and gas industry and links to microbiologically influenced corrosion. *International Biodeterioration & Biodegradation*, 126, 169–176. ISSN 0964-8305, 10.1016/j.ibiod.2016.11.019

Eroini, V., Andfinsen, H., and Mitchell, A. 2015. Investigation, Classification and Remediation of Amorphous Deposits in Oilfield Systems, SPE 173719-MS. Society of Petroleum Engineers.

Gareth, J., Edwards, T., Wright, A., Broadhurst, M., and Newton, C. 2009. Learning lessons from forensic investigations of corrosion failures. *Proceedings of the Institution of Civil Engineers: Civil Engineering*, 162, 5, 18–24.

ISO 8407. 2009. *Corrosion of Metals and Alloys — Removal of Corrosion Products from Corrosion Test Specimens.* International Organization for Standardization, Geneva, Switzerland.

Kagarise, K., Eckert, R., and Vera, J. 2017. *The Importance of Deposit Characterization in Mitigating UDC and MIC in Dead Legs, Paper 9128, Corrosion 2017.* NACE International, Houston, TX.

Kotu, S. P., and Eckert, R. B. 2019. A framework for conducting analysis of microbiologically influenced corrosion failures. *Inspectioneering Journal* 25, 4.

Lahme, S., Mand, J., Longwell, J., Smith, R., and Enning, D. 2020. *Severe corrosion of carbon steel in oil field produced water can be linked to methanogenic archaea containing a special type of [NiFe] hydrogenase. bioRxiv 2020.07.23.219014.* 10.1101/2020.07.23.219014

Larsen, J., Skovhus, T. L., Saunders, A. M., Højris, B., and Agerbæk, M., 2008. *Molecular Identification of MIC Bacteria from Scale and Produced Water: Similarities and Differences, Corrosion 2008 Paper 08652.* NACE International, Houston, TX.

Lee, J., and Little, B. 2017. *Diagnosing Microbiologically Influenced Corrosion in the Oil and Gas Industry in Microbiologically Influenced Corrosion in the Upstream Oil and Gas Industry* (1st ed.) eds. T. L. Skovhus, D. Enning, and J. S. Lee, CRC Press, Taylor and Francis, New York, https://doi.org/10.1201/9781315157818.

Little, B. J., Blackwood, D. J., Hinks, J., Lauro, F. M., Marsili, E., Okamoto, A., Rice, S. A., Wade, S. A., and Flemming, H.-C. 2020. Microbially influenced corrosion—Any progress? *Corrosion Science*, 170, 108641. ISSN 0010-938X, 10.1016/j.corsci.2020.108641

McNeil, M. B., and Little, B. J. 1990. Mackinawite formation during microbial corrosion, *Corrosion*, 46, 7, 599–600.

NACE Standard Test Method TM0212. 2018. *Detection, Testing, and Evaluation of Microbiologically Influenced Corrosion on Internal Surfaces of Pipelines.* NACE International, Houston, TX.

Papavinasam, S. 2014. *Chapter 4 - The Main Environmental Factors Influencing Corrosion in Corrosion Control in the Oil and Gas Industry.* Gulf Professional Publishing, Houston, TX, USA, pp. 179–247. ISBN 9780123970220, 10.1016/B978-0-12-397022-0.00004-2

Pitblado, R., Fisher, M., and Benavides, A. 2011. *Linking Incident Investigation to Risk Assessment. Presented at the Mary Kay O'Connor Process Safety Conference*, College Station, Texas.

Skovhus, T. L., and Eckert, R. B. 2014. Practical Aspects of MIC Detection, Monitoring and Management in the Oil and Gas Industry. *CORROSION 2014*, San Antonio, TX.

Starosvetsky, J., Starosvetsky, D., and Armon, R. 2007. Identification of microbiologically influenced corrosion (MIC) in industrial equipment failures. *Engineering failure analysis* 14, 8, 1500–1511.

Stoecker, J. 1984. *Guide for the Investigation of MIC, Materials Performance.* NACE International, Houston, TX, p. 48.

Wolodko, J., Haile, T., Khan, F., Taylor, C., Eckert, R., Hashemi, S., Ramirez, A., and Skovhus, T. L. 2018. *Modeling of Microbiologically Influenced Corrosion (MIC) in the Oil and Gas Industry - Past, Present and Future, Paper 11398, Corrosion 2018,* NACE International, Houston, TX.

Wrangham, J. B., and Summer, E. J. 2013. *Planktonic Microbial Population Profiles Do Not Accurately Represent Same Location Sessile Population Profiles, Corrosion 2013 Paper 2780.* NACE International, Houston, TX.
Zintel, T. P., Kostuck D. A., and Cookingham, B. A., 2003. *Evaluation of Chemical Treatments in Natural Gas Systems vs. MIC and Other Forms of Internal Corrosion Using Carbon Steel Coupons, Corrosion 2003 Paper 03574.* NACE International, Houston, TX.

4 Analytical Methods for MIC Assessment

Torben Lund Skovhus
VIA University College

Richard B. Eckert
DNV GL USA

CONTENTS

4.1 INTRODUCTION

This chapter provides an overview of the microbiological, chemical, and metallurgical analytical methods used to provide information in support of corrosion threat assessment and failure analysis, including in particular the threat of MIC. These methods are commonly used in forensic examinations of failed metallic and

DOI: 10.1201/9780429355479-5

non-metallic engineering components to discern whether microorganisms played a role in the damage mechanism. Wherever possible, references to further details about these methods are provided for readers seeking a deeper understanding.

It should be clear from the onset that, to date, there is currently no single unique test that can positively determine that MIC has occurred, or that it is actively occurring. The need for multiple lines of evidence to diagnose MIC has been made clear since early research by Pope (1988, 1990, 1992) for the Gas Research Institute and is still true today, as described by Little and Lee (2014). While there is promising research proceeding to elucidate genetic characteristics of samples that indicate when MIC is underway (Lahme et al. 2020), these "biomarkers" will undoubtedly, at least initially, be specific to certain types or groups of microorganisms known to be involved in MIC under specific environmental conditions. Currently, a diagnosis of MIC is supported by information and evidence about the chemical conditions in the environment, the microorganisms living in the environment – specifically on surfaces where corrosion has occurred, and metallurgical information about the material and corrosion products identified. Further, a complete picture of the factors leading to MIC requires the integration of information about operating conditions, design, construction, and mitigation history. It is with this combination of information that a solid diagnosis for MIC can be supported.

To collect information from these different categories requires the use of analytical methods in various scientific disciplines, as described in Figure 4.1. Each of the methods listed in Figure 4.1 under the microbiological, chemical, and corrosion/materials headings has strengths and limitations in terms of the data produced, speed, sensitivity, reproducibility, cost, training required to use it, and the way in which the data from any given method can be interpreted.

This chapter will introduce each of these methods and provide some context as to how the data can be applied.

4.2 MICROBIOLOGICAL TOOLS

While there are a large variety of testing and analytical methods that can be used to characterize the microbiological conditions in a given environment, they all produce

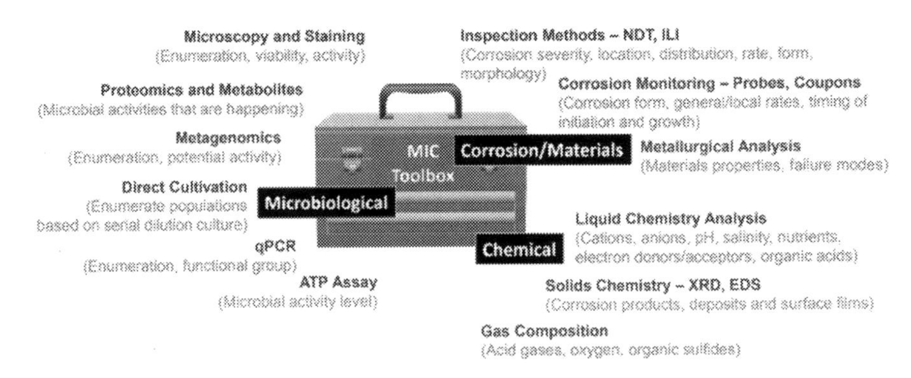

FIGURE 4.1 The analytical tools used for MIC identification including chemical, microbiological, and metallurgical methods.

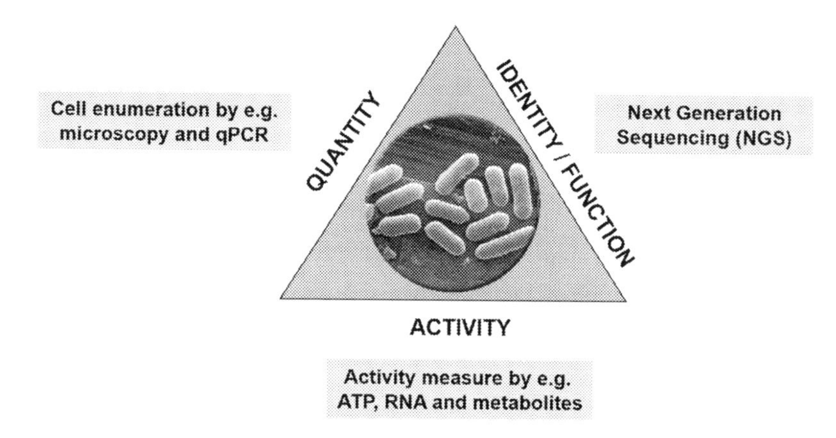

FIGURE 4.2 Combination of complementary microbiological techniques proposed by Skovhus et al. (2007). Modified version of the original figure with permission by the author.

some form of information about the diversity or types of microorganisms that are present, their numbers (enumeration), or activity. When used together, these three categories of microbiological information are far more informative than having data from only one category. This idea is illustrated in Figure 4.2.

There are several reviews that discuss the ever-evolving state of the art in microbiological diagnostic testing (Skovhus and Eckert 2014; Little et al. 2020) and a comparative summary of some commonly used methods is presented by NACE International in the standard "Detection, Testing, and Evaluation of Microbiologically Influenced Corrosion on Internal Surfaces of Pipelines" (NACE Standard Test Method TM0212 2018). Some of the important points to consider about each method and the data produced, include the following:

- Sensitivity, minimum detection limit
- Accuracy, repeatability, and bias
- Quantitative or qualitative results
- Sample volume and type of preservation required
- Interferences from other chemical species or oil in the sample
- False positives and false negatives
- Field or laboratory method (special handling requirements for reagents if field method)
- Time required to perform test and obtain results
- Cost and complexity; special analytical equipment needed and its maintenance
- Output format of data – relative ease of data interpretation and application

Four major groups of microbiological methods are discussed here including those based on enzyme measurement, microscopic examination, cultivation in media, and molecular microbiological methods (MMM) based on DNA. Most often, the ideal data for MIC failure investigation and diagnosis is obtained from the use of multiple microbiological methods that collectively provide a better insight on system microbiology than any single method, because the data from multiple methods help support each other. Further details on the application of these methods can be found in the case studies included in this book.

4.2.1 Enzyme-Based Methods

Being living organisms, microorganisms produce enzymes that aid in various metabolic activities. Such enzymes for which field tests have been created include adenosine triphosphate (ATP), adenosine phosphosulfate (APS) reductase, and hydrogenase. Modifications of the ATP assay exist that include measurement of adenosine monophosphate (AMP). The concentration of these enzymes in a sample is believed to be proportional to the concentration and activity of the microorganisms that produce them. These techniques can be used to estimate the concentration of certain types of microorganisms based on activity; however, they do not determine which microorganisms are present (active or inactive). Care must be taken when interpreting the results of any of these microbiological characterization techniques since the mere presence of microorganisms, or even specific types of bacteria, does not indicate that MIC is occurring. Further, microorganisms may be inactive for a number of reasons at the time of analysis; thus, the activity may not accurately represent their condition in the field. ATP methods are described in multiple standards and literature including ASTM D7687 (2017), NACE Standard Test Method TM0212 (2018), and by Skovhus and Højris (2018).

4.2.2 Microscopy Methods

Microscopy methods (Holmkvist et al. 2011) are used to examine samples directly to determine the overall numbers of microorganisms present (e.g., using various staining techniques), or activity and abundance using fluorescent in situ hybridization (FISH) probes. Depending on the method of sample preparation, the detection limit for all microorganisms present in a liquid sample, including viable, non-viable, and dead cells, is approximately 10^3 cells per mL unless the sample is concentrated by filtration. Biological stains such as acridine orange (N,N,N',N'-tetramethylacridine-3,6-diamine), fluorescein isothiocyanate (FITC), and 4', 6-diamidino-2-phenylindole (DAPI) are often used for epifluorescent microscopy.

Fluorescent probes have been developed to "label" specific groups of microorganisms, or to distinguish live vs. dead cells in a sample. FISH probes are used to identify and quantify certain species and groups of microorganisms; however, organic and inorganic materials in the sample can interfere with examination through autofluorescence. DAPI is often used in combination with FISH analysis, to help distinguish the total cell count from the number of cells that are labeled using the FISH probes. Confocal laser scanning microscopy (CLSM) is another means of microscopic imaging of stained biofilms that can provide a three-dimensional rendering of cell distribution within a biofilm a few microns in thickness if no organic or inorganic materials are also present. Microscopic methods for MIC failure analysis have seen infrequent application because of the limited data that may be produced, without regard to cell viability or identification, and due to interferences from oil and solid particles.

4.2.3 Culture-Based Methods

Serial dilution of liquid culture media using the most probable number (MPN) method has been the method most used for detection of microorganisms in the oil and gas industry for decades and is still widely used by the industry as described in standard TM0194 (NACE Standard Test Method TM0194 2014). Several studies have shown the limitations of this method (Larsen et al. 2008; Mitchell et al. 2012; Jensen et al. 2013; Skovhus et al. 2017) and for that reason it is slowly being phased out by major oil and gas operators. The MPN method gives a measure of the culturable fraction of the microorganisms in a sample at a given salt concentration (i.e., total dissolved solids [TDS]), growing at a given incubation temperature for a given period of time using the nutrients provided in the medium. Since no culture medium can approximate the complexity of a natural environment, even under ideal circumstances, liquid culture provides favorable growth conditions for only 1 to 10% of the natural population and many microorganisms in oil and gas systems are not culturable using traditional media. The MPN method does not provide a direct measure of the potential for MIC (Jensen et al. 2013; Skovhus et al. 2007).

4.2.4 Molecular Microbiological Methods (MMM)

Molecular microbiological methods are microbiological analysis techniques that do not rely on the cultivation of microorganisms; rather, they commonly rely on the detection of a DNA, RNA, or protein-based markers. Since the broader introduction of MMM to the oil and gas industry in about 2007 (Skovhus et al. 2007), their use has developed at a rapid rate and the benefits of their results have been demonstrated by numerous publications (Zhu et al. 2003; Gittel et al. 2009; Whitby et al. 2013; Skovhus et al. 2014, 2017). A detailed review of applied MMM was published by Muyzer and Marty in "Applications of Molecular Microbiological Methods" (2014). From an industrial perspective, a valuable resource with further details on the applicability of MMM in the oil and gas industry is the Energy Institute (EI) report "A practical evaluation of 21st century microbiological techniques for the upstream oil and gas industry" (Juhler et al. 2012). The report compared a range of traditional oil industry microbiological techniques (including MPN) with MMM to help advance the use of the most appropriate technology for microbiological assessment in the oil and gas industry. The EI report examined a total of six common analytical methods on four different types of samples from injection seawater, produced crude, pigging debris, and corrosion products/biofilm. The pros and cons of each of the methods are discussed in the EI technical report.

4.2.4.1 DNA-Based Methods

Polymerase chain reaction (PCR) and quantitative PCR (qPCR) are molecular microbiological methods used to amplify a single or few copies of a DNA sequence across several orders of magnitude, generating millions of copies. The qPCR method enumerates genetic markers rather than individual cells by applying a modified PCR method and may be applied to estimate total cell numbers or quantify specific microbial groups (e.g., surfate reducing prokaryotes (SRP) or methanogenic archaea). Because qPCR targets the DNA in all prokaryotes, the qPCR method measures living, inactive, and dead microorganisms whose DNA has not yet degraded. qPCR may be used to quantify the total number of microorganisms or a specific genus/species or functional group of microorganisms in nearly any type of sample, including produced fluids, oil/emulsion, and solids. The qPCR method does not underestimate microorganisms in a sample as is the case with culture methods.

Next generation sequencing (NGS) involves the extraction of DNA from all microorganisms contained in a liquid or solid sample. Specific genes of interest from the extracted DNA can then be amplified via PCR for further analysis. Most commonly, the gene of interest is the 16S rRNA gene, a gene that is present in all bacteria and archaea. Subsequent sequencing of the 16S rRNA gene on one of a number of available platforms (i.e., specific methodologies and instruments) generates sequence libraries that are compared to public or private DNA sequence databases to identify specific microorganisms in the sample. 16S rRNA gene sequencing offers a phylogenetic survey on the diversity of a single ribosomal gene (Rudkjøbing et al. 2014). Particularly useful is the combination of sequencing of 16S rRNA genes for identification of all microorganisms and their relative fractions in a microbial community, combined with the quantification of total bacteria and archaea via qPCR. Sequencing data are best used for comparing samples from different locations or samples collected from the same locations over a period of time to observe changes in the overall community composition. Lomans et al. (2016) discussed a complementary approach for integrating data from qPCR and NGS in a case study, demonstrating how data from different molecular methods can support each other.

The shotgun metagenomic sequencing approach is used to examine thousands of microorganisms from a sample in parallel and comprehensively sample all genes, with the intent of providing insight into biodiversity and microbial function. Shotgun sequencing also allows for the detection of low abundance members of microbial communities.

4.2.4.2 "OMICS" Methods

Bioinformatics is a broad term that encompasses the application of physical sciences, mathematics, physics, microbiology, and statistics to begin to associate meaning to massive volume of genetic and other data that can be collected from microbiomes. Analytical approaches (i.e., OMICS methods, Proco 2020) such as metagenomics, proteomics, and metabolomics, are also supported by "bioinformatics" since the data produced by these methods is referenced to a number of large online databases to help interpret the massive volume of raw data produced. Schreiber (2014) describes the general workflow of a metagenomic study through the application of three different metagenomic approaches and "meta-analysis" of microbial community gene expression using metatranscriptomics and metaproteomics.

Metabolomics seeks to understand the metabolic processes occurring within the microbiome and those of specific microorganisms by analyzing the different molecules they produce. Identifying the primary and intermediary metabolites present in the environment helps scientists understand "who is doing what" in the microbiological consortium. Ultra-high performance liquid chromatography/high-resolution mass spectrometry are methods commonly used for metabolite analysis. One of the challenges of this method is separating microbially produced molecules from complex organic compounds in crude oil samples.

Proteomics is the study of the proteins (gene products) expressed by microorganisms in response to the environmental conditions that are present, including energy sources, physical conditions and other microbiological activities occurring within a microbial consortium. The rationale for employing proteomics is the idea that the final product of a gene better describes its function than identifying the gene itself. For example, real time (RT)-qPCR can identify RNA that could be used in gene expression, but without identifying whether it was actually used, where proteomics can identify the protein a gene actually expressed. Methods applied to proteomics include the use of antibodies (immunoassays), mass spectrometry, and 2D electrophoresis (Graves and Haystead 2002). The application of proteomics to industry and MIC has been fairly limited thus far due to the complexity of the methods used and challenges in isolation of proteins; however, it is more broadly applied in bioremediation.

Metagenomics refers to the use of genomic technologies and bioinformatics tools to assess the genetic content of entire communities of microorganisms based on DNA (Handelsman 2004). As mentioned previously, analytical methods that supply data for metagenomic analysis include qPCR, RT-qPCR, NGS, and 16S rRNA high throughput and shotgun gene sequencing. A summary of microbiological methods discussed here is provided in Table 4.1.

Bioinformatics uses the integration of different types of microbiological and OMICS data and metadata to gain new perspectives on the way in which microorganisms behave as a community in a specific environment. As these methods become more accessible to industry it is likely that a better understanding of the MIC process will result. For example, a combined metagenomic and metabolomic approach was used to investigate differences in corrosion severity in two similar North Sea oil production pipelines (Bonifay et al. 2017). This study identified microbial communities with dramatically different metabolic and genomic signatures, with notable differences in the complexity and abundance of metabolites, the potential to utilize nitrate or sulfate as terminal electron acceptors, and the ability to degrade hydrocarbons.

In summary, microbiological tools can help answer questions about MIC in the following ways:

TABLE 4.1
Summary of Microbiological Methods

Method	Basis for Measurement	Sample Preservation	Quantitative?	Analysis Time	Information Yielded	Benefits and Limitations
ATP – Adenosine Triphosphate	Enzyme assay	Immediate measurement on site or preservation in lysis buffer for up to 7 days (in cooler)	Yes	Hours	Microbial equivalents, translated to total active cells	Interference issues are common Cannot distinguish between relevant and irrelevant microorganisms Lower detection limit about 1 x 10^3
Hydrogenase	Enzyme assay	Immediate chemical preservation is required	Yes	Hours	Activity of hydrogenase enzyme in sample	Hydrogenase enzyme has been implicated in MIC Not all MIC mechanisms involve hydrogenase enzyme
Adenosine phosphosulfate (APS) reductase	Immunological detection of the enzyme APS reductase	Immediate analysis	Semi-quantitative	Hours	Estimated number of SRB	Rapid test kit for estimation of SRB MIC by microorganisms other than SRB is possible
Microscopy based methods FISH, DAPI, etc.	Specific fluorescent staining of targeted microorganisms	Immediate analysis or chemical fixation	Yes	1 to 10 days	Measurement of specific species or active vs. inactive cells or total number of cells	Need for optimization of protocol based on sample matrix Interference issues common (e.g. autofluorescence of hydrocarbon) Spatial relationship between microorganisms in biofilm can be visualized Lower detection limit about 1 x 10^3 unless sample is concentrated by centrifuge

(Continued)

TABLE 4.1 (Continued)
Summary of Microbiological Methods

Method	Basis for Measurement	Sample Preservation	Quantitative?	Analysis Time	Information Yielded	Benefits and Limitations
qPCR	DNA amplification using PCR then quantification of specific genes (DNA) of interest	Freezing of sample or chemical preservation. Immediate analysis possible with field-based kits.	Yes	1 to 10 days	Depends on primers used. Can measure microbial functional genes, specific bacteria or total bacteria/archaea	DNA extraction procedures used largely affect the results; oil or solid matrix samples require more knowledge/skill to extract effectively. Adequate primer design important to guarantee broad coverage. Lower detection limit about 1 x 10^3. May lead to false negatives if sample contains uncommon microorganisms; in other words, the method only measures DNA for the primers used.
Microbial community analysis – NGS	Amplified and sequenced 16S rRNA genes are compared to public data bases	Freezing of sample or chemical preservation	No, provides ratio of different microorganisms or relative distribution	2 to 8 weeks	List of all microorganisms present in analyzed sample on a genera, phyla or species level	Adequate primer design is important Expert evaluation required due to complexity of data Useful for comparing between samples over time or from different locations; can apply PCA analysis.
-OMICs Metagenomics, Proteomics, metabolomics	Measurement of DNA, metabolic chemical species, proteins, etc.	Various depending on type of analysis to be performed and	Yes, but may not be directly	Weeks	Study of genetic material, proteins, or metabolites recovered directly from environmental samples	Researchers can analyze microbial diversity and also identify new proteins, enzymes, and biochemical pathways

(Continued)

TABLE 4.1 (Continued)
Summary of Microbiological Methods

Method	Basis for Measurement	Sample Preservation	Quantitative?	Analysis Time	Information Yielded	Benefits and Limitations
		the stability of the analytes	related to cell numbers		Characteristic by-products of microbial activity	Does require specialized knowledge and is time consuming. Not a routine method for monitoring, rather a research tool
Serial dilution – MPN Methods	Culture-based method using liquid media or gel.	Immediate or storage of sample in cooler for up to 48 hours	Yes	7 to 28 days	Cultivatable numbers within specific microbial functional groups	Only a small number of microorganisms can be grown in liquid media Culture medium salinity and other chemical parameters, and incubation temperature, must match field conditions Can only detect culturable microorganisms. Organisms that produce positive test results may not be active in service due to inavailability of nutrients.

Examples of Questions That Microbiological Analysis Methods Can Help Answer

Category	Questions
Cause determination	Which microorganisms are most prevalent in the overall environment and more specific to the corrosion damage?
	Which microorganisms are active? Are their nutritional requirements met by the environment?
	Which corrosion mechanism processes can be supported by the activity of the microorganisms that are present? How do they work together as a community?
Mitigation selection	Is the most effective chemical mitigation going to be based on corrosion inhibitor or biocide?
	How important will it be to pig the pipeline or clean the exposed surface to reduce the threat of corrosion?
Monitoring system selection	What type(s) of microbiological methods will need to be used to monitor mitigation effectiveness?
	What types of samples will provide the most reliable microbiological monitoring data?

4.3 CHEMICAL COMPOSITION ANALYSIS TOOLS

Chemical composition analysis provides important information to support the investigation and assessment of MIC and abiotic corrosion mechanisms. In general, the purpose of chemical analysis is to characterize the chemistry of the environment in the bulk fluid phase(s) and on the metal surface, ideally in areas of corrosion and areas without corrosion. Composition data provides insight on the resources available to microorganisms as energy sources, electron donors and acceptors, and also the by-products of microbial metabolism and resultant corrosion products. Importantly, understanding the chemical composition of an environment is also crucial for determining the abiotic corrosion threats that are present, e.g., dissolved carbon dioxide, hydrogen sulfide and oxygen, the pH, scaling tendency, salinity, conductivity, and organic acids that are present.

In multi-phase systems, such as crude production, hydrocarbons, water, and gas phases are present concurrently during gathering operations. Two important considerations for multiphase systems are that 1) the solubility (and thus concentration) of chemical species varies between the phases in equilibrium at temperature and pressure, and 2) multiphase flow in a pipeline is more complex than for a pipeline transporting gas or water only, and the flow rate and flow regime can have profound effects on corrosion. Organic and inorganic solids are yet another form of sample whose composition needs to be evaluated in consideration of the composition of the other phases present.

Finally, after collecting a sample from a pressurized system and potentially exposing it to atmospheric pressure and air, significant changes can occur rapidly, skewing the test results. When a sample of produced water is collected from a crude gathering system with carbon dioxide and hydrogen sulfide gas, the following changes occur very quickly unless preventive measures are taken:

1. Acid gases are released, raising the pH, and affecting the alkalinity of the sample.
2. The pH change impacts solubility of chemical species and can lead to solid precipitation.
3. Oxygen is entrained during sampling, reacting with dissolved (ferrous) iron and sulfide, oxidizing both species.

Different strategies for preventing or reducing changes to sample composition include the use of specific sample collection procedures and equipment to maintain pressure and prevent oxygen ingress, chemical preservation, and field testing of constituents that can change rapidly after sample collection.

Since field tests need to be relatively uncomplicated, produce timely results, and not require large equipment, there sometimes are trade-offs in terms of accuracy, precision, and resolution when performing field tests. Field tests are often performed on oilfield water samples for pH, dissolved CO_2 and H_2S, bicarbonate and carbonate alkalinity, and soluble iron. Several manufacturers provide test kits for these species that are suited for field use. For cooling water, steam and boiler water, a wider variety of field test kits is available. Since many field tests use a titration method, darkly colored samples can pose a problem for interpretation and hydrocarbons also present challenges.

4.3.1 Analysis of Liquids

The analysis of produced fluids (water and crude) or injected waters for enhanced oil recovery (EOR) or can provide information useful in determining the cause of corrosion. Some of the tests that can be performed in oilfield waters are dissolved ionic solids (usually referred as total dissolved solids (TDS)), total suspended solids (TSS), dissolved gases, and residual oil (oil-in-water). A complete water analysis including total alkalinity, cations, anions, specific gravity, and organic acids is often performed for general characterization. In oilfield-produced brines, TDS range from less than 10 g/L to over 350 g/L, of which salt (NaCl) often constitutes 80% or more, along with calcium, magnesium, and potassium salts. Troublesome cations from a scaling perspective are calcium (Ca^{2+}), magnesium (Mg^{2+}), barium (Ba^{2+}), and strontium (Sr^{2+}). Ferrous iron (Fe^{2+}) and manganese levels may indicate the dissolution of carbon steel, if no oxidizing species (e.g., sulfide, oxygen) are present to cause precipitation of ferric iron solids. Commonly encountered anions include bromide (Br^-), chloride (Cl^-), sulfate (SO_4^{2-}), bicarbonate (HCO_3^-), and sulfide (S^-). Ion chromatography is often used for analysis of anions and organic acids, while inductively coupled plasma spectroscopy (ICP) may be used for analysis of cations. The presence of oil-water emulsions can complicate the analysis.

TSS are organic and inorganic solids found in petroleum waters and seawater. These may be particles of metal oxides from well casings, mineral scales e.g., calcium carbonate, or oxidized iron or manganese. Other suspended solids may be silt, sand, clay, or biological material. These particles are often collected on a 0.45 μm membrane filter for identification or samples with heavy solids loading may be first filtered using coarse filter papers and solvent washing to remove residual oil. The filtered solids are the analyzed separately, as discussed later in this section. Separation of inorganic sediments from paraffin or heavy crude can be difficult.

The dissolved gases of greatest concern for corrosion in the oil and gas industry are hydrogen sulfide (H_2S), carbon dioxide (CO_2), and oxygen (O_2). Oil reservoirs can become sour over time through the activity of sulfate reducing bacteria (SRB) in the producing formation, from contamination during drilling or well workovers, and/or from injected seawater breakthrough. H_2S concentrations in water and oil from souring can reach concentrations up to hundreds of milligrams per liter. Carbon dioxide forms carbonic acid (H_2CO_3) when dissolved in water. It is one of the main contributors to production system corrosion. Dissolved oxygen is rarely present in produced fluids unless entrained from leakage or from seawater injection. If oxygen is present in oilfield fluids, it is perhaps the most serious corrodent due to the normally high chloride levels, compounded effects of acid gases and potential for oxidizing sulfide to elemental sulfur.

In terms of the role of liquid composition with regard to MIC, the objective of the analysis is to determine whether the energy sources and electron donors/acceptors needed for metabolism by the microorganisms are present, and whether the salinity, pH, and redox conditions present are suitable for the growth of those microorganisms. For abiotic corrosion mechanisms, liquid composition also plays an important role, with pH, alkalinity, chloride content, and organic acids being key parameters for determining the likely corrosion mechanism, in concert with dissolved gases.

4.3.2 Gases

Analysis of gas phase composition is important because the species of concern from a corrosion standpoint in the gas phase will be dissolved in equilibrium with the oil and water phases present at pressure and temperature. A portable chromatograph may be used, if available; however, gas composition is often measured using gas chromatography (GC) in the laboratory. For accurate test results, samples for laboratory GC analysis must be collected at system pressure in stainless steel cylinders that have been cleaned and purged with an inert gas prior to sample collection. For analysis of organic sulfur and sulfides, samples are collected in lined cylinders to prevent the sulfides from reacting with the cylinder itself and affecting the measurement. Water vapor dewpoint of gas can be measured in the field using portable equipment or in the laboratory. For field testing of the gas phase, stain tube technologies have been used for decades; however, users are cautioned to be aware of interferences and condensation of liquids when venting high pressure gas. The condensed water (and sometimes hydrocarbon) can invalidate the measurements taken with water vapor stain tubes. ASTM Standards D4888 (2015), D4984 (2016) and D4810 (2006) provide guidance for stain tube analysis of water vapor, carbon dioxide, and hydrogen sulfide, respectively.

4.3.3 Solids

Solids exist in many forms and can be found in various ways, such as a commingled solid deposit from pigging a pipeline, a scraping of corrosion products collected from within a pit, fine powder found on a filter element in a gas system or solid debris separated during produced water processing. Solids found on a corroded surface may or may not reflect the conditions that caused the corrosion if the environment has changed since the corrosion occurred. The composition of solid deposits can sometime reflect conditions far upstream, such as with fine iron sulfide compounds (i.e., black powder; Powell et al. 2018) that are transported in a gas pipeline for very long distances, in gas that no longer contains hydrogen sulfide at the downstream sample point. Particle size, density, and shape analysis can provide information on the distribution of these characteristics within a population of diverse solid particles. NACE International Publication 61114 (2014) published a detailed report on the multiple ways solids can affect UDC and the selection of mitigation measures.

Solids and surface deposits are also relevant to MIC investigation. For example, Larsen and Hilbert (2014) investigated under deposit corrosion in production tubulars and discussed the significance of deposit composition and microbiological communities at different depths within the layers of solid deposits inside pipelines. Wrangham and Summer (2013) also emphasized the need to focus on surfaces and their research demonstrated that planktonic microbial population profiles did not accurately represent sessile population profiles same location. Stewart (2012) noted, as many have others, that for microorganisms, surfaces are a preferential location for life as they offer a more suitable environment for their survival, growth, protection, and reproduction and favor synergistic interactions within the consortium. Therefore, focusing on analysis of solids associated with surfaces can provide valuable information for diagnosing MIC.

Solids on a surface can influence abiotic corrosion as well. As described in a NACE report (2014) on the subject, the term *under-deposit corrosion* (UDC) is used to describe any corrosion associated with deposits that could be caused by a variety of underlying mechanisms. In the oil and gas industry, several types of deposits can be encountered individually or in combination, including:

- Inorganic deposits, including material transported from the producing formation (sand, silt, salts, etc.), corrosion products (iron oxides, -sulfides, and -carbonates), and precipitates of water (calcium, barium, magnesium-carbonate/-sulfate scales);
- Organic components from production (asphaltenes, waxes, etc.); and
- Organic components formed due to microbiological activity (exopolysacharrides, organic sulfur compounds, etc.).

Solids are typically analyzed in two forms, i.e., solids removed from a surface or solids attached to a surface. Solids can be removed from surfaces a variety of ways;

however, when sampling it is advisable to remember that solids may exhibit heterogeneities both in thickness and lateral distribution/concentration, particularly as involved with corrosion. Standard best practice for solids analysis relative to corrosion investigation is to sample deposits both within the corroded material and away from the corrosion to help identify differences and the chemical species that were particularly associated with the corrosion. Filtration of liquids can also be used to collect suspended solids; however, their composition only represents a point in time whereas solid accumulations generally occur over some period of time.

Bulk inorganic solids can generally be analyzed by dissolving the samples in acid and using ICP. Two other methods commonly used for laboratory analysis of deposits removed from surfaces and in-situ are scanning electron microscopy (SEM) – Energy-dispersive Spectroscopy (EDS) and X-ray powder diffraction (XRD).

4.3.3.1 Energy Dispersive X-Ray Spectroscopy (EDS)

SEM-EDS is likely the most widely used microanalysis technique in failure investigation, providing semi-quantitative elemental results (i.e., not compounds). EDS is conducted in combination with SEM to allow analysis of specific portions of a sample or specific particles. EDS uses an x-ray detector positioned near the sample that can simultaneously count and determine the energy of emitted x-rays; it is particularly helpful for characterizing concentrations of specific elements in one area of a sample, e.g., corroded areas vs. uncorroded. Since it generally must be performed under a high vacuum, the sample must be free of volatile organic material and dried. Environmental SEM (ESEM) can accommodate wet samples for analysis, but the EDS capabilities are limited for some samples. Samples of dried corrosion products or deposits are typically mounted on a carbon stub for analysis. If the sample is highly non-conductive, it may be coated with an atomically thin layer of carbon or gold-palladium to improve imaging; these elements are then disregarded in the compositional analysis. The size of the SEM chamber and range of x-y motion must be considered when determining sample size if a corrosion deposit is to be examined in-situ on the component or in a metallographic specimen. Modern EDS technology approaches quantitative analysis capabilities, given the right hardware, software and use of standards for calibration. Finally, EDS is often performed before XRD to provide the XRD analyst some idea of the materials that may be present. EDS will also detect/measure the elemental composition of amorphous materials, as may be found in some cases with sulfur and carbon in particular, where XRD will not characterize non-crystalline materials.

4.3.3.2 X-Ray Powder Diffraction

XRD is a common technique performed on solid corrosion products or surface deposits removed from a sample since it can identify crystalline phases in unknown materials. Identification of mineral compounds, rather than elements or ions, is useful for understanding the corrosion reactions that have taken place and to help identify corrosion mechanisms. Typically, only a small sample volume is required, i.e., a few grams. Micro-XRD is a technique that is gaining popularity as it allows SEM visualization of the sample and targeted XRD of a specific area.

One critical drawback for data interpretation is that oxides present in XRD results of corrosion products can result from exposure of the sample to atmosphere, rather than indicate the presence of oxygen in the system. Further, many iron sulfide species oxidize upon air exposure and change to other forms of sulfur. Eroini et al. (2015) described a sampling protocol for solids collected from offshore oil production that used a nitrogen purge to prevent oxidation prior to analysis. To eliminate oxidation of the sample as a source of contamination, even sample processing prior to XRD would need to be performed in an inert environment, such as a glove box with a nitrogen or other inert gas purge. Often, because samples are known to oxidize, the iron oxides are discounted to some extent when interpreting the data, particularly when the operating environment of their origin is known to be anoxic.

4.3.3.3 Mixed Phase Solids

For samples that contain different phases and various organic and inorganic materials, e.g., sludge from a cleaning pig run, EDS and XRD may not accurately describe the full nature of the sample if the water and hydrocarbons are removed. This situation is where wet chemical methods can be helpful. Wet chemistry is a general term that describes laboratory analysis techniques such as gravimetry (weighting), titrimetric (volume), and numerous separation techniques. Methods vary widely with sample type and composition and often require some experience to assemble the right analytical pieces. For example, solid organic phase samples are characterized by the percent weight of sample that is dissolved in water, acetone, toluene, and hexane. The inorganic percent weight of a mixed solid sample is sometime determined by combustion, i.e. heating to a high temperature to evaporate water and burn off the organic phase. Determining the percent water in a sample that is mostly solid can be performed through Karl Fischer moisture analysis as described in ASTM E203 (2006). Samples collected from pigging often contain multiple phases that require multiple extraction and analytical steps to characterize relative to the overall sample composition. Kagarise et al. (2017) discussed the importance of characterizing solid and sludge samples from pipelines in order to select the most appropriate means of mitigation.

4.3.4 Surface Characterization Tools

For solids that must be analyzed in-situ on a sample, various surface microanalytical methods can be employed. Surface analysis samples could originate from coupons, bioprobes, pipe cutouts, failed components, ILI verification cutouts, cross sections, laboratory simulation test specimens, metallographic specimens, or scale layers. The object of surface microanalytical methods is to chemically characterize solids on a surface, particularly those that may only be present as a thin film or a diffused layer into the metal. Samples used for these techniques are typically sectioned to a small size, sometimes only a few centimeters square.

The most common methods for chemical characterizations of surfaces are shown in Table 4.2. The techniques to be applied depend on the type and size of the

TABLE 4.2
Characteristics of Surface Analysis Techniques

	Composition Information Yielded	Analysis Depth	Analysis Area	Reported Detection Limits
EDS	Elemental	<5 μm	<1 μm	<1 atomic %
WDS	Elemental	<5 μm	>1 μm	<0.1 atomic %
AES	Elemental	<5 nm	>100 nm	<0.5 atomic %
XPS	Elemental, chemical structure	<5 nm	>10 μm	<0.1 atomic %
FTIR	Chemical structure	<5 μm	>10 μm	<100 ppm

sample, the desired depth and area of the analysis, and the information sought. In many cases, a combination of analytical techniques may be required to evaluate the physical and chemical nature of the surface under study.

Wavelength dispersive spectroscopy (WDS) is another method used with an SEM (like EDS). Where EDS measurements are based on the energy level of the emitted x-rays, WDS is based on the wavelength of the emitted x-rays. The WDS method offers advantages in certain applications, particularly those where sulfur and molybdenum need to be resolved.

Auger Electron Spectroscopy (AES) is a microanalytical method that offers greater sensitivity in terms of the depth of material analyzed since the technique can measure species within a few nanometers of the surface. Thus, the technique is most ideal for evaluation of contaminated surfaces and thin coatings. Depth profiling of composition can be accomplished by stepwise progressive removal of material from a surface using an ion beam, usually of argon ions, followed by Auger analysis.

X-ray photoelectron spectroscopy (XPS), also known as ESCA (Electron Spectroscopy for Chemical Analysis), is a surface analysis technique that can be used to determine both the elemental and chemical composition of the outermost atomic layers of a solid material. With the exception of H and He, all elements can be detected.

Fourier transform infrared spectroscopy (FTIR) is a nondestructive micro-analytical spectroscopic technique. The analysis results are principally qualitative, but also quantitative information regarding the composition and state of the material can also be obtained. FTIR can be used to identify small amounts of organic materials on surfaces and allows chemical mapping of the surface as well. Organic compounds are best characterized if the FTIR spectra can be compared to those of known compounds of interest.

In summary, the types of questions that can be answered through chemical analysis include the following:

Examples of Questions That Chemical Analysis Methods Can Help Answer

Category	Questions
Cause determination	To what extent could the chemical composition of the operating environment contribute to abiotic corrosion? Which microorganisms would be supported by the nutrients present in the environment? Were the chemical condition more or less constant or were there episodic changes? Did these episodes affect biotic or abiotic corrosion in some way? Does the composition of solids/deposits point to a particular corrosive chemical species? Is that chemical species involved in microbial metabolism?
Mitigation selection	Are surface deposits present that promote a biofilm and prevent adequate contact with biocide treatment? Is the primary corrosion mechanism best mitigated with an inhibitor, biocide, oxygen scavenger, or some other chemical? What chemical conditions that promote corrosion in this case can be mitigated by operational measures (filtration, separation, dehydration, scrubbing, etc.)? How do chemical conditions in the environment affect the distribution and effectiveness of chemical treatment?
Monitoring system selection	Are there chemical parameters present that correlate with increased corrosion or MIC activity that can be used as a key performance indicator? Are there microbial metabolites produced that can be used as a monitoring measure? If solid deposits are involved, what type of monitoring can be performed to track the growth or removal of deposits? How will the effectiveness of mitigation be measured?

4.4 METALLURGICAL AND CORROSION CHARACTERIZATION TOOLS

Metallurgical characterization, as described here, includes metallographic analysis of the metal/material being corroded, and broader characterization describing the way that corrosion damage is manifested in an asset, e.g., the size, distribution, and severity of corrosion damage and its relationship to operating conditions and design. It is one thing to examine a single leak in a pump station piping system and yet another to visualize the distribution of corrosion damage across various piping in dead legs in the entire pump station through risk-based inspection (RBI). In this example, characterization of the leak location could provide information regarding how the local corrosion damage interacted with the microstructure of the metal or whether welds or defects were present originally, while inspection of the piping system could help determine, e.g., if piping that rarely experienced flow was more likely to have corrosion. This type of information is particularly helpful for root cause analysis (RCA). SEM-EDS and other surface methods characterization methods, as previously detailed, are also considered metallurgical tools since they are used to study material degradation mechanisms, such as environmental cracking, dealloying, sensitization, etc., in addition to MIC.

The following section starts at a high level with inspection methods and corrosion monitoring techniques, followed by methods that examine corrosion with more detail, including metallographic examination, corrosion failure analysis, and RCA.

4.4.1 Inspection Methods

A multitude of inspection methods and technologies are available for corrosion damage characterization and monitoring on assets. NACE Technical Report 3T199 (2012) describes an important distinction between direct and indirect inspection and monitoring methods as follows:

> "Direct techniques are those that measure parameters directly affected by metal loss caused by the corrosion or erosion process."

> "Indirect techniques are those that provide data on parameters that either affect, or are affected by, the corrosiveness of the environment or the products of the corrosion process."

Direct techniques include nondestructive inspection (radiography, ultrasonic inspection, etc.), in-line inspection (ILI) of pipelines and tethered logging of well lines, and visual inspection of the internal and external surfaces of piping when they are exposed. Direct techniques can be either intrusive (coupons, sand and erosion probes, etc.) or non-intrusive, such as electrical field mapping (EFM) or permanently mounted ultrasonic sensors.

Automated ultrasonic inspection (AUT) and digital radiography (DRT) are commonly applied technologies for assessing the presence of physical wall loss on piping and equipment. Each technique and its method of application needs to be suitable for the material type, the nature/form of damage and the severity of damage that may be significant from an integrity standpoint. Although inspection is a valuable tool, in reality it can't be applied to all equipment all of the time, therefore inspection programs need to be prioritized to focus on the assets with the highest risk. In the oil and gas industry, RBI programs are frequently based on American Petroleum Institute standards API RP 580 (2016) and API RP 581 (2016), API 580 being a joint API/ANSI standard. RBI is intended to prioritize inspection programs based on risk, i.e., the potential severity of each threat and the consequences of a leak or rupture. MIC has not been previously differentiated as a threat that is specifically considered in RBI ranking; however, recently Skovhus et al. (2018) and then Abilio et al. (2019) have proposed assessment methods to rank the potential severity of MIC for the purposes of providing input to RBI programs.

The integration of inspection data with other MIC assessment information is quite valuable for determining the contributing causes to the environments that promote MIC. Particularly when operation and design information are available, a better picture of the reasons a certain location is more susceptible to MIC become evident. While the need to integrate inspection data with failure data appears to be common sense, some organizations still lack communications between the inspection/integrity and the corrosion control functions of the company that allow such data exchange to occur.

4.4.2 Corrosion Monitoring

Corrosion coupons and electronic probes (e.g., electric resistance [ER] probes) are a form of direct monitoring of corrosion phenomena; however, translation of coupon data or probe data to corrosion damage on a pipeline system is not a one-to-one association for several reasons. One advantage of coupons, however, is that they provide a surface exposed to the process fluid from which biofilms can be sampled. DNV GL Standard ST F101 (2017) guides users that corrosion probes and monitored spools are primarily for detection of changes in fluid corrosivity and are not applicable for verification of the integrity of the pipeline – a very important point! Inspection and monitoring methods can be viewed as illustrating the conditions leading to corrosion growth over various period of time:

- Inspection and ILI: conditions occurring over months or years.
- Coupons: conditions occurring over weeks or months.
- Probes: conditions by the minute, hour or day, depending on how data are collected.

A combination of these technologies provides the optimum monitoring data, demonstrating short-term responses to changes, for example in chemical treatment, and also showing the long-term effectiveness of corrosion control activities.

Indirect monitoring, by collection of gas, liquid and solid samples for microbiological and chemical characterization is an important part of an internal corrosion monitoring program, largely because these parameters indicate the types of corrosion mechanisms that could occur, or that are occurring. The real value from sampling and analysis comes from trending and comparison of large data sets collected over time; not from isolated samples. Additionally, most of these monitored indirect parameters should not be used as key performance indicators (KPIs) because they only demonstrate a potential for corrosion; not that corrosion has occurred in most cases. Regardless, data from liquid, solid, and gas samples collected from a failed system over time can provide data that is helpful for determining the corrosion mechanism and contributing factors to the corrosive environment, if the samples have been handled and analyzed correctly.

4.4.3 Metallurgical Analysis and Failure Analysis

Metallographic cross sections of corroded components are typically prepared and examined using a metallograph to visualize details of the corrosion morphology, e.g., whether grain boundaries or inclusions were involved or the thermomechanical processing of the metal during manufacturing influenced or helped initiate the corrosion. Often, cross sections of corrosion pits are prepared, taking care to preserve the deposits present within the pit and minimize contamination. Metallographic examination of these mounted cross sections may show layers of deposits and/or corrosion products that represent different operating condition or stages of growth of the pit. It is quite helpful to examine the preserved deposits using SEM-EDS to characterize elemental chemical differences between different features and whether a certain chemical species was involved with the corrosion.

Deposit morphology can sometimes provide useful data; however, morphology alone cannot be used as conclusive evidence that MIC has occurred. For example, the appearance of tubercles has often been (incorrectly) attributed to MIC. Ray et al. (2010) noted that the presence of tubercles on carbon steel cannot be used to conclude localized corrosion directly under the tubercles or a role for bacteria in their formation. Others have documented tubercle formation in drinking water systems with no corrosion present beneath them. Little and Lee (2014) emphasized the need for connecting corrosion products with the operating environment, noting that, "In all cases, the mineralogy of MIC depends on the microorganism, the nature of the substratum, and the environment."

While SEM-EDS are invaluable for corrosion product analysis, in some cases, carefully preserved corrosion deposits may be observed using SEM to contain evidence of iron oxidizing bacteria, particularly iron oxide covered straight and twisted stalks and sheaths. These delicate structures are fragile and easily damaged in handling, but sometimes they are observable above 1,000x SEM magnification. The presence of these "fossils" suggests that opportunistic iron oxidizers may have benefitted from the iron in solution near the corroding metal.

4.4.3.1 Corrosion Failure Analysis

Corrosion failure investigation and root cause analysis (RCA) are useful tools for asset owners and operators who are seeking ways to optimize performance, control costs and reduce risk. When it comes to corrosion, operators are sometimes reluctant to invest in determining the cause of corrosion in the interest of quickly making repairs and maintaining throughput, particularly when economic conditions are challenging. In the United States, regulated gas and liquid hydrocarbon pipeline operators are required to maintain incident investigation procedures as a part of their Operations and Maintenance manual, although these are normally viewed as being applicable only in the event of significant incidents rather than being used whenever corrosion is discovered. Certainly, if previous corrosion failures have been analyzed properly and documented, this provides a useful resource for investigating future corrosion issues in similar assets.

Understanding the reason that internal or external corrosion has occurred (beyond only after major incidents) is necessary for optimizing mitigation and prevention measures, and in most cases, worth the relatively minor investment required for a technically valid corrosion investigation. To control costs, the degree of analytical rigor can be made proportional to the severity of the corrosion or the level of risk associated with the asset. Anytime that corrosion is found on a pipeline, an opportunity presents itself to clearly understand the mechanism causing the corrosion and the contributing factors that supported it; leading to the ability to better manage the threat of corrosion in the future and extension of asset life.

The main objective of a corrosion failure analysis is to reliably determine the mechanistic cause. While there are many potential pipeline corrosion mechanisms, they can be generally divided between biotic (caused or promoted by microorganisms and/or their activities) and abiotic (corrosion in the absence of any direct microbiological contribution). Abiotic mechanisms, for example, include corrosion of steel in an electrolyte in the presence of dissolved carbon dioxide, hydrogen sulfide or oxygen, or galvanic effects. The definition of biotic corrosion is more complicated because microbiological activities may be intracately intertwined with chemical/electrochemical processes that cause corrosion. Biotic and abiotic processes are also affected by complex transformations of chemical species that have growth, inhibitory or synergistic effects on microbiological activities that cause MIC. When establishing the mechanistic cause of corrosion damage, each potential mechanism should be considered, and use the available information to assess the role of each mechanism.

It may appear that determining the true extent to which microorganisms contribute to corrosion in a failure investigation could be time consuming and costly, and some may take the position that the lack of plausible abiotic corrosion mechanisms in a given situation conveys MIC by default. Fortunately, given advances in the characterization of biofilms using MMM, a sound investigation of MIC can still be produced using a logical, methodical, and analytical approach.

In summary, metallurgical analysis can help answer the following questions related to MIC failure investigation:

Examples of Questions That Metallurgical Analysis and Inspection Methods Can Help Answer

Category	Questions
Cause determination	Are the elements and mineral compounds in the corrosion products indicative of an environment produced by the bacteria found in biofilms for the asset? Is the distribution and severity of corrosion damage in the asset indicative of an operating condition and/or design element that promotes corrosion or even a specific corrosion mechanism? Did the material composition, metallurgical condition or fabrication method increase the potential for corrosion damage? Are metallurgical defects the cause of damage and not corrosion?
Mitigation selection	What parameters can be changed in terms of design or operation that are likely to reduce the potential for MIC in susceptible locations? How can mitigation measures be applied (e.g. chemical treatment, pigging, cleaning, sand jetting, etc.) to the asset to mitigate MIC? Based on inspection and monitoring data, where do mitigation measures need to be considered beyond the location that was analyzed or to other similar assets?
Monitoring system selection	Based on the location of the corrosion, what parameters can be monitored as leading indicators of corrosion or MIC potential? Which technologies can be applied to provide sufficient insight as to when mitigation is effective or when additional measures are required?

4.5 DATA INTEGRATION AND ANALYSIS

More than for any other corrosion mechanism, the diagnosis of MIC requires integration of microbiological, chemical, metallurgical, operational, and design data and information. While this requirement has been repeated in the literature for decades now, the practicality of performing the integration is a process that is largely undocumented, relying heavily on the experience of the experts involved in the failure analysis. The transdisciplinary nature of MIC further requires specialized knowledge in multiple technical and scientific areas, including metallurgy, microbiology, production chemistry, asset operation and maintenance, integrity management, flow assurance, etc. to fully integrate different data sets. Further, there is a lack of well-documented MIC case histories that include data from all of these different categories.

One challenge to having available a well-documented library of MIC failure analyses is the lack of widely adopted, industry consensus, step-by-step procedures specifically for MIC failure investigation. So far, only limited guidance is available. For example, NACE TM0212 for internal MIC and NACE TM0106 for external

TABLE 4.3
Integration of Four Types of Data to Identify the Role of MIC

Chemical Conditions	**Microbiological Conditions**
1. Abiotic threats; pH, corrosive gases 2. Microbial energy sources 3. Electron acceptors for microorganisms 4. Surface conditions vs. bulk phase differences 5. Presence of deposits 6. Episodic events: • Oxygen ingress • Water upsets • Contamination	1. Compare corroded with uncorroded areas for microbial: • Diversity • Activity • Abundance 2. Compare microbial function relative to corrosion mechanisms, chemistry and operating conditions: • Acid production • Direct iron oxidation • Elemental sulfur/sulfide production • Supports intermediate electrochemical reaction
Corrosion and Metallurgical Information	**Design and Operation Information**
1. Material properties 2. Corrosion products – XRD 3. Elemental composition – EDS 4. Distribution of minerals/elements relative to corrosion damage 5. Morphology and rate of damage 6. Known mechanisms for given alloy in given environment	1. Water wet vs. oil wet 2. % time water-wetted 3. Down time with no flow or dead leg 4. Solid accumulations present 5. Temperature range 6. Mitigation measures applied or missed 7. Operational problems 8. Missed maintenance 9. Design shortcomings

MIC of pipelines provide general information about sampling and test methods that can be used to support corrosion investigations, along with some guidelines as follows:

- Merely detecting viable microorganisms in liquid or solid samples associated with corrosion does not prove that MIC has occurred.
- For any given sample, the actual number of microorganisms determined by any analytical technique is far less significant than trends that may be observed among samples from a series of locations or over a period of time.
- Analysis of the data should demonstrate that microorganisms and their activities provide the predominant influence over the corrosion mechanism present in the pipeline, as opposed to abiotic mechanisms.

ASTM G161 (2018) provides a useful overview of the corrosion failure analysis process and a checklist that can be used when collecting information that will be used to support the corrosion failure analysis, as do the appendices of the two aforementioned NACE standards. In addition, Kotu and Eckert (2019) described a process for integrating multiple lines of evidence for MIC investigation. The need for integration of data from multiple categories is summarized in Table 4.3.

REFERENCES

Abilio, A., Eckert, R., Skovhus, T. L., Wolodko, J. 2019. Modeling of Microbiologically Influenced Corrosion (MIC) for Risk-Based Inspection (RBI) in the Oil and Gas Industry: Screening Influential Parameters, *Poster Presentation at ISMOS 7*, http://www.ismos-7.org/wp-content/uploads/2019/06/Abstractbook_ISMOS7, p. 48.

API RP 580. 2016. Risk-Based Inspection, 3rd ed. American Petroleum Institute, Washington, DC.

API RP 581. 2016. Risk-Based Inspection Methodology, 3rd ed. American Petroleum Institute, Washington, DC.

ASTM D4810. 2006. *Standard Test Method for Hydrogen Sulfide in Natural Gas Using Length-of-Stain Detector Tubes*. ASTM International, West Conshohocken, PA.

ASTM D4888. 2015. *Standard Test Method for Water Vapor in Natural Gas Using Length-of-Stain Detector Tubes*. ASTM International, West Conshohocken, PA.

ASTM D4984. 2016. *Standard Test Method for Carbon Dioxide in Natural Gas Using Length-of-Stain Detector Tubes*. ASTM International, West Conshohocken, PA.

ASTM D7687. 2017. *Standard Test Method for Measurement of Cellular Adenosine Triphosphate in Fuel and Fuel-Associated Water With Sample Concentration by Filtration*. ASTM International, West Conshohocken, PA.

ASTM E203. 2006. *Standard Test Method for Water Using Volumetric Karl Fischer Titration*. ASTM International, West Conshohocken, PA.

ASTM G161. 2018. *Standard Guide for Corrosion-Related Failure Analysis*. ASTM International, West Conshohocken, PA.

Bonifay, V., Wawrik, B., Sunner, J., Snodgrass, E. C., Aydin, E., Duncan, K. E., Callaghan, Amy V., Oldham, A., Liengen, T., Beech, I., 2017. Metabolomic and metagenomic analysis of two crude oil production pipelines experiencing differential rates of corrosion. *Frontiers in Microbiology* 8. 10.3389/fmicb.2017.00099

DNV GL Standard ST F101. 2017. Submarine Pipeline Systems. Veritasveien 1 1363, Oslo, Norway.

Eckert, R., Skovhus, T. L. 2018. Advances in the application of molecular microbiological methods in the oil and gas industry and links to microbiologically influenced corrosion. *Int. Biodeterior. Biodegradation.* 126:169–176. ISSN 0964-8305, 10.1016/j.ibiod.2016.11.019

Enning, D., Garrelfs, J. 2014. Corrosion of iron by sulfate-reducing bacteria: New views of an old problem. *Appl. Environ. Microbiol.* 80(4): 1226–1236. 10.1128/AEM.02848-13

Eroini, V., Anfindsen, H., Mitchell, A. F., 2015. Investigation, classification and remediation of amorphous deposits in oilfield systems. *Soc. Petroleum. Eng.* 3. 10.2118/173719-MS

Gittel, A., Sørensen, K. B., Skovhus, T. L., Ingvorsen, K., Schramm, A. (2009). Prokaryotic community structure and activity of sulfate reducers in production water from high-temperature oil reservoirs with and without nitrate treatment. *Appl. Environ. Microbiol.* 75: 7086–7096.

Graves, P. R., Haystead, T. A. J. 2002. Molecular biologist's guide to proteomics. *Microbiol. Mol. Biol. Rev.* 66(1): 39–63. 10.1128/MMBR.66.1.39-63.2002

Handelsman, J. 2004. Metagenomics: Application of genomics to uncultured microorganisms. *Microbiol. Mol. Biol. Rev.* 68(4): 669e685. 10.1128/MMBR.68.4.669-685.2004

Holmkvist, L., Ostergaard, J. J., Skovhus, T. L. 2011. Which microbial communities are present? Using fluorescence in situ hybridisation (FISH): Microscopic techniques for enumeration of troublesome microorganisms in oil and fuel samples. In *Applied Microbiology and Molecular Biology in Oilfield Systems*, 55–61, eds.Whitby, C. and Skovhus, T. L. Springer, Netherlands. DOI 10.1007/978-90-481-9252-6

Jensen, M. L., Jensen, J., Lundgaard, T., Skovhus, T. L. (2013). Improving Risk Based Inspection with Molecular Microbiological Methods. *Corrosion 2013, Paper C2013-0002247.* NACE International, Houston, TX.

Juhler, S., Vance, I., Skovhus, T.L. 2012. *A Practical Evaluation of 21st Century Microbiological Techniques for the Upstream Oil and Gas Industry.* Energy Institute, London. ISBN: 9780852936382

Kagarise, K., Eckert, R., Vera, J. 2017. *The Importance of Deposit Characterization in Mitigating UDC and MIC in Dead Legs, Paper 9128, Corrosion 2017.* NACE International, Houston, TX.

Kotu, S. P., Eckert, R. B. 2019. A framework for conducting analysis of microbiologically influenced corrosion failures. *Inspectioneering J.* 25(4).

Lahme, S., Mand, J., Longwell, J., Smith, R., Enning, D. 2020. Severe corrosion of carbon steel in oil field produced water can be linked to methanogenic archaea containing a special type of [NiFe] hydrogenase. *Appl. Environ. Microbiol.* 87(3): e01819–e01820. 10.1101/2020.07.23.219014

Larsen J., Hilbert L. R. 2014. *Investigation into Under Deposit Corrosion in Halfdan Production Tubulars, Corrosion 2014-Paper 3746.* NACE International, Houston, TX.

Larsen, J., Rasmussen, K., Pedersen, H., Sørensen, K., Lundgaard, T., Skovhus, T. L., 2010. *Consortia of MIC Bacteria and Archaea Causing Pitting Corrosion in Top Side Oil Production Facilities. CORROSION 2010- Paper 10252.* NACE International, Houston, TX.

Larsen, J., Skovhus, T. L., Saunders, A. M., Højris, B., Agerbæk, M. 2008. *Molecular Identification of MIC Bacteria from Scale and Produced Water: Similarities and Differences Corrosion 2008-Paper 08652.* NACE International, Houston, TX.

Lee, J., Little, B. 2017. *Diagnosing Microbiologically Influenced Corrosion in the Oil and Gas Industry in Microbiologically Influenced Corrosion in the Upstream Oil and Gas Industry*, eds Skovhus, T. L., Enning, D. and Lee, J. S. CRC Press, Taylor and Francis, Boca Raton, London, New York.

Lenhart, T. R., Duncan, K. E., Beech, I. B., Sunner, J. A., Smith, W., Bonifay, V., Biri, B., Suflita, J. M. 2014. Identification and characterization of microbial biofilm communities associated with corroded oil pipeline surfaces. *Biofouling* 30(7): 823e835. 10.1080/08927014.2014.931379

Little, B. J., Blackwood, D. J., Hinks, J., Lauro, F. M., Marsili, E., Okamoto, A., Rice, S. A., Wade, S. A., Flemming, H.-C. 2020. Microbially influenced corrosion—Any progress? *Corros. Sci.* 170: 108641. ISSN 0010-938X, 10.1016/j.corsci.2020.108641

Little, B. J., Lee, J. 2014. Microbiologically influenced corrosion: An update. *Int. Mater. Rev.* 59(7): 384–393. 10.1179/1743280414Y.0000000035

Lomans, B. P., de Paula, R., Geissler, B., Kuijvenhoven, C. A. T., Tsesmetzis, N. 2016. *Proposal of Improved Biomonitoring Standard for Purpose of MIC Risk Assessment. SPE-179919-MS.* Society of Petroleum Engineers, Richardson, TX.

Mand, J., Park, H. S., Okoro, C., Lomans, B. P., Smith, S., Chiejina, L., Voordouw, G. 2016. Microbial methane production associated with carbon steel corrosion in a nigerian oil field. *Front. Microbiol.* 6. 10.3389/fmicb.2015.01538, Article 1538

Mitchell, A. F., Liengen, T., Anfindsen, H., Molid, S. 2012. *Experience of Molecular Monitoring Techniques in Upstream Oil and Gas Operations.* NACE International, Houston, TX. NACE-2012-1756

Muyzer, Gerard, Marty, Florence. 2014. Chapter 1 - Molecular methods in microbiologically influenced corrosion: Research, monitoring and control. In *Applications of Molecular Microbiological Methods*, eds. Skovhus, T. L., Caffrey, S. and Hubert, C. Caister Academic Press, UK. ISBN 978-1-908230-31-7, https://www.caister.com/ammm

NACE International Publication 61114. 2014. *Under Deposit Corrosion (UDC) Testing and Mitigation Methods in the Oil and Gas Industry.* NACE International, Houston, TX.

NACE Standard Test Method TM0194. 2014. *Field Monitoring of Bacterial Growth in Oil and Gas Systems.* NACE International, Houston, TX.

NACE Standard Test Method TM0212. 2018. *Detection, Testing, and Evaluation of Microbiologically Influenced Corrosion on Internal Surfaces of Pipelines.* NACE International, Houston, TX.

NACE Technical Report 3T199. 2012. *Techniques for Monitoring Corrosion and Related Parameters in Field Applications.* NACE International, Houston, TX.

Pope, D. H. 1988. *Guide to the Investigation of Microbial Corrosion in Gas Industry Facilities.* Gas Research Institute, Chicago.

Pope, D. H. 1990. *GRI-90/0299 Field Guide: Microbiologically Influenced Corrosion (MIC): Methods of Detection in the Field.* Gas Research Institute, Chicago, IL.

Pope, D. H. 1992. *Microbiologically Influenced Corrosion (MIC) II: Investigation of Internal MIC and Testing Mitigation Measures.* Gas Research Institute, Chicago.

Powell, D. E., Winters R. H., Mercer, M. A., 2018. *Field Guide for Managing Iron Sulfide (Black Powder) Within Pipelines or Processing Equipment.* NACE International, Houston, TX. ISBN-13:978-1575903835

Proco, L. 2020. The era of 'omics' technologies in the study of microbiologically influenced corrosion. *Biotechnol. Lett.* 42: 341–356. 10.1007/s10529-019-02789-w

Ray, R. I., Lee, J. S., Little, B. J., Gerke, T. L. 2010. The anatomy of tubercles: a corrosion study in a fresh water estuary. *Mater. Corros.* 61(12): 993–999. 10.1016/B978-0-444-63228-9.00005-X

Reservoir Microbiology Forum (https://www.energyinst.org/whats-on/search/events-and-training?meta_eventId=62111B)

Rudkjøbing, V. B., Wolff, T. Y., Skovhus, T. L. 2014. Chapter 16: Quantitative real-time polymerase chain reaction (qPCR) methods for abundance and activity measures. In *Applications of Molecular Microbiological Methods,* eds. Skovhus, T. L., Caffrey, S. and Hubert, C. Caister Academic Press, UK. ISBN 978-1-908230-31-7

Schreiber, L. 2014. Chapter 13. Metagenomic analysis of microbial communities and beyond. In *Applications of Molecular Microbiological Methods,* eds. Skovhus, T. L., Caffrey, S. and Hubert, C.Caister Academic Press, UK. ISBN 978-1-908230-31-7

Skovhus, T. L., Andersen, E. S., Hillier, E. 2018. *Management of Microbiologically Influenced Corrosion in Risk-Based Inspection Analysis.* SPE Production & Operations. 10.2118/179930-PA

Skovhus, T. L., Caffrey, S., Hubert, C. 2014. *Molecular Methods and Applications in Microbiology.* Horizon Press, Norfolk, UK. ISBN 978-1-908230-31-7

Skovhus, T. L., Eckert, R. 2014. *Practical Aspects of MIC Detection, Monitoring and Management in the Oil and Gas Industry, Corrosion 2014 Paper 3920.* NACE International; Houston, TX.

Skovhus, T. L., Eckert, R., Rodrigues, E. 2017. Management and control of microbiologically influenced corrosion (MIC) in the oil and gas industry - Overview and a North Sea case study. *J. Biotechnol.* 256: 31–45. 10.1016/j.jbiotec.2017.07.003

Skovhus, T. L., Højris, B. 2018. *Chapter 8 - Adenosine Triphosphate (ATP) measurement technology in Microbiological Sensors for the Drinking Water Industry.* IWA Publishing, London. ISBN 13: 9781780408682

Skovhus, T. L. Højris, B., Saunders, A. M., Thomsen, T. R., Agerbæk, M., Larsen, J. 2007. *SPE 109104 Practical Use of New Microbiology Tools in Oil Production.* Society of Petroleum Engineers.

Sooknah, R., Papavinasa, S., Revie, R. W., 2007. *Modelling the Occurrence of Microbiologically Influenced Corrosion, Corrosion 2007 Paper 07515.* NACE International, Houston, TX.

Stewart P. S. 2012. Mini-review: Convection around biofilms. *Biofouling.* 28: 187–198.

Stoecker, J. 1984. Guide for the investigation of MIC. *Matls. Perf.* 48e55.

Suflita, J., Atkas, D., Duncan, K., Oldham, A. 2012. Molecular tools to track bacteria responsible for fuel deterioration and microbiologically influenced corrosion. *Biofouling.* 28(9): 1003e1010. 10.1080/08927014.2012.723695

Whitby, C., Skovhus, T. L., 2011. *Applied Microbiology and Molecular Biology in Oil Field Systems.* Springer Publisher, Switzerland AG. ISBN 978-90-481-9251-9, p. 17.

Whitby, C., Skovhus, T. L., Passman, F. J. (2013). *ISMOS-3: Third International Symposium on Applied Microbiology and Molecular Biology in Oil Systems – Preface to special issue. Book chapter in Special Issue – 3rd International Symposium on Applied Microbiology and Molecular Biology in Oil Systems,* Vol. 81. Elsevier, 1–146.

Wrangham, J. B., Summer, E. J. 2013. *Planktonic Microbial Population Profiles Do Not Accurately Represent Same Location Sessile Population Profiles, Corrosion 2013-Paper 2780.* NACE International, Houston, TX.

Zhu, Xiang Y., Lubeck, J., Kilbane II, J. J. 2003. Characterization of microbial communities in gas industry pipelines. *Appl. Environ. Microbiol.* 69(9): 5354e5363. 10.1128/AEM.69.9.5354-5363.2003

5 Microbiologically Influenced Corrosion Mechanisms

Jason S. Lee and Treva T. Brown
Naval Research Laboratory

Brenda J. Little
B.J. Little Corrosion Consulting, LLC

CONTENTS

5.1 INTRODUCTION

Microbiologically influenced corrosion (MIC) is a collective term for all mechanisms by which the presence or activities of microorganisms influence the kinetics of corrosion reactions. Mechanisms for MIC can be attributed to the interactions of three basic components: 1) causative microorganisms, 2) susceptible materials, and 3) environments (electrolytes) capable of supporting life (Little et al. 2020). This chapter will provide an overview of the three components and potential MIC mechanisms for different metals and alloys.

DOI: 10.1201/9780429355479-6

5.2 CAUSATIVE MICROORGANISMS

Causative microorganisms of MIC include microflora from the three kingdoms of life: bacteria, eukaryota, and archaea. Liquid culture kits have been developed for serial dilution to extinction for the enumeration of groups of microorganisms, typically bacteria, associated with MIC mechanisms, such as sulfate-reducing bacteria (SRB) and acid-producing bacteria (APB). Liquid culture can detect only a very small percentage of naturally occurring bacteria and more importantly, no one has ever demonstrated a relationship between numbers of microorganisms and the likelihood that MIC has or will occur. Molecular microbiological techniques can more accurately identify microorganisms associated with electrolytes and corrosion products, provided appropriate primers are used (Skovhus, Caffrey, and Hubert 2014).

MIC has historically been associated with bacteria in biofilms, which are polymeric matrices of living and dead microbial cells and their extracellular exudates. Species from all three kingdoms can form biofilms and multi-kingdom biofilms have been reported for MIC (Stamps et al. 2020). Metabolic and chemical reactions that occur within biofilms determine the interfacial chemistries that control MIC mechanisms. Biofilm research has provided useful insights into biofilm complexities. Royer et al. (2003) described biofilms as "dynamic constructs" in which heterogeneity is both spatial and temporal. For example, biofilms do not typically provide uniform coverage of a surface, but are instead spatially heterogeneous. Water channels for electrolyte/nutrient distribution can form and dissipate within biofilms, leading to temporal heterogeneity. Naturally occurring biofilms contain multiple species with mutualistic relationships. Symbiotic relationships between groups of microorganisms that depend entirely on each other for energy conservation and growth have been identified within natural biofilms (Davey and O'Toole 2000). For example, acetogens oxidize propionate and butyrate to acetate, molecular (H_2) and/or formate that can then be used by archaeal methanogens or SRB. Some SRB couple the oxidation of organic compounds and H_2 to sulfate reduction. Methanogens, archaea capable of growth on H_2 and CO_2, are often found in biofilms with H_2-consuming SRB (Plugge et al. 2011).

All stages of biofilm formation are influenced by hydrodynamics. Fluid flow in a given system influences the transport of cells to surfaces, transfer, and reaction rates within biofilms, as well as detachment, i.e., erosion and sloughing. Dense biofilms can form as a result of high shear stress or as a result of starvation (Beyenal and Lewandowski 2002). vanLoosdrecht et al. (1995) discussed the effects of substrate loading, shear stress, and growth rate on biofilm structure. Immediately after attachment, microorganisms initiate production of extracellular polymeric substances (EPS), e.g., proteins, polysaccharides, nucleic acids, and lipids, that assist formation of microcolonies and microbial films. EPS bridge negatively charged bacteria to both negatively and positively charged surfaces. Metal ions (e.g., Ca^{2+}, Cu^{2+}, Mg^{2+}, and Fe^{2+}) interact with anionic functional groups e.g., carboxyl, phosphate, sulfate, glycerate, pyruvate, and succinate groups associated with EPS. EPS-bound metal ions can act as electron shuttles providing novel redox reaction pathways and gradients in biofilms (Beech and Sunner 2004).

Low-flow and stagnant conditions encourage cell attachment and biofilm development. Oxygen gradients can be established in quiescent electrolytes and in biofilms, providing microaerobic and anaerobic conditions required for some MIC mechanisms, such as iron oxidation and sulfide production, respectively.

One phenomenon related to biofilm formation on passive alloys is predictable, namely ennoblement, an increase of the corrosion potential (E_{corr}) (Little, Lee, and Ray 2008). Ennoblement is not a mechanism for MIC. Instead, E_{corr} ennoblement can increase the probability for pitting and crevice corrosion initiation and propagation of some passive alloys. Ennoblement is particularly important for those alloys for which the pitting potential is close to E_{corr}, for example 300 series stainless steels. Ennoblement has been observed in fresh, estuarine, and marine waters. Attempts to relate ennoblement in marine waters to a single microbiologically mediated reaction within biofilms have failed. Martin et al. (2007) demonstrated that for a particular metal, the extent of ennoblement varied between estuarine and marine locations and that the extent of ennoblement cannot be used to predict an increased likelihood of localized corrosion. Ennoblement in fresh water has been attributed to microbial manganese deposition and will be discussed in detail in Section 5.3.

MIC literature includes lists of microorganisms, particularly bacteria, that are referred to as "corrosive" or "problematic." without reference to the specific electrolyte in which a MIC mechanism could be maintained. The chemical composition of the electrolyte determines the survival and growth of specific microorganisms, leading to specific MIC mechanisms (see Section 5.3).

5.3 ELECTROLYTES

All microorganisms have minimum and optimum growth conditions, including water activity (a_w), pH, pressure, ionic strength, temperature, electron donors/acceptors, carbon sources, and other nutrients. Microorganisms cannot grow at $a_w = 1$ (pure water) in the absence of electron donors/ acceptors and nutrients. However, many microorganisms can adapt to a range of growth conditions that are less than optimum. Waters composed of suitable forms of carbon (C), nitrogen (N), phosphorus (P), and sulfur (S) can support microbial growth. In general, microorganisms can use a variety of electron acceptors for respiration, including oxygen, sulfate, nitrate, nitrite, carbon dioxide, Fe^{3+}, Mn^{+4}, and Cr^{+6}. Some microorganisms can use multiple electron acceptors, depending on availability (Richter, Schicklberger, and Gescher 2012).

Laboratory media and experimental conditions are designed to supply all of the requirements for optimum microbial growth or to demonstrate a particular mechanism. In natural and industrial settings, some microorganisms exploit environments with water activity (a_w) from 0.650 to 0.999, however, the majority of microorganisms cannot multiply below 0.900 a_w. (Stevenson et al. 2015). Maintenance or inhibition of localized corrosion is based on the ratio of aggressive ions: inhibiting ions, for example chloride:nitrate (Cl^-:NO_3^-) (Leckie and Uhlig 1966, Little 2003). Microorganisms influence the corrosivity of electrolytes during growth and respiration simply by the assimilatory uptake of nutrients (e.g., SO_4^{2-},

NO_3^-, PO_4^{3-}, NO_2^-) and dissimilatory reduction of electron acceptors (e.g., SO_4^{2-}, NO_3^-, NO_2^-, CrO_4^{2-}). For example, one of the arguments used to recommend Yucca Mountain, Nevada, as a site for storage of Alloy 22 (UNS N06022) metal canisters of nuclear waste was the nitrate-rich waters associated with the mountain (Kehler, Ilevbare, and Scully 2001). Nitrate in groundwater was identified as a naturally occurring corrosion inhibitor for localized corrosion for tightly creviced Alloy 22 in the presence of high Cl^- concentrations (Kehler, Ilevbare, and Scully 2001). However, Little (2003) suggested that nitrate concentration would be much reduced in the presence of growing and respiring microorganisms.

One dramatic example of electrolyte alteration during microbial growth is the breakdown and consumption of commercial corrosion inhibitors. Many compounds typically added to electrolytes as corrosion inhibitors, e.g., phosphates, aliphatic amines, nitrites, nitrates, and chromates can be degraded/assimilated by microorganisms, decreasing their concentration and effectiveness while increasing microbial populations (Little 2003). Concentrations and types of anions required for corrosion inhibition in Cl^--containing media are specific for both metals and environments. Sulfates can inhibit Cl^--induced pitting of stainless steels, but are aggressive toward mild steel. To be fully effective, inhibitor anions must be present in a certain minimum concentration. At concentrations below the critical value, inhibitive anions may act aggressively and stimulate breakdown of the oxide films. Cooke, Hughes, and Poole (1995) reported that potassium chromate (K_2CrO_4) was ineffective as a corrosion inhibitor in an electricity generating station because of microbial reduction of the hexavalent soluble chromium (Cr^{+6}) to insoluble Cr^{3+}. Cr^{+6} oxyanions are mobile in aqueous solutions and Cr^{+6} is typically transported to sites of localized corrosion where it is reduced and irreversibly adsorbed. In the presence of Cr^{+6}-reducing microorganisms, Cr^{3+} precipitates in association with the microbial cells. Lee et al. (2012) reported that a microbial consortium of the fungus *Aspergillus niger* and bacteria influenced the corrosion-inhibiting effectiveness of Cr^{+6} leached from a chromate (CrO_4^{-2})-containing coating exposed in an artificial medium designed to replicate the composition of human urine. In the absence of microorganisms, Cr^{+6} migrated from coatings to corroding sites on aluminum alloy A2024-T3. In the presence of microorganisms, Cr^{+6} was removed from solution and was associated with the cell mass. Pitting in A2024-T3 was suppressed by Cr^{+6} in exposures without microorganisms compared to severe pitting measured in the presence of microorganisms with the same initial Cr^{+6} concentration. Many bacteria and fungi can carry out the reduction of Cr^{+6} to Cr^{3+} (Bennett et al. 2013, Chen and Hao 1998).

5.4 SUSCEPTIBLE METALS AND ALLOYS

The susceptibly of metals and alloys is directly related to the propensity of the metal oxide to be derivatized (sulfide production), removed (metal ion reduction or acid production), or inhibited (under deposit corrosion) by microbiologically mediated reactions. MIC has been reported for carbon/mild steel, 300 series austenitic stainless steels, Cu, Cu/Ni alloys, Ni/Cu alloys, and Al alloys (Little and Lee 2007). MIC failures have not been reported for titanium alloys and Ni/Cr alloys. Increasing

Mo content in stainless steels above 6 wt% has been shown to significantly decrease susceptibility to MIC (Felder and Stein 1994).

5.5 MECHANISMS

Mechanisms for MIC are the result of specific metal/microorganism/electrolyte interactions. The specificity of these interactions makes it impossible to generalize a mechanism for MIC. MIC can cause pitting, crevice corrosion, enhanced erosion corrosion, enhanced galvanic corrosion. and cracking of some materials (Little and Lee 2007). In addition, several investigators have demonstrated that the outcome of corrosion experiments using identical microorganism-metal combinations varies with electrolyte composition (Jigletsova et al. 2004, Javed et al. 2017, Mehanna et al. 2010).

5.5.1 Sulfide-Producing Microorganisms

Sulfate-reducing bacteria (SRB) are the microorganisms most closely identified with MIC. SRB use the sulfate ion as the terminal electron acceptor, producing hydrogen sulfide in anaerobic environments. SRB are typically classified as anaerobes, but some species can survive exposure to aerobic environments (Beech and Sunner 2004). SRB are ubiquitous in natural and industrial waters. The metabolic activity of SRB causes accumulation of sulfide near metal surfaces. Other bacteria can produce sulfides by the reduction of elemental sulfur or thiosulfate. The inclusive term that includes all sulfide-producing bacteria and archaea is "sulfide-producing prokaryotes (SPP)"; opposed to "*sulfate*-reducing prokaryotes." Not all sulfide-producing microorganisms use sulfate as the terminal electron acceptor (Escoffier et al. 2001). Jia et al. (2019) suggested the term *sulfate-reducing microbes*, but that term excludes those microorganisms that produce sulfides via alternative reduction pathways.

While diverse sulfide-producing bacteria and archaea are recognized as important contributors to sulfide production in natural and industrially relevant biofilms, most research related to MIC mechanisms has been performed using pure cultures of SRB. Mechanisms attributed to SRB, include cathodic depolarization, sulfide derivatization of metal oxides, embrittlement related to hydrogen uptake, and electrogenic reactions

5.5.2 Cathodic Depolarization Theory

The cathodic depolarization theory (CDT) is based on the notion that removal/consumption of surface-absorbed cathodic hydrogen by hydrogenase-expressing SRB could accelerate the cathodic reaction leading to increased corrosion rates of steel. Blackwood (2018) provided a thorough review of the CDT and concluded that removal of adsorbed hydrogen was not a mechanism that contributed to MIC. Specifically, the rate-limiting step for hydrogen evolution (cathodic reaction) is the *absorption* of hydrogen to the steel surface – not *desorption* as required by CDT. In addition, hydrogen-consuming SRB have no effect on corrosion rates when compared to rates measured in the presence of non-hydrogen consuming SRB (Dinh et al. 2004, Mori et al. 2010).

5.5.3 Sulfide Derivatization of Metal Oxides

McNeil and Odom (1994) developed a model for predicting SRB-influenced corrosion based on thermodynamics, i.e., the likelihood that a metal oxide will react with microbiologically produced sulfide. If the reaction to convert the metal oxide to a metal sulfide has a positive Gibbs free energy, the reaction will not take place and sulfides will accumulate at the biofilm/metal interface. If the Gibbs free energy for that reaction is negative, a reaction with the metal oxide and sulfide will proceed. The model predicts that the oxides on copper alloys and low alloy steels will be derivatized and that titanium and most stainless steels will be immune to reactions with sulfide. The model is limited to thermodynamic predictions as to whether or not a reaction will take place and does not consider metal toxicity to the organisms, tenacity of the resulting sulfide or other factors that influence the reaction.

Theoretically, uniform sulfide films can inhibit corrosion of vulnerable metals and alloys (Enning and Garrelfs 2014). However, natural biofilms are patchy. In addition, metal sulfide layers on carbon steel and copper surfaces are not uniformly tenacious. The result is a potentially patchy distribution of cathodic sulfide-covered areas adjacent to anodic bared areas, leading to localized micro-galvanic corrosion (Jia et al. 2019).

Any discussion of mechanisms attributed to MIC resulting from sulfide production under anaerobic conditions should acknowledge that in the presence of oxygen, sulfide minerals undergo oxidation and corrosion rates increase dramatically (Lee and Characklis 1993). Hamilton (2003) concluded that oxygen was a requirement for aggressive SRB-influenced corrosion. In his unifying theory, he acknowledged sulfate as the terminal electron acceptor in anaerobic SRB respiration, but he identified oxygen as the terminal electron acceptor in the corrosion reaction. Lee et al. (2012) demonstrated that low levels of dissolved oxygen (DO) in natural seawater (100 ppb) influenced biodegradation pathways for alternative plant-derived fuels and subsequent SRB influenced corrosion of carbon steel. The absolute limit for DO in "anaerobic" artificial metal media has not been established. Jia et al. (2019) suggest that laboratory experiments with pure cultures of SRB should be conducted in media containing less than 40 ppb DO to minimize abiotic corrosion.

5.5.4 Hydrogen Damage

Hydrogen uptake in steels can cause multiple failure mechanisms in steel including hydrogen embrittlement, stress-corrosion cracking, hydrogen-induced stress-cracking (HISC), and blistering. Labanowski et al. (2019) demonstrated that SRB increased the susceptibility of cathodically protected 2205 duplex stainless steel (DSS) to HISC. Experiments conducted with an SRB, *Desulfovibrio desulfuricans,* in Postgate B medium demonstrated that the combination of cathodic polarization and SRB increased the hydrogen uptake of the DSS, leading to brittle fracture failure. Wang et al. (2017) evaluated the influence SRB on the blistering of a cathodically protected pipeline steel (X80) in a simulated soil solution. Hydrogen permeation in the steel increased three-fold with SRB present compared to sterile control. In both the Labanowski et al. (2019) and the Wang et al. (2017) studies, the effect of SRB on the steels was minimal at open circuit potential (OCP), i.e., under conditions of no polarization.

5.5.5 Electrogenic Reactions

Electrogenic microorganisms can transfer electrons to extracellular electron acceptors. Venzlaff et al. (2013) demonstrated electron uptake from steel by *Desulfopila corrodens* grown in an artificial seawater (ASW) medium buffered by CO_2/$NaHCO_3$, and provided with an anoxic headspace of CO_2/N_2 (10:90, v/v). The ASW medium contained typically 28 mM sulfate as an electron acceptor and no oxidizable organic substrates. They concluded that the "specially adapted, highly corrosive SRB" derived energy directly from the iron substratum. Their observations led to the conclusion that in their studies, starved cells were more aggressive to carbon steel than well-fed cells. Multiple mechanisms for electron exchange between microorganisms within biofilms have been suggested. Electron transfer relies on surface bound proteins (e.g., c-type cytochromes and heme proteins) and/or pili to pass electrons from outside the cells to the cytoplasm. Pili formation has been related to both oxygen limitation (Gorby et al. 2009) and nutrient deprivation (Sherar et al. 2011). Mediated electron transfer depends on redox active electron shuttles. Redox-active electron transfer mediators such as formate, 2-amino-3-carboxy-1, 4-naphtoquinone (ACNQ), flavin, riboflavin, and nicotinamide adenine dinucleotide can be secreted by some microorganisms. Exopolymers secreted by microorganisms can contain uronic acids that are electron acceptors. Stimulation of electron exchange has been presented (Gu 2012) as a mechanism for converting non-corrosive biofilms into corrosive biofilms. Enning and Garrelfs (2014) concluded that while all SRB can influence corrosion through H_2S formation if sulfate and suitable electron donors are present, only a few are capable of electrogenic reactions.

Li et al. (2018) suggested a more general occurrence of elemental iron as a source of electrons. They suggested that gradients in biofilms create a lack of carbon sources at the metal biofilm interface creating starved SRB that switch to elemental iron as the electron donor for maintenance energy. Li et al. (2020) reported that a consortium of SRB growing at the bottom of a biofilm formed on carbon steel in ATCC 1249 culture medium used Fe^0 as an electron donor when 90% of the carbon in the starting medium had been depleted.

5.6 METAL OXIDIZING AND REDUCING MICROORGANISMS

Oxidation of Fe^{2+} to Fe^{3+} by iron-oxidizing bacteria (FeOB) and mechanisms for MIC are related to pH. Extreme acidophilic FeOB grow optimally within pH 1–3; moderate acidophiles, pH 3–6 and neutrophiles, and pH 6–8. MIC mechanisms have been attributed to both extreme acidophilic and neutrophilic FeOB. Extreme acidophiles will be discussed under microbial acid production (see section 5.7). Neutrophilic FeOB such as *Gallionella* sp., *Mariprofundus ferrooxydans*, and *Sideroxydans lithotrophicus,* oxidize Fe_{aq}^{2+} to Fe_{ppt}^{3+} to obtain energy in microaerobic environments (<50 μM DO) as higher DO concentrations result in abiotic oxidization of Fe_{aq}^{2+}. Deposition of Fe_{ppt}^{3+} by neutrophilic FeOB can occur in association with polymers, especially negatively charged acidic polysaccharides, e.g., extracellular stalks, sheaths, dreads, and granules (Kato et al. 2015, Chan et al. 2011). The resulting dense deposits of bacteriogenic oxides (BIOS) made up of live cells, dead cells, and extracellular material, in addition to Fe_{ppt}^{3+} can cause aggressive pitting (1 mm per month) for stainless steels, containing 17.5 to 20% chromium (Cr) exposed in oxygenated Cl^{-}-containing media. The mechanism most often cited for the observed pitting is under deposit corrosion or differential aeration cells, i.e., a small anode fixed by BIOS deposits surrounded by a large cathode (Rzepa et al. 2016, Little and Lee 2007). Extensive reviews of this phenomenon and resulting anodic chemical reactions have been published (Lee and Little 2019). FeOB initiate the conditions for the under deposit corrosion but pit propagation depends on the subsequent chemical reactions and are alloy-specific. Stainless steels containing 6% or more molybdenum are not vulnerable to this type of attack. Attempts to kill the organisms within BIOS deposits using biocides will not prevent pit propagation. The microaerobic conditions required for significant microbial iron oxidation can be created by stagnation of aerobic electrolytes. Early reports of MIC failures due to FeOB were consistently reported for stagnant water often associated with hydrotest procedures (Tatnall 1981, Kobrin 1976).

There have been challenges to the under deposit corrosion mechanism for FeOB influenced corrosion of 300 series stainless steels (reviewed in (Lee and Little 2019). Dense deposits produced by FeOB have long been proposed as probable locations for SRB growth and activity (Emerson 2018). Others (Chamritski et al. 2004, Newman et al. 1986) suggested that pitting in 304 SS could be influenced by thiosulfate. The problem with any SRB or sulfur-assisted pitting mechanism is the inability of investigators to demonstrate SRB or sulfur compounds in association with BIOS.

Many bacteria and fungi can catalyze Mn oxidation (Mn^{2+} to Mn^{+4}) (Sutherland, Wankel, and Hansel 2018). Mn^{2+} is soluble while the oxidized forms, e.g., Mn_2O_3, MnOOH, Mn_3O_4, MnO_2, are insoluble. As a result of microbial action, manganese oxides can form on submerged materials in fully oxygenated waters with manganese levels as low as 6 ppb (Mathiesen and Frantsen 2008). Since 1994, there have been numerous published case histories of corrosion in natural or industrial waters due to biomineralized MnO_2 on copper (Dickinson and Pick 2002), AISI 304, AISI 316L, and martensitic alloy 1.4313 (Fe-13Cr-4Ni) (Linhardt 1994, 1996a, 1996b, 1997, 1998, 2000, Linhardt and Nichtawitz 2003, Linhardt and Mori 2004, Mathiesen and Frantsen 2008). Linhardt and Nichtawitz (2003) observed a relationship between duration of water stagnation, extent of manganese oxide formation and the severity of corrosion attack.

The development of oxygen concentration cells (under deposit corrosion) similar to those described for FeOB, has also been cited as the mechanism for pitting in stainless steels ≤20% Cr in the presence of MnOB in fresh river water (Dickinson and Lewandowski 1996), with one notable difference. Microbial deposition of manganese causes a shift of the E_{corr} in the positive direction, i.e., ennoblement. Dickinson, Lewandowski, and Geer (1996) demonstrated a 6% surface coverage of manganese oxide deposits increased the open circuit potential of 316L SS by 500 mV vs SCE.

Dissimilatory iron and/or manganese reduction occurs in several microorganisms, including in anaerobic and facultative aerobic bacteria. Inhibitor and competition experiments demonstrate that Mn^{+4} and Fe^{3+} are efficient electron acceptors similar to nitrate in redox ability, capable of out-competing electron acceptors of lower potential, such as sulfate or carbon dioxide (Myers and Nealson 1988). Dissimilatory manganese reduction has not been reported in association with MIC. Dissimilatory iron reducing bacteria (IRB) are heterotrophic facultative anaerobes, capable of using oxygen aerobically or reducing ferric to ferrous under anaerobic conditions. Royer et al. (2003) suggest that the IRB are an expanding group of microorganisms, however, *Shewanella* sp. and *Geobacter* sp are most often cited in the context of MIC. The reductive dissolution of iron oxides, particularly those associated with mild steels is acknowledged as an obvious MIC mechanism (Little et al. 1998). However, Videla et al. (2008) suggested that the role of IRB in MIC was "controversial" based on reports that *Shewanella oneidensis* could inhibit corrosion of steel (Dubiel et al. 2002). Royer et al. (2003) examined the conceptual model for corrosion inhibition of steel suggested by Dubiel et al. (2002). The model includes consumption of oxygen by *S. oneidensis* and reduction of Fe^{3+} to Fe^{2+} as contributors to corrosion. However, the model suggests that Fe^{2+} diffusion into the electrolyte may scavenge oxygen from the water and form a "protective shield against oxygen attack on the steel surface," i.e., corrosion inhibition.

5.7 ACID PRODUCTION

Acid-producing bacteria (APB), ferment organic compounds to low molecular weight organic acids such as acetic, formic, and/or lactic acids. The impact of acidic metabolites is intensified when they are trapped at the biofilm/metal interface. APB in a biofilm can produce a pH up to 2 pH units lower than the pH in the bulk electrolyte (Jia et al. 2019). Such an environment can inhibit the formation of protective films. Organic acids may force a shift in the tendency for corrosion to occur and their impact depends on the properties of the metal oxide on which they are produced. Pope (1990) conducted a study of buried carbon steel gas pipelines and concluded that APB were more important to the corrosion than SRB.

Several species of bacteria produce inorganic acids such as nitric, nitrous, sulfuric, sulfurous or carbonic acids. *Acidithiobacillus ferrooxidans,* an acidophilic FeOB, oxidizes Fe^{2+} to Fe^{3+} and reduced inorganic sulfur compounds to H_2SO_4 under aerobic conditions. Under acidic conditions ($pH < 3.5$) Fe^{3+} is soluble ($Fe_{aq}{}^{3+}$), leading to the rapid corrosion of some alloys, with $Fe_{aq}{}^{3+}$ acting as an oxidant (Inaba et al. 2019). A study measuring the corrosion rate of C1010 steel with and without *A. ferrooxidans* reported that corrosion increased 3–6 times in the biotic medium over that of an abiotic acidic medium (Wang et al. 2014). The accelerated corrosion did not require biofilm formation or direct cell contact with the corroding steel. Fe_{aq}^{3+} continuously produced by *A. ferrooxidans* was responsible for the increased corrosion. Dong et al. (2018) attributed corrosion of a super austenitic stainless steel to acid production by *A. caldus*. *Thiobacillus,* capable of growth at pH 1, produces sulfuric acid as a result of sulfide oxidation. Sulfuric acid is associated with concrete deterioration and iron rebar corrosion in sewer systems (Islander et al. 1991).

Fungi degrade organic materials and can also secrete organic acids such as formic, acetic, and propionic acids. Acid production by fungi has been reported as the cause of localized pitting for carbon steel walls in ship holds (Stranger-Johannessen 1984), 2024 T-6 aluminum in aircraft (Lavoie et al. 1997), carbon steel wire rope (Little et al. 1995), and sheathed carbon steel tendons (Little and Staehle 2001), exposed to carbon-rich substrates, such as cereal grains, hydraulic fluid, wood and lubricating grease, respectively. There is extensive literature detailing organic acid production and corrosion by *Hormoconis resinae* growing in the water phase of fuel/water mixtures, particularly aviation fuel (Videla et al. 1993, McNamara et al. 2005). Microbial growth and subsequent corrosion depend on fuel chemistry, a_w and the susceptibility of the metal/alloy. Stamps et al. (2020) determined that a mixed microbial community of filamentous fungi and APB were responsible for steel corrosion in underground storage tanks containing B20, a blended fuel containing 20% biodiesel. They further concluded, "Microbial contamination and proliferation in biodiesel poses a risk to fuel storage infrastructure worldwide" (Stamps et al. 2020).

5.8 SUMMARY

Identification of plausible mechanisms for MIC in natural or industrially relevant electrolytes is essential for diagnosis and control. However, the significance of individual mechanisms attributed to MIC is controversial and can vary with small changes in the electrolyte composition, for example aggressive anions:inhibiting anion ratios. The demonstration and quantification of MIC mechanisms have been largely established in laboratory settings with optimized growing conditions for the putative microorganisms. Most mechanisms have been evaluated individually, without considering the possibility of simultaneous or sequential multiple reactions and influences within biofilms. Mechanisms for MIC are not mutually exclusive. Archaea and fungi are acknowledged contributors to biofilm chemistry, but their contribu tions to MIC have received much less attention than mechanisms attributed to bacteria.

REFERENCES

Beech, W. B., and J. Sunner. 2004. "Biocorrosion: towards understanding interactions between biofilms and metals." *Current Opinion in Biotechnology* 15 (3):181–186. doi: 10.1016/j.copbio.2004.05.001.

Bennett, R. M., P. R. F. Cordero, G. S. Bautista, and G. R. Dedeles. 2013. "Reduction of hexavalent chromium using fungi and bacteria isolated from contaminated soil and water samples." *Chemistry and Ecology* 29 (4):320–328. doi: 10.1080/02757540.2013.770478.

Beyenal, H., and Z. Lewandowski. 2002. "Internal and external mass transfer in biofilms grown at various flow velocities." *Biotechnology Progress* 18 (1):55–61. doi: 10.1021/Bp010129s.

Blackwood, Daniel. 2018. "An electrochemist perspective of microbiologically influenced corrosion." *Corrosion and Materials Degradation* 1 (1):59–76. doi: 10.3390/cmd1010005.

Chamritski, I. G., G. R. Burns, B. J. Webster, and N. J. Laycock. 2004. "Effect of iron-oxidizing bacteria on pitting of stainless steel." *Corrosion* 60 (7):658–669.

Chan, C. S., S. C. Fakra, D. Emerson, E. J. Fleming, and K. J. Edwards. 2011. "Lithotrophic iron-oxidizing bacteria produce organic stalks to control mineral growth: implications for biosignature formation." *Isme Journal* 5 (4):717–727. doi: 10.1038/ismej.2010.173.

Chen, J. M., and O. J. Hao. 1998. "Microbial chromium (VI) reduction." *Critical Reviews in Environmental Science and Technology* 28 (3):219–251. doi: 10.1080/10643389891254214.

Cooke, V. M., M. N. Hughes, and R. K. Poole. 1995. "Reduction of chromate by bacteria isolated from the cooling water tower of an electrical generating station." *Journal of Industrial Microbiology & Biotechnology* 14 (3–4):323–328.

Davey, M. E., and G. A. O'Toole. 2000. "Microbial biofilms: from ecology to molecular genetics." *Microbiology and Molecular Biology Reviews* 64 (4):847. doi: 10.1128/Mmbr.64.4.847-867.2000.

Dickinson, W. H., and Z. Lewandowski. 1996. "Manganese biofouling and the corrosion behavior of stainless steel." *Biofouling* 10 (1–3):79–93.

Dickinson, W. H., Z. Lewandowski, and R. D. Geer. 1996. "Evidence for surface changes during ennoblement of type 316L stainless steel: Dissolved oxidant and capacitance measurements." *Corrosion* 52 (12):910–920.

Dickinson, W. H., and R. W. Pick. 2002. "Manganese-dependent corrosion in the electric utility industry." CORROSION / 2002, Denver, CO, April 7–11.

Dinh, H. T., J. Kuever, M. Mubmann, A. W. Hassel, M. Stratmann, and F. Widdel. 2004. "Iron corrosion by novel anaerobic microorganisms." *Nature* 427:829–832.

Dong, Y. Q., B. T. Jiang, D. K. Xu, C. Y. Jiang, Q. Li, and T. Y. Gu. 2018. "Severe microbiologically influenced corrosion of S32654 super austenitic stainless steel by acid producing bacterium Acidithiobacillus caldus SM-1." *Bioelectrochemistry* 123:34–44. doi: 10.1016/j.bioelechem.2018.04.014.

Dubiel, M., C. H. Hsu, C. C. Chien, F. Mansfeld, and D. K. Newman. 2002. "Microbial iron respiration can protect steel from corrosion." *Applied and Environmental Microbiology* 68 (3):1440–1445. doi: 10.1128/Aem.68.3.1440-1445.2002.

Emerson, D. 2018. "The role of iron-oxidizing bacteria in biocorrosion: a review." *Biofouling* 34 (9):989–1000. doi: 10.1080/08927014.2018.1526281.

Enning, D., and J. Garrelfs. 2014. "Corrosion of iron by sulfate-reducing bacteria: New views of an old problem." *Applied and Environmental Microbiology* 80 (4):1226–1236. doi: 10.1128/Aem.02848-13.

Escoffier, S., J. L. Cayol, B. Ollivier, B. K. C. Patel, M. L. Fardeau, P. Thomas, and P. A. Roger. 2001. "Identification of thiosulfate- and sulfur-reducing bacteria unable to reduce sulfate in ricefield soils." *European Journal of Soil Biology* 37 (3):145–156. doi: 10.1016/S1164-5563(01)01079-2.

Felder, C. M., and A. A. Stein. 1994. "Microbiologically influenced corrosion of stainless steel weld and base metal - four year tests results." CORROSION / 94, Baltimore, MD.

Gorby, Y. A., S. Yanina, J. S. McLean, K. M. Rosso, D. Moyles, A. Dohnalkova, T. J. Beveridge, I. S. Chang, B. H. Kim, K. S. Kim, D. E. Culley, S. B. Reed, M. F. Romine, D. A. Saffarini, E. A. Hill, L. Shi, D. A. Elias, D. W. Kennedy, G. Pinchuk, K. Watanabe, S. Ishii, B. Logan, K. H. Nealson, and J. K. Fredrickson. 2009. "Electrically conductive bacterial nanowires produced by Shewanella oneidensis strain MR-1 and other microorganisms (vol 103, pg 11358, 2006)." *Proceedings of the National*

Academy of Sciences of the United States of America 106 (23):9535–9535. doi: 10.1073/pnas.0903426106.

Gu, T. 2012. "New understandings of biocorrosion mechanisms and their classifications." *Journal of Microbial and Biochemical Technology* 4 (4). doi: 10.4172/1948-5948.1000e107.

Hamilton, W. A. 2003. "Microbially influenced corrosion as a model system for the study of metal microbe interactions: a unifying electron transfer hypothesis." *Biofouling* 19 (1):65–76. doi: 10.1080/0892701021000041078.

Inaba, Y., S. Xu, J. T. Vardner, A. C. West, and S. Banta. 2019. "Microbially influenced corrosion of stainless steel by acidithiobacillus ferrooxidans supplemented with pyrite: importance of thiosulfate." *Applied and Environmental Microbiology* 85 (21). doi: ARTN e01381-19. 10.1128/AEM.01381-19.

Islander, R. L., J. S. Devinny, F. Mansfeld, A. Postyn, and S. Hong. 1991. "Microbial ecology of crown corrosion in sewers." *Journal of Environmental Engineering-Asce* 117 (6):751–770. doi: 10.1061/(Asce)0733-9372(1991)117:6(751).

Javed, M. A., W. C. Neil, G. McAdam, and S. A. Wade. 2017. "Effect of sulphate-reducing bacteria on the microbiologically influenced corrosion of ten different metals using constant test conditions." *International Biodeterioration & Biodegradation* 125:73–85. doi: 10.1016/j.ibiod.2017.08.011.

Jia, R., T. Unsal, D. K. Xu, Y. Lekbach, and T. Y. Gu. 2019. "Microbiologically influenced corrosion and current mitigation strategies: a state of the art review." *International Biodeterioration & Biodegradation* 137:42–58. doi: 10.1016/j.ibiod.2018.11.007.

Jigletsova, S. K., V. B. Rodin, N. A. Zhirkova, N. V. Alexandrova, and V. P. Kholodenko. 2004. "Influence of nutrient medium composition on the direction of microbiologically influenced corrosion." CORROSION / 2004, New Orleans, LA, March 28–April 1.

Kato, S., M. Ohkuma, D. H. Powell, S. T. Krepski, K. Oshima, M. Hattori, N. Shapiro, T. Woyke, and C. S. Chan. 2015. "Comparative genomic insights into ecophysiology of neutrophilic, microaerophilic iron oxidizing bacteria." *Frontiers in Microbiology* 6. doi: ARTN 1265. 10.3389/fmicb.2015.01265.

Kehler, B. A., G. O. Ilevbare, and J. R. Scully. 2001. "Crevice corrosion stabilization and repassivation behavior of Alloy 625 and Alloy 22." *Corrosion* 57 (12):1042–1065.

Kobrin, G. 1976. "Corrosion by microbiological organisms in natural-waters." *Materials Performance* 15 (7):38–43.

Labanowski, J., T. Rzychon, W. Simka, and J. Michalska. 2019. "Sulfate-reducing bacteria-assisted hydrogen-induced stress cracking of 2205 duplex stainless steels." *Materials and Corrosion-Werkstoffe Und Korrosion* 70 (9):1667–1681. doi: 10.1002/maco.201910802.

Lavoie, D. M., B. J. Little, R. I. Ray, K. R. Hart, and P. A. Wagner. 1997. "Microfungal degradation of polyurethane paint and corrosion of aluminum alloy in military helicopters." CORROSION / 97, New Orleans, LA.

Leckie, H. P., and H. H. Uhlig. 1966. "Environmental factors affecting critical potential for pitting in 18-8 stainless steel." *Journal of the Electrochemical Society* 113:1262–1267.

Lee, W. C., and W. G. Characklis. 1993. "Corrosion of mild steel under anaerobic biofilm." *Corrosion* 49(3):186–199.

Lee, J. S., and B. J. Little. 2019. "A mechanistic approach to understanding microbiologically influenced corrosion by metal-depositing bacteria." *Corrosion* 75 (1):6–11. doi: 10.5006/2899.

Lee, J. S., R. I. Ray, B. J. Little, K. E. Duncan, A. L. Oldham, I. A. Davidova, and J. M. Suflita. 2012. "Sulphide production and corrosion in seawaters during exposure to FAME diesel." *Biofouling* 28 (5):465–478.

Lee, J. S., R. I. Ray, B. J. Little, and J. T. Stropki. 2012. "Fate of Cr+6 from a coating in an electrolyte with microorganisms." *Journal of the Electrochemical Society* 159 (11):C530–C538. doi: 10.1149/2.057211jes.

Li, Y. C., S. Q. Feng, H. M. Liu, X. K. Tian, Y. Y. Xia, M. Li, K. Xu, H. B. Yu, Q. P. Liu, and C. F. Chen. 2020. "Bacterial distribution in SRB biofilm affects MIC pitting of carbon steel studied using FIB-SEM." *Corrosion Science* 167. doi: ARTN 108512. 10.1016/j.corsci.2020.108512.

Li, Y. C., D. K. Xu, C. F. Chen, X. G. Li, R. Jia, D. W. Zhang, W. Sand, F. H. Wang, and T. Y. Gu. 2018. "Anaerobic microbiologically influenced corrosion mechanisms interpreted using bioenergetics and bioelectrochemistry: a review." *Journal of Materials Science & Technology* 34 (10):1713–1718. doi: 10.1016/j.jmst.2018.02.023.

Linhardt, P. 1994. "Manganese oxidizing bacteria and pitting of turbine components made of CrNi steel in a hydroelectric power plant." *Werkstoffe und Korrosion* 45 (2):79–83.

Linhardt, P. 1996a. "Failure of chromium-nickel steel in a hydroelectric power plant by manganese-oxidizing bacteria." In *Microbially Influenced Corrosion of Materials*, edited by E. Heitz, H. C. Flemming, and W. Sand, 221–230. Berlin, Heidelberg: Springer-Verlag.

Linhardt, P. 1996b. "Pitting of stainless steel in freshwater influenced by manganese oxidizing microorganisms." *Biodeterioration and Biodegradation* 133:77–83.

Linhardt, P. 1997. "Corrosion of metals in natural waters influenced by manganese oxidizing microorganisms." *Biodegradation* 8 (3):201–210. doi: 10.1023/A:1008294003160.

Linhardt, P. 1998. "Electrochemical identification of higher oxides of manganese in corrosion relevant deposits formed by microorganisms." *Electrochemical Methods in Corrosion Research Vi, Pts 1 and 2* 289-2:1267–1274. doi: 10.4028/www.scientific.net/MSF.289-292.1267.

Linhardt, P. 2000. "Corrosion processes in the presence of microbiologically deposited manganese oxides." CORROSION / 2000, Orlando, FL.

Linhardt, P., and G. Mori. 2004. "MIC by manganese oxidizers in a paper mill." CORROSION / 2004, New Orleans, LA, March 28–April 1.

Linhardt, P., and A. Nichtawitz. 2003. "MIC in hydroelectric powerplants." CORROSION / 2003, San Diego, CA.

Little, B. J. 2003. "A perspective on the use of anion ratios to predict corrosion in Yucca Mountain." *Corrosion* 59 (8):701–704.

Little, B. J., D. J. Blackwood, J. Hinks, F. M. Lauro, E. Marsili, A. Okamoto, S. A. Rice, S. A. Wade, and H. C. Flemming. 2020. "Microbially influenced corrosion-Any progress?" *Corrosion Science* 170. doi: ARTN 108641. 10.1016/j.corsci.2020.108641.

Little, B. J., and J. S. Lee. 2007. "Microbiologically influenced corrosion." In *Wiley Series in Corrosion,*edited byR. Winston Revie. Hoboken, New Jersey: John Wiley and Sons, Inc.

Little, B. J., J. S. Lee, and R. I. Ray. 2008. "The influence of marine biofilms on corrosion: a concise review." *Electrochimica Acta* 54 (1):2–7.

Little, B. J., R. I. Ray, K. R. Hart, and P. A. Wagner. 1995. "Fungal-induced corrosion of wire rope." *Materials Performance* 34 (10):55–58.

Little, B. J., and R. W. Staehle. 2001. "Fungal influenced corrosion in post-tension structures." *Interface* 10 (4):44–48.

Little, B. J., P. A. Wagner, K. R. Hart, R. I. Ray, D. M. Lavoie, K. Nealson, and C. Aguilar. 1998. "The role of biomineralization in microbiologically influenced corrosion." *Biodegradation* 9:1–10.

Martin, F. J., S. C. Dexter, M. Strom, and E. J. Lemieux. 2007. "Relations between seawater ennoblement selectivity and passive film semiconductivity on Ni-Cr-Mo alloys." CORROSION / 2007, Nashville, TN.

Mathiesen, T., and J. E. Frantsen. 2008. "Unusual corrosion failures of stainless steel in low chloride waters." CORROSION / 2008, New Orleans, LA, March 16–20.

McNamara, C. J., T. D. Perry, R. Leard, K. Bearce, J. Dante, and R. Mitchell. 2005. "Corrosion of aluminum alloy 2024 by microorganisms isolated from aircraft fuel tanks." *Biofouling* 21 (5–6):257–265. doi: 10.1080/08927010500389921.

McNeil, M. B., and A. L. Odom. 1994. "Thermodynamic prediction of microbiologically influenced corrosion (MIC) by sulfate-reducing bacteria (SRB)." In *Microbiologically Influenced Corrosion Testing*, edited byJ. R. Kearns and B. J. Little, 173–179. Philadelphia, PA: ASTM.

Mehanna, M., R. Basseguy, M. L. Delia, and A. Bergel. 2010. "Geobacter sulfurreducens can protect 304L stainless steel against pitting in conditions of low electron acceptor concentrations." *Electrochemistry Communications* 12 (6):724–728. doi: 10.1016/j.elecom.2010.03.017.

Mori, K., H. Tsurumaru, and S. Harayama. 2010. "Iron corrosion activity of anaerobic hydrogen-consuming microorganisms isolated from oil facilities." *Journal of Bioscience and Bioengineering* 110 (4):426–430. doi: 10.1016/j.jbiosc.2010.04.012.

Myers, C., and K. H. Nealson. 1988. "Bacterial manganese reduction and growth with manganese oxide as the sole electron acceptor." *Science* 240:1319–1321.

Newman, R. C., W. P. Wong, and A. Garner. 1986. "A mechanism of microbial pitting in stainless-steel." *Corrosion* 42 (8):489–491. doi: 10.5006/1.3583056.

Plugge, C. M., W. Zhang, C. M. Johannes, and A. J. M. Stams. 2011. "Metabolic flexibility of sulfate-reducing bacteria." *Frontiers in Microbiology* 2. doi: 10.3389/fmicb.2011.00081.

Pope, D. H. 1990. *GRI-90/0299 Field Guide: Microbiologically Influenced Corrosion (MIC): Methods of Detection in the Field.* Chicago, IL: Gas Research Institute.

Richter, K., M. Schicklberger, and J. Gescher. 2012. "Dissimilatory reduction of extracellular electron acceptors in anaerobic respiration." *Applied and Environmental Microbiology* 78 (4):913–921. doi: 10.1128/Aem.06803-11.

Royer, R. A., R. F. Unz, Brian A. Dempsey, and William D. Burgos. 2003. "Dissimilatory metl reducing bacteia in biogeochemistry and corrosion." CORROSION / 2003, San Diego, CA.

Rzepa, G., G. Pieczara, A. Gawel, A. Tomczyk, and R. Zalecki. 2016. "The influence of silicate on transformation pathways of synthetic 2-line ferrihydrite." *Journal of Thermal Analysis and Calorimetry* 125 (1):407–421. doi: 10.1007/s10973-016-5345-6.

Sherar, B. W. A., I. M. Power, P. G. Keech, S. Mitlin, G. Southam, and D. W. Shoesmith. 2011. "Characterizing the effect of carbon steel exposure in sulfide containing solutions to microbially induced corrosion." *Corrosion Science* 53 (3):955–960. doi: 10.1016/j.corsci.2010.11.027.

Skovhus, T. L., S. M. Caffrey, and C. R. J. Hubert, eds. 2014. *Applications of Molecular Microbiological Methods.* UK: Caister Academic Press.

Stamps, B. W., C. L. Bojanowski, C. A. Drake, H. S. Nunn, P. F. Lloyd, J. G. Floyd, K. A. Emmerich, A. R. Neal, W. J. Crookes-Goodson, and B. S. Stevenson. 2020. "In situ linkage of fungal and bacterial proliferation to microbiologically influenced corrosion in B20 biodiesel storage tanks." *Frontiers in Microbiology* 11. doi: ARTN 167. 10.3389/fmicb.2020.00167.

Stevenson, A., J. A. Cray, J. P. Williams, R. Santos, R. Sahay, N. Neuenkirchen, C. D. McClure, I. R. Grant, J. D. R. Houghton, J. P. Quinn, D. J. Timson, S. V. Patil, R. S. Singhal, J. Anton, J. Dijksterhuis, A. D. Hocking, B. Lievens, D. E. N. Rangel, M. A. Voytek, N. Gunde-Cimerman, A. Oren, K. N. Timmis, T. J. McGenity, and J. E. Hallsworth. 2015. "Is there a common water-activity limit for the three domains of life?" *ISME Journal* 9 (6):1333–1351. doi: 10.1038/ismej.2014.219.

Stranger-Johannessen, M. 1984. "Fungal corrosion of the steel interior of a ship's hold." In *Biodeterioration VI. Proceedings*, 218–223. Slough, UK: CAB International Mycological Institute.

Sutherland, K. M., S. D. Wankel, and C. M. Hansel. 2018. "Oxygen isotope analysis of bacterial and fungal manganese oxidation." *Geobiology* 16 (4):399–411. doi: 10.1111/gbi.12288.

Tatnall, R. E. 1981. "Case-histories - bacteria induced corrosion." *Materials Performance* 20 (8):41–48.

van Loosdrecht, M. C. M., D. Eikelboom, A. Gjaltema, A. Mulder, L. Tijhuis, and J. J. Heijnen. 1995. "Biofilm structures." *Water Science and Technology* 32 (8):35–43. doi: 10.1016/0273-1223(96)00005-4.

Venzlaff, H., D. Enning, J. Srinivasan, K. J. J. Mayrhofer, A. W. Hassel, F. Widdel, and M. Stratmann. 2013. "Accelerated cathodic reaction in microbial corrosion of iron due to direct electron uptake by sulphate-reducing bacteria." *Corrosion Science* 66 (1):88–96.

Videla, H. A., P. S. Guiamet, S. DoValle, and E. H. Reinoso. 1993. "Effects of fungal and bacterial contaminants of kerosene fuels on the corrosion of storage and distribution systems." In *A Practical Manual on Microbiologically Influenced Corrosion*, edited by G. Kobrin, 125–139. Houston, TX: NACE International.

Videla, H. A., S. Le Borgne, C. Panter, and R. S. K. Raman. 2008. "MIC of steels by iron reducing bacteria." NACE CORROSION / 2008, New Orelans, LA, March 16–20.

Wang, H., L. K. Ju, H. Castaneda, G. Cheng, and B. M. Z. Newby. 2014. "Corrosion of carbon steel C1010 in the presence of iron oxidizing bacteria Acidithiobacillus ferrooxidans." *Corrosion Science* 89:250–257. doi: 10.1016/j.corsci.2014.09.005.

Wang, D., F. Xie, M. Wu, D. X. Sun, X. Li, and J. Y. Ju. 2017. "The effect of sulfate-reducing bacteria on hydrogen permeation of X80 steel under cathodic protection potential." *International Journal of Hydrogen Energy* 42 (44):27206–27213. doi: 10.1016/j.ijhydene.2017.09.071.

6 Iron to Gas: The Mechanisms behind Methanogen-Induced Microbiologically Influenced Corrosion (Mi-MIC) and Their Importance for the Industry and Infrastructure

Sherin Kleinbub, Annie Biwen An-Stepec, and Andrea Koerdt
Bundesanstalt für Materialforschung und -prüfung (BAM)

CONTENTS

DOI: 10.1201/9780429355479-7

6.1 PART I: MECHANISMS BEHIND MI-MIC

6.1.1 BACKGROUND

Iron is among the most distributed metals in our society. It is widely used in the industrial and socially relevant infrastructure, energy and manufacturing industry, including oil and gas pipelines, water systems and storage facilities or medical devises. Despite its benefits e.g., being an inexpensive material with high malleability, the main disadvantage of iron is its susceptibility to corrosion. The corrosion of metal causes enormous economic damage. It is estimated that the direct cost of corrosion, such as maintenance or mitigation, in developed countries ranges between 2% and 3% of the gross domestic product (Enning and Garrelfs 2014), which for instance encompasses 276 billion dollars in the United States annually (Uchiyama et al. 2010).

Further costs, for instance environmental remediation or loss of production, have not been calculated. Iron corrosion is mainly an electrochemical process combining the oxidation of the metal (anodic reaction) and the reduction of a suitable reactant (cathodic reaction). Electrons are released from the metal forming aqueous ferrous ions. Under aerobic conditions, the electron acceptor is usually molecular oxygen, the responsible agent for the formation of various hydroxides, or so-called "rust" on the metal surface. In environments where oxygen is not available, oxygen can be replaced by protons (H^+; Equation 6.1) from dissociated water (Enning and Garrelfs 2014).

$$Fe^0 + 2H^+ \rightarrow Fe^{2+} + H_2;\ \Delta G^{0\prime} = -10.6\ \text{kJ/mol}\ Fe^0 \tag{6.1}$$

Under those abiotic conditions where oxygen or another suitable electron acceptor is missing, the reaction represented in Equation 6.1 is very slow or even absent (Hamilton 2003). However, metal corrosion can be accelerated in the presence of microorganisms, also known as microbiologically influenced corrosion (MIC). MIC occurs in various aqueous environments, including marine systems, water treatment plants, sewage plants, and particularly for energy infrastructures such as in underground oil and gas pipelines, crude oil reservoir tanks, and water systems (Kato 2016, Loto 2017, Jia et al. 2019). It is estimated that MIC contributes up to 20% of the overall corrosion costs which is assessed to be in the range of billons of U.S. dollars and is not restricted to one type of metal. Next to iron, also cast iron, mild steel, stainless steel, and aluminum alloys, etc. can be affected (Dinh et al. 2004, Mori et al. 2010, Uchiyama et al. 2010, Enning and Garrelfs 2014).

The high significance and the enormous costs are only two reasons why scientists have been working on MIC for almost 100 years. One of the first proceedings on the topic of MIC already existed in 1934 and was published by Wolozogen Kühr and van der Vlugt. They described MIC as an "electrobiological process of biochemical and chemical nature" (von Wolzogen Kühr and van der Vlugt 1964). In the 1960s and early

1970s, MIC received increased attention aided by the advancements in cultivation methods. From this time on, it was assumed for a long time that sulfate-reducing bacteria (SRB) were mainly, sometimes even exclusively, responsible for MIC in anaerobic environments. This assumption was supported by the metabolite hydrogen sulfide (HS^-), which can be corrosive per se (Daniels et al. 1987). Furthermore, the remarkably high corrosion rate of SRB was explained by the theory of cathodic depolarization (CDT). The basic principle of CDT is the removal of hydrogen (H_2) from the cathodic region on the iron surface by H_2 consuming microorganisms e.g., through hydrogenases, which is applicable for SRB. It has been assumed that the removal of H_2 accelerates the normally abiotic reaction (Equation 6.1) and thus leads to increased corrosion. Since the publication of the CDT, it has been heavily discussed. Culture-based experiments showed that bacterial consumption of cathodic H_2 did not significantly accelerate iron corrosion compared to abiotic controls (Dinh et al. 2004, Mori et al. 2010, Enning and Garrelfs 2014, Deutzmann et al. 2015, Tsurumaru et al. 2018). Furthermore, it is becoming more and more evident that the real bottleneck of the CDT is not the removal of H_2 from the system but rather the accumulation of protons on the metal surface which drives the reaction (Venzlaff et al. 2013).

One important finding, achieved by Daniels et al., was the discovery that methanogenic archaea (MA) contribute to metal corrosion more significantly than initially thought (Daniels et al. 1987). Besides that, surface analysis techniques, electrochemical methods, and microscopic analyzations such as atomic force microscopy (AFM), the measurement of current transients or scanning electron microscopy (SEM) coupled with energy-dispersive analysis of X-rays (EDAX) have found increased applications for metal surface and corrosion products analyses (Videla and Herrera 2005). More recently, significant progress in molecular, electrochemical and microscopy techniques have been obtained. For example, confocal laser scanning microscopy (CLSM) is used frequently for the analysis and characterization of (corrosive) biofilms. In combination with microelectrodes, CLSM analysis allows the correlation of oxygen profiles within the biofilm structure. Another important achievement was the development of specific DNA and RNA probes that can be used for the identification of single microorganisms or whole communities involved in MIC (Videla and Herrera 2005).

The most studied microorganisms in the context of MIC are still SRB and thus the mechanism of how SRB corrode metals is well understood. In the presence of ferrous ions and HS^-, the corrosive iron sulfide (FeS) is formed on the metal surface, generating a localized corrosion cell (Enning 2012, Enning and Garrelfs 2014). More recent studies even demonstrated iron corrosion by the SRB *Desulfovibrio ferrophilus* IS5 through direct electron uptake have corrosion rates with up to 0.9 mm/yr under laboratory conditions (Deng et al. 2015, Enning and Garrelfs 2014).

Other MIC-inducing microorganisms are acetogenic bacteria (AB). AB produce organic acids as their final metabolic product. The organic acids can lower local pH to <3, resulting in an increased corrosion rate (Gu 2014). Additionally, an acidic environment can prevent the formation of protective corrosion product films (Xu et al. 2016, Jia et al. 2019). Iron-oxidizing bacteria (IOB) that oxidize Fe^{2+} to Fe^{3+} are also involved in metal corrosion. This reaction accelerates the abiotic formation of Fe^{2+} (Equation 6.1) resulting in pitting corrosion and thus in an increased corrosion rate (Wang et al. 2014).

One group of microorganisms that receives more and more attention regarding MIC are methanogenic archaea (MA). Originally it was thought that MA participate in the corrosion process by only consuming cathodic H_2 on the metal surface, which causes cathodic depolarization of the metal (von Wolzogen Kühr and van der Vlugt 1964, Daniels et al. 1987). However, more recent studies demonstrated that MA corrode metals in a direct manner, most likely by the withdrawal of electrons from the iron surface (Dinh et al. 2004, Mori et al. 2010). This was particularly demonstrated on *Methanococcus maripaludis* Mic1c10 and *M. maripaludis* KA1, both isolated from a crude-oil storage tank (Mori et al. 2010, Uchiyama et al. 2010), and the *Methanobacterium*-affiliated strain IM1, isolated from a marine sediment (Dinh et al. 2004). The fact that *Methanobacterium*-affiliated strain IM1 can extract electron from iron, even with a higher methane production than *M. maripaludis* species, shows that this feature is not only restricted to the order of *Methanococcales* (Uchiyama et al. 2010). The abilities of *M. maripaludis* spp. Mic1c10 and KA1 to corrode iron seem to be strain specific since other *M. maripaludis* strains e.g., strain JJ^T or strain S2 are not corrosive (Mori et al. 2010, Uchiyama et al. 2010). In the following sections of this chapter, we will focus on the methanogen-induced microbiologically influenced corrosion (Mi-MIC) and the suspected mechanisms behind the underestimated corrosion potential of these microorganisms.

6.1.2 Mi-MIC Mechanisms

MA are a phylogenetically diverse group of microorganisms within the phylum Euryarchaeota and have an important key function in the final stage of anaerobic, microbial degradation of biomass. During the process of methanogenesis, oxidized carbon e.g., CO_2 serves as a terminal electron acceptor. According to literature, MA can be grouped in two ways. First, depending on the substrate used, H_2, formate, acetate, or methylated compounds (three classes can be distinguished) or in metabolic-physiological dependence (on whether cytochromes are present).

6.1.2.1 Classification Depending on the Substrate

MA are categorized into three different classes according to their substrates used for methanogenesis (Table 6.1). The first class is those of acetoclastic methanogens represented by *Methanosarcina* and *Methanosaeta* species. Acetoclastic methanogens utilize mainly acetate and pyruvate for methanogenesis. The second class belongs to methylotrophic methanogens. Only species in the order of *Methanosarcinales* can use methylated compounds for CO_2 reduction. These compounds include e.g., methanol, methylated amines, or methylated sulfides. The third class is those of hydrogenotrophic methanogens, where most of MA belong to. Many hydrogenotrophic methanogens can utilize formate for methanogenesis. However, their major electron donor is H_2 (Equation 6.2) (Liu and Whitman 2008).

$$4H_2 + CO_2 \rightarrow CH_4 + 2\ H_2O;\ \Delta G^{0'} = -136\ \text{kJ} \tag{6.2}$$

TABLE 6.1
Classification of Methanogens according to Their Substrates

	Acetoclastic Methanogens	Methylotrophic Methanogens	Hydrogenotrophic Methanogens
Order/Genus	Methanosarcinales (e.g., *Methanosarcina; Methanosaeta*)	Methanobacteriales (*Methanosphaera*) Methanosacinales (e.g., *Methanosarcina, Methanohalophilus*)	Methanobacteriales (e.g., *Methanobacterium, Methanobrevibacter*) Methanococcales (e.g., *Methanococcus, Methanocaldococcus*) Methanomicrobiales (e.g., *Methanomicrobium, Methanospirillum*) Methanopyrales (*Methanopyrus*)

6.1.2.2 Classification on Metabolic-Physiological Dependence

In this type of categorization, the assignment is straight-forward, depending on the occurrence of cytochromes. Hydrogenotrophic methanogens, i.e., MA that convert H_2 and CO_2 and/or formate to methane, lack cytochromes. Of the cytochrome containing MA that are specialized on acetate, methanol and/or methylamine, only a few can grow on H_2 and CO_2. However, the H_2 partial pressure is an important feature that differentiates the two types of MA. A characteristic feature of hydrogenotrophic methanogens is their ability to grow at H_2 partial pressures below 10 Pa, while methylotrophic methanogens, for example, need H_2 partial pressures above 10 Pa for growth (Thauer et al. 2010).

Currently, only hydrogenotrophic microorganisms (without cytochromes) could be identified as corrosive. This can be explained by the low H_2 partial pressure present on the metal surface. In competition with H_2-consuming organisms such as SRB, MA-containing cytochromes would have no chance to compete.

Unlike MIC that is caused by SRB, only few studies are available that focus on Mi-MIC. Additionally, there is still a lot of controversy about the mechanism behind Mi-MIC and the interaction of MA with the metal surface is still not solved. In the following section, the different suspected mechanisms of MIC in anaerobic conditions, will be discussed focusing on the mechanism of Mi-MIC.

In fact, there is no general mechanism for MIC. Instead, MIC occurs rather as an interaction of several different aspects that variates depending on the microorganism and the community. Due to the unexplained mechanisms, different classifications, classes, and sub-classes have been described in the literature. The different terminology is comprehensive and causes often more confusion than benefits; especially since basically two main types can be derived from all described types: indirect and direct MIC.

As the term *indirect MIC* already implies, an indirect way of metal corrosion can be caused by microbially produced metabolic products or chemical substances. Some microorganisms produce, secrete, or export substances that can attack metal and lead to corrosion. Well-known examples are e.g., acids from fermentative microorganisms (AB, Figure 6.1) or quite classical H_2S from SRBs. This mechanism is presented, for example, by Enning et al., as "chemical-microbiologically influenced corrosion" (CMIC) (Enning and Garrelfs 2014) and from Jia et al., as metabolite-MIC (M-MIC) (Jia et al. 2019).

CMIC usually does not occur in MA, which produce exclusively the non-corrosive methane as their main metabolite. Thus Mi-MIC occurs most likely via the second type of MIC: the direct MIC. In this case MIC occurs through the direct attack of the iron surface due to the withdrawal of electrons and is termed as "electrical microbiologically influenced corrosion" (EMIC) (Enning and Garrelfs 2014) or extracellular electron transfer MIC (EET-MIC) (Jia et al. 2019). According to Jia et al., EMIC can be further differentiated, depending on if the cell is in direct contact to the iron surface as in direct electron transfer (DET-MIC/direct EMIC) (Jia et al. 2019) or whether the electron withdrawal is mediated via enzymes or chemical compounds without direct cellular contact to the metal as mediated electron transfer (MET-MIC)/ indirect EMIC (Figure 6.1) (Jia et al. 2019). To simplify the diverse terminologies regarding the different types of MIC-mechanisms, the terms *direct* and *indirect EMIC* will be used throughout this chapter.

In direct EMIC, the cell attaches with different cell-bound surface structures to the iron surface. The probably best-known structures are conductive pili, which

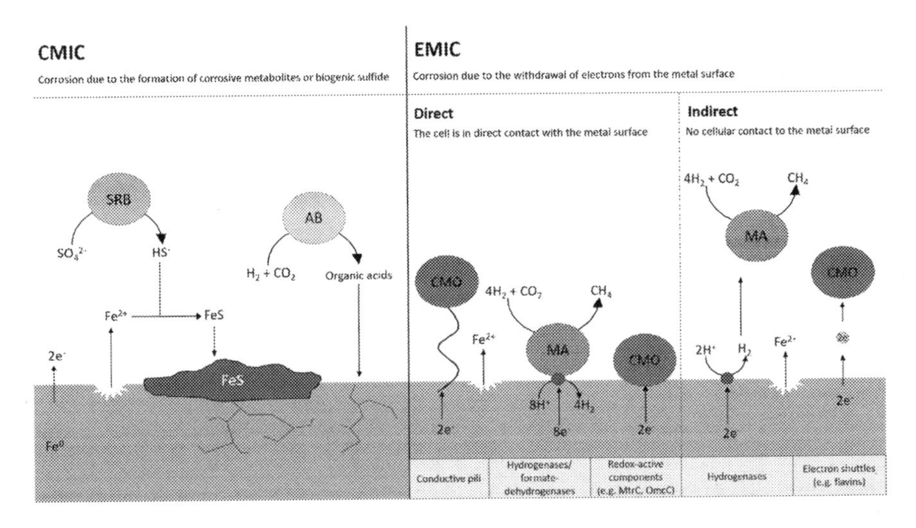

FIGURE 6.1 Schematic representation of the different types of MIC. MIC can be either caused chemically (CMIC) or electrically (EMIC). EMIC is further differentiated between a direct uptake of electrons for example via conductive pili or redox active components or an indirect electron uptake for example via electron shuttles. SRB: sulfate-reducing bacteria; AB: acetogenic bacteria; CMO: general corrosive microorganisms (e.g., *Geobacter sulfurreducens* or *Shewanella oneidensis* MR-1); MA: methanogenic archaea.

mediate cell-surface and cell-cell interactions for electron transfer for instance found in *Geobacter sulfurreducens* (Reguera et al. 2005). Another example is *S. oneidensis* MR-1 whose conductive pili can transport electrons with rates up to 109/s at 100 mV along micrometer-length scales (El-Naggar et al. 2010). Recently, Wagner and colleagues reported that the archaellum of the hydrogenotrophic methanogen *Methanospirillum hungatei* is electrically conductive as well, with an even higher electron flow compared to *G. sulfurreducens* (Walker et al. 2019). These results imply that conductive cell appendages such as e-pili or e-archaella could also play a role in MA-mediated electron uptake. In addition to cell appendances, cell surface associated enzymes such as hydrogenases or formate dehydrogenases are thought to be involved in MIC. The hydrogenases or formate dehydrogenases interact with the iron surface and convert the electrons derived from the iron to H_2 or formate, respectively (Figure 6.1). H_2 and formate can then further used by MA for methane formation (Deutzmann et al. 2015, Tsurumaru et al. 2018). However, an *M. maripaludis* mutant carrying marker less in-frame deletions of all five catabolic hydrogenases and a deletion of the anabolic hydrogenase was still able to grow on iron as a sole electron source. Interestingly, the same mutant was unable to grow when H_2 and CO_2 was provided as energy sources (Lohner et al. 2014). This observation provided direct evidence for a H_2 independent electron uptake mechanism by *M. maripaludis*. Instead, the electrons needed for CO_2 reduction may be provided by outer membrane redox proteins, which may interact with the iron surface (Uchiyama et al. 2010, Enning and Garrelfs 2014, Lohner et al. 2014). This principle is already known from other metal corroding bacteria. Examples for such redox proteins are OmcC or MtrC that have been reported in *Geobacter* sp. and *S. oneidensis*, respectively (Lohner et al. 2014).

In indirect EMIC, microorganisms release enzymes or redox-active proteins such as flavins that act as "electron shuttles" between the metal and the cell (Figure 6.1). In accordance with the current state of knowledge and similar to direct EMIC, the enzymes are primary hydrogenases and formate dehydrogenases. The secreted enzymes interact with the iron surface and catalyze the formation of molecular H_2 or formate, respectively, using the electrons derived from Fe^0. The generated H_2 and formate is then consumed by the MA. First evidence in the involvement of (extracellular) hydrogenases showed a *M. maripaludis* Mic1c10 culture that grew with Fe^0 as electron donor. Even though the growth and H_2 production stagnated, iron corrosion proceeded continuously (Mori et al. 2010). Deutzmann and colleagues confirmed this assumption further. They used spent cell-free culture medium of a *M. maripaludis* strain (derivative of strain S2) that revealed molecular H_2 and formate formation in a significant higher rate than the abiotic control (Deutzmann et al. 2015). Recently, a 12 kb chromosomal gene region, termed the "MIC island," was identified in the corrosive *M. maripaludis* strains OS7 and KA1 (Tsurumaru et al. 2018). Mutants, lacking the "MIC island" were incapable of using electrons directly from Fe^0. The "MIC island" encodes genes for the large and small subunit of a [NiFe] hydrogenase that is thought to be the causative enzyme of Mi-MIC. Therefore, the hydrogenase was termed "MIC hydrogenase" (Tsurumaru et al. 2018). In addition, the "MIC island" encodes additional genes potentially for the

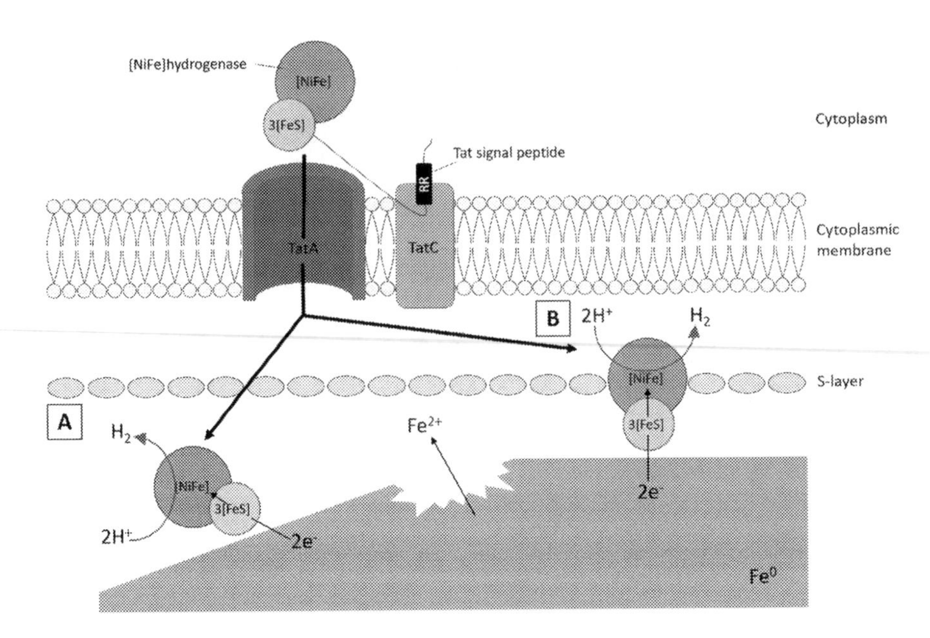

FIGURE 6.2 Suspected corrosion mechanism of *M. maripaludis* sp. The MIC-hydrogenase is first secreted through the cytoplasmic membrane via the Tat secretion pathway. In a second step, the hydrogenase is transported through the S-layer in a yet-unknown mechanism. Then the hydrogenase adheres to the iron surface, converting the electrons deriving from the metal and the protons present in the aqueous phase to H_2. The H_2 is then consumed (A). Alternatively, the hydrogenase is incorporated in the S-layer and the whole cell is in contact to the iron (B). Figure adapted according and Palmer and Berks (2012).

secretion of the hydrogenase via the Tat pathway and a gene for the hydrogenase maturation protease. The resulting model of the Mi-MIC mechanism proposed by Tsurumaru et al. is depicted in Figure 6.2. In the first step, the "MIC hydrogenase" is translocated through the cytoplasmic membrane via the Tat secretion pathway. Then the hydrogenase is either secreted through the S-layer in a yet-unknown mechanism (Figure 6.2A) or is located on the external S-layer surface (Figure 6.2B). In both scenarios, the hydrogenase contacts the iron surface directly and uses the electrons from Fe^0 for the generation of molecular H_2, which is subsequently converted to methane by MA (Tsurumaru et al. 2018). The fact, that spent culture filtrate of wildtype *M. maripaludis* strains OS7 and KA1 induces H_2 formation in contrast to the spent culture filtrate of the mutants, indicates that the MIC hydrogenases most likely is secreted into the medium (Lahme et al. 2020). Due to the instability of the "MIC island", *M. maripaludis* strains OS7 and KA1 lose their corrosivity upon prolonged cultivation with of H_2 and CO_2 (Tsurumaru et al. 2018). Comparative genomics revealed that the loss of the "MIC island" in the mutant strains or H_2-grown compared to the iron-grown wildtype is due to an excision from the genome by homologous recombination (Tsurumaru et al. 2018).

Similar to *M. maripaludis*, the litho-autotrophic *Methanobacterium*-affiliated strain IM1 utilizes electrons from cathodes for methane formation (Dinh et al. 2004,

Beese-Vasbender et al. 2015). However, IM1 cells seems to attach quite strong to iron while *M. maripaludis* forms a rather loose biofilm on the metal surface. This observation may be a first indication that the corrosion mechanisms within the MA could be different from each other. However, further investigations are required to gain an improved understanding in the mechanisms of IM1.

Another, if not the most important aspect regarding MIC is the formation of biofilms that significantly influences metal deterioration (Loto 2017, Jia et al. 2019). In a biofilm, the microorganisms are embedded in a self-produced matrix that influences the metal surface in different manners. Microorganisms within the biofilm can create localized microenvironments caused by their metabolic activity eventually resulting in localized pH changes, or alteration in organic and inorganic species that lead to electrochemical reactions (Loto 2017). The activity of secreted enzymes, such as the previously mentioned hydrogenases or formate dehydrogenase, contribute significantly to the corrosion by acceleration of the cathodic reactions (Beech and Sunner 2004). So far, there are almost no studies available that focus on methanogenic biofilms. For a better understanding of Mi-MIC, it's crucial to gain more information of the development and composition of methanogenic biofilms, corrosion potential and corrosion products of Mi-MIC. For various reasons, MA have been underestimated for decades in their contribution to MIC. The two most obvious reasons were 1) their low corrosion potential under laboratory conditions and 2) the suspected corrosion products that are formed.

1. *Corrosion rates:*

Corrosion rates of published MA pure cultures ranged so far from 0.02 mm/yr for *M. maripaludis* KA1 to a maximum of 0.07 mm/yr for *M. thermolithotrophicus* (Daniels et al. 1987, Mori et al. 2010). By comparison, SRB values can reach up to 0.7–0.9 mm/yr and are classified as severely corrosive microorganisms (Enning 2012, Enning and Garrelfs 2014). However, a recent study carried out by An et al. (2020) showed that MA can also achieve higher and relevant corrosion rates up to 0.31 mm/yr. These high corrosion rates were achieved by an environmental simulating flow system instead of static batch cultures. This new test system simulates the corrosion conditions under more realistic conditions as they are also found in e.g., real oil and gas pipelines. This flow system allowed a doubling of the corrosion rate compared to a static IM1 culture (0.15 mm/yr). The highest corrosion rate achieved and published to date for IM1 was 0.52 mm/yr (An et al. 2020).

2. *Corrosion products:*

The postulated and only published corrosion products of MA explain the low corrosion rates which are achieved so far. In contrast to SRB, which produce the electrically conductive FeS as a corrosion product, it has been assumed that MA produce mainly/only siderite ($FeCO_3$), an electrically non-conductive compound (Kip et al. 2017). This conclusion was drawn based on a study with *M. maripaludis* KA1 by Uchiyama et al. (2010), which described siderite was the only corrosion product (Uchiyama et al. 2010). However, different physical studies showed that the formation of corrosion products can be very complex. The stability of siderite in an aqueous environment varies, for example depending on the iron and carbonate ion concentration (Barker et al. 2018). In other low-sulfate and CO_2-rich environments, the formation of semi-conducting magnetite (Fe_3O_4) and chukanovite ($Fe_2(OH)_2CO_3$) was observed (Joshi 2015, In 't Zandt et al. 2019). Furthermore, environmental factors such as high temperature, low pH, increased salinity and flow velocity disrupt the nucleation process of siderite, causing the dissolution of Fe-CO complexes and the formation of other compounds, i.e., iron phosphates (Barker et al. 2018). Therefore, the composition of the corrosion products formed in Mi-MIC is still an open question and needs further investigation.

6.2 PART II: MI-MIC AND THE INDUSTRY

6.2.1 Background

Methanogens are found in a wide range of industries and environments, including hydrothermal vents, biogas plants, petroleum, marine sediments, wastewater treatment, rumen, and many others (Calderon et al. 2013, Liu and Whitman 2008, Zehnder 1988, An et al. 2013). The production of pure methane by MA exhibit high industrial and ecological potential (Enzmann et al. 2018). Here, we will discuss the positive and negative impacts of methanogens on the industry. In particularly, the consequences of Mi-MIC on the industry and current corrosion monitoring methods.

6.2.2 Methanogens in the Industry

Methanogens occur naturally in the environment and could be co-habitating with other organisms, i.e., bacteria, plants, or animals. There are six well-defined methanogenic orders: *Methanobacteriales, Methanococcales, Methanomicrobiales, Methanosarcinales, Methanopyrales* and *Methanocellales* (Calderon et al. 2013). In the renewable energy industry, MA are often explored for their biogas potential. By 2015, over 15,000 biogas plants were installed in Europe with a total energy production capacity of 8.73 GW (Balussou 2018). In biogas plants, large polymers are hydrolyzed into sugars and amino acids that undergo fermentation into acetate, H_2 and CO_2, which are substrates for methanogenesis (Enzmann et al. 2018, De Vrieze et al. 2012). Members of *Methanosarcina* are known to be the dominant MA in biogas plants (De Vrieze et al. 2012), where they often form syntrophic relationships with AB (Nakayama et al. 2019). Syntrophic relationships between microorganisms are classified into mediated-interspecies electron transfer (MIET) or direct-interspecies electron transfer (DIET). In one study on DIET anaerobic digestion, *Methanosaeta* aggregated with *Geobacter* to reduce CO_2 to CH_4, by using electrons shuttled from *Geobacter* through direct pili contact (Lovley 2017, Yin and Wu 2019). Other conductive materials, such as magnetite and activated carbon, can also be used for DIET pathway (Lovley 2017, Nakayama et al. 2019). The presence of conductive materials can compensate for the absence of electrically-conductive pili or cytochromes (Yin and Wu 2019). *Methanosarcina barkeri,* which is a known syntrophic partner of *Geobacter*, lack both electrically conductive pili and membrane-bound c-type cytochrome (Holmes et al. 2018). In a DIET-grown culture with *Geobacter metallireducens*, *M. barkeri* had increased gene expressions of the F_{420}-H_2-dehydrogenase (Fpo) and the deterodisulfide reductase (HdrABC) (Holmes et al. 2018). Furthermore, DIET-grown *M. barkeri* have low affinity to H_2, suggesting intracellular H_2-transfer is more unlikely (Holmes et al. 2018).

To study biogas formation, different types of bioreactors are used, including continuously stirred tank reactor (CSTR), up-flow anaerobic sludge blanket reactor (UASB) and expanded granular sludge bed bioreactor (EGSB) (Bensmann et al. 2016, Nakayama et al. 2019). Bioreactors are used to study gas production based on biomass-substrate contact in terms of velocity and kinetics (Bensmann et al. 2016, Nakayama et al. 2019, Koch et al. 2016). Similarly, MA are also involved in wastewater treatment and studied by using flow reactors (Calderon et al. 2013). The removal of acetate and other organic matter by MA decreases the amount of organic sludge while aerating the system with produced methane (Enzmann et al. 2018). However, the presence of MA are also considered undesirable in wastewater treatment (Calderon et al. 2013) due to the formation of archaeal biofilms. Based on several studies of methanogenic-wastewater reactors, granule-forming *Methanosaeta concilii* initiates the colonization of AB and other microorganisms, creating a stable biofilm structure that leads to biofouling (Calderon et al. 2013). Biofouling in anaerobic sewage systems is further enhanced by microbial syntrophy between sulfate-reducing bacteria and methanogens (Auguet et al. 2015).

6.2.3 Methanogens in Oil and Gas Industries

As part of the oil, subsurface, and water supplies, methanogens are found across the oil and gas industry. Methanogens are regarded as both detrimental and beneficial for the industry as they can cause biofouling, oil biodegradation, methanogen-induced microbiologically influenced corrosion (MI-MIC) or be used for microbial-enhanced oil recovery (MEOR) (Head and Gray 2016, Grigoryan and Voordouw 2008). Here, we will briefly discuss the roles methanogens exhibit in the petroleum industry.

Conventional oil recovery is defined as oil extraction, typically light oil, through conventional methods or by relying on natural geological pressure (Goodwin et al. 2014). The main challenge associated with methanogens in conventional oil fields is oil degradation (Gray et al. 2009, 2010). Methanogenic oil degradation is a well-known phenomenon and it is known to contribute to the rise of heavy oil (Gray et al. 2009, 2010). Lighter hydrocarbons are easier to biodegrade as opposed to long and complex hydrocarbon chains (Head and Gray 2016, Gieg et al. 2008). Furthermore, biodegrading MA could grow under thermophilic conditions (up to 80 °C) (Grigoryan and Voordouw 2008). Hydrogenotrophic methanogens generally dominate in both low and high temperature oil reservoirs (Gray et al. 2010), which are also involved in Mi-MIC (Kato 2016, Dinh et al. 2004, Tsurumaru et al. 2018, An et al. 2020). As discussed earlier, hydrogenotrophic methanogens are involved in biocorrosion through different pathways. In several reported corrosion cases, methanogens are present as part of the biofilm microbial community, which will be discussed further in the following section.

In unconventional heavy oil reservoirs, methanogens are important for biotechnology associated with microbial enhanced oil recovery (MEOR) (Tucker et al. 2015, Head and Gray 2016, Foght et al. 2017, Gieg et al. 2008). As reservoir pressure depletes in unconventional oil fields, effective, ecological and economical re-pressurization methods are desirable. Methanogens can form syntrophic relationships with hydrocarbon degraders that convert complex compounds into acetate or H_2 for methanogenesis (Gieg et al. 2008). In shale reservoirs, methanogens are not generally considered to be involved in biocorrosion (An et al. 2019). Due to the oil recovery process, shale reservoirs are typically extremely saline that limits the growth of non-halophilic microorganisms (An et al. 2019, Borton et al. 2018). Microbial community analyses of several shale reservoirs revealed halophilic methanogens, particularly *Methanohalophilus* to be the dominant methanogen (An et al. 2017, 2019, Borton et al. 2018, Daly et al. 2016). *Methanohalophilus* is a halophilic, mesophilic, methylotrophic methanogen that utilizes methylated compounds, such as trimethylamine, for methanogenesis (An et al. 2019). It was proposed that *Methanohalophilus* may contribute to the increase in ammonium concentration by converting methylamines or glycine betaine into methane (An et al. 2019). However, *Methanohalophilus* may indirectly contribute to biocorrosion in shale reservoirs by providing osmotic solutes, i.e., glycine betaine, for corrosive halophiles (Daly et al. 2016). But the potential of halophilic methanogens to produce large amounts of methane using methylated compounds may contribute to reservoir re-pressurization for enhanced oil recovery (Tucker et al. 2015).

6.2.4 Risk of Mi-MIC in the Industry

Currently, *Methanococcus maripaludis* strains OS7, KA1, Mic1c10 isolated from Japan, *Methanobacterium*-affiliated IM1 and Baltic-sea *Methanosarcina*, are the best-known iron-corroding methanogens (Dinh et al. 2004, Kato 2016, Mori et al. 2010, Uchiyama et al. 2010, Tsurumaru et al. 2018, Palacios et al. 2019, An et al. 2020). Though the specific mechanisms behind the electron uptake pathways remain unclear, they exhibit the abilities to utilize iron as the sole electron donor. In the industry, methanogenic microbial communities are often detected in corrosive field samples, some are closely related to corrosive isolates. Pipeline cut-outs obtained from a Canadian oil field undergoing oxygen scavenger sodium bisulfite treatment, showed high fractions of *Methanobacteriaceae* in the pipeline-associated samples indicating their corrosion potential (Park et al. 2011). Pipeline pigging samples obtained from a Nigerian oil field were dominated primarily by MA (79.5 to 85.6%) (Mand et al. 2016) with high proportions of *Methanobacterium*. The high proportions of MA in the pipeline-associated solids further indicated an involvement of methanogens in pipeline corrosion (Mand et al. 2016). In a low-sulfate Nigerian crude oil transporting pipeline, higher corrosion rates were positively correlated with methane production (Okoro et al. 2016a). Microbial community analyses conducted on the pipeline solids showed high proportions of

Methanobacterium, *Methanoculleus*, *Methanosaeta*, and *Methanolobus* (Okoro et al. 2016b). Under high temperature (>50°C), such as the one reported by Liang et al. 2014, thermophilic *Methanotermobacter* formed a syntrophic relationship with thiosulfate-reducing bacteria yielding in high corrosion rates (0.052 ± 0.012 mm/yr) (Liang et al. 2014). A similar relationship was observed in a co-culture isolated from a high temperature pipeline inner surface (Davidova et al. 2012). The fermentative *Thermococcus* strain was tightly aggregated with the thermophilic methanogen, the versatile metabolic activities of the co-culture led to pipeline corrosion even under elevated temperatures (Davidova et al. 2012). Another case study conducted on a cut-out of a North Sea Halfdan oil field pipeline, with corrosion rates between 0.3 mm/yr up to 2.6 mm/yr, showed presences of both sulfate-reducing bacteria (SRB) and methanogens (Larsen et al. 2010). Based on the qPCR results, SRB and methanogen were spatially separated on the pipeline surface, where the hydrogenotrophic methanogens were more abundant closer to the metal surface and SRB near the transportation fluids (Larsen et al. 2010). However, any associated DIET between SRB and methanogen and their contributions to biocorrosion remain unclear and require further investigation. Overall, methanogens exhibit high corrosion potential in a variety of environments that require further research on identifying the specific pathways involved.

6.2.5 Corrosion Monitoring of Mi-MIC

6.2.5.1 Standard Laboratory Method

Standard laboratory corrosion measurements of methanogens rely on the traditional serum bottle experiments. Typically, the metal of interest, i.e., carbon steel, is placed into anaerobic medium and sealed with rubber stopper. For methanogenic incubations, serum bottle experiments allow a constant formation of methane gas without environmental disturbances that can be easily quantified. Currently, most experiments conducted on Mi-MIC use serum bottles enrichments. The corrosion rates are then calculated based on weight loss and a uniform corrosion rate is generated. However, several factors are limited in serum bottle experiments, such as substrate and nutrient availabilities. To combat substrate availability, i.e., surface area limitations, iron granules can be used, which is conducted by (Mori et al. 2010) on iron-utilizing *Methanococcus maripaludis* (Mori et al. 2010). However, specific corrosion rates and corrosion products of iron granules can only be estimated based on dissolved iron concentrations. Furthermore, the corrosion products formed in serum bottles are less representative of natural conditions. For instance, the crystallization of siderite nucleates is sensitive to several external factors, i.e., pH, flow, and salinity (Joshi 2015), and these conditions are difficult to replicate in serum bottles.

6.2.5.2 Flow Systems for Mi-MIC Monitoring

Flow systems are typically implemented in junction to serum bottle experiments, particularly for sulfate-reducing bacteria (Hubert et al. 2003, Pinnock et al. 2018). Flow systems of SRB are easy to maintain due to the usage of an aqueous electron acceptor and an oxygen-reducing metabolic product. However, flow systems for MA are difficult to operate: 1) system pressure changes due to methane production,

2) ingress of oxygen, 3) continuously measure methane, and 4) electron acceptor availability. Establishing a flow-system for iron-corroding MA would offer new opportunities to measure Mi-MIC under more natural conditions with a continuous influx of nutrients (An et al. 2020). Furthermore, flow systems can be analyzed sectionally to characterize the associated corrosion products and corrosion rate calculations (An et al. 2020). Flow systems also offer opportunities to implement different corrosion inhibitor and biocide treatment that mimics field applications (Duncan et al. 2014), which is currently lacking on Mi-MIC. Recently, a studied published on Mi-MIC using a once-flow through system yielded high corrosion rates (up to 0.52 mm/yr) for *Methanobacterium*-IM1, compared to 0.15 mm/yr in serum bottles (An et al. 2020). Overall, the study of Mi-MIC under flow condition showed high corrosion potentials by MA that offers more industrial relevance than typical serum bottle experiments.

6.3 PART III: SUMMARY AND FUTURE RESEARCH DIRECTIONS OF MI-MIC

Overall, this chapter provided an overview of the current understandings of Mi-MIC. However, the study of Mi-MIC is still at its infancy and much of the pathways involved are largely unknown. It is important to note that methanogens capable of iron corrosion offers a unique physicochemical opportunity that can be explored by a wide range of industries. In particularly, with the increased concerns on climate change and emphases on "green fuel", electro-active methanogens are both beneficial and detrimental to the industries. To further our understandings of Mi-MIC and the associated mechanisms, innovative monitoring techniques capable of integrating field parameters such as portable flow cells, are needed. Overall, increased attention on Mi-MIC in the industry is needed for necessary mitigation and biotechnological advances that will be beneficial for all.

REFERENCES

An, Dongshan, et al. 2013. "Microbial community and potential functional gene diversity involved in anaerobic hydrocarbon degradation and methanogenesis in an oil sands tailings pond." *Genome/National Research Council Canada = Génome / Conseil national de recherches Canada* 56:612–618. doi: 10.1139/gen-2013-0083.

An, B. A., et al. 2017. "Control of sulfide production in high salinity bakken shale oil reservoirs by Halophilic bacteria reducing nitrate to nitrite." *Frontiers in Microbiology* 8:1164. doi: 10.3389/fmicb.2017.01164.

An, B. A., et al. 2019. "Halophilic methylotrophic methanogens may contribute to the high ammonium concentrations found in shale oil and shale gas reservoirs." *Frontiers in Energy Research* 7:1–13. doi: 10.3389/fenrg.2019.00023.

An, B. A., et al. 2020. "Iron to gas: Versatile multiport flow-column revealed extremely high corrosion potential by methanogen-induced microbiologically influenced corrosion (Mi-MIC)." *Frontiers in Microbiology* 11:527. doi: 10.3389/fmicb.2020.00527.

Auguet, O., et al. 2015. "Changes in microbial biofilm communities during colonization of sewer systems." *Applied and Environmental Microbiology* 81 (20):7271–7280. doi: 10.1128/AEM.01538-15.

Balussou, David. 2018. "An analysis of current and future electricity production from biogas in Germany." *Doktors der Ingenieurwissenschaften (Dr.-Ing.)*, Karlsruhe Institute of Technology (KIT).

Barker, Richard, et al. 2018. "A review of iron carbonate (FeCO3) formation in the oil and gas industry." *Corrosion Science* 142:312–341. doi: 10.1016/j.corsci.2018.07.021.

Beech, Iwona B., and Jan Sunner. 2004. "Biocorrosion: Towards understanding interactions between biofilms and metals." *Current Opinion in Biotechnology* 15 (3):181–186. doi: 10.1016/j.copbio.2004.05.001.

Beese-Vasbender, Pascal F., et al. 2015. "Selective microbial electrosynthesis of methane by a pure culture of a marine lithoautotrophic archaeon." *Bioelectrochemistry* 102:50–55. doi: 10.1016/j.bioelechem.2014.11.004.

Bensmann, Astrid, et al. 2016. "Diagnostic concept for dynamically operated biogas production plants." *Renewable Energy* 96:479–489. doi: 10.1016/j.renene.2016.04.098.

Borton, Mikayla A., et al. 2018. "Coupled laboratory and field investigations resolve microbial interactions that underpin persistence in hydraulically fractured shales." *Proceedings of the National Academy of Sciences* 115:E6585–E6594. doi: 10.1073/pnas.1800155115.

Calderon, K., et al. 2013. "Archaeal diversity in biofilm technologies applied to treat urban and industrial wastewater: Recent advances and future prospects." *International Journal of Molecular Sciences* 14 (9):18572–18598. doi: 10.3390/ijms140918572.

Daly, Rebecca A., et al. 2016. "Microbial metabolisms in a 2.5-km-deep ecosystem created by hydraulic fracturing in shales." *Nature Microbiology* 1:16146. doi: 10.1038/nmicrobiol.2016.146.

Daniels, Lacy, et al. 1987. "Bacterial methanogenesis and growth from CO2 with elemental iron as the sole source of electrons." *Science* 237:509–511. doi: 10.1126/science.237.4814.509.

Davidova, Irene A., et al. 2012. "Involvement of thermophilic archaea in the biocorrosion of oil pipelines." *Environmental Microbiology* 14:1762–1771. doi: 10.1111/j.1462-2920.2012.02721.x.

De Vrieze, J., et al. 2012. "*Methanosarcina*: The rediscovered methanogen for heavy duty biomethanation." *Bioresource Technology* 112:1–9. doi: 10.1016/j.biortech.2012.02.079.

Deng, Xiao, et al. 2015. "Electron extraction from an extracellular electrode by *Desulfovibrio ferrophilus* strain IS5 without using hydrogen as an electron carrier." *Electrochemistry* 83 (7):529–531. doi: 10.5796/electrochemistry.83.529.

Deutzmann, J. S., et al. 2015. "Extracellular enzymes facilitate electron uptake in biocorrosion and bioelectrosynthesis." *mBio* 6 (2). doi: 10.1128/mBio.00496-15.

Dinh, Hang T., et al. 2004. "Iron corrosion by novel anaerobic microorganisms." *Nature* 427:829–832. doi: 10.1038/nature02321.

Duncan, Kathleen E., et al. 2014. "The effect of corrosion inhibitors on microbial communities associated with corrosion in a model flow cell system." *Applied Microbiology and Biotechnology* 98 (2):907–918. doi: 10.1007/s00253-013-4906-x.

El-Naggar, Mohamed Y., et al. 2010. "Electrical transport along bacterial nanowires from *Shewanella oneidensis* MR-1." *Proceedings of the National Academy of Sciences* 107:18127–18131.

Enning, Dennis. 2012. "Bioelectrical corrosion of iron by lithotrophic sulfate-reducing bacteria." PhD Thesis, University of Bremen.

Enning, Dennis, and Julia Garrelfs. 2014. "Corrosion of iron by sulfate-reducing bacteria: New views of an old problem." *Applied and Environmental Microbiology* 80:1226–1236. doi: 10.1128/AEM.02848-13.

Enzmann, F., et al. 2018. "Methanogens: Biochemical background and biotechnological applications." *AMB Express* 8 (1):1. doi: 10.1186/s13568-017-0531-x.

Foght, J. M., et al. 2017. "The microbiology of oil sands tailings: Past, present, future." *FEMS Microbiology Ecology* 93 (5). doi: 10.1093/femsec/fix034.

Gieg, L. M., et al. 2008. "Bioenergy production via microbial conversion of residual oil to natural gas." *Applied and Environmental Microbiology* 74 (10):3022–3029. doi: 10.1128/AEM.00119-08.

Goodwin, Anthony R. H., et al. 2014. "Conventional oil and gas." In *Future Energy*, 19–52.

Gray, N. D., et al. 2009. "Biogenic methane production in formation waters from a large gas field in the North Sea." *Extremophiles* 13 (3):511–519. doi: 10.1007/s00792-009-0237-3.

Gray, N. D., et al. 2010. "Methanogenic degradation of petroleum hydrocarbons in subsurface environments." In *Advances in Applied Microbiology*, 137–161.

Grigoryan, Alexander, and Gerrit Voordouw. 2008. "Microbiology to help solve our energy needs: Methanogenesis from oil and the impact of nitrate on the oil-field sulfur cycle." *Annals of the New York Academy of Sciences* 1125:345–352. doi: 10.1196/annals.1419.004.

Gu, Tingyue. 2014. "Theoretical modeling of the possibility of acid producing bacteria causing fast pitting biocorrosion." *Journal of Microbial & Biochemical Technology* 6 (2). doi: 10.4172/1948-5948.1000124.

Hamilton, W. A. 2003. "Microbially influenced corrosion as a model system for the study of metal microbe interactions: A unifying electron transfer hypothesis." *Biofouling* 19 (1):65–76. doi: 10.1080/0892701021000041078.

Head, Ian M., and Neil D. Gray. 2016. "Microbial biotechnology 2020; microbiology of fossil fuel resources." *Microbial Biotechnology* 9:626–634. doi: 10.1111/1751-7915.12396.

Holmes, D. E., et al. 2018. "Electron and proton flux for carbon dioxide reduction in Methanosarcina barkeri during direct interspecies electron transfer." *Frontiers in Microbiology* 9:3109. doi: 10.3389/fmicb.2018.03109.

Hubert, Casey, et al. 2003. "Containment of biogenic sulfide production in continuous up-flow packed-bed bioreactors with nitrate or nitrite." *Biotechnology Progress* 19:338–345. doi: 10.1021/bp020128f.

In 't Zandt, Michiel H., et al. 2019. "High-level abundances of *Methanobacteriales* and *Syntrophobacterales* may help to prevent corrosion of metal sheet piles." *Applied and Environmental Microbiology* 85 (20):e01369–e01319. doi: 10.1128/aem.01369-19.

Jia, Ru, et al. 2019. "Microbiologically influenced corrosion and current mitigation strategies: A state of the art review." *International Biodeterioration & Biodegradation* 137:42–58. doi: 10.1016/j.ibiod.2018.11.007.

Joshi, Gaurav. 2015. "Elucidating sweet corrosion scale." PhD Thesis, The Univeristy of Manchester.

Kato, Souichiro. 2016. "Microbial extracellular electron transfer and its relevance to iron corrosion." *Microbial Biotechnology* 9:141–148. doi: 10.1111/1751-7915.12340.

Kip, Nardy, et al. 2017. "Methanogens predominate in natural corrosion protective layers on metal sheet piles." *Scientific Reports* 7:11899. doi: 10.1038/s41598-017-11244-7.

Koch, Sabine, et al. 2016. "Predicting compositions of microbial communities from stoichiometric models with applications for the biogas process." *Biotechnology for Biofuels* 9:1–16. doi: 10.1186/s13068-016-0429-x.

Lahme, Sven, et al. 2020. "Severe corrosion of carbon steel in oil field produced water can be linked to methanogenic archaea containing a special type of [NiFe] hydrogenase." *bioRxiv*:2020.07.23.219014. doi: 10.1101/2020.07.23.219014.

Larsen, Jan, et al. 2010. "Consortia of MIC bacteria and archaea causing pitting corrosion in top side oil production facilities." *NACE International* 2010 (Paper No. 10252).

Liang, Renxing, et al. 2014. "Roles of thermophilic thiosulfate-reducing bacteria and methanogenic archaea in the biocorrosion of oil pipelines." *Frontiers in Microbiology* 5:1–12. doi: 10.3389/fmicb.2014.00089.

Liu, Y., and W. B. Whitman. 2008. "Metabolic, phylogenetic, and ecological diversity of the methanogenic archaea." *Annals of the New York Academy of Sciences* 1125:171–189. doi: 10.1196/annals.1419.019.

Lohner, Svenja T., et al. 2014. "Hydrogenase-independent uptake and metabolism of electrons by the archaeon *Methanococcus maripaludis*." *ISME Journal* 8:1673–1681. doi: 10.1038/ismej.2014.82.

Loto, C. A. 2017. "Microbiological corrosion: Mechanism, control and impact—a review." *The International Journal of Advanced Manufacturing Technology* 92:4241–4252. doi: 10.1007/s00170-017-0494-8.

Lovley, D. R. 2017. "Happy together: Microbial communities that hook up to swap electrons." *ISME Journal* 11 (2):327–336. doi: 10.1038/ismej.2016.136.

Mand, Jaspreet, et al. 2016. "Microbial methane production associated with carbon steel corrosion in a Nigerian oil field." *Frontiers in Microbiology* 6:1–12. doi: 10.3389/fmicb.2015.01538.

Mori, Koji, et al. 2010. "Iron corrosion activity of anaerobic hydrogen-consuming microorganisms isolated from oil facilities." *Journal of Bioscience and Bioengineering* 110:426–430. doi: 10.1016/j.jbiosc.2010.04.012.

Nakayama, Cristina Rossi, et al. 2019. "Improved methanogenic communities for biogas production." In *Improving Biogas Production*, 69–98.

Okoro, C. C., et al. 2016a. "Molecular analysis of microbial community structures in Nigerian oil production and processing facilities in order to access souring corrosion and methanogenesis." *Corrosion Science* 103:242–254. doi: 10.1016/j.corsci.2015.11.024.

Okoro, C. C., et al. 2016b. "The effects of Tetrakis-hydroxymethyl phosphonium sulfate (THPS), nitrite and sodium chloride on methanogenesis and corrosion rates by methanogen populations of corroded pipelines." *Corrosion Science* 112:507–516. doi: 10.1016/j.corsci.2016.08.018.

Palacios, P. A., et al. 2019. "Baltic Sea methanogens compete with acetogens for electrons from metallic iron." *ISME Journal* 13 (12):3011–3023. doi: 10.1038/s41396-019-0490-0.

Palmer T., and Berks B. C. 2012. "The twin-arginine-translocation (Tat) protein export pathway." *Nature Reviews Microbiology* 10: 483-496. doi: 10.1038/nrmicro2814.

Park, Hyung Soo, et al. 2011. "Effect of sodium bisulfite injection on the microbial community composition in a brackish-water-transporting pipeline." *Applied and Environmental Microbiology* 77:6908–6917. doi: 10.1128/AEM.05891-11.

Pinnock, Tijan, et al. 2018. "Use of carbon steel ball bearings to determine the effect of biocides and corrosion inhibitors on microbiologically influenced corrosion under flow conditions." *Applied Microbiology and Biotechnology* 102 (13):5741–5751. doi: 10.1007/s00253-018-8974-9.

Reguera, Gemma, et al. 2005. "Extracellular electron transfer via microbial nanowires." *Nature* 435 (7045):1098–1101. doi: 10.1038/nature03661.

Thauer, Rudolf K., et al. 2010. "Hydrogenases from methanogenic archaea, nickel, a novel cofactor, and H_2 storage." *Annual Review of Biochemistry* 79:507–536. doi: 10.1146/annurev.biochem.030508.152103.

Tsurumaru, Hirohito, et al. 2018. "An extracellular [NiFe] hydrogenase mediating iron corrosion is encoded in a genetically unstable genomic island in *Methanococcus maripaludis*." *Scientific Reports* 8:15149. doi: 10.1038/s41598-018-33541-5.

Tucker, Y. T., et al. 2015. "Methanogenic archaea in marcellus shale: A possible mechanism for enhanced gas recovery in unconventional shale resources." *Environmental Science & Technology* 49:7048–7055. doi: 10.1021/acs.est.5b00765.

Uchiyama, T., et al. 2010. "Iron-corroding methanogen isolated from a crude-oil storage tank." *Applied and Environmental Microbiology* 76 (6):1783–1788. doi: 10.1128/AEM.00668-09.

Venzlaff, Hendrik, et al. 2013. "Accelerated cathodic reaction in microbial corrosion of iron due to direct electron uptake by sulfate-reducing bacteria." *Corrosion Science* 66:88–96. doi: 10.1016/j.corsci.2012.09.006.

Videla, H. A., and L. K. Herrera. 2005. "Microbiologically influenced corrosion: Looking to the future." *International Microbiology* 8 (3):169–180.

von Wolzogen Kühr, C. A. H., and L. S. van der Vlugt. 1964. "Graphitization of cast iron as an electro-biochemical process in anaerobic soils. Translation No. 1021 of the original article C. A. H. von Wolzogen Kühr, L. S. van der Vlugt (1934). "De grafiteering van gietijezer als electrobiochemish process in anaerobe gronden."" *Water* 18 (16):147–165.

Walker, D. J. F., et al. 2019. "The archaellum of *Methanospirillum hungatei* is electrically conductive." *mBio* 10 (2). doi: 10.1128/mBio.00579-19.

Wang, Hua, et al. 2014. "Corrosion of carbon steel C1010 in the presence of iron oxidizing bacteria Acidithiobacillus ferrooxidans." *Corrosion Science* 89:250–257. doi: 10.1016/j.corsci.2014.09.005.

Xu, D., et al. 2016. "Mechanistic modeling of biocorrosion caused by biofilms of sulfate reducing bacteria and acid producing bacteria." *Bioelectrochemistry* 110:52–58. doi: 10.1016/j.bioelechem.2016.03.003.

Yin, Q., and G. Wu. 2019. "Advances in direct interspecies electron transfer and conductive materials: Electron flux, organic degradation and microbial interaction." *Biotechnology Advances* 37 (8):107443. doi: 10.1016/j.biotechadv.2019.107443.

Zehnder, A. J. B. 1988. *Biology of anaerobic microorganisms*. New York: John Wiley and Sons Inc.

Part II

MIC Failure Analysis Case Studies

7 Failure Investigation of a Leak in the Offshore Water Injection System

Torben Lund Skovhus
VIA University College

Øystein Bjaanes and Bjarte Lillebø
DNV GL Norway

Jo-Inge Lilleengen
VIA University College

CONTENTS

DOI: 10.1201/9780429355479-9

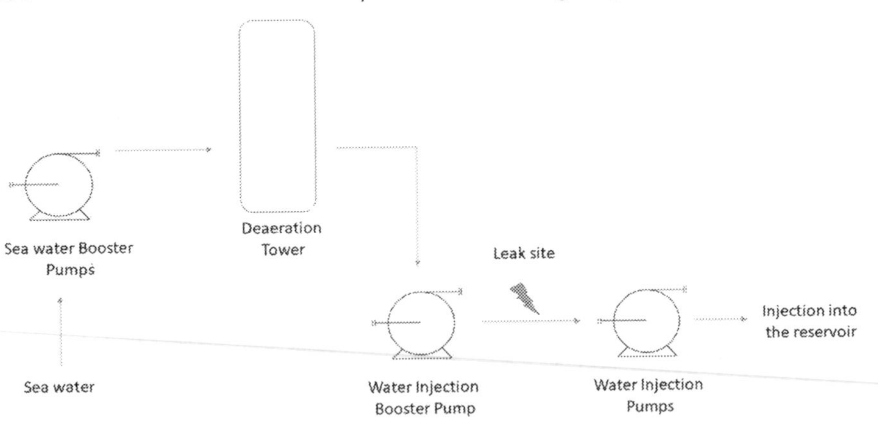

FIGURE 7.1 Location of the leak in the offshore process system.

7.1 INTRODUCTION

Both treated seawater and produced water is applied to keep up the pressure of oil reservoirs during secondary oil recovery (water flooding of the reservoir). Seawater is filtered, treated with biocides, and deaerated before it is pumped into the reservoir (SP0499-2012-SG 2012). On some occasions, produced water from the reservoir is commingled with treated seawater for injection into the reservoir, which is the case in the current study.

The oil and gas industry has seen several cases in the past where both produced- and injection water systems have suffered from severe corrosion in general and microbiologically influenced corrosion (MIC) in particular (Skovhus and Eckert 2014, Skovhus et al. 2017a, 2017b). If not managed well, in time, such corrosion threats could potentially compromise the safety of the workers, the integrity of the assets, and the marine environment. Figure 7.1 shows a schematic of where in the process the investigated failure analysis took place.

7.2 CORROSION IN THE WATER INJECTION SYSTEM

A number of leaks have been experienced in the water injection system on the offshore oil-producing platform in the past few years. Most of these have been wrapped, and the spools are still in operation.

One of the leaking spools was replaced August 3rd, 2017. The leaking spool (and a neighboring spool section) was made available for testing and failure investigation. The purpose of performing testing and failure investigation was to identify the direct and underlying causes for the leak, and to recommend mitigating actions to prevent further degradation and future leaks in the water injection system.

7.2.1 System Overview and Materials Information

Part of an 8" pipe was replaced August 3rd, 2017. The spool was located downstream from a water injection pump (Figure 7.1). The water injected through this pipe was mainly produced water (from the reservoir), but may also have contained treated seawater (filtrated, deaerated, and treated with biocide). The leak location is shown in Figure 7.2.

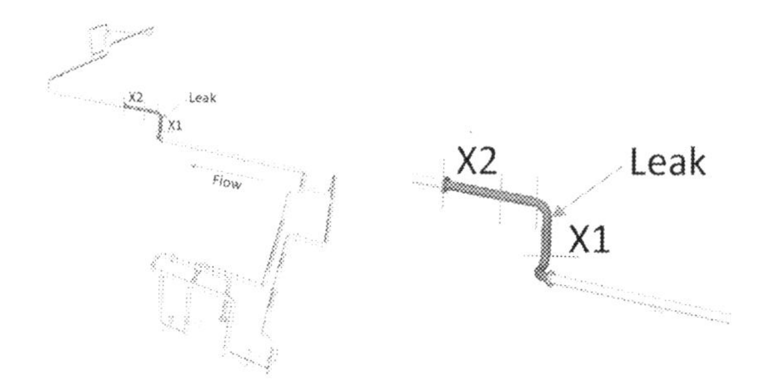

FIGURE 7.2 3D sketch of the examined pipe spool with flow direction and leakage location indicated. X1 and X2 indicates the two sections of piping recovered for failure analysis.

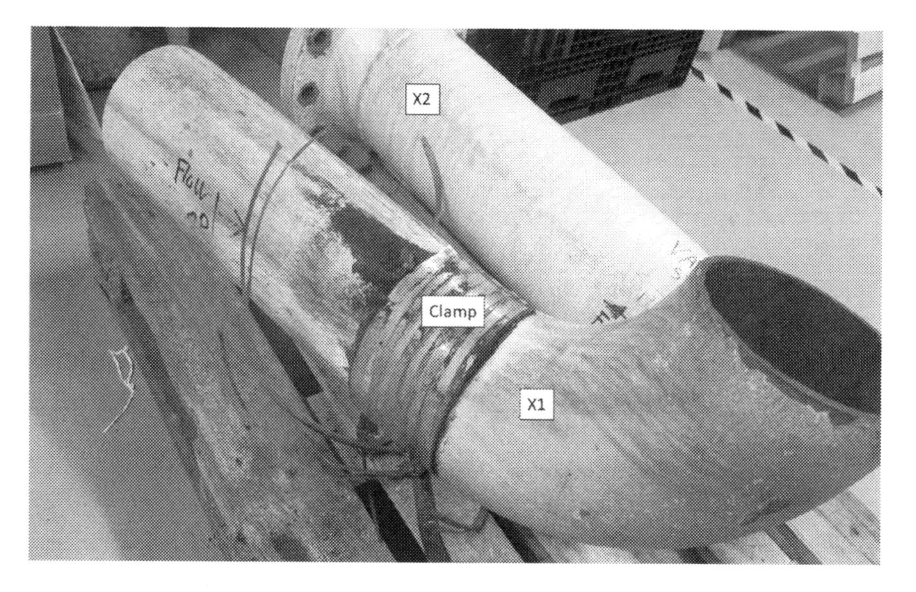

FIGURE 7.3 Pipe samples denoted X1 and X2, respectively. Note that the leak point on X1 is wrap repaired by a glued rubber plate fixed to the pipe using metallic strips. X1: Spool section with reported pinhole leakage. X2: Spool section from 2.5 m downstream the leakage location, including flange.

Two sections of the pipe spool, each approximately 1 m long, were cut, packed, and sent to shore and thereafter to the failure investigation laboratory. The two sections were denoted X1 and X2, respectively. The two sections are illustrated in Figure 7.3.

The piping is a seamless carbon steel pipe (ASTM A106 grade B). This is a steel pipe for high-temperature service. The nominal pipe dimension is 8" (219.1 mm), and the wall thickness is 8.18 mm.

7.2.2 Process and Operational Information

The pipe section under investigation has been in use for 18 years. In the beginning of oil production, the pipe was mainly transporting treated seawater (95%) and some produced water (5%). Over 14 years, this was gradually changed to 35% seawater and 65% produced water. Treated seawater has a temperature of 10–25°C and produced water has a maximal temperature of 78°C. The exact flow velocity and temperature of the pipe was not provided as part of the failure investigation data input. However, it is estimated that the temperature would have increased from 10 to 20°C to over 50°C over the 18 years of operation.

The seawater was treated with sodium hypochlorite to prevent marine growth in the seawater system. Chlorine concentration in the seawater system will have varied between 0.5 and 2 ppm. Oxygen was removed in the deaerator and was kept below 30 ppb. Scale inhibitor, biocide, and antifoam were also applied. Raw seawater from the North Sea contains relatively high concentrations of sulfate, chloride, sodium, magnesium, calcium, and potassium.

Produced water from the various fields are mixed and treated in the produced water deoiling system, prior to be used as injection water. Typically, > 95% of the produced water is re-injected. Water samples from the producing wells are collected regularly and analyzed for their ion composition. This aims to detect signs of injection water breakthrough, and to assess the environment. There have been no signs of major injection water breakthrough. Water samples from the wells are presently not used for evaluating corrosion and scale formation. The produced waters from the five oil fields are very similar and have the composition as listed in Table 7.1.

TABLE 7.1
Water Chemistry Data from the Five Oil Fields (Produced Water from the Wells)

Parameter (Unit)	Field A	Field B	Field C	Field D	Field E
Sodium (mg/l)	27,500	30,000	28,100	24,900	30,000
Calcium (mg/l)	4,700	4,800	5,300	3,420	4,800
Magnesium (mg/l)	500	875	420	380	875
Barium (mg/l)	42	92	21	83	NA
Strontium (mg/l)	610	650	660	400	650
Potassium (mg/l)	250	250	230	150	250
Iron (mg/l)	20	41	11	NA	NA
Chloride (mg/l)	54,000	57,000	57,000	36,500	57,000
Sulphate (mg/l)	36	14	67	20	14
Bicarbonate (mg/l)	230	190	270	175	NA
pH	6,7	6,5	6,9	6,3	6,5
CO_2 in gas (mol%)	1	1	1	3	2,5
H_2S (cm^3/m^3)	1	1	1	1	1

NA: Not Analyzed

7.3 MATERIALS AND METHODS APPLIED IN THE FAILURE INVESTIGATION

7.3.1 Overview of Sampling Procedure

Sampling of the spool sections X1 and X2 was performed in three stages, as outlined in Figure 7.4.

FIGURE 7.4 The sampling process offshore, at the shore base and further sampling processing at two specialist laboratories onshore.

The first stage was cutting the sections out of the existing pipework offshore. The second stage was sampling for microbiological analyses at the shore base, and finally the third stage was onshore investigation and further sub-sampling at two specialist laboratories (materials and microbiology). To allow for the extraction of spool section X1 and X2, and in order to detail the preservation requirements (for sampling microbiological samples) after cutting offshore, detailed instructions were developed for cutting, marking and preservation. The detailed instructions are provided in Chapter 24. Figure 7.5 shows the parts that were shipped to the two specialist laboratories in stage 2 of the sampling process (shore base sampling).

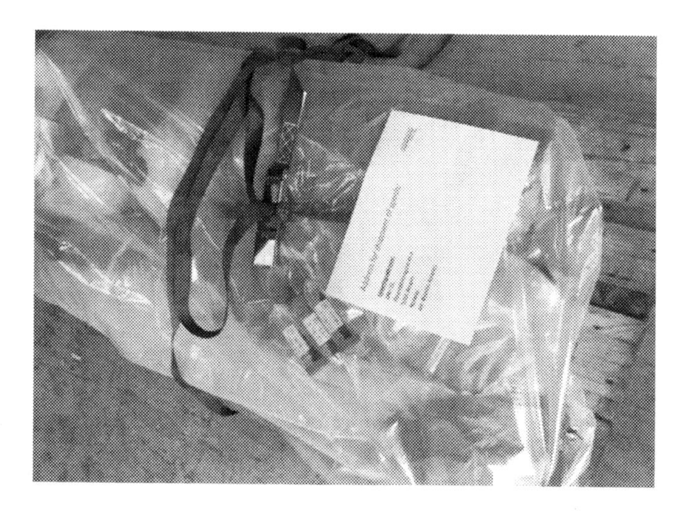

FIGURE 7.5 Overview of three sub-samples taken (tubes with red lids) from the spool sections for microbial tests (microbiological laboratory). The two spool sections X1 and X2 were wrapped in clean plastic and send to the metallurgical laboratory.

7.3.2 Sampling for Microbiological Analyses

A total of three sub-samples were taken for microbiological analysis. Sampling onshore for microbiological analyses is further described in Chapters 24. The following parameters were analyzed on the three samples.

- Total numbers of *Bacteria* and *Archaea*
- Total sulfate-reducing bacteria (SRB)
- Total sulfate-reducing archaea (SRA)
- Three groups of methanogenic Archaea

Prokaryotic numbers of total *Bacteria* and total *Archaea* were enumerated by quantitative Polymerase Chain Reaction (qPCR) targeting the gene for 16S rRNA, which is present in all *Bacteria* and *Archaea*. Sulfate-reducing bacteria (SRB) and archaea (SRA) were quantified by qPCR targeting the dsrAB gene, which is present in all sulfate-reducing prokaryotes (SRP). Methane-producing Archaea were quantified using qPCR targeting the gene encoding the Methyl-Coenzyme M Reductase (MCR) genes of *Methanothermococcus*, *Methanocaldococcus,* and *Methanosarcinales*.

7.3.3 Visual Inspection, Metallographic Examination, and Chemical Analyses

Wrap repairs were gently removed, and leak points and other corroded regions were inspected visually through various methods, including the use of stereo microscope. Cross sections were made through the leak points and corroded regions in order to examine corrosion deposits, perform chemical analysis of corrosion deposits, base material and weld metal, and to document the material mictrostructure.

7.4 RESULTS FROM THE FAILURE INVESTIGATION

On August 3rd, 2017, the two spool sections were cut out offshore, marked, and preserved in accordance with the detailed instructions (see Chapter 24). A specialist was present at the shore base upon arrival of the spools and took samples for microbiological analysis August 11th, 2017, and handed them over to the microbiological laboratory the same day for the microbiological analysis. The pipe spools were sent to the materials laboratory for further examination.

7.4.1 Visual Inspection Results

When the spools X1 and X2 arrived at the materials laboratory, they were visually examined, and documented primarily through photography.

7.4.1.1 Visual Examination of Spool Section X1

The wrap repair was gently removed in order to reveal the leak point, as shown in Figure 7.6.

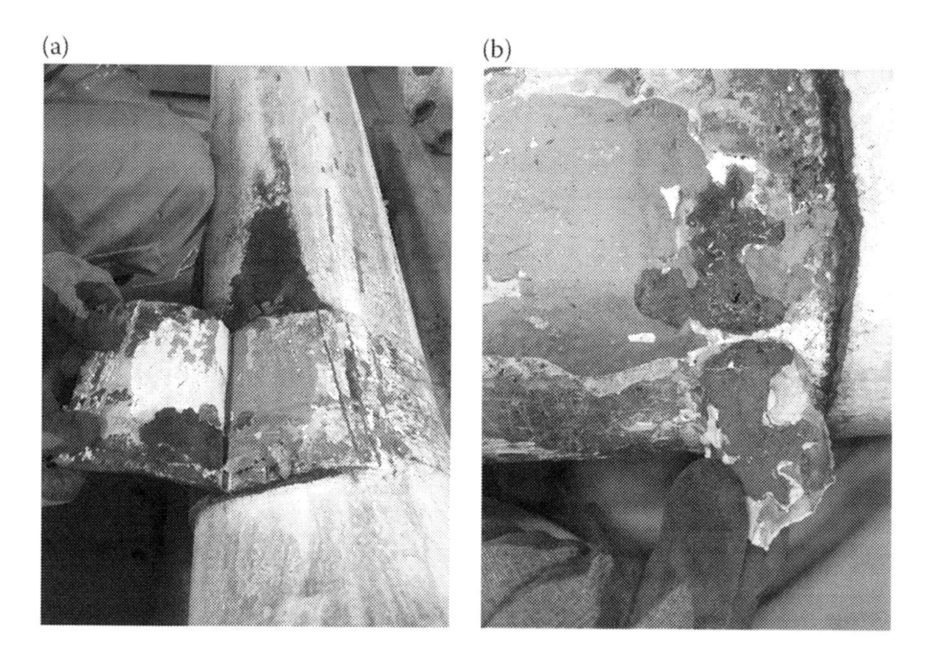

FIGURE 7.6 Wrap repair being removed to reveal the leak point (A). Leak point visible after removal of wrap repair (B).

Close-up pictures of the leak point are shown in Figure 7.7.

The leak point was located approximately 20 mm from the weld between the vertical pipe and the vertical to horizontal bend, where the flow direction was from the pipe to the bend, as shown in Figure 7.8.

Significant corrosion was observed in an approximately 25 mm broad band along the weld root on the bend side of the weld, ref. Figures 7.9 to 7.11.

Another corrosion pit/cavity was observed in the same region as the leak point. It was also observed that the weld root has not been ground, and that there is a significant geometric transition at the weld root.

7.4.1.2 Visual Examination of Spool Section X2 (Horizontal Section Downstream X1)

A reference area from spool section X2 was also inspected and examined, as shown in Figure 7.12.

The internal surface was generally corroded, but there was significantly less general corrosion along the weld, although some corrosion pits were observed, ref. Figure 7.13.

When removing the outer, brownish corrosion layer, a blackish corrosion layer was revealed (Figure 7.14).

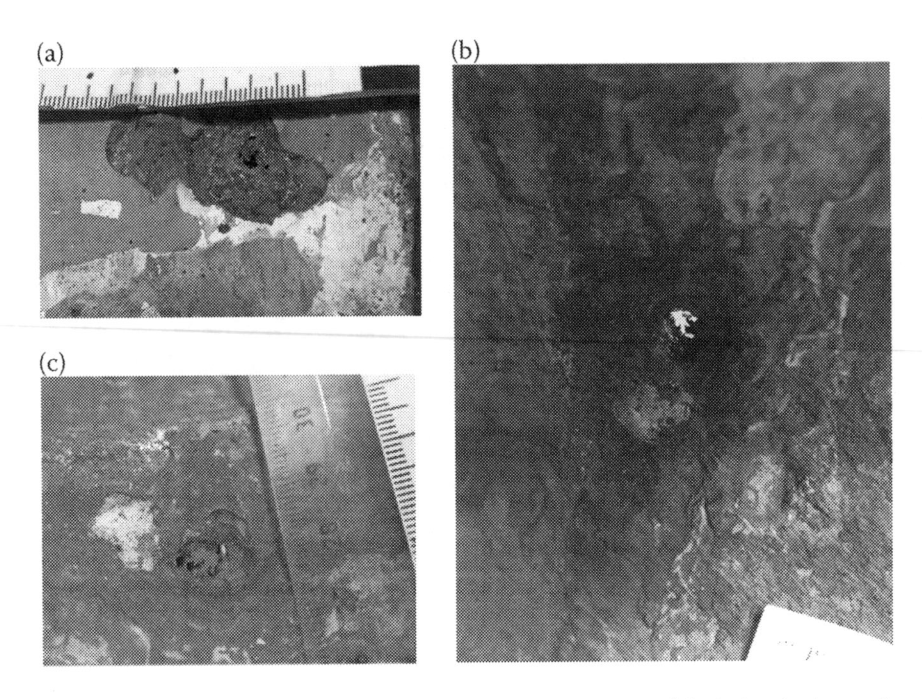

FIGURE 7.7 Close-up of the leak point externally (A). Close-up of the leak point internally (B and C).

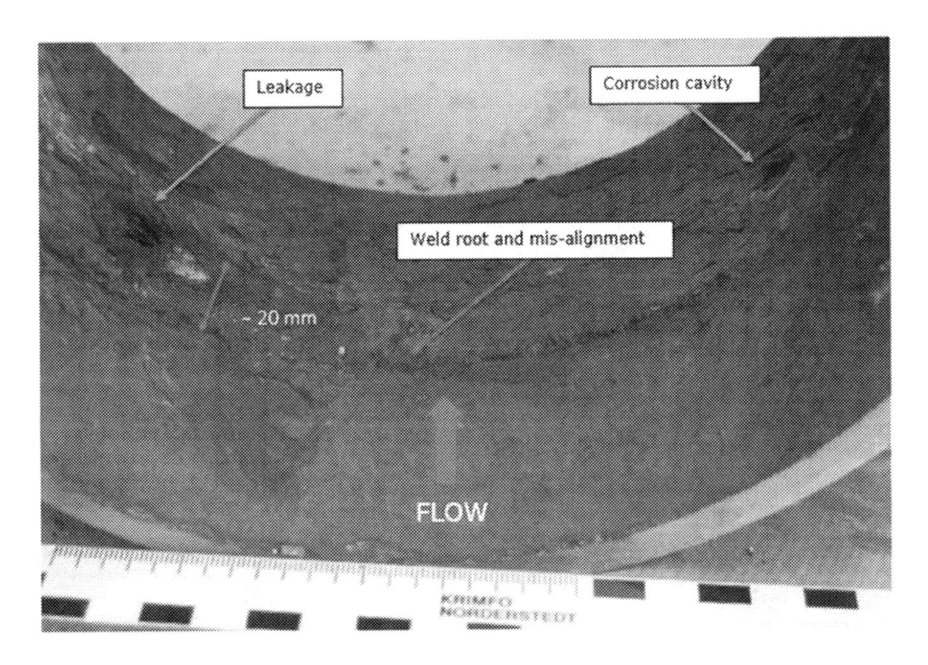

FIGURE 7.8 Close-up photo of internal area comprising the leakage.

FIGURE 7.9 Close-up photo of a local corrosion cavity approximately 20 mm from the weld root as also shown in Figure 7.8.

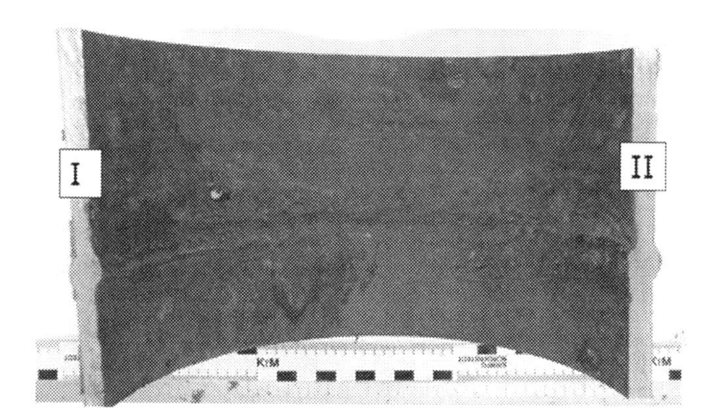

FIGURE 7.10 Area close to weld after partly rinsing with Hexamine.

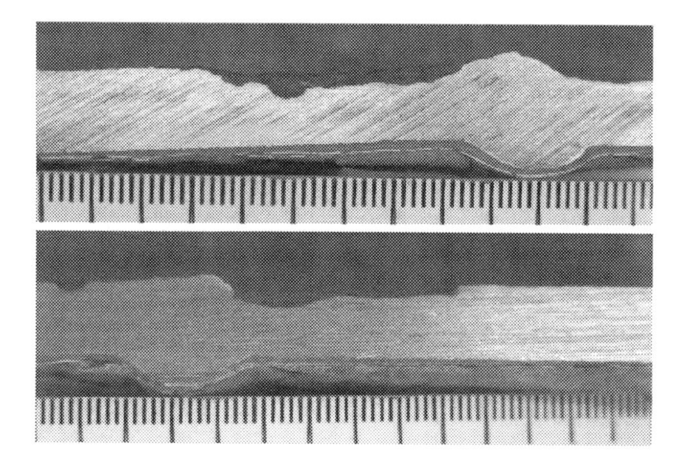

FIGURE 7.11 Cross section cut I (top) and II (bottom) in Figure 7.10, indicate the corroded area on the bend side of the weld (approx. 30 mm from center of weld root).

(a)

(b) (c)

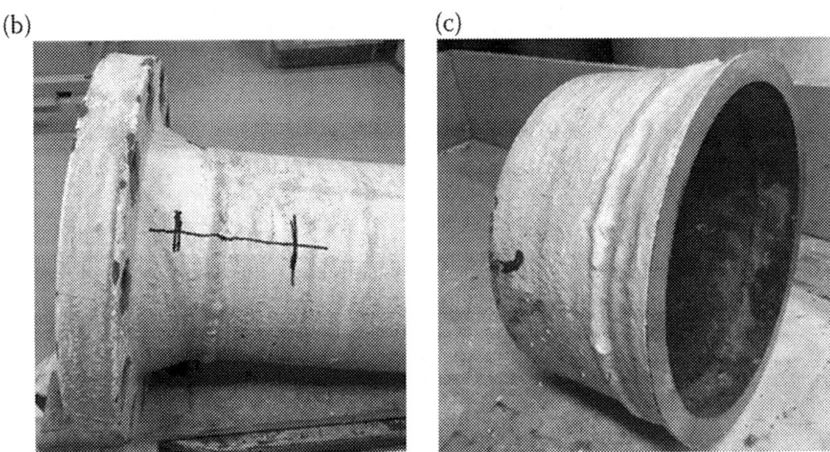

FIGURE 7.12 Spool section X2, flow direction indicated on the pipe (A). Indication of sectioning of the weld on spool section X2 (B). Spool section X2 after sectioning (C).

7.4.2 Metallographic Examination Results

A metallographic sample was prepared for microstructure examinations. The sample comprised the leak point and weld zone as indicated in Figures 7.15 and 7.16.

The microstructure observed in different locations on the sample is presented in Figure 7.17 (locations for microstructure analysis are seen in Figure 7.16).

The microstructure of both the pipe and bend base material consists of ferrite and pearlite as expected in a carbon steel.

7.4.3 Chemical Analysis of the Base Materials

A sample from the pipe and bend material, respectively, in spool section X1 was subject to spectrographic analysis to determine the chemical composition of the base materials. The results are shown in Tables 7.2 and 7.3. The requirements, as described in ASTM A106, are shown in Table 7.4.

FIGURE 7.13 Representative internal surface along the weld. General corrosion and some corrosion pits along the weld.

FIGURE 7.14 Area inside sample X2 including pipe-to-flange weld. When removing outer corrosion layer, a blackish surface appeared.

The chemical composition of both the pipe and bend material is in accordance with the specification. The pipe material, however, has higher content of alloying elements including Mn, Cr, and Mo. These elements, in general, contribute to improving the corrosion resistance properties of the material.

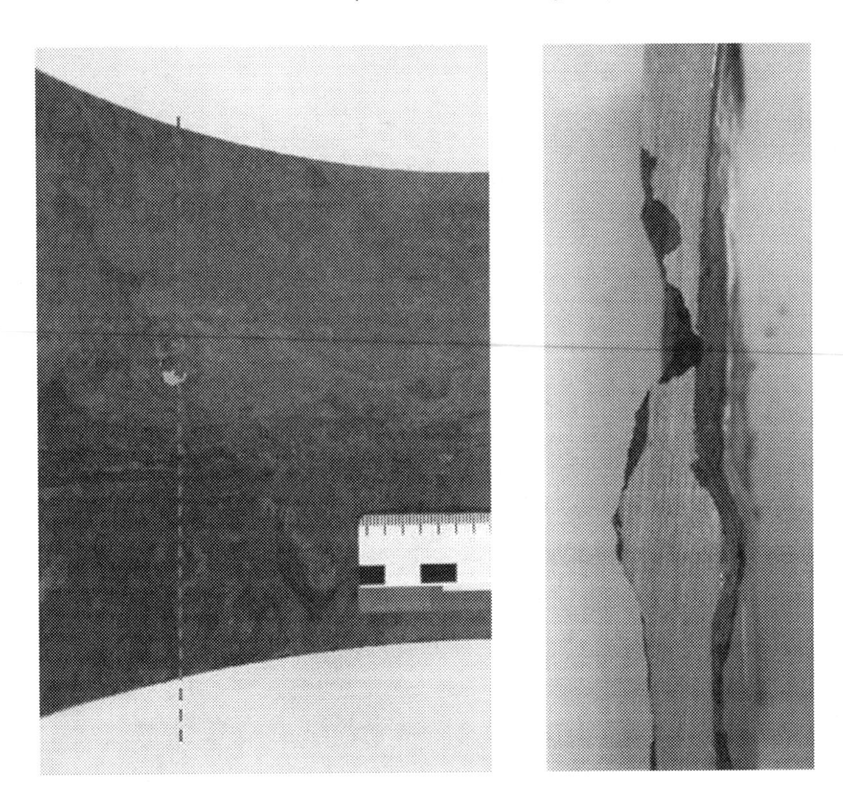

FIGURE 7.15 Metallographic sample (cross section) cut out for micro examinations (X1 through leak site).

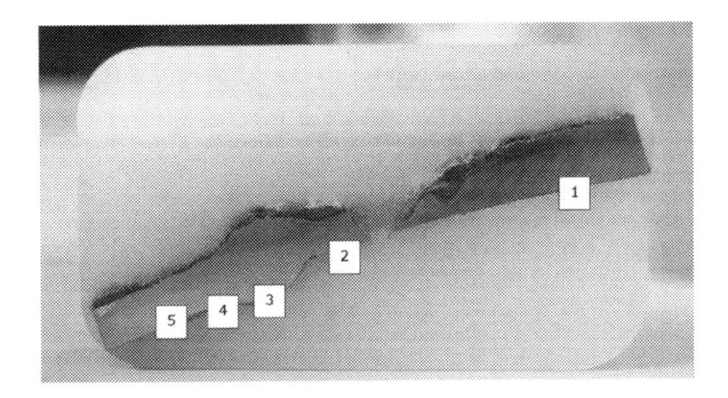

FIGURE 7.16 Metallographic sample was grinded, polished, and etched prior to microscope examination (numbers indicate areas for microstructure analysis, Figure 7.17).

7.4.4 Chemical Analysis of the Corrosion Products

The chemical composition of the corrosion products present on the surface was analyzed by the Energy Dispersive Spectrometer (EDS) unit in the Scanning

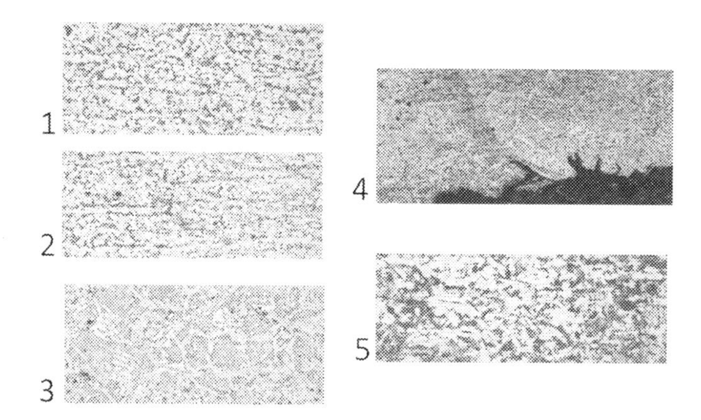

FIGURE 7.17 The microstructure in the bend consist of pearlite and ferrite, x200, ref. position 1 in Figure 7.16. The microstructure in the bend close to the weld, x200, ref. position 2 in Figure 7.16. Weld metal, X200, ref. position 3 in Figure 7.16. Microstructure in fusion line between pipe and weld, X100, ref. position 4 in Figure 7.16. Microstructure in pipe consists of pearlite and ferrite, X200, ref. position 5 in Figure 7.16.

TABLE 7.2
Chemical Analysis of Sample X1 (Pipe)

Element	**C**	**Si**	**Mn**	**P**	**S**	**Cr**	**Ni**	**Mo**	**Ti**
Wt%	0.185	0.20	0.67	0.015	0.014	0.10	0.19	0.04	<0.003
Element	Nb	Cu	Co	N	Sn	W	V	Al	Fe
Wt%	<0.005	0.25	<0.005	0.009	0.050	<0.01	<0.005	0.025	98.21

TABLE 7.3
Chemical Analysis of Sample X1 (Bend)

Element	C	Si	Mn	P	S	Cr	Ni	Mo	Ti
Wt%	0.172	0.21	0.54	0.007	0.003	0.04	0.03	<0.01	<0.003
Element	Nb	Cu	Co	N	Sn	W	V	Al	Fe
Wt%	<0.005	0.074	<0.005	0.007	0.007	<0.01	<0.005	0.004	98.85

Electron Microscope (SEM). Parts of the brownish corrosion product were removed, revealing the blackish corrosion product underneath. Chemical analysis was performed on the blackish surface (analyses 1 and 2) and the brownish surface (analyses 3 and 4), as illustrated in Figures 7.18 to 7.21, and the results are presented in Table 7.5.

TABLE 7.4
ASTM A106

Element	C	Si	Mn	P	S	Cr	Ni	Mo	Ti
Wt%	Max 0.30	-	0.29–1.06	Max 0.035	Max 0.035	Max 0.40	Max 0.40	Max 0.15	-
Element	Nb	Cu	Co	N	Sn	W	V	Al	Fe
Wt%	-	Max 0.40	-	-	-	-	Max 0.08	-	Balance

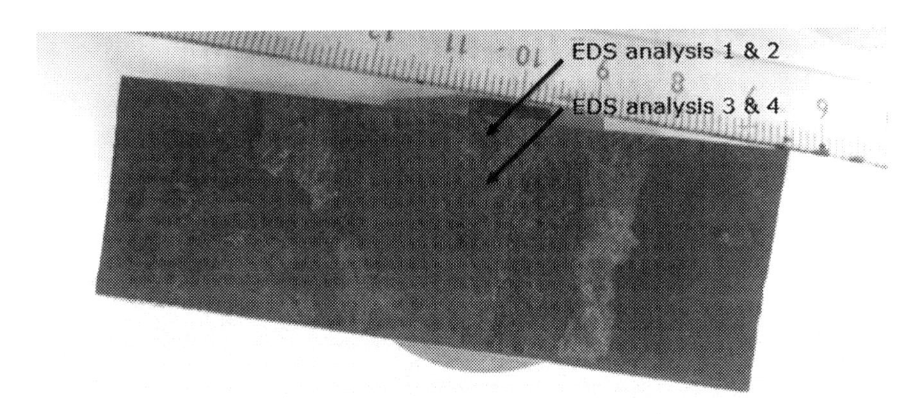

FIGURE 7.18 Section prepared for EDS analysis (in area indicated by arrows).

FIGURE 7.19 Cross section comprising weld root and corroded zone on bend side.

7.4.5 Microbiological Results

The onsite specialist extracted sample A, B, and C for microbiological analysis August 11th, 2017, in accordance with the detailed instruction in Chapter 24. The samples are shown in Figure 7.22.

Sample A was taken as close to the leak hole as possible and scraped out near/in the welding zone. Sample B was taken at the opposite end of the spool piece that

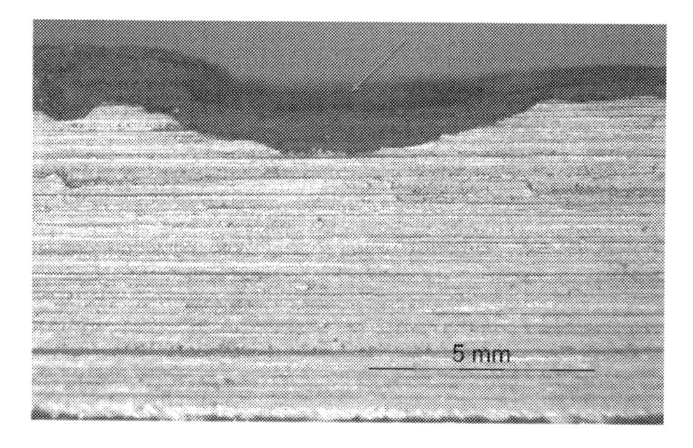

FIGURE 7.20 Area subject to chemical analyses of corrosion products in SEM (dark bottom layer, and rust-colored top layer), top layer is partly removed, ref. arrow.

FIGURE 7.21 SEM image: areas subject to chemical analyses of corrosion products in SEM, ref. 18 to Figure 7.20. Top layer is partly removed, ref. arrows (left: analyses 1 and 2, right: analyses 3 and 4).

contained the leak hole; not in a welding zone. Sample C was taken at the other spool sample (X2, reference spool) near the flange—not in a welding zone.

In the three samples from the inner surface of an injection water spool *low* numbers of bacteria were detected. The total number of bacteria identified in the samples were 7.7×10^4 cells/gram in sample A, 2.5×10^3 cells/gram in sample B, and below the detection limit in sample C. Sample A contained total archaea of 6.0×10^3 cells/gram dry weight. In samples B and C the level of total archaea was below the detection limit. No SRA nor SRB were detected in any of the samples.

TABLE 7.5
Chemical Analysis Obtained by EDS Unit in the Scanning Electron Microscope

Analysis*	Elements (mass %)							
	C	O	Na	Si	S	Cl	Ca	Fe
1	9.5	40.9	–	1.1	0.6	3.3	–	44.6
2	9.3	40.3	–	1.2	0.6	1.7	–	47.0
3	16.6	38.2	1.8	5.8	2.8	1.6	1.4	31.9
4	19.2	29.4	1.7	2.6	7.3	0.7	1.8	37.4

Note:
* Samples 1 and 2 were taken on the blackish corrosion products, while samples 3 and 4 were taken on the brownish corrosion products.

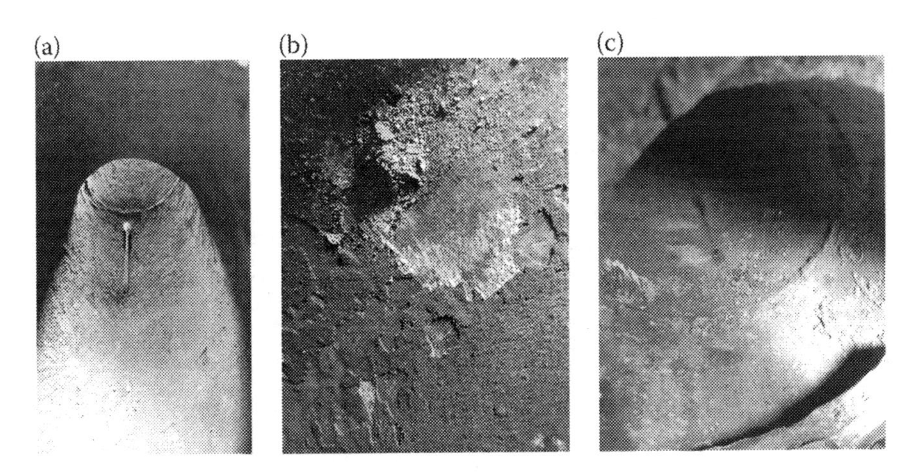

FIGURE 7.22 Location for sample A, close to leak point in X1. Location for sample B, upstream the leak point in spool section X1. Location for sample C, downstream the leak point in spool section X2.

Methanogens were only detected in Sample A, in numbers of 4.4×10^3 cells/gram dry weight. The sample showing presence of methanogens was closest to the leak point.

The samples were old when retrieved and had been exposed to air and light for an extended period, which leads to rapid degradation of the biomaterial. It is recommended that microbial samples are retrieved immediately as standard procedure upon system failures. This allows proper microbial assessment if needed.

7.5 DISCUSSION

The produced water and the water injection system on the platform are two connected systems where significant internal corrosion has developed, and where a number of through wall thickness corrosion developments has led to leak of injection water (produced water and treated sea water). Prior to entering the water injection system, the produced water has been through the degasser where the majority of the CO_2 is removed at low pressure. Hence, significant CO_2 corrosion should not be expected in the water injection system.

The injection water is intended to be oxygen free. Presence of oxygen in the injection water may lead to corrosion in the injection system. Sources of oxygen may be the low pressure part of the system (related to the degasser) and the water injection pumps (e.g., leaking seals on the suction side). Based on experience from other offshore installations, there is reason to believe that some oxygen may enter the water injection system. However, low levels of oxygen in the water injection system is not considered a likely cause of the significant corrosion experienced in this study.

Microbiological activity is known to be a potential source of significant corrosion in offshore process systems, including produced water and water injection systems. Microbiologically influenced corrosion (MIC) may potentially lead to significant localized corrosion rates. Regular analysis to monitor the presence and content of microbiological activity at the platform has been performed, and the reported levels of bacteria have been interpreted to conclude that the threat related to MIC is medium.

The microbiological analysis performed as part of this examination showed that sample A contained bacteria; 7.7×10^4 cells/gram dry weight and sample B contained 2.5×10^3 cells/gram dry weight. This means that bacteria were present, but in low abundance in both samples. Sample C contained values below the detection limit of the method, and was not analyzed further.

Samples A and B were further analyzed for microorganisms often seen causing MIC (SRP and methanogens). These analyses showed that sample A contained total archaea (6.0×10^3 cells/gram dry weight) and methanogens (4.4×10^3 cells/gram dry weight) but no SRB and SRA in levels that could be quantified by the qPCR method.

Sample B contained neither archaea, SRA, SRB, nor methanogens in levels that could be quantified by the qPCR method.

From these results it was concluded that sample A contained the highest abundance of bacteria and archaea as compared to samples B and C. Sample A was taken closest to the leak point, in the welding zone (this was not the case for sample B and C). None of the three samples contained SRB/SRA in amounts that could be detected by the qPCR method. Sample A was the only sample that contained moderate levels of methanogens (archaea). Methanogens have been found in equivalent produced water systems in the North Sea and found to promote MIC and pitting corrosion (Larsen et al. 2010, Skovhus et al. 2012).

Equations 7.1 and 7.2 below (illustrated in Figure 7.23) represent two potential contributors to the MIC observed in the offshore water injection systems:

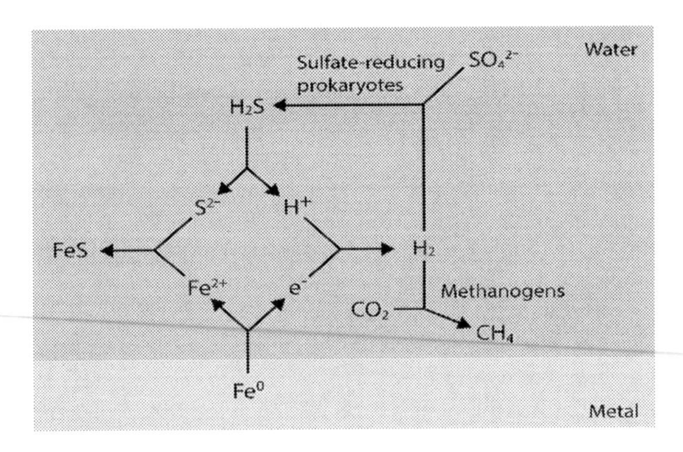

FIGURE 7.23 Illustration of sulphate reduction and methane production as possible contributors to MIC (Skovhus et al. 2017a, 2017b).

$$\text{Sulphate reduction: } 4Fe + SO_4^{2-} + 3HCO_3^- + H_2O \; - > FeS + 3FeCO_3 + 5OH^- \quad (7.1)$$

$$\text{Methane production: } 4Fe + 5HCO_3^- + 2H_2O \; - > 4FeCO_3 + CH_4 + 5OH^- \quad (7.2)$$

It is most likely that Eq. 7.2, methane production, was the driving factor of MIC causing the pinhole that led to a leak in the pipe spool system. However, as we analyzed samples taken from the system several years after the pitting initiation may have started, it cannot be excluded that sulphate reduction, ref. Eq. 7.1, may have taken place at an earlier point in time due to the activity of SRBs. The blackish scale observed, which most likely contains iron sulphides, may have its origin from HS- in the water phase, and may be an indication that SRBs have contributed to the corrosion development at an earlier stage.

The observed corrosion was located in the bend material, in a band along the weld root. Information received during the examination indicated that there seemed to be significant corrosion development in a number of the bends in the water injection system, while less corrosion had been observed in the pipe sections of the inspected spools.

The chemical analysis performed on one bend and one pipe indicate that the pipe material contains significant higher levels of alloying elements which contribute toward improving the corrosion resistance, e.g., Cr, Mn, Mo. Whether this may have contributed to more aggressive corrosion in the bends is, however, difficult to conclude.

The weld root upstream of the corrosion band and the leak point represent a significant geometric curvature, which may have impacted on the local flow condition over and just after the weld root. As the flow may influence significantly on the formation and maintenance of surface films and corrosion products on the steel surface, the local flow conditions just downstream of the weld root may have contributed to the localized corrosion development. The corroded band evolved beyond the heat-affected zone (HAZ), hence, the metallographic characteristics are not believed to have a significant impact on the corrosion development. Note that equivalent corrosion along the weld in the base material has been reported in the literature (Alabbas 2017).

7.6 CONCLUSIONS AND RECOMMENDATIONS

Based on the performed examination, the following can be concluded, as also summarized in Table 7.6:

- Microbiologically influenced corrosion (MIC) is most likely the cause of the localized corrosion, leading to leak in the examined pipe spools from the water injection system. Methanogens have most likely played an active role in driving the localized pitting corrosion. However, it cannot be excluded that SRBs have also contributed to the observed corrosion.
- CO_2 and O_2, to the extent that they are present in the injection water, may have contributed to the general corrosion in the system, but cannot explain the localized corrosion leading to leaks in the water injection system.
- The examined bend material had lower content of alloying elements which enhance corrosion resistance, e.g., Cr, Mo, and Mn, than the pipe material. This may have contributed to the more aggressive corrosion attacks experienced in the bends compared to the pipe sections.
- The geometric transition at the weld root may have contributed to the severe corrosion attack in an approximately 25 mm band downstream of the weld root. The local flow conditions may have impacted on the local flow velocity

TABLE 7.6
Factors Contributing to the Determination of MIC

Chemical Conditions	**Microbiological Conditions**
• Mix of sea water and formation water • Low concentration of H_2S and CO_2 • Corrosive compounds in the water (Cl^-, SO_4^{2-}) • No (observed) oxygen • pH < 7	• Both bacteria and archaea measured close to the leak position • Methanogens measured close to the leak position • Samples were dry when they arrived at shore base
Corrosion and Metallurgical Information	**Design and Operating Information**
• Localized corrosion and pits • Scale formation and deposits • Presence of sulfur in corrosion product	• Mixing of produced water and treated seawater over time • Dead end piping observed in the system • Increased from 10 to over 50°C during 18 years of operation

and therefore the formation and maintenance of surface films and corrosion products.

In order to reduce future corrosion in the injection water system going forward, the following mitigating activities are recommended:

- It is recommended to establish an optimized program for biocide injection as early as possible in the process.

The injection of corrosion inhibitor may also be considered if the above recommendation is not successful; alternatively, corrosion inhibitor could be included as part of the biocide injection program. When it is possible to access the separator, the electrostatic coalescer or the degasser, measurement of the microbiological activity in the sediments and internal cleaning of these vessels is recommended. Monitoring of the amount of CO_2 and O_2 in the injection water should also be considered, to verify that these are not at a level which could significantly contribute to future corrosion development.

REFERENCES

Alabbas, F. M. 2017. MIC cases histories in oil, gas and associated operations. In Skovhus T. L., Lee, J. and Enning, D. *Microbiologically influenced corrosion in the upstream oil & gas industry* (1st ed.). CRC Press, New York.

ASTM A106. 2015. Standard specification for seamless carbon steel pipe for high-temperature service. Pennsylvania: ASTM International. https://www.astm.org/Standards/A106.

Larsen, J., Rasmussen, K., Pedersen, H., Sørensen, K., Lundgaard, T. and Skovhus, T. L. 2010. Consortia of MIC bacteria and archaea causing pitting corrosion in top side oil production facilities. *Corrosion 2010, paper 12779.* Houston, TX: NACE International.

Skovhus, T. L. and Eckert, R. 2014. *Practical aspects of mic detection, monitoring and management in the oil & gas industry. Corrosion 2014, paper C2014-3920.* Houston, TX: NACE International.

Skovhus, T. L., Eckert, R. and Rodrigues, E. 2017a. Management and control of microbiologically influenced corrosion (MIC) in the oil and gas industry - Overview and a North Sea case study. *Journal of Biotechnology* 256, 31–45. 10.1016/j.jbiotec.2017.07.003.

Skovhus, T. L., Holmkvist, L., Andersen, K., Pedersen, H. and Larsen, L. 2012. MIC risk assessment of the halfdan oil export spool, SPE-155080, *SPE International Conference and Exhibition on Oilfield Corrosion*, Aberdeen, UK, 28–29 May 2012.

Skovhus. T. L., Lee, J. and Little, B. 2017b. Predeomiant MIC Mechanisms in the Oil and Gas Industry. In Skovhus T. L., Lee, J. and Enning, D. *Microbiologically influenced corrosion in the upstream oil & gas industry* (1st ed.). CRC Press, New York.

SP0499-2012-SG. 2012. Corrosion control and monitoring in seawater injection systems. Houston, TX: NACE International. https://store.nace.org/sp0499-2012-2.

8 Failure Analysis for Internal Corrosion of Crude Oil Transporting Pipelines

Mohita Sharma
University of Calgary

Trevor Place
Enbridge

Nicolas Tsesmetzis
Shell International Exploration and Production, Inc.

Lisa M. Gieg
University of Calgary

CONTENTS

8.1 INTRODUCTION

The service life of a crude oil transporting pipeline depends on interactions with its biotic and abiotic surroundings and the characteristics of the material used for

DOI: 10.1201/9780429355479-10

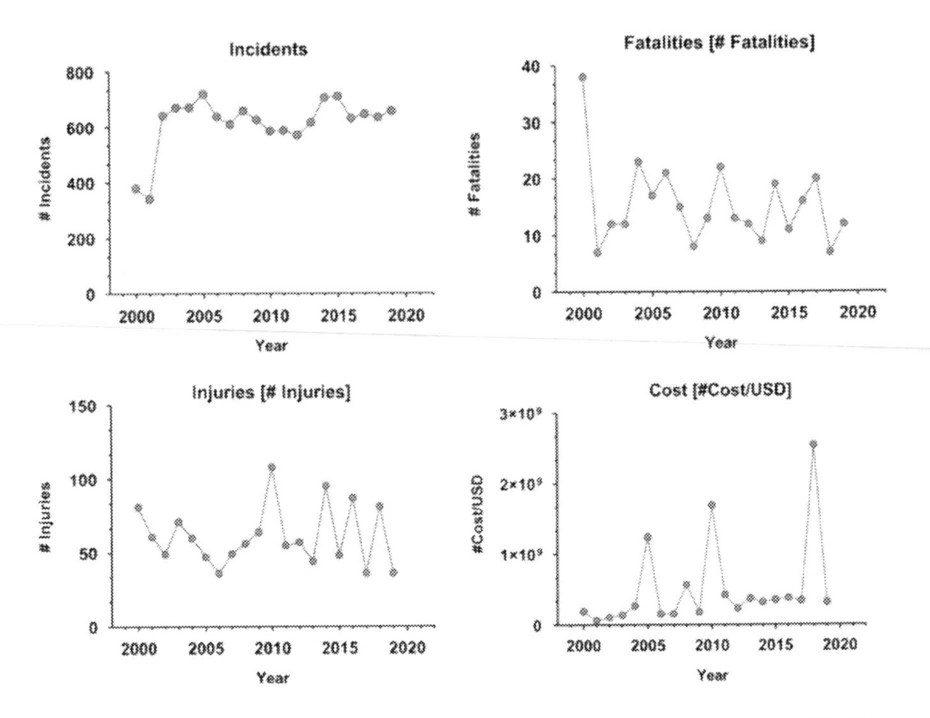

FIGURE 8.1 Pipeline incident 20-year trends of (a) number of incidents (b) number of fatalities, (c) number of injuries, and (d) cost related to pipeline failures accidents in the USA. Data provided by Pipeline and Hazardous Materials Safety Administration. Accessed June 2020 from: https://www.phmsa.dot.gov/data-and-statistics/pipeline/pipeline-incident-20-year-trends

its construction. Regular predictions, monitoring, and mitigation methods are applied to pipeline systems to maintain integrity, though failures occasionally occur (Koch et al. 2016). Unexpected corrosion failures can lead to substantial downtime, environmental, safety, health-related consequences, and economic losses due to the efforts and costs needed to repair/replace such issues (Koch et al. 2016; Skovhus et al. 2017). Figure 8.1 shows the statistics across 20 years of pipeline incidents, fatalities, injuries, and cost of corrosion/annum, calculated from monitoring of the wide network of gas transmission and hazardous liquid pipelines in the USA (Figure 8.2).

Microbiologically influenced corrosion (MIC), also known as biocorrosion, can be an important contributor to some of these corrosion failures. MIC-related issues are more difficult to identify because it is not a stand-alone reason for failure and is usually accompanied by one or more abiotic mechanisms that can occur due the surrounding chemical environment and physical conditions (Shi et al. 2011; Tidwell et al. 2017; Little et al. 2020). This leads to unexpected additional operational costs for replacement of equipment and pipelines in addition to local environmental damage that can happen due to leakages. The studies of MIC mechanisms, microbial monitoring, and prediction of MIC using modeling as well as development of new mitigation methods have been active areas of research in recent years (Videla and Herrera 2005; Voordouw et al. 2016; Rajala et al. 2017; Vigneron et al. 2016, 2018; Bagaria et al. 2019; Sharma and Voordouw 2016; Sharma et al. 2016,

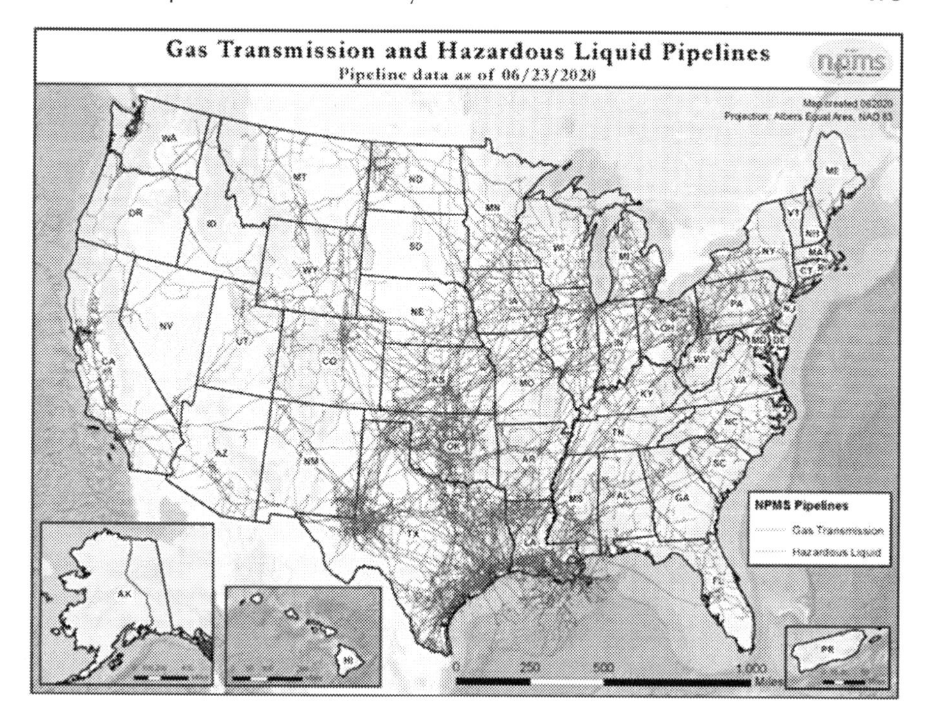

FIGURE 8.2 Network of gas transmission and hazardous liquid pipelines in the United States. Accessed June 2020 from: https://www.phmsa.dot.gov

2017, 2018a, 2018b, 2019). Though corrosion is an electrochemical process that requires an anode, a cathode, a metal surface, and an electrolyte for it to propagate, microorganisms can participate in influencing this rate of corrosion in many ways (Kip and Veen 2015; Lanneluc et al. 2015). For example, microorganisms can alter the thermodynamics and reaction kinetics at the anode and/or cathode, alter the matter/solution interface, influence the redox potential with many interrelated mechanisms, and produce metabolites that lead to the generation of corrosive agents such as iron sulfides (Shi et al. 2011; Enning and Garrelfs 2014; Liu et al. 2017, 2018). Further, many types of microorganisms are now known to directly consume electrons at the steel surface to drive respiration through a process known as electrical MIC, or EMIC (Enning and Garrelfs 2014; Kato 2016).

Many different factors can contribute to pipeline corrosion and corrosion related pipeline failure. These factors can be broadly categorized as chemical (into which microorganisms fall as they facilitate chemical reactions), physical, or environmental factors (Figure 8.3). The interplay between these various factors can give rise to what may be termed a "corrosion mechanism." For instance, the presence of water (chemical-abiotic factor) at a crevice (a structural factor) can give rise to chemical gradients that exacerbate corrosion – this is termed "crevice corrosion." Similarly, the presence of hard particulate in a fast-flowing pipeline can give rise to "erosion-corrosion." Duration of exposure, structural properties of the pipe including geometry and size, and operating pressure will then affect the end stage of a corrosion failure (pinhole leak, large leak, or rupture).

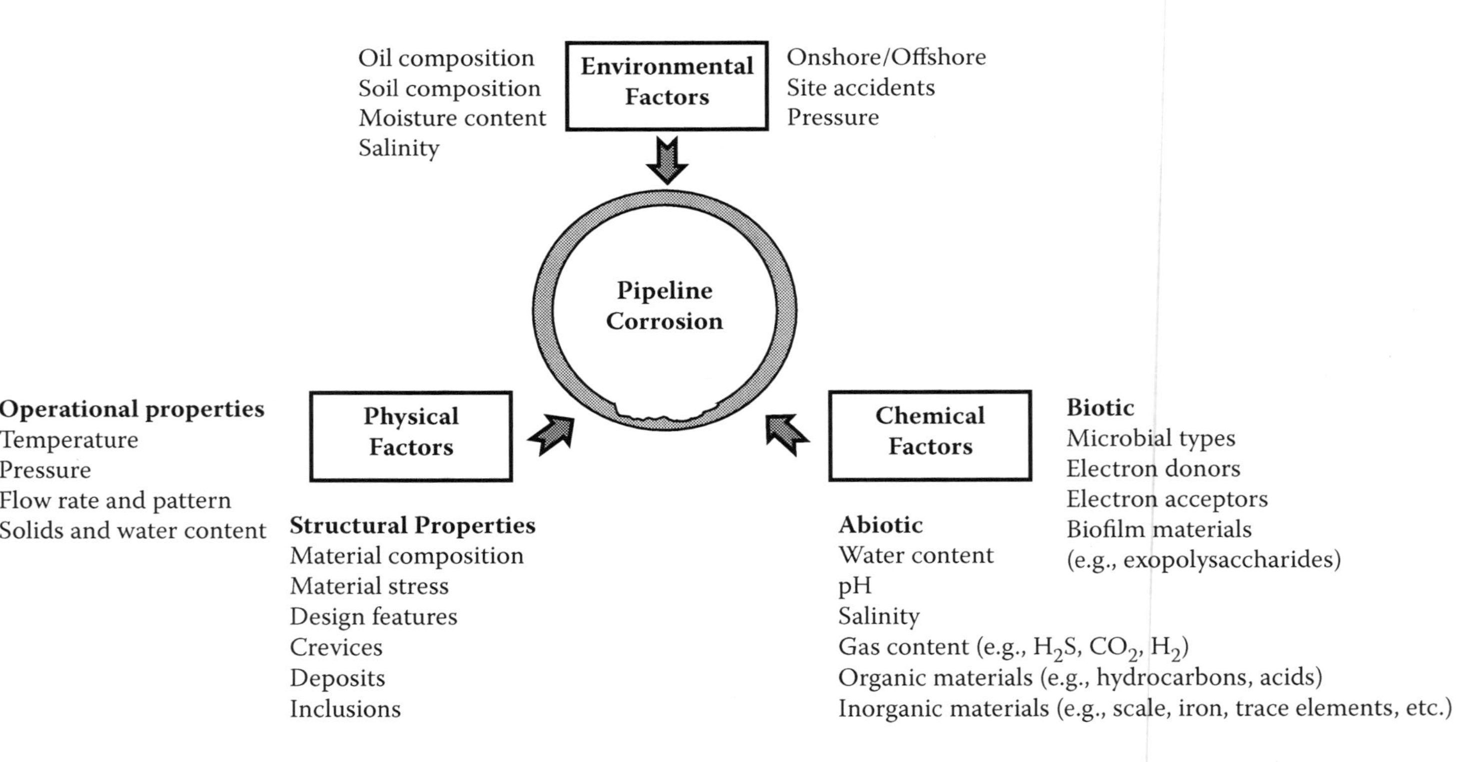

FIGURE 8.3 Factors that can influence corrosion of a metal pipeline (Adapted from Ossai et al. 2015; Sharma and Voordouw 2016).

Another possible mechanism is MIC. Systems with abundant microbial populations and ineffective microbial management and those experiencing periods of stagnation or low flow conditions and moderate temperatures that permit microbial growth and activity are often susceptible to MIC (Eckert 2016; Little et al. 2020). In these low-flow or no-flow areas, deposits consisting of a mixture of sand and other solids, water, crude oil, salts, and microorganisms can accumulate on the pipe surface, potentially leading to a damaging phenomenon referred to as under deposit corrosion (UDC) (Vera et al. 2012; Eckert 2016). These areas of localized corrosive deposits are often difficult to monitor and difficult to clean using conventional methods like pigging (Bensman 2016; Eckert 2016; Kagarise et al. 2017). Once deposited in such zones, microorganisms can grow, form biofilms (surface attached colonies of microorganisms) and secrete metabolites that increase localized corrosion or pitting of surfaces, can de-alloy metals, and can cause galvanic corrosion, stress corrosion, and hydrogen-induced cracking which can impact both internal and external surface of the pipelines (Eckert 2016; Suarez et al. 2019). Higher biological activity mainly due to sulfate-reducing prokaryotes (SRP include both sulfate-reducing bacteria and sulfate-reducing archaea) activity has also been directly associated with H_2S production and consequently higher corrosion rates of carbon steel infrastructure like pipelines (Vera et al. 2012; Gonzalez et al. 2014; Enning and Garrelfs 2014; Vigneron et al. 2018). However, it should be noted that many other types of microorganisms such as acid-producing bacteria (fermenters), iron-oxidizing/reducing bacteria, nitrate-reducing bacteria, manganese-oxidizing bacteria, sulfide-oxidizing bacteria, methanogens, and others also directly/indirectly influence the rate of corrosion and cause pipeline damage. Several of these other types of microorganisms involved in MIC are described in recent reviews (Kip and Veen 2015; Enning and Garrelfs 2014; Sharma and Voordouw 2016; Kato 2016; Vigneron et al. 2018; Suarez et al. 2019).

Linking together microbiological activities and communities, the chemical environment, and operating conditions to assess whether failures are due to UDC and MIC using a multidisciplinary approach is still an important knowledge gap (Hashemi et al. 2018). As such, there have been few reports documenting a "multiple-line-of-evidence" approach wherein microbiological assessments are considered alongside chemical and metallurgical analysis for evaluating failure mechanisms, particularly for failures in dead leg segments (Larsen and Hilbert 2014; Bensman 2016; Suarez et al. 2019). In this chapter, we present the results of a case study wherein we had a unique opportunity to evaluate a failure occurring on low flow portion of a crude oil transmission using a holistic, multiple-line-of-evidence approach. In addition to metallurgical analysis, we describe how samples can be collected at and around the vicinity of a leak so that microbiological analyses can be effectively incorporated into a failure analysis procedure to determine whether MIC plays a significant role.

8.2 CASE STUDY: FAILURE ANALYSIS OF A PIN HOLE LEAK AT A DEAD LEG SEGMENT OF A CRUDE OIL TRANSPORTING PIPELINE

A perforation was observed in a carbon steel (API 5L, Gr. B) crude oil transporting pipeline between a 34-inch diameter tee and pipe which was in operation for

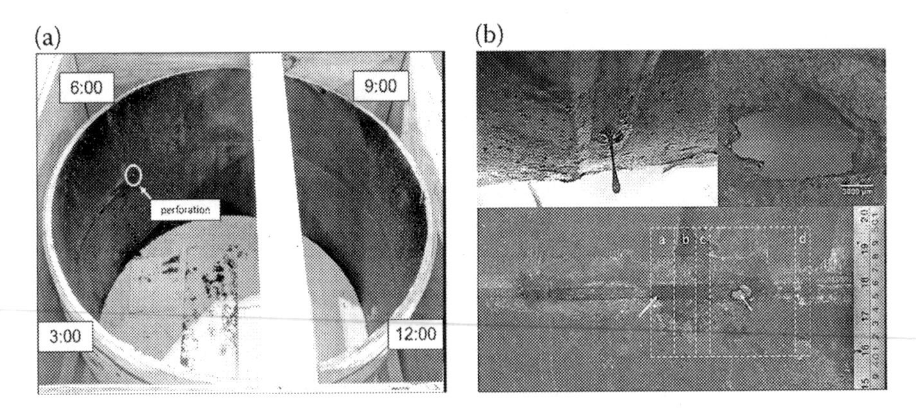

FIGURE 8.4 (A) Photo showing the location of the pinhole leak at the six o'clock position on a bypass segment of a crude oil transporting pipeline. (B) The perforation leak as seen from different positions and at different scales.

13 years (Figure 8.4). The maximum operating pressure (MOP) for this segment of pipe was 275 psi and it was operating below this limit at the time of failure. This portion of the pipeline was normally used for tank-to-tank transfers and was operated approximately once per month. No signs of cracking and brittle damage were observed in the region of failure. The geometry of the pipe at the leak site promoted deposits and liquid accumulation along the weld root. This led to the proposition that the failure mechanism may be UDC, mainly due to the low/no-flow nature of this pipe segment. As such, the location was hypothesized to be susceptible to microbiological biofilm formation with high metabolic activity, hence was selected for carrying out multiple analyses (microbiological, chemical, and metallurgical) to assess whether the failure was at least partly due to MIC.

8.2.1 Methods

The various analyses conducted to assess the mechanisms for failure of the pipe segment are summarized in Table 8.1. Upon discovering the leak through a routine inspection, containment was immediately applied, and the section of the pipe was isolated and drained. Within four days of the leak discovery, the pipe segment was cut out and sampled immediately for microbiological analyses, and the pipe was then subject to several other analyses.

8.2.1.1 Microbiological and Chemical Analyses

Samples were collected from four different locations from the dead leg pipe cutout. Location 1 (L1) samples were collected directly from the leak location (which was at the six o'clock position), while Location 2 (L2) samples were collected from a corroded area immediately adjacent to the leak location. This latter location was sampled because leaks that occur under pressure can wash away microorganisms at the leak location; adjacent locations will often have similar pitting as the leak location thus can still reflect the chemical and microbiological environment at the

TABLE 8.1
Tests Performed with the Pipe Cutout and with Unpreserved and Chemically Preserved Samples Removed from Various Locations on the Surface of the Pipe in Order to Gather Multiples Lines of Evidence for This Failure Analysis (MIC) Study

Pipe Cutout Analysis	Unpreserved Samples	Chemically Preserved Samples
• Visual examination • Metallography and light optical microscopy (LOM) • Scanning electron microscopy (SEM) • Energy-dispersive spectroscopy (EDS) • Chemical analysis of weld and bulk metal • X-ray diffraction (XRD) • Atomic emission spectroscopy (AES) • Mechanical testing • Tensile testingCharpy V-notch (CVN)	• pH • NaCl (mM equivalents) • Volatile organic acids (acetate, propionate, butyrate, mM) • Ammonium (mM) • Total iron (mM) • Sulfate, nitrate, nitrite (mM) • Sulfide (mM) • X-ray diffraction (XRD)	DNA extraction and PCR amplification of the 16S rRNA gene to determine microbial community composition

leak. Location 3 (L3) samples were collected from a non-corroded segment of the pipe at the six o'clock position, and location 4 (L4) samples consisted of sediments that were found on the inner surface of the pipeline away from the leak site and in a non-corroded area. Samples were collected using sterile implements (sterilized using ethanol) and placed into sterile mason jars. Four samples were collected from each location and placed into separate mason jars; two jars were amended with 70% isopropanol that was added as a chemical preservative to capture the microbial community at the time of sampling, while the other two jars remained unpreserved to be used for chemical analysis and establishing laboratory incubations. When sampling for microbial community analysis using molecular methods (such as the 16S rRNA gene analysis that was done here, see below), it is critical to either process samples immediately (e.g., extract DNA in the field), or to chemically preserve the samples so that the microbial community composition does not shift between the time samples are collected and transported to the lab. Several studies have now shown that microbial communities can change rapidly following sample collection (Kilbane 2014; Sharma and Huang 2019; De Paula et al. 2018; Rachel and Gieg 2020), underlining the importance of field preservation. All samples were also placed on ice and shipped to the laboratory.

Once the samples were received in the lab (still cold), the non-preserved samples were transferred to an anaerobic hood containing 90% N_2 and 10% CO_2, and the preserved samples were used to determine microbial community composition based on sequencing the 16S rRNA gene (Sharma et al. 2019). DNA was extracted from each chemically preserved replicate from each of the four sampled locations using published protocols (Sharma et al. 2019), then subject to PCR amplification of the 16S rRNA gene. To ensure representative coverage irrespective of the amplified region, three different primer sets were used, targeting the V3–V4, V4–V5, and V6–V8 hypervariable regions of the 16S rRNA gene; here, we present the results obtained using the V3–V4 primers. Samples were sequenced on an Illumina MiSeq platform, then processed through QIIME 2 (version 2019.1; Bolyen et al. 2019) to obtain the microbial community composition of the different samples. The unpreserved samples collected from the four locations on the pipe were used for chemical analyses. The samples were prepared for "water chemistry" analysis by mixing the sample 1:1 with sterile water, then analyzing the water extract for volatile fatty acids (acetate, propionate, butyrate) and anions (sulfate, nitrate, nitrite) by HPLC, pH, salinity (in mM equiv. NaCl), and using spectrophotometric assays for determining ammonium, sulfide, and total iron concentrations (all methods described in Sharma et al. 2017). The samples collected from locations 1, 2, and 3 were also analyzed by XRD in order to determine the composition of putative corrosion products in the samples. While incubations were also prepared from the non-preserved samples, the results are not shown here but are included in a separate publication (Sharma et al. 2020).

8.2.1.2 Metallurgical Analyses

The pipe cutout was sent for surface analysis, microscopic analysis, and mechanical testing to a third-party laboratory. Surface and microscopic analyses were performed using visual inspection of the pipe, light optical microscopy, scanning electron microscopy, energy dispersive spectroscopy (EDS), X-ray diffraction (XRD), and atomic emission spectroscopy (AES). Mechanical testing was done using tensile testing and the Charpy V-notch method.

8.3 RESULTS AND DISCUSSION

8.3.1 Analysis of the Pipe Cutout

Following sample collection for microbiological analysis and initial cleaning of the affected pipe segment, a visual examination showed that the perforation was at the six o'clock position (Figure 8.4A). Based on the local wall thickness of 9.5 mm and the 13-year service life of the pipe segment, an average corrosion rate (CR) of 0.75 mm/yr was established. A closer microscopic inspection and pictures of the perforation from different angles are shown in Figure 8.4B. Different colored and textured corrosion products (powdery, chalky, and flaky to dense) were observed in this section of the pipe, with severe thickness loss.

For further examination of the internal surface of the pipe, cleaning was performed using tetrachloroethylene of the internal surface of the pipe to remove crude oil stuck on the internal surface of the pipe. Metallography and light optical microscopy (LOM) of one of the areas close to the perforation marked as sample a in Figure 8.5A showed the failure of external coating in this area that may have allowed the fluid accumulation in the area between the metal and the coating. After

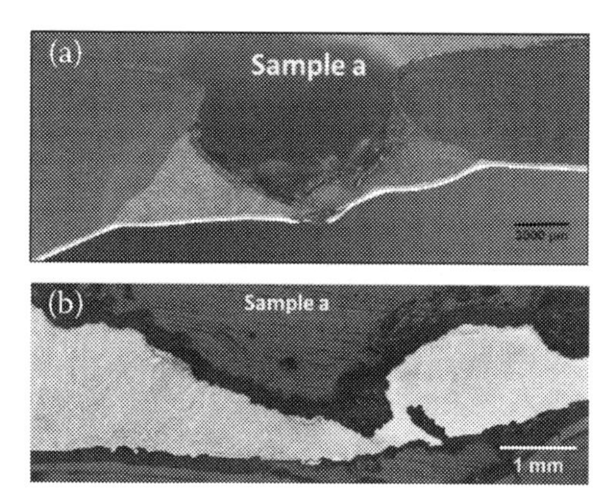

FIGURE 8.5 Micrograph of cross section of sample marked as "Sample a" before (A) and after etching (B). Here corrosion has degraded almost the entire weld bead and the external coating has failed.

etching of this area, as shown in Figure 8.5B, it was evident that corrosion propagated primarily from the inside of the pipe to the outer surface and that there was a tiny void at the outer surface – possibly due to a metal inclusion, weld defect, or external corrosion.

Scanning electron microscopy in backscattered mode was used to study the internally corroded (Figure 8.6A,B,C) and an outer section of the pipe (Figure 8.6D). The inner surfaces of the pipe section showed the predominance of Fe, Si, S, Cl, and Mn in varying proportions, while section D showed a predominance of Fe, Si, Al, and Ca and various other elements in trace concentrations, mainly due to exposure to many external environmental factors on this section of the pipe. The presence of S and Cl on the inner surface is consistent with corrosion processes common to crude oil production and transportation systems containing trace amounts of naturally occurring brine (from a geologic source) and H_2S (which can be from both biogenic or geologic sources). The scarcity of these elements in the compounds on the outer surface of the pipe demonstrate the corrosion mechanism at the internal and external surface of the pipe were likely not the same. The results of mechanical tests were compared to API 5L requirements for line pipe steels as a reference for relative strength and toughness. The chemical analysis and mechanical testing (tensile testing and Charpy V-notch testing) of the bulk and weld metal were within the range of industry standards (data not shown) of Grade X52 and Grade B, suggesting that changes in metal composition or a mechanical deformity could not explain the failure.

8.3.2 Microbiological and Chemical Analyses

The non-preserved samples collected for microbiological analysis were also used for measuring several chemical analytes and for the presence of possible corrosion products using XRD. Chemical analyses of water extracts of each duplicate sample collected from each location, and results of the individual analyses and the averages for each analyte are shown in Figure 8.7.

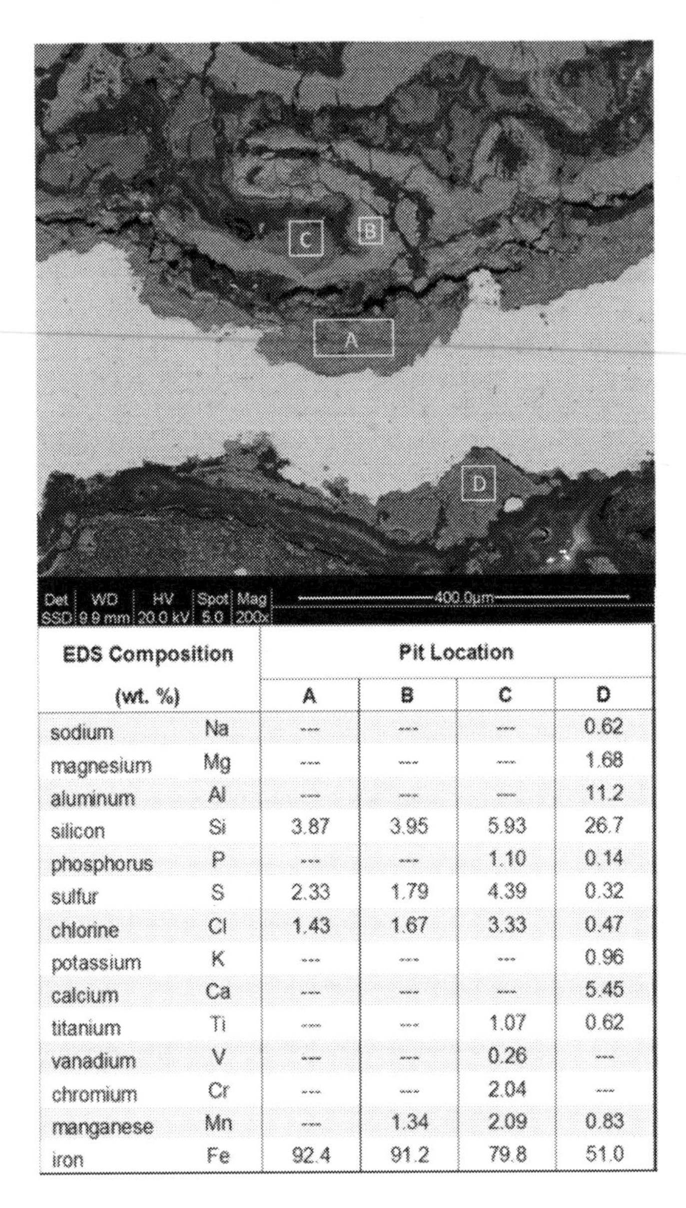

EDS Composition (wt. %)		Pit Location			
		A	B	C	D
sodium	Na	---	---	---	0.62
magnesium	Mg	---	---	---	1.68
aluminum	Al	---	---	---	11.2
silicon	Si	3.87	3.95	5.93	26.7
phosphorus	P	---	---	1.10	0.14
sulfur	S	2.33	1.79	4.39	0.32
chlorine	Cl	1.43	1.67	3.33	0.47
potassium	K	---	---	---	0.96
calcium	Ca	---	---	---	5.45
titanium	Ti	---	---	1.07	0.62
vanadium	V	---	---	0.26	---
chromium	Cr	---	---	2.04	---
manganese	Mn	---	1.34	2.09	0.83
iron	Fe	92.4	91.2	79.8	51.0

FIGURE 8.6 Scanning electron microscopy (SEM) and energy dispersive spectroscopy (EDS) analysis of corrosion deposits where A, B, and C represent internal and D represents an external area on the surface of the pipe.

For some measurements, there were large differences in the duplicate measurements, presumably due to the heterogeneous nature of the samples, however similarities and differences between sampling locations could be discerned for several analytes. All samples were characterized by similar average pH values in the neutral range (pH 6.5–7). The average salinities were higher in locations 1 and 2

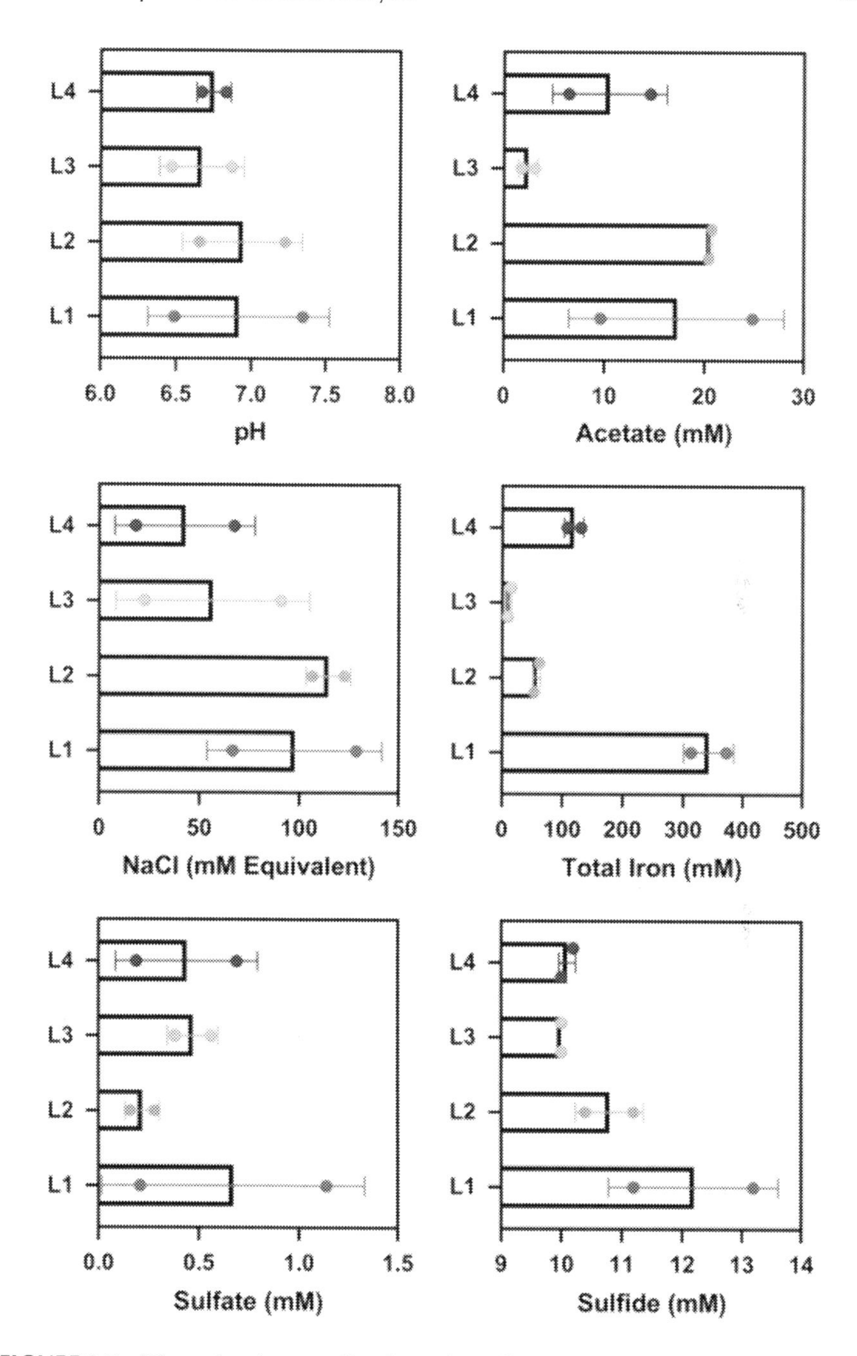

FIGURE 8.7 Water chemistry profile of samples collected from location 1 (L1), location 2 (L2), location 3 (L3), and location 4 (L4). The dots represent the data recorded for each of the duplicate samples collected and analyzed from each location and the horizontal bar represents the average value for the duplicate samples.

(98 and 115 mM equivalents of NaCl, respectively) compared to locations 3 and 4 (57 and 42 mM equivalents of NaCl, respectively), potentially contributing to the corrosion observed at the leak-associated sites compared to the non-corroded areas. However, the overlapping error bars associated with the duplicate salinity measurements amongst all samples makes it difficult to attribute salinity as a factor contributing to corrosion at the leak site. Propionate, butyrate, nitrate, and nitrite were not detected in any of the samples. Sulfate was detected in all samples, but due to error, differences could not be discerned between samples (Figure 8.7). However, there were significant differences measured for acetate, total iron, and sulfide values measured in the non-corroded location 3 samples compared to the leak location samples (locations 1 and 2). The concentration of total iron was lowest in the non-corroded Location 3 sample (13 mM), compared to higher iron levels measured in leak location 1 (343 mM) and adjacent to leak location 2 (58 mM). Increased amounts of total iron in the leak-associated samples suggested increased iron metal dissolution, aligning with the location of the pipe perforation. Average acetate concentrations were also significantly higher in the leak-associated samples (17 mM and 21 mM for locations 1 and 2, respectively), compared to the non-corroded location 3 sample (2 mM). Increased amounts of acetate, a by-product of many metabolic reactions, including acid-producing fermentation reactions, in the leak location samples suggest that enhanced microbial activity occurred in these locations, positively correlating with the leak site. Dissolved sulfide values were also significantly higher in leak-associated location 1 and 2 samples (averaging 11.5 mM) compared to non-corroded location 3 (10 mM) (Figure 8.7), suggesting enhanced activity of sulfide-producing microorganisms at the leak location. Location 4 samples, which consisted of sediments collected from a section of the pipe near to the leak site, also showed elevated levels of total iron (119 mM) and acetate (10 mM) compared to the non-corroded location 3 samples (Figure 8.7). Though these were collected from a non-corroded segment of the cutout near the leak site, pipe sediments are typically an accumulation of solids that originated from different locations. Thus, these kinds of samples can contain corrosion product and other chemicals, a likely explanation for the elevated levels of iron and acetate measured in the location 4 samples. XRD measurements were also conducted to help provide some insight into the mechanism of corrosion, as some products can be indicative of microbial corrosion rather than (or in addition to) chemical corrosion. Using this analysis, the non-corroded location 3 sample showed the presence of quartz (55–65%), calcite (1–10%), paraffin (5–15%), mackinawite (1–10%) and other unidentified compounds. In contrast, samples collected from the corrosion-affected locations 1 and 2 showed a comparatively high abundance of siderite ($FeCO_3$: 45–55% in location 1 and 20–30% in location 2). Higher amounts of calcite were also detected in these corroded area samples. Iron (1–10%) was also detected in the location 2 sample. The higher abundance of siderite ($FeCO_3$) in the samples at or next to the leak site can be indicative of the action of corrosive microorganisms known to directly attack the surface of metal and withdraw electrons from their surface (e.g., via EMIC). This phenomenon is known to lead to the generation of excess ferrous ions, which can combine with carbonate to form $FeCO_3$. Siderite has been associated with biological iron cycling and is reported to

be produced by direct enzymatic reduction of Fe^{3+} by SRP (Pye et al. 1990; Usher et al. 2015). Notably, no FeS could be detected in any sample by XRD analysis. As FeS is a common, corrosive product observed in MIC due to the action of SRP (Enning and Garrelfs 2014), its absence suggests that this type of microbial metabolism did not contribute to the failure in this case study.

8.3.3 Microbial Community Analysis Using the Preserved Samples Collected from Locations 1–4

Although traditional, culture-based methods used for the detection and enumeration of sulfate-reducing bacteria, acid-producing bacteria, and general heterotrophic bacteria in oil and gas samples have been widely used, a major limitation of this approach is that they identify less than 1% of the microbial population present in a given sample (Konopka 2009). This is because most microorganisms are not cultivable in traditionally designed growth media and their electron donor/acceptor capabilities can be very different from what is provided through these growth media. Alternatively, the use of molecular microbiological methods that do not rely on cultivation are better able to capture the diversity of microorganisms within a given sample (Eckert and Skovhus 2018). Beale et al. (2016) provided a comprehensive review on the recent progresses made in a number of fields that have used omics-based applications to improve the fundamental understanding of biofilms and MIC processes. In the current study, 16S rRNA gene sequencing was thus used to identify the kinds of microorganisms associated with the collected samples.

Results procured from the bioinformatics analysis of samples amplified using the V3–V4 set of primers are shown as a heat map in Figure 8.8. Notably, all samples were comprised of microorganisms that were indicative of an anoxic environment. All samples were comprised predominantly of fermentative microorganisms (e.g., known to produce organic acids such as acetate, including acetogens), methanogens, sulfur-/thiosulfate-reducers, and known biofilm-forming organisms – all metabolisms that are known to play a role in MIC (Kip and Veen 2015; Sharma and Voordouw 2016; Vigneron et al. 2016, 2018). Although the relative abundances of most of the microbial taxa identified were similar across all locations (Figure 8.8), there were a few key differences in the microbial community compositions in samples collected from leak-associated locations 1 and 2 versus those collected from non-corroded locations 3 and 4. Most notably, the relative abundances of *Acetobacterium*, an acetate-producing anaerobe, was substantially higher (Figure 8.8). This observation correlates with the higher average concentrations of acetate found in locations 1 and 2 (18 mM) versus locations 3 and 4 (6.5 mM). Acetogens such as *Acetobacterium* have been previously shown to be involved in MIC by producing acetate from H_2 and CO_2, which can subsequently be used by other microorganisms (Mand et al. 2014) and have been detected in other oilfield facilities were MIC was a concern (Duncan et al. 2009). Further, some acetogenic species have also reported to be EMIC organisms that can directly use iron as an electron donor, facilitating microbial corrosion (Kato 2016). Another major difference in microbial community composition was an increased abundance of *Actinobacteria*. These organisms are obligate anaerobes able to attach to various surfaces; they have been previously associated with small nodular

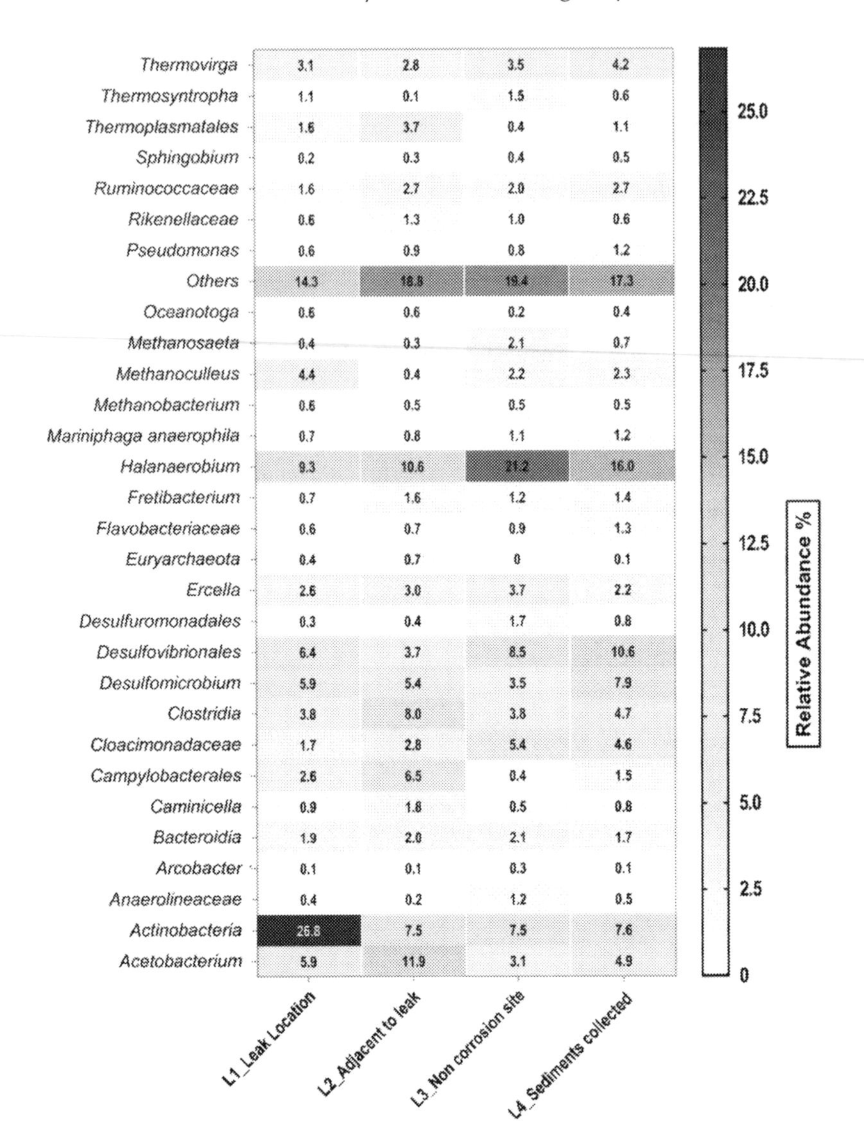

FIGURE 8.8 Heat map showing the percent relative abundance of microbial taxa in the chemically preserved samples collected from the four locations on the pipe segment.

structures in corroding systems (Chen et al. 2013), as well as with corroded water-transporting pipelines (Park et al. 2011). In addition, members of the *Actinobacteria* have been found in offshore and onshore oil production facilities in Nigeria that experienced corrosion and souring issues (Okoro et al. 2014). Some members of the *Actinobacteria*, such as *Coriobacteriaceae,* may also contribute to MIC either directly by fermenting a variety of substances to organic acids, or indirectly in conjunction with methanogens on biocathodes for generating methane from CO_2 by aiding in electron uptake (Wojcieszak et al. 2017; Kobayashi et al. 2017). Though knowledge is still limited, some members of the identified taxa are known as electrical MIC (EMIC) microbes that have the ability to use electrons directly from carbon steel to drive microbial reduction reactions (such as acetogenesis or methanogenesis). These processes are known to yield $FeCO_3$ as a major corrosion product, which also aligned closely with the increased abundance of this mineral at leak-associated locations 1 and 2. Notably, sulfate-reducers (SRP) were found in low relative abundance in microbial community profiles at or near the leak site, in agreement with the XRD measurements showing a lack of FeS minerals at the leak-associated locations.

8.4 CONCLUSIONS

A pinhole leak that occurred on a bypass segment of a crude oil transporting pipeline was investigated using a multi-faceted approach that included microbiological analysis alongside chemical and metallurgical analysis. As this pipe segment experienced intermittent low flow, it was hypothesized that the failure was due to UDC and potentially MIC. The failure allowed for the opportunity to collect samples for microbiological analysis immediately following the pipe cutout. Sampling in several locations (leak site, adjacent to the leak site, and at non-corroded areas) allowed for comparative analyses with respect to microbial community composition and the chemical environment that could potentially indicate the involvement of microorganisms. The physical/metallurgical analysis revealed that the geometry of the pipe segment was conducive to accumulation of sediment in the vicinity of the leak. While the weld metal was more strongly affected by corrosion than the base metal the high localization and rate of corrosion (~0.75 mm/yr) could not be explained by a purely abiotic corrosion mechanism. The XRD results showing an enrichment of $FeCO_3$ at/near the leak site, elevated levels of acetate (17–20 mM) and total dissolved iron (304 mM) at/near the leak site compared to the non-leak location (2 mM acetate, 13 mM iron). Further, the dominant microbial taxa present in the leak-associated samples compared to non-corroded samples showed that fermentative organisms producing organic acids, biofilm-forming organisms, and possibly organisms able to withdraw electrons directly from the steel contributed to the corrosion at the bypass segment. This case exemplified how a microbial analysis alongside other more commonly used metallurgical analysis can help to pinpoint MIC as at least an important contributor to dead-leg failures, all of which supported the hypothesis that the failure was due to UDC and MIC.

Chemical conditions

- Dissolved H_2S: 0 (typical range 0 to 20 mg/l)
- BS&W or Water Cut: 0.35% (typical range 0 to 0.5%)
- The chemical analysis and mechanical testing (tensile testing and Charpy V-notch testing) of the surface of bulk and weld metal pipe-cutout were within the range of industry standards of grade

Microbiological conditions

- MIC and/or EMIC mechanisms are suspected to have resulted in elevated levels of $FeCO_3$, acetate, and salinity at and adjacent to the point of failure, all of which can accelerate UDC
- Samples collected from the four locations in and around the failure site showed a predominantly anaerobic microbial community, comprised of fermentative

X52 and Grade B eliminating the possibility of metal composition or mechanical deformity of the pipe and weld as the cause of failure.

- The unpreserved sludge scraped from the four locations on the failure pipe was also used for water chemistry and XRD analysis of corrosion products.
    a. Increased concentrations of total iron (200 mM), acetate (18 mM), and sulfide (11.5 mM) were found in corrosion-affected locations 1 and 2 as compared to non-affected locations 3 and 4 (iron: 65 mM; acetate: 6.5 mM; sulfide: 10 mM).
    b. Iron compounds that are known to form abiotically and via microbiologically influenced corrosion (MIC) such as siderite ($FeCO_3$) were found in comparatively high amounts corrosion-affected locations 1 and 2 but not in location 3 samples (non-corroded area), indicating the possible role of microorganisms in corrosion.

microorganisms (e.g., known to produce organic acids such as acetate, including acetogens), methanogens, sulfur-/thiosulfate-reducers, and known biofilm-forming organisms – all metabolisms that are known to play a role in MIC.

- Some members of the identified taxa are known to have the ability to use electrons directly from carbon steel to drive microbial reduction reactions (such as sulfate reduction, acetogenesis, or methanogenesis). These processes are known to yield $FeCO_3$ as a major corrosion product, as seen in the XRD analysis.

Corrosion and metallurgical information

- Age of the pipeline: 13 years
- Material: API 5L Steel, grade B, 9.5 mm w.t.
- No signs of cracking and brittle damage were observed in the region of failure
- Ovality of the pipe resulted in increased concavity at 6 o'clock position that may have resulted in accumulation of liquids, other deposits, and biofilm formation at the lowest point of the pipe.
- Failure of external coating in this area close to the perforation may have allowed the fluid accumulation in the area between the metal and the coating
- Deposits rich in Fe-, Si-, S,- Cl-, and Ca-enabled UDC cells to form along the bottom of the pipe.
- Texture of the corrosion product ranged from powdery, chalky, or flaky to dense, matted, or glassy. The range in colors and textures indicates changing conditions and suggests multiple corrosion mechanisms.

Design and operation information

- Perforations were seen on the bypass/dead-leg line for booster pumps of a crude oil transporting pipeline.
- This line has laminar flow with tank-to-tank transfer with a frequency of less than a month.
- Leak occurred at the 6 o'clock position on a dead leg segment
- Pressure at time of collection: 1.01 bar (typical operating pressure range 1.7 to 2 bar)
- Temperature: sample temperature at collection was 25°C (typical operating temperature range 10–38°C)
- Maximum Operating Pressure (MOP): The MOP of a pipeline in a given location based on its design factors and hydrotest pressures was 275 psi.

REFERENCES

Bagaria, H., M. Sharma, R. Gutierrez, A. Oliver, J. Sargent, T. Place, et al. 2019. Synergistic effect of biocide and corrosion inhibitor in mitigation of microbiologically influenced corrosion in wet parked crude oil pipelines. NACE Northern Area Western Conference, paper no. 03-1419.

Beale, D. J., A. V. Karpe, S. Jadhav, T. H. Muster, and E. A. Palombo. 2016. Omics-based approaches and their use in the assessment of microbial-influenced corrosion of metals. *Corrosion Reviews* 34: 1–15.

Bensman, L. 2016. Dead leg internal corrosion management. NACE CORROSION 2016, paper no. 7715.

Bolyen, E., J. R. Rideout, M. R. Dillon, N. A. Bokulich, C. C. Abnet, G. A. Al-Ghalith, et al. 2019. Reproducible, interactive, scalable and extensible microbiome data science using QIIME 2. *Nature Biotechnology* 37: 852–857.

Chen, L., R. B. Jia, and L. Li. 2013. Bacterial community of iron tubercles from a drinking water distribution system and its occurrence in stagnant tap water. *Environmental Science: Processes & Impacts* 15: 1332–1340.

De Paula, R., C. St. Peter, A. Richardson, J. Bracey, E. Heaver, K. Duncan, et al. 2018. DNA sequencing of oilfield samples: impact of protocol choices on the microbiological conclusions. NACE CORROSION 2018, paper no. 11662.

Duncan, K. E., L. M. Gieg, V. A. Parisi, R. S. Tanner, S. G. Tringe, J. Bristow, et al. 2009. Biocorrosive thermophilic microbial communities in Alaskan North Slope oil facilities. *Environmental Science and Technology* 43: 7977–7984.

Eckert, R. B. 2016. Field Guide to Internal Corrosion Mitigation and Monitoring for Pipelines. NACE International, Houston, TX, p. 343.

Eckert, R. B., and T. L. Skovhus. 2018. Advances in the application of molecular microbiological methods in the oil and gas industry and links to microbiologically influenced corrosion. *International Biodeterioration and Biodegradation* 126: 169–176.

Enning, D., and J. Garrelfs. 2014. Corrosion of iron by sulfate-reducing bacteria: new views of an old problem. *Applied and Environmental Microbiology* 80: 1226–12360.

Gonzalez, J. E., N. J. R. Kraakman, Y. C. Wang, L. Vorreiter, J. Cesca, and T. Nguyen. 2014. Sydney Water's corrosion and odour management tool: a new process for selecting the most cost-effective corrosion and odour measures. *Water* 41: 65–71.

Hashemi, S. J., N. Bak, F. Khan, K. Hawboldt, L. Lefsrud, and J. Wolodko. 2018. Bibliometric analysis of microbiologically influenced corrosion (MIC) of oil and gas engineering systems. *Corrosion* 74(4), 468–486.

Kagarise, C., R. B. Eckert, and J. R. Vera. 2017. The importance of deposit characterization in mitigating UDC and MIC in dead legs. NACE CORROSION 2017, paper no. 9128.

Kato, S. 2016. Microbial extracellular electron transfer and its relevance to iron corrosion. *Microbial Biotechnology* 9: 141–148.

Kilbane, J. 2014. Effect of sample storage conditions on oilfield microbiological samples. NACE CORROSION 2014, paper no. 3788.

Kip, N., and J. A. van Veen. 2015. The dual role of microbes in corrosion. *ISME Journal* 9: 542.

Kobayashi H., Q. Fu, H. Maeda, and K. Sato. 2017. Draft genome sequence of novel *Coriobacteriaceae* sp. strain EMTCatB1, reconstructed from the metagenome of a thermophilic electromethanogenic biocathode. *Genome Announcements* 5: e00022–e00017.

Koch, G., J. Varney, N. Thompson, O. Moghissi, M. Gould, and J. Payer. 2016. *NACE IMPACT – International Measures of Prevention, Application, and Economics of Corrosion Technologies – Study*. NACE International, Houston, TX.

Konopka, A. 2009. What is microbial community ecology? *ISME Journal* 3: 1223–1230.

Lanneluc, I., M. Langumier, R. Sabot, M. Jeannin, P. Refait, and S. Sablé. 2015. On the bacterial communities associated with the corrosion product layer during the early stages of marine corrosion of carbon steel. *International Biodeterioration & Biodegradation* 99: 55–65.

Larsen, J., and L. R. Hilbert. 2014. Investigation into Under Deposit Corrosion in Halfdan Production Tubulars, *NACE CORROSION 2014*, paper no. 3746, NACE International, Houston, TX, USA.

Little, B. J., D. J. Blackwood, J. Hinks, F. M. Lauro, E. Marsili, A. Okamoto, et al. 2020. Microbially influenced corrosion – any progress? *Corrosion Science* 170: 108641.

Liu, T., Y. F. Cheng, M. Sharma, and G. Voordouw, G. 2017. Effect of fluid flow on biofilm formation and microbiologically influenced corrosion of pipelines in oilfield produced water. *Journal of Petroleum Science and Engineering* 156: 451–459.

Liu, H., M. Sharma, J. Wang, Y. F. Cheng, and H. Liu. 2018. Microbiologically influenced corrosion of 316L stainless steel in the presence of *Chlorella vulgaris*. *International Biodeterioration & Biodegradation* 129: 209–216.

Mand, J., H. S. Park, T. R. Jack, and G. Voordouw. 2014. The role of acetogens in microbially influenced corrosion of steel. *Frontiers in Microbiology* 5: 268.

Okoro, C., S. Smith, L. Chiejina, R. Lumactud, D. An, H. S. Park, et al. 2014. Comparison of microbial communities involved in souring and corrosion in offshore and onshore oil production facilities in Nigeria. *Journal of Industrial Microbiology & Biotechnology* 41: 665–678.

Ossai, C. I., B. Boswell, and I. J. Davies. 2015. Pipeline failures in corrosive environments – A conceptual analysis of trends and effects. *Engineering Failure Analysis* 53: 36–58.

Park, H. S., I. Chatterjee, X. Dong, S. H. Wang, C. W. Sensen, S. M. Caffrey, et al. 2011. Effect of sodium bisulfite injection on the microbial community composition in a brackish-water-transporting pipeline. *Applied and Environmental Microbiology* 77: 6908–6917.

Pye, K., J. A. D. Dickson, N. Schiavon, M. L. Coleman, and M. Cox. 1990. Formation of siderite-Mg-calcite-iron sulphide concretions in intertidal marsh and sandflat sediments, North Norfolk, England. *Sedimentology* 37: 325–343.

Rachel, N. M., and L. M. Gieg. 2020. Preserving microbial community integrity in oilfield produced water. *Frontiers in Microbiology* 11: 581387.

Rajala, P., M. Bomberg, M. Vepsäläinen, and L. Carpén. 2017. Microbial fouling and corrosion of carbon steel in deep anoxic alkaline groundwater. *Biofouling* 33: 195–209.

Sharma, M., D. An, K. Baxter, M. Henderson, L. Edillon, and G. Voordouw. 2016. Understanding the role of microbes in frequent coiled tubing failures. NACE CORROSION 2016, paper no. 7815.

Sharma, M., D. An, T. Liu, T. Pinnock, F. Cheng, and G. Voordouw. 2017. Biocide-mediated corrosion of coiled tubing. *PLoS One* 12: e0181934.

Sharma, M., J. Handy, D. An, G. Voordouw, and L. M. Gieg. 2019. Characterization of microbiologically influenced corrosion potential in nitrate injected produced waters. NACE CORROSION 2019, paper no. 13198.

Sharma N., and W. Huang. 2019. Rapid in-field collection and ambient temperature preservation of corrosion-related microbial samples for downstream molecular analysis. In: *Oilfield Microbiology*, Eds. T. L. Skovhus and C. Whitby. CRC Press, Boca Raton, FL.

Sharma, M., H. Liu, S. Chen, F. Cheng, G. Voordouw, and L. M. Gieg. 2018a. Effect of selected biocides on microbiologically influenced corrosion caused by *Desulfovibrio ferrophilus* IS5. *Scientific Reports* 8: 1–12.

Sharma, M., H. Liu, N. Tsesmetzis, J. Handy, A. Kapronczai, T. Place, et al. 2020. Diagnosing microbiologically influenced corrosion at a crude oil pipeline facility leak site – a holistic approach. *Corrosion* 62: 13.

Sharma, M., P. Menon, J. Voordouw, Y. Shen, and G. Voordouw. 2018b. Effect of long term application of tetrakis (hydroxymethyl) phosphonium sulfate (THPS) in a light oil-producing oilfield. *Biofouling* 34: 605–617.

Sharma, M., and G. Voordouw. 2016. MIC detection and assessment - A holistic approach. In: *Microbiologically Influenced Corrosion in the Upstream Oil and Gas Industry*, Eds. T. L. Skovhus, D. Enning, and J. S. Lee. CRC Press, Boca Raton, FL, pp. 177–212.

Shi, X., N. Xie, and J. Gong. 2011. Recent progress in the research on microbially influenced corrosion: a bird's eye view through the engineering lens. *Recent Patents on Corrosion Science* 1: 118–131.

Skovhus, T. L., R. B. Eckert, and E. Rodrigues. 2017. Management and control of microbiologically influenced corrosion (MIC) in the oil and gas industry – Overview and a North Sea case study. *Journal of Biotechnology* 256: 31–45.

Suarez, E., L. Machuca, and K. Lepkova. 2019. The role of bacteria in under-deposit corrosion in oil and gas facilities: a review of mechanisms, test methods and corrosion inhibition. *Corrosion & Materials* February: 80–87.

Tidwell, T. J., R. De Paula, Z. Broussard, and V. V. Keasler. 2017. Mitigation of severe pitting corrosion caused by MIC in a CDC biofilm reactor. NACE CORROSION 2017, paper no. 9604.

Usher, K. M., A. H. Kaksonen, D. Bouquet, K. Y. Cheng, Y. Geste, P. G. Chapman, et al. 2015. The role of bacterial communities and carbon dioxide on the corrosion of steel. *Corrosion Science* 98: 354–365.

Vera, J. R., D. Daniels, and M. H. Achor. 2012. Under deposit corrosion (UDC) in the oil and gas industry: a review of mechanisms, testing and mitigation. NACE CORROSION 2012, paper no. C2012-0001379.

Videla, H. A., and L. K. Herrera. 2005. Microbiologically influenced corrosion: looking to the future. *International Microbiology* 8: 169.

Vigneron, A., E. B. Alsop, B. Chambers, B. P. Lomans, I. M. Head, and N. Tsesmetzis. 2016. Complementary microorganisms in highly corrosive biofilms from an offshore oil production facility. *Applied and Environmental Microbiology* 82: 2545–2554.

Vigneron, A., I. M. Head, and N. Tsesmetzis. 2018. Damage to offshore production facilities by corrosive microbial biofilms. *Applied Microbiology and Biotechnology* 102: 2525–2533.

Voordouw, G., P. Menon, T. Pinnock, M. Sharma, Y. Shen, A. Venturelli, et al. 2016. Use of homogeneously-sized carbon steel ball bearings to study microbially-influenced corrosion in oil field samples. *Frontiers in Microbiology* 7: 351.

Wojcieszak, M., A. Pyzik, K. Poszytek, P. S. Krawczyk, A. Sobczak, L. Lipinski, et al. 2017. Adaptation of methanogenic inocula to anaerobic digestion of maize silage. *Frontiers in Microbiology* 8: 1881.

9 Failure Analysis of Microbiologically Influenced Corrosion in Storage Tanks Containing B20 Biodiesel

Bradley S. Stevenson
University of Oklahoma

Blake W. Stamps
Air Force Research Laboratory and UES Inc.

James G. Floyd
University of Oklahoma

Caitlin L. Bojanowski and Wendy J. Goodson
Air Force Research Laboratory

CONTENTS

DOI: 10.1201/9780429355479-11

9.1 INTRODUCTION

In the mid- to late 2000s, the U.S. Department of Defense (DOD) used record amounts of fuel. High, fluctuating fuel prices, foreign oil dependencies, and subsequent government mandates prompted the DOD to increase their use of alternative fuels such as biodiesel. The U.S. Air Force (USAF) set aspirational goals to increase their use of alternative fuels by 10% each year through 2015 ("Air Force Infrastructure Energy Plan" 2010). To meet alternative fuel usage goals, the USAF began to use diesel fuel containing 20% biodiesel (B20 biodiesel) in non-emergency, non-tactical, ground vehicles.

Currently, outside of the DOD, the consumption of biodiesel is on the rise worldwide, helped in part by the global effort to remove sulfur compounds from petroleum distillate transportation fuels (e.g., diesel). These sulfur compounds are linked to emissions that are harmful to the environment and human health ("Health Assessment Document For Diesel Engine Exhaust (Final 2002)" 2002, Ingersoll et al. 2003). Since 2006, the U.S. Environmental Protection Agency (EPA) has mandated the use of Ultra Low Sulfur Diesel (ULSD, < 15 ppm sulfur; 40 CFR 80 Subpart 1), whereas the current European Union standard is < 10 ppm sulfur (Euro 5; EN 590:2009).

Removing over 97% of the organosulfur compounds from diesel also has some negative consequences. Mainly, the organosulfur compounds removed from ULSD have critical lubricative properties, requiring additives to restore lost lubricity (Hoekman et al. 2012). One additive that restores the lost lubricity of ULSD is biodiesel, composed of straight-chain fatty acid methyl esters (FAME) (Hazrat, Rasul, and Khan 2015). FAME for biodiesel is produced by the esterification of vegetable oils and animal fats with alcohols, usually methanol or ethanol. The U.S. EPA allows up to 5% biodiesel v/v in ULSD, while the EU allows up to 7% v/v (EN 590:2009). In addition to its use as a fuel additive, biodiesel is also used as a fuel extender (i.e., blended fuels with ULSD at higher percentages). Blended fuel, containing up to 20% v/v biodiesel (i.e., B20 biodiesel; 20% biodiesel, 80% ULSD) can reduce the impact of diesel-burning engines on harmful emissions and the global carbon budget, while not requiring modification of existing storage infrastructure and engines (Özener et al. 2014).

Despite the benefits of biodiesel as an ULSD additive or extender, its introduction correlates with anecdotal reports of tank contamination and corrosion (US EPA, OLEM 2016). According to the Air Force Petroleum Office (personal communication), soon after the USAF began adopting B20 biodiesel, many bases reported issues with particulates in fuel. Fuel fouling was recurrent at many bases even after the removal of contaminated fuel and cleaning of storage tanks by pressure washing and physical removal of biomass Wendy Goodson, personal communication. Fuels can become fouled with waxes and other non-biological compounds (Jakeria, Fazal, and

Haseeb 2014, Leung, Koo, and Guo 2006), but microbiological contamination is an ever-present threat.

Another potentially serious threat posed by microorganisms to tank integrity is microbiologically influenced corrosion (MIC). Soon after the adoption of ULSD (most of which contained up to 5% biodiesel), owners of underground storage tanks (UST) began reporting rapid and severe corrosion of metal components (US EPA, OLEM 2016). Observations suggested that corrosion may be common and severe on metal surfaces in the upper vapor space of underground storage tanks (UST), an area that was not known to be prone to corrosion before 2007 (Nelson, Lapara, and Novak 2010, Sowards and Mansfield 2014, Williamson et al. 2015). Vapor phase corrosion resulted in flaking of metallic surfaces; flakes resembling coffee grounds began to appear in fuel filters and in fuel samples removed from UST bottoms. Furthermore, it appeared many operators were not aware of the presence or severity of corrosion within their UST systems (US EPA, OLEM 2016). At least 83% of the tanks inspected by the EPA had visible corrosion that was moderate or severe with regard to the presence of turbercles and their distribution, yet less than 25% of owners reported knowledge of corrosion in their UST systems. It would appear that severe and rapid corrosion has been occurring in USTs storing ULSD across the country since soon after the fuel was introduced in 2006 (US EPA, OLEM 2016). In a study of 42 USTs containing ULSD, the most common predictive (but not statistically significant) factors for moderate to severe corrosion were particulates (i.e., corrosion products) and water entrained in the fuel.

Fuel storage tanks are often contaminated with microorganisms (largely bacteria and fungi) that might enter in several ways. As atmospheric storage tanks, they are not air-tight. By design, the tank headspace and outside atmosphere interact to compensate for changing temperatures and air pressure, or when fuel is dispensed. Suspended microorganisms (cells or spores) or those attached to airborne particles can be carried in through the venting system required to maintain and equilibrate pressure with the atmosphere. Microorganisms may also enter the tank with any liquid water that may infiltrate through poorly sealed closures. Lastly, microorganisms can gain access to the storage tank as contaminants of instrumentation that enters the tank during routine operation or maintenance (Jia et al. 2019). As evidence of communication between the storage tank and outside environment, samples at one USAF base in the spring contained abundant tree pollen detected in molecular analyses. Any immigration of microorganisms represents a contamination threat, one in which the stored fuel can act as a suitable oxidizable substrate for growth.

Here, we present a failure analysis and case study of B20 biodiesel USTs to illustrate how a fuel storage tank may become contaminated with microorganisms and water. Proliferation of the microorganisms can lead to degradation of the fuel, production of acids and extensive biofilms, contributing to MIC.

9.2 FAILURE ANALYSIS

We attempted to determine whether reports of fuel fouling at several USAF facilities were biologically derived, and if so, whether this represented an increased corrosion risk. The data presented here represent a failure analysis of contaminated B20 biodiesel tanks at a single USAF base in the southeastern United States. The focus of these analyses was to determine 1) whether a diesel/biodiesel storage tank was contaminated with microbiological biomass, 2) what impact this biomass had on fuel quality, and 3) how this microbial contamination affected corrosion in the tank. Based on these findings, a broader, more comprehensive longitudinal study was later carried out that further identified the microbiota in fuels and measured the corrosion occurring in multiple tanks at two geographically distinct locations (Stamps et al. 2020).

9.2.1 Characterizing Biomass Fouling of Fuel and Surfaces

A first, critical step in failure analysis of MIC in USTs is to establish the location, nature, extent, and composition of biomass in the tank. The two B20 biodiesel USTs that were the focus of this failure analysis were located at a USAF base in the southeastern United States. Both tanks were located at the same filling station and shared lines that led to two dispensers. The tanks were 10,000 gallon, double-walled, fiberglass USTs, with topside openings for a manway, automatic tank gauge (ATG), and fill tube.

Fuel management personnel recollected previous and recurring incidents of fuel fouling, or "sludge," that required replacement of fuel. Samples of fuel from these tanks were taken using a tank sampler, commonly known as a Bacon Bomb (Figure 9.1A). This device allows samples to be taken from the bottom of tanks upon contact of the tank bottom with a piston mechanism, or at any intermediary depth by triggering the piston with a fuel-resistant string. While fuel samples from the dispensers and upper levels of the storage tank were "clear and bright," bottom samples from the tanks were cloudy and contained water, an intermediate "sludge" layer, and fuel (Figure 9.1B). The presence of cloudy fuel and sludge at the interphase between water and fuel could be biofilm/biomass, entrained water, or fatty acid waxes

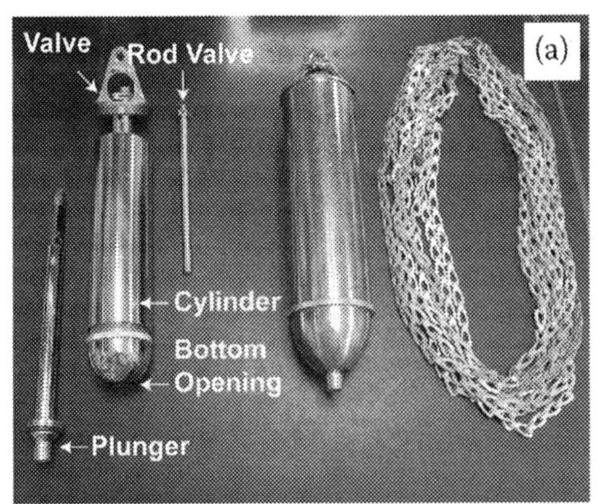

FIGURE 9.1 Example of common fuel sampling device known as a "Bacon Bomb" (A). Bottom fuel samples recovered by the Bacon Bomb were often cloudy and heavily fouled with obvious separation of fuel and water layers (B, left) or were more heavily clouded (B, right). Image credit: Wendy Goodson, Air Force Research Laboratory.

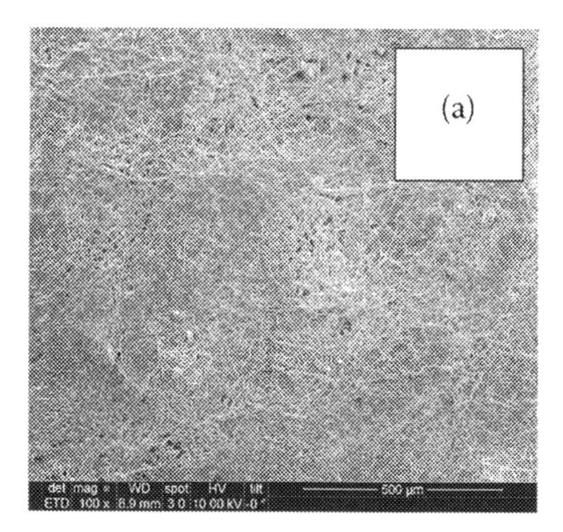

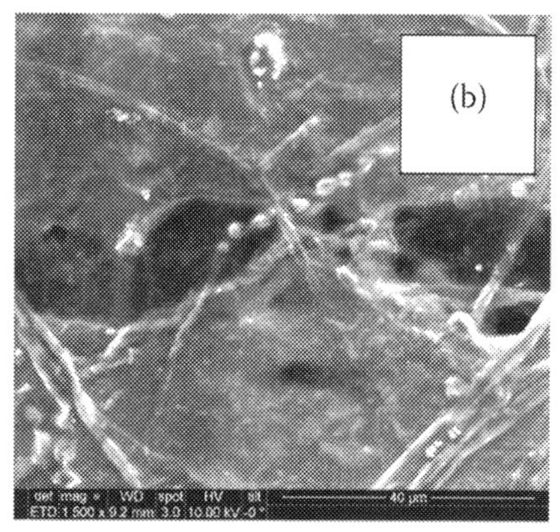

FIGURE 9.2 Scanning electron microscopy images of biofilms recovered from environmental biofilm samples (A). Upon closer examination, fungal hyphae and microbial morphologies were visible (B). Image credit: Pamela Lloyd, UES, Inc.

(Passman 2013). Scanning Electron Microscopy (SEM) imaging of the interphase biomass showed abundant fungal hyphae (Figure 9.2), confirming biological fouling.

Biomass from the fuel and biofilms attached to the ATG and walls of the tank were readily quantifiable and contained an abundance of fungal morphology (discussed in detail below). Universal amplification of small subunit rRNA produced both 16S bacterial rRNA, and 18S rRNA that was primarily fungal. Quantitative PCR results suggested that the population density of these fungal communities was quite high and varied considerably over time (Figure 9.3). In fuel samples, the number of 18S rRNA gene copies varied from less than 10^2 copies/L, to over 10^6 copies/L fuel. "Receipt" fuels were fuel samples taken directly from the fuel delivery truck, prior to being stored within any tank. These fuel samples were apparently devoid of contamination, with almost no quantifiable 18S rRNA genes and very few detectable microbial taxa (Figure 9.3). These receipt samples suggest that large numbers of microorganisms did not come from delivered fuels but could have resulted from microbiota immigrating through either air or water ingress and subsequent, significant growth in the tank.

The two most abundant and commonly observed fungal families in fuel and biofilm samples were the *Saccharomycetaceae* and *Trichocomaceae* (including operational taxonomic units, or OTUs, belonging to the unclassified *Eurotiomycetes*, to which the *Trichocomaceae* belong). While this initial study lasted only four months, these two families of fungi were also observed in a longer, more comprehensive study of fuels and biofilm samples taken both at this base and another in the southwestern United States (Stamps et al. 2020). Even though members of the *Saccharomycetaceae* were abundant in Tank 4 biofilms and Tank 3 fuels (Figure 9.3), we focused on the *Trichocomaceae*, which was more abundant when microbial fouling was greatest, especially in Tank 3. Tank 3 had greater amounts of generalized and pitting corrosion than Tank 4, and filamentous fungi were repeatedly visible (Figure 9.3).

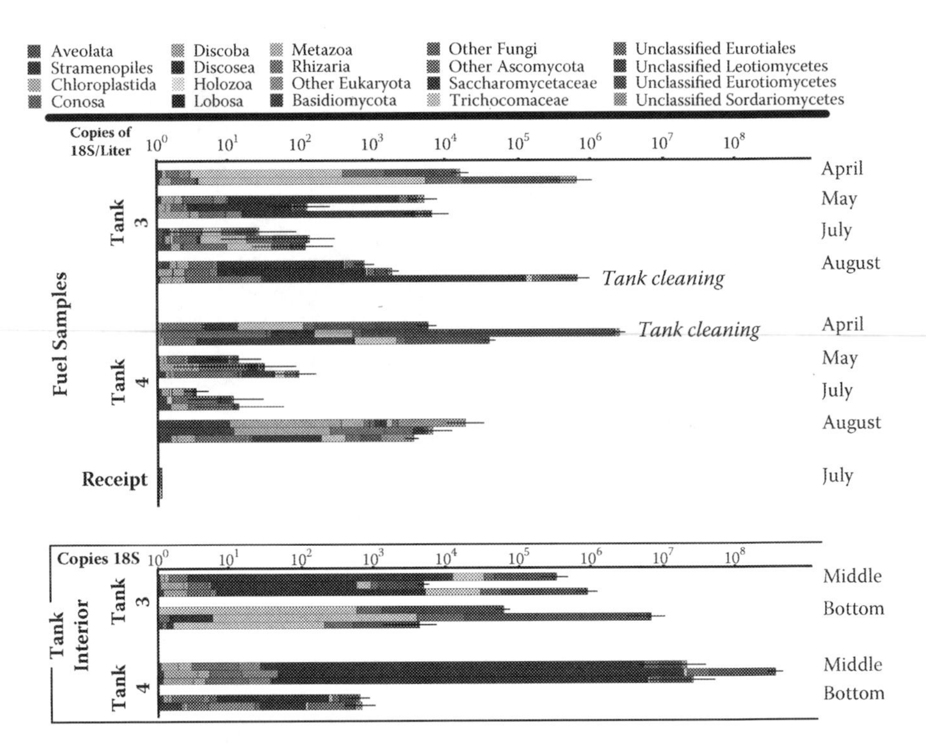

FIGURE 9.3 18S rRNA gene bar chart, normalized to rRNA gene copy number as estimated by quantitative PCR. Error bars shown are for triplicate technical replicates of qPCR, and individual bars at each time point are true biological replicate samples taken in the field.

Using qPCR estimations of fungal community density, the level of fouling in fuel was both dynamic and recurring. This is to be expected in operational fuel systems, in which fuel in the UST is periodically replenished with "clean" fuel received from a production facility, and where UST contamination events result in the implementation of various mitigation strategies. Fuel samples from Tank 4 in Figure 9.3 were taken just prior to the removal of contaminated fuel and pressure washing of the interior surfaces. Within as little as five months, the fungal population density had increased to nearly the same levels seen before the tank cleaning. Fungal population density in Tank 3 was also initially higher in April. About 250 gallons of contaminated fuel was removed from the bottom of Tank 3 with a "stinger" vacuum line at the same time that Tank 4 was emptied and cleaned. Over the course of the five months shown, the fungal population density increased to a level that prompted the disposal of all fuel and a tank cleaning.

9.2.2 Characterization of Fuel Degradation

A profile of the various FAMEs in B20 biodiesel was determined for a fouled fuel sample taken from Tank 3, and a "receipt" fuel sample not exposed to storage conditions using gas chromatography coupled to mass spectrometry (GC-MS).

Major differences were observed in the concentration of FAME components in the fuel (measured as peak height, Figure 9.4). Specifically, receipt fuel had higher concentrations of palmitic, oleic, and linoleic acid methyl esters when compared to the fuel at the bottom Tank 3. The differences between fuel in Tank 3 and the receipt fuel was likely due to the microbial community in Tank 3 metabolizing the FAME compounds, decreasing their overall abundance within the measured GC-MS trace. This initial GC-MS analysis focused on the FAMEs present within the fuel; however, degradation of FAME results in the generation of fatty acids that increase fuel acidity, which elevates the risk of corrosion (Andrade 2016, Stamps et al. 2020). Variations in fuel composition are certainly possible between deliveries, potentially explaining why palmitoleic, heptadecenoic, and steric acid methyl esters were elevated in the Tank 3 fuel sample relative to the receipt sample (Figure 9.4). To add a causative link to the correlation between microbial activity and shifts in fuel composition observed in the field, laboratory-based experiments with fuel-degrading microorganisms and fuel of the same composition were carried out.

A fungal isolate obtained from a USAF base in the Southwest United States but matching the abundant Trichocomaceae found within Tank 3 at the southeastern USAF base (Figure 9.3) was used in a fuel degradation assay. This isolate, *Byssochlamys* sp. SW2 showed preference for palmitic and linoleic acid methyl esters (Andrade 2016). Another common fuel contaminant, *Wickerhamomyces anomalus* was also isolated and tested, again showing the ability to degrade FAME,

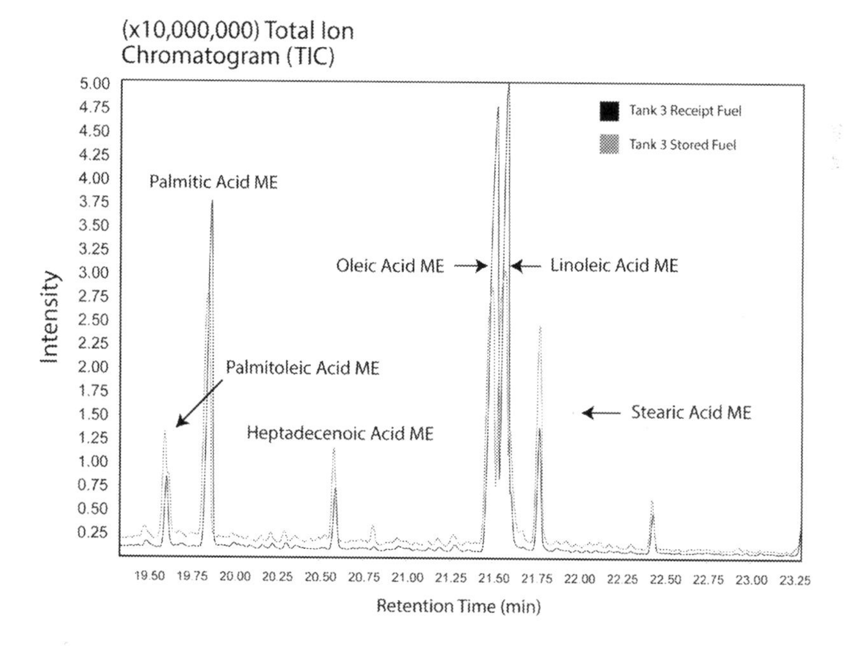

FIGURE 9.4 Fatty Acid Methyl Ester (FAME) profile of a B20 fuel sample taken prior to storage in SE Tank 3 (Black line,"Tank 3 Receipt Fuel) and fuel after exposure to the microbiota present within the tank (Red line, "Tank 3 Stored Fuel").

but also hydrocarbons from the ULSD (Andrade 2016). By combining field-based studies with careful laboratory experimentation performed with environmental isolates, it was possible to confirm that the abundant microorganisms present in B20 tanks can preferentially degrade FAME, resulting in increased fuel acidity.

9.2.3 Assessment of Corrosion Threat

The overall goal of any microbiome driven failure analysis should be to link specific microbiological populations, their abundance, and their activities to rates and locations of corrosion. It is important, therefore, to assess the level of corrosion that is presumed to be MIC vs abiotic corrosion mechanisms. The extent of corrosion, presumed to be MIC, in the storage tanks was measured over the course of one year using 1010 carbon steel witness coupons. These coupons were attached to racks lowered to the bottom, exposing them to fuel and biomass very near the bottom as well as about 18 inches off the bottom of the UST (Figure 9.5).

The coupons were pre-weighed and cleaned with isopropyl alcohol prior to their attachment to the rack. Once installed, the rack remained undisturbed until sampling occurred at roughly three-month intervals. The results of this survey were previously reported by Stamps et al. (2020) and showed that corrosion measured by mass loss was less than 100 mg at both sampling locations with the exception of one tank at SW which experienced greater than 200 mg mass loss per coupon. Calculation of corrosion rates (in mils per year, or MPY) using profilometry data was more informative, as much higher MPY were identified in pitting compared to generalized surface corrosion. Corrosion rates of over 7 MPY were observed in the tank discussed as a part of this failure analysis (Stamps et al. 2020), which was higher than those observed for other tanks in the system (>3 and 4 MPY, respectively.)

An important parameter needed to identify and measure pitting corrosion, is the definition of a "pit." In our study, pits were defined as any points that were greater than 20 μm below the mean surface average. Maximum pit depth and total pitted area were then calculated from profilometry surface analyses. To more firmly establish pitting corrosion as a generalizable term, more discussion is needed to establish unified standards, ensuring that measurements made across studies and research groups can be directly compared.

9.3 DISCUSSION

Throughout this failure analysis, we were able to measure the fate of contaminated B20 biodiesel and its impact on the UST. Even after physical tank cleaning, fungal biomass quickly recovered, and the composition of the extant community shifted in a way that was suggestive of a "bloom" of a few fungi. Two abundant members of the fungal community were isolated and were capable of growing on B20 biodiesel as their sole carbon and energy source, preferentially degrading major FAME components of the fuel (Andrade 2016). This increase in living fungal biomass also elevated fuel acidity, and finally, increased the threat of corrosion, as demonstrated in a one-year survey (Stamps et al. 2020).

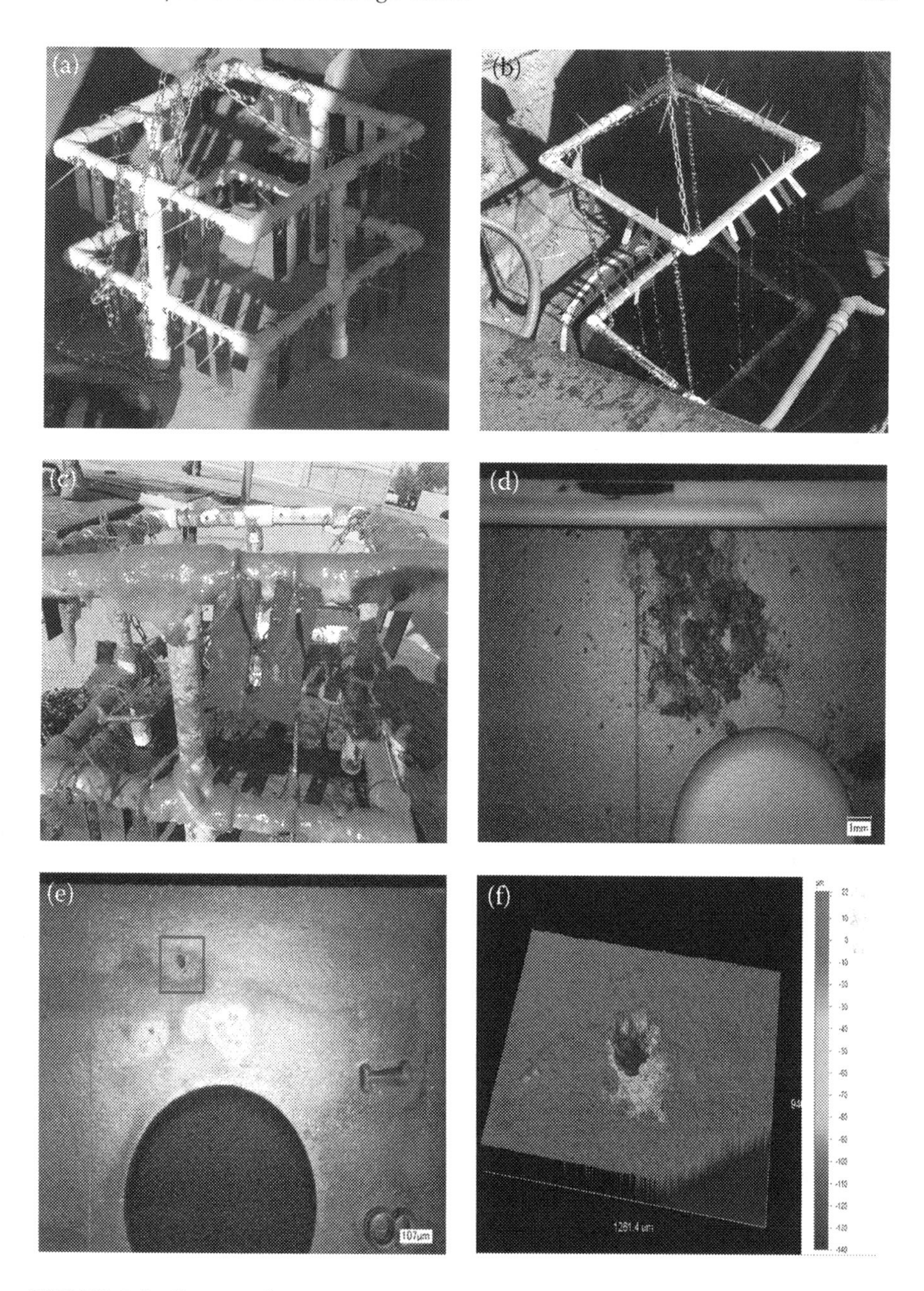

FIGURE 9.5 Images of coupon *in situ* sampling "rigs" prior to incubation (A) and after exposure to the fuel tank microbiome (B 3 months C) 12 months. Clear, viscous biofilms are present across heavily fouled tanks (C). Biofilms on individual coupons (D) and be peeled carefully away, and pits caused by corrosion (E) are visible to the naked eye, and quantifiable by profilometry (F). Image credit (D–F) Carrie Drake, Air Force Research Laboratory and UES, Inc.

Most critical to the initiation of fouling within a tank, and indeed required by all life, is the presence of water. Biodiesel is significantly more hygroscopic than ULSD, accumulating more water during storage (Fregolente, Fregolente, and Maciel 2012). The amount of biodiesel in ULSD is directly corelated with the amount of water that can be obtained from the atmosphere above the fuel and remain entrained in the fuel itself. Operators noted large quantities of water when fouling was worst, which was observable in fuels taken from these tanks (Figure 9.1B). Without water, microorganisms would have no habitat. Microorganisms can exist in microdroplets of water suspended within hydrocarbon fuels, and have even been shown to emulsify fuel, but even more water could support a larger population of microorganisms (Meckenstock et al. 2014). Additionally, metabolism of the fatty acid chains that occurs through beta oxidation would produce copious amounts of water as a by-product. Tank operators have long known the importance of removing water from their storage systems, regardless of how difficult it can be to do so, but biodiesel represents a compounded risk of water accumulation as biological activity generates water as a by-product, and more is accumulated by the hydroscopic nature of the fuel.

In addition to the production of water, microbial metabolic activity has a direct, and quantifiable impact on fuel acidity. The amount of acid that is produced as a by-product of microbial metabolism has an impact on the corrosivity of the fuel, causing generalized corrosion of metal surfaces that are in contact with the fuel and the atmosphere of the storage and distribution system. Acidic microbial metabolites can also become volatilized producing additional corrosion (Williamson et al. 2015, US EPA, OLEM 2016). Further degradation of methanol and fatty acids produced during the oxidation of FAME, produces smaller, more volatile organic acids (e.g., acetic, lactic, propionic acid) and CO_2. These volatile acids can enter the vapor phase in the headspace of the storage tank, produce highly acidic localized conditions when they condense on surfaces, and cause severe corrosion of any metal in these spaces. The fuel tank discussed within this review had considerable increases in fuel acidity (measured in mg KOH/g B20), and at the sampled location, the greatest amount of observable corrosion as previously reported (Stamps et al. 2020). This increase in fuel acidity over time matches another study in which increased acidity, and corrosivity was identified in ULSD tanks across the United States (US EPA, OLEM 2016). Multiple organic acids including acetic, formic, and propionic acid were detected in the headspace of sampled USTs that exhibited varying levels of corrosion (US EPA, OLEM 2016). While this result does not fully confirm direct biological activity, fuel acidity measurements should be considered a standard measure for fuel quality for any biodiesel storage system, using well-established ASTM protocols (D27 Committee 2020). While abiotic fuel hydrolysis could also increase acidity, any increase outside of ASTM specification ("ASTM D6751-20, Standard Specification for Biodiesel Fuel Blend Stock (B100) for Middle Distillate Fuels," 2020, 6751–20) should be cause for concern. Ideally, a test that could rapidly distinguish between abiotic increases in fuel acidity and those caused by If elevated levels of fuel acidity are detected, tank cleaning, fuel polishing, or even complete replacement of fuel should be considered to decrease the threat of corrosion.

Generalized corrosion occurs whenever metal surfaces are exposed to moisture in an aerobic atmosphere. However, the extensive biofilms found attached to surfaces in the B20 storage tanks can increase the acidity in the immediate environment under the biofilm and represent a greater corrosion risk (Stamps et al. 2020). Microbial biomass on the surface of a metal component also allows for a difference in oxygen concentration. The resulting oxygen concentration corrosion cell is established as electrons from oxygenated areas are pulled from the anoxic regions under the biofilm (Norsworthy 2014). Once a biofilm is established, this area becomes a cathode leading to the solubilization of Fe^{2+} under the biofilm, producing a pit. As discussed previously, pitting corrosion was the greater risk to the observed USTs. This type of corrosion is, by its nature, more difficult to identify as it is highly localized. Advancements in visual tank inspection via robotics could assist but such monitoring is time consuming for day-to-day operators of USTs. The use of witness coupons *in situ* is the simplest method to identify pitting corrosion as coupons can be easily removed and inspected at regular time intervals. Likely, the best approach is to perform *in situ* incubation of witness coupons in the pilot deployment of a new or existing biofuel, identify the risks to infrastructure, and devise means of mitigating or reducing these risks prior to broad scale deployment of a new fuel in which the risks of corrosion are unknown.

Currently, there are no standards that are broadly agreed upon for biological sampling and identification within USTs. Molecular surveys of microbial communities establish whom is the likely causative agent of tank contamination, although the diversity of organisms, both fungal (as presented above) and bacterial that can contaminate a biodiesel UST make these conclusions difficult. Advances in small subunit rRNA gene primers allow for the even amplification of all three domains of life (Parada, Needham, and Fuhrman 2016) although the resolution of the rRNA gene for fungi is lower than that of bacteria, and should primarily fungal contamination be suspected, internally transcribed spacer (ITS) sequencing can provide higher resolution (Lindahl et al. 2013). Care must also be taken to prepare sufficient negative controls and extraction blanks as some samples, such as receipt fuels, can be extremely low in biomass. Low biomass samples can produce spurious results if contamination is not controlled for (Salter et al. 2014) but there are bioinformatic analysis methods to rapidly and accurately "decontaminate" low biomass amplicon and metagenomic samples to produce reliable results (Davis et al. 2018). Better still would be real-time metagenomic (or transcriptomic) sequencing to provide in-field results more akin to those in commonly used ATP or fuel acidity tests. Such sequencing approaches are becoming more common in forensics and other built environments (Sim and Chapman 2019, Nygaard et al. 2020) but further validation on fuel samples is required. Instead of sequencing or other costly, complicated methods, simply the presence of water could be a proxy (e.g., the presence of water likely indicates that life will be present within the tank), but measurements of water can be unreliable. Many fuel storage tanks have an automatic tank gauge (ATG), designed to provide the operator with an indication of the level of fuel in the tank and the presence of any water. Yet, the buildup of biomass on the ATG will impair its function. The proliferation of microorganisms at the fuel-water interphase can, therefore, prevent the ATG from detecting the water in the tank, exacerbating the issue. Even with high fuel turnover (i.e., rapid use of fuel) the risk of water buildup may be reduced through constant emptying and replenishment but biofilms will remain on surfaces such as the ATG and tank walls ready to "infect" the next batch of fuel. Whether through advanced methods such as DNA or RNA sequencing, or more simply through the detection of water or increased acidity in fuel, more attention must be given to monitoring biodiesel for microbial growth. These measurements also should not be taken from topside fuel dispensers, as the openings in the tanks are often 18" or more above the tank floor. Unless tank bottoms are sampled, operators may miss a growing and insidious problem until it is too late, and a difficult to remove sludge layer is present. Storage tanks and dispensers containing ULSD with up to 5–7% biodiesel or B20 biodiesel blends need to be inspected and tested regularly, preferably from the tank bottoms. Any indication of microbial contamination should be taken seriously as a sign of corrosion risk. Regular sampling of the fuels, biomass, and water on the bottom of storage tanks would give a critical initial indication of contamination.

Once enough microbial biomass or particulate matter is present, stored fuels need to be "polished" by removing particulates and water via filtration. More commonly, fuels are simply disposed of, which is expensive, wasteful, and time-consuming. If the contamination is extensive, the storage tank will also need to be cleaned, typically through mechanical means such as power washing, or in many cases by the use of biocides such as Biobor (Biobor Fuel Additives, Houston TX) without any physical removal of biomass. This act of cleaning is akin to washing one's mouth with mouthwash without brushing one's teeth. While it may temporarily "freshen" a tank, without physical removal of biomass, the problem will only persist. Biomass removal is itself not complete and any biomass left behind can be sufficient to re-establish widespread contamination over time, as evidenced in this failure analysis (Figure 9.3).

Establishing more than anecdotal connections between microbiological contamination of fuels and the risk of corrosion will provide tank operators a set of guidelines to determine the next course of action. Corrosion monitoring via witness coupons is not a practical solution for front-line UST operators, yet the approach is viable for performing risk assessments of new, next generation biofuels prior to mass implementation. While materials compatibility testing, chemical stability, and resistance to fuel oxidation were performed on biodiesel and biodiesel blends (Jain and Sharma 2010) and fuel oxidation was correlated to subsequent corrosion, no one group thought to perform the same stability tests in the presence of microbiological flora known to inhabit fuel systems. Our work suggests that during the development and scale-up of next generation biofuels, real-world storage conditions should be tested for up to a year in a "best worst-case" storage environment, to allow for biological contamination to take root, if possible. Cooperative studies between industry, military, and academia, such as the one described here, allow for each partner to contribute their skills and expertise required to tackle the issue of biofouling and corrosion in fuels.

9.4 EXPERIMENTAL METHODS

9.4.1 Sampling Fuel, Biomass and Biofilms

In diesel storage tanks, microbial biomass is abundant wherever water can be found, which is at the bottom. Fuel samples were taken at each time point from the bottom of each tank using a 500 mL Bacon bomb fuel sampler (Thermo Fisher Scientific, Hampton, NH). Fuel samples were also taken from dispenser nozzles and fuel that was retained from the original delivery (i.e., receipt fuel). All fuel samples were transferred to sterile 1 L glass bottles and kept at room temperature until processed. Approximately 1 L of each fuel sample was filtered through a 45 mm dia, 0.22 µm nominal pore size polyether sulfone bottle-top filter (Steritop™, MilliporeSigma) to collect any biomass in the fuel, which was used in Stamps et. al (2020). The filtered fuel was collected into sterile 1 L glass bottles and used for acid-index measurements and GC-MS analysis (discussed below).

The filters were cut into quarters using a sterile scalpel. Three of these filter quarters were placed into individual ZR BashingBead™ Lysis tubes containing 0.7 mL (dry volume) of 0.5 mm ZR BashingBead™ lysis matrix (Zymo Research Corp., Irvine, CA) and 750 µL Xpedition™ Lysis/Stabilization Solution (Zymo), for a total of three technical replicates per sampling. These samples were homogenized for 30 seconds on-site using a sample cup attached to a cordless reciprocating saw, transported overnight at room temperature, and stored at −20 °C until needed. The fourth quarter of the filter was placed in 1 mL of sterile phosphate buffered saline (PBS) in a 2 mL screw-cap microcentrifuge tube and kept at 4°C until used to determine the most-probable number (MPN) of viable cells in less than 24 hours. The microcentrifuge tube was then vortexed to resuspend cells that were trapped on the filter for 10 seconds. Following this MPNs were set up using 96 well plates and inoculated on Hestrin Schramm medium and incubated at room temperature for 72 hours.

Biofilms from tank were collected during tank cleaning operation. Nylon flocked swabs (Therapak Corp, Los Angeles, CA) were used to remove biofilm from surfaces of coupons, walls of the storage tank. The tip of each swab was cut off, placed into ZR BashingBead™ Lysis Tubes, and then prepared as described previously.

9.4.2 Quantifying Microbial Biomass

One of the more challenging measurements to make from these fuel systems was an accurate quantification of microbial biomass. As mentioned previously, the microorganisms were not evenly distributed throughout the fuel or on surfaces. Replicate samples were used to estimate biomass concentration in several ways. The number of viable cells was estimated using a modified MPN approach with the filter or swab samples suspended in PBS. The biomass was suspended by vortex at full speed for one minute. Five 50 uL aliquots of each suspension were transferred to individual wells across the top row of a microtiter plate. Sterile PBS was added to the sixth well as a sterility control. With a multi-channel pipet, ten-fold serial dilutions were made into sterile PBS in each row beneath, creating a range of dilution factors of 10^0 to 10^7. A 10 uL aliquot of each dilution was transferred onto the surface of Hestrin Schramm (HS) agar medium (per L: 20 g glucose, 5 g yeast extract, 5 g peptone, 2.7 g Na_2HPO_4, 1.15 g citric acid, 7.5 g Agar; pH adjusted to 6.0 with diluted HCl or NaOH) (27), and incubated at 25°C for 7 days. Any visible colonies from each aliquot were scored as positive for growth and used to calculate MPN (cells/mL) of the fuel.

The abundance of fungi in fuel and in biomass samples was also estimated via qPCR of the 18S rRNA gene (Liu et al. 2012) as a proxy for biomass. Based on the apparent abundance of fungal morphologies in SEM images, estimates of bacterial and archaeal biomass using 16S rRNA gene qPCR was not conducted. Briefly, 1 μL of extracted DNA template (described in section 9.4.3 below) was added to a master mix containing 1.8 μM of the primers FungiQuantF (5'-GGR AAA CTC ACC AGG TCC AG-3') and FungiQuantR (5'-GSW CTA TCC CCA KCA CGA-3'), 225 nM of the TaqMan® probe (6FAM 5'-TGGTGCATGGCCGTT-3' MGBNFQ), and 1 X using Platinum Quantitative PCR SuperMix-UDG with ROX (Thermo Fisher Scientific Inc). The reaction was carried out under the following conditions: 3 minutes at 50 °C for UNG 80 treatment, 10 minutes at 95 °C for Taq activation, 50 cycles of 15 seconds at 95 °C for denaturation and 1 minute at 65 °C for annealing. The Ct-value for each reaction was determined using a manual Ct threshold of 0.10 and automatic baseline detection. Linear 18S rRNA gene standards were prepared by amplifying a known isolate using the primers (*Tolypocladium inflatum*) in seven 50 μL reactions and purifying and pooling the final product using an 0.9 X concentration of Agencourt AMPure XP paramagnetic beads (Beckman Coulter, Inc.), and two successive 75% EtOH washes. Standards were quantified using the QuBit BR assay (Thermo Fisher Scientific Inc.). Standard concentrations were 20 ng/μL, 2 ng/μL, 0.2 ng/μL, 20 pg/μL, 2 pg/μL, 0.2 pg/μL, and 20 fg/μL. Each standard was frozen once, and never refrozen. The number of copies was estimated using the approximate length of the amplified 18S fragment of ≈750 bp.

9.4.3 Identifying the Fuel Microbiome

Samples of fuel and biofilms were homogenized with Xpedition™ buffer (Zymo Research Corp.), which lysed the cells and preserved the DNA. These samples were kept at –20°C until needed. After thawing, each sample was homogenized for an additional 30 seconds using a BioSpec Mini-BeadBeater-8 (Biospec Products Inc., Bartlesville, OK). DNA extractions were performed per manufacturer specifications using the Zymo Xpedition™ Kit (Zymo Research Corp.).

Eukaryotic small subunit (SSU) ribosomal RNA (rRNA) gene fragments were amplified from each DNA extraction using a single set of PCR primers that spanned the SSU rRNA gene between V3 to V5 hypervariable regions between position 566 and 1,200 (*Saccharomyces cerevisiae* numbering). These primers were selected to evenly amplify a large percentage of the eukaryotic microbial community (Hadziavdic et al. 2014). Amplified gene libraries were sequenced using a Pacific Biosciences RS II sequencing instrument, utilizing concensus circular sequencing (CCS) at the Duke Center for Genomic and Computational Biology. Quality control, demultiplexing, and operational taxonomic unit (OTU) clustering at 97% sequence similarity were performed as previously described (Stamps et al. 2016) within QIIME 1 (Caporaso et al. 2010), with the modification of using mothur (Schloss et al., 2009) to assign taxonomy against the SILVA database (Quast et al. 2013) (r128) formatted for use with mothur (Schloss et al. 2009).

9.4.4 Measuring Microbial Degradation of B20 Biodiesel

The biodegradation of B20 *in situ* by mixed microbial communities was measured using gas chromatography mass spectroscopy. It was possible to determine the impact of microbial metabolism on fuel *in situ* because fuel had been taken directly from the delivery before it was placed in the storage tank (i.e., holdback fuel). The differential abundances of FAME and other components in holdback fuel compared to fuel samples taken from the storage tank were used to identify the extent and preference of FAME degradation.

The chemical composition of the B20 biodiesel samples was determined by GC/MS using a Shimadzu QP 2010 SE (Shimadzu Corporation, USA). Each sample was diluted 1:200 with hexane prior to injection. A volume of 1 μL was injected via autosampler with a split ratio of 1:10. Injection started at 300°C, the oven was at 40°C with a 0.5-minute hold, which increased to 320°C at a rate of 10°C min^{-1}. Chemical components were separated with a Restek Column Rxi 5Sil with dimensions: 30 m, 0.25 mm ID, 0.25 μm. High purity helium was used as a carrier gas at a linear velocity of 36.8 cm s^{-1}. Mass spectra were analyzed in scan mode with the following parameters: interface at 320°C, ion source 200°C, solvent cut of 2 minutes, event time of 0.25 seconds and scan speed of 2000. Each Total Ion Chromatogram (TIC) was processed using the software LabSolutions version 4.20 (Shimadzu Corporation, USA). Peaks were identified using the mass spectra library NIST version 14 and verified using reference standards for FAME (Supelco® 37 Component FAME Mix, Sigma Aldrich, USA) and B20 (Diesel:Biodiesel (80:20) Blend Standard, RESTEK, USA). Major alkane and FAME peaks were identified by the NIST library replicates and underwent destructive sampling of triplicates at each time point.

REFERENCES

"Air Force Infrastructure Energy Plan." 2010. US Air Force. https://www.dm.af.mil/Portals/99/Docs/Infrastructure.pdf?ver=2016-02-22-172103-900.

Andrade, Oderay. 2016. "Characterization of Fungal Contaminants in B20 Biodiesel Storage Tanks and Their Effect on Fuel Composition." *University of Oklahoma*. https://shareok.org/handle/11244/44854.

"ASTM D6751-20, Standard Specification for Biodiesel Fuel Blend Stock (B100) for Middle Distillate Fuels." 2020. ASTM International. Accessed July27. http://www.astm.org/cgi-bin/resolver.cgi?D6751-20.

Caporaso, J. Gregory, Justin Kuczynski, Jesse Stombaugh, Kyle Bittinger, Frederic D. Bushman, Elizabeth K. Costello, Noah Fierer, et al. 2010. "QIIME Allows Analysis of High-Throughput Community Sequencing Data." *Nature Methods* 7 (5). Nature Publishing Group: 335–336. doi:10/d63hsf.

D27 Committee. 2020. "Guide for Sampling, Test Methods, and Specifications for Electrical Insulating Oils of Petroleum Origin." *ASTM International*. Accessed July27. doi: 10.1520/D0117-18.

Davis, Nicole M., Diana M. Proctor, Susan P. Holmes, David A. Relman, and Benjamin J. Callahan. 2018. "Simple Statistical Identification and Removal of Contaminant Sequences in Marker-Gene and Metagenomics Data." *Microbiome* 6 (1): 226. doi:10/gfxzx5.

Fregolente, Patricia Bogalhos Lucente, Leonardo Vasconcelos Fregolente, and Maria Regina Wolf Maciel. 2012. "Water Content in Biodiesel, Diesel, and Biodiesel–Diesel Blends." *Journal of Chemical & Engineering Data* 57 (6). American Chemical Society: 1817–1821. doi:10/f33jtr.

Hadziavdic, Kenan, Katrine Lekang, Anders Lanzen, Inge Jonassen, Eric M. Thompson, and Christofer Troedsson. 2014. "Characterization of the 18S RRNA Gene for Designing Universal Eukaryote Specific Primers." *PLOS ONE*9 (2). Public Library of Science: e87624. doi:10/f5w5r4.

Hazrat, M. A., M. G. Rasul, and M. M. K. Khan. 2015. "Lubricity Improvement of the Ultra-Low Sulfur Diesel Fuel with the Biodiesel." *Energy Procedia*, Clean, Efficient and Affordable Energy for a Sustainable Future: The 7th International Conference on Applied Energy (ICAE2015), 75 (August): 111–117. doi:10/gdm3h9.

"Health Assessment Document For Diesel Engine Exhaust (Final 2002)." 2002. Reports & Assessments EPA/600/8-90/057F. U.S. Environmental Protection Agency: Office of Research and Development, National Center for Environmental Assessment, Washington Office, Washington, DC. https://cfpub.epa.gov/ncea/risk/recordisplay.cfm?deid=29060.

Hoekman, S. Kent, Amber Broch, Curtis Robbins, Eric Ceniceros, and Mani Natarajan. 2012. "Review of Biodiesel Composition, Properties, and Specifications." *Renewable and Sustainable Energy Reviews* 16 (1): 143–169. doi:10/bkmcjb.

Ingersoll, Christine M., Donald E. Roland, Amanda M. Kacuba, Christina L. Wilkinson, and József M. Berty. 2003. "Removal of Sulfur Oxides from Diesel Exhaust Gases." *Environmental Progress* 22 (3): 199–205. doi:10/djhxvj.

Jain, Siddharth, and M. P. Sharma. 2010. "Stability of Biodiesel and Its Blends: A Review." *Renewable and Sustainable Energy Reviews* 14 (2): 667–678. doi:10/fkzcvv.

Jakeria, M. R., M. A. Fazal, and A. S. M. A. Haseeb. 2014. "Influence of Different Factors on the Stability of Biodiesel: A Review." *Renewable and Sustainable Energy Reviews* 30 (February): 154–163. doi:10/f5sqdj.

Jia, Ru, Tuba Unsal, Dake Xu, Yassir Lekbach, and Tingyue Gu. 2019. "Microbiologically Influenced Corrosion and Current Mitigation Strategies: A State of the Art Review." *International Biodeterioration & Biodegradation* 137 (February): 42–58. doi:10/gg55xz.

Leung, D. Y. C., B. C. P. Koo, and Y. Guo. 2006. "Degradation of Biodiesel under Different Storage Conditions." *Bioresource Technology* 97 (2): 250–256. doi:10/fn6c6k.

Lindahl, Björn D., R. Henrik Nilsson, Leho Tedersoo, Kessy Abarenkov, Tor Carlsen, Rasmus Kjøller, Urmas Kõljalg, et al. 2013. "Fungal Community Analysis by High-Throughput Sequencing of Amplified Markers – A User's Guide." *The New Phytologist* 199 (1): 288–299. doi:10/f2z2nx.

Liu, Cindy M., Sergey Kachur, Michael G. Dwan, Alison G. Abraham, Maliha Aziz, Po-Ren Hsueh, Yu-Tsung Huang, et al. 2012. "FungiQuant: A Broad-Coverage Fungal Quantitative Real-Time PCR Assay." *BMC Microbiology* 12 (1): 255. doi:10/gb47tb.

Meckenstock, Rainer U., Frederick von Netzer, Christine Stumpp, Tillmann Lueders, Anne M. Himmelberg, Norbert Hertkorn, Philipp Schmitt-Kopplin, et al. 2014. "Water Droplets in Oil Are Microhabitats for Microbial Life." *Science* 345 (6197). American Association for the Advancement of Science: 673–676. doi:10/f6bq24.

Nelson, Denice K., Timothy M. Lapara, and Paige J. Novak. 2010. "Effects of Ethanol-Based Fuel Contamination: Microbial Community Changes, Production of Regulated Compounds, and Methane Generation." *Environmental Science & Technology* 44 (12): 4525–4530. doi:10/dc3qtx.

Norsworthy, R. 2014. "1 - Understanding Corrosion in Underground Pipelines: Basic Principles." In *Underground Pipeline Corrosion*, edited by Mark E. Orazem, 3–34. Woodhead Publishing. doi:10.1533/9780857099266.1.3.

Nygaard, Anders B., Hege S. Tunsjø, Roger Meisal, and Colin Charnock. 2020. "A Preliminary Study on the Potential of Nanopore MinION and Illumina MiSeq 16S RRNA Gene Sequencing to Characterize Building-Dust Microbiomes." *Scientific Reports* 10 (1). Nature Publishing Group: 3209. doi:10/gg55zh.

Özener, Orkun, Levent Yüksek, Alp Tekin Ergenç, and Muammer Özkan. 2014. "Effects of Soybean Biodiesel on a DI Diesel Engine Performance, Emission and Combustion Characteristics." *Fuel* 115 (January): 875–883. doi:10/gg55xv.

Parada, Alma E., David M. Needham, and Jed A. Fuhrman. 2016. "Every Base Matters: Assessing Small Subunit RRNA Primers for Marine Microbiomes with Mock Communities, Time Series and Global Field Samples." *Environmental Microbiology* 18 (5): 1403–1414. doi:10/f8n8hp.

Passman, F. J. 2013. "Microbial Contamination and Its Control in Fuels and Fuel Systems Since 1980 – A Review."*International Biodeterioration & Biodegradation*, Special Issue: 3rd International Symposium on Applied Microbiology and Molecular Biology in Oil Systems, 81 (July): 88–104. doi:10/f4zf47.

Quast, Christian, Elmar Pruesse, Pelin Yilmaz, Jan Gerken, Timmy Schweer, Pablo Yarza, Jörg Peplies, and Frank Oliver Glöckner. 2013. "The SILVA Ribosomal RNA Gene Database Project: Improved Data Processing and Web-Based Tools." *Nucleic Acids Research* 41 (D1): D590–D596. doi:10/gfb6mr.

Salter, Susannah J., Michael J. Cox, Elena M. Turek, Szymon T. Calus, William O. Cookson, Miriam F. Moffatt, Paul Turner, et al. 2014. "Reagent and Laboratory Contamination Can Critically Impact Sequence-Based Microbiome Analyses." *BMC Biology* 12 (1): 87. doi:10/f6wj9s.

Schloss, Patrick D., Sarah L. Westcott, Thomas Ryabin, Justine R. Hall, Martin Hartmann, Emily B. Hollister, Ryan A. Lesniewski, et al. 2009. "Introducing Mothur: Open-Source, Platform-Independent, Community-Supported Software for Describing and Comparing Microbial Communities." *Applied and Environmental Microbiology* 75 (23): 7537. doi:10/fqcv8t.

Sim, Justin, and Brendan Chapman. 2019. "In-Field Whole Genome Sequencing Using the MinION Nanopore Sequencer to Detect the Presence of High-Prized Military Targets." *Australian Journal of Forensic Sciences* 51 (sup 1). Taylor & Francis: S86–S90. doi:1 0/gg55zg.

Sowards, Jeffrey W., and Elisabeth Mansfield. 2014. "Corrosion of Copper and Steel Alloys in a Simulated Underground Storage-Tank Sump Environment Containing Acid-Producing Bacteria." *Corrosion Science* 87 (October): 460–471. doi:10/gg55xx.

Stamps, Blake W., Caitlin L. Bojanowski, Carrie A. Drake, Heather S. Nunn, Pamela F. Lloyd, James G. Floyd, Katelyn A. Emmerich, et al. 2020. "In Situ Linkage of Fungal and Bacterial Proliferation to Microbiologically Influenced Corrosion in B20 Biodiesel Storage Tanks." *Frontiers in Microbiology* 11. Frontiers. doi:10/ggpcxg.

Stamps, Blake W., Christopher N. Lyles, Joseph M. Suflita, Jason R. Masoner, Isabelle M. Cozzarelli, Dana W. Kolpin, and Bradley S. Stevenson. 2016. "Municipal Solid Waste Landfills Harbor Distinct Microbiomes." *Frontiers in Microbiology* 7. Frontiers. doi:1 0/gftjhx.

US EPA, OLEM. 2016. "Investigation of Corrosion-Influencing Factors in Underground Storage Tanks with Diesel Service." Reports and Assessments EPA 510-R-16-001. https://www.epa.gov/ust/investigation-corrosion-influencing-factors-underground-storage-tanks-diesel-service.

Williamson, Charles H. D., Luke A. Jain, Brajendra Mishra, David L. Olson, and John R. Spear. 2015. "Microbially Influenced Corrosion Communities Associated with Fuel-Grade Ethanol Environments." *Applied Microbiology and Biotechnology* 99 (16): 6945–6957. doi: 10/f7xp26.

10 Elemental Sulfur Corrosion

A Case Study

Katherine M. Buckingham and Richard B. Eckert
DNV GL USA

CONTENTS

10.1 CASE STUDY OVERVIEW

A crude oil operator experienced aggressive internal corrosion on a recently constructed pipeline. The pipeline carried crude that was transported by truck from tanks at small producer sites in the region and also crude from other large gathering systems. The crude was unprocessed, other than for removal of most of the bulk produced water phase and degassing at atmospheric pressure. Dissolved H_2S in the water phase was typically zero. The corrosion occurred despite efforts to prevent and mitigate internal corrosion through regular pigging, monitoring, and the use of corrosion inhibitors and biocide. The corrosion was identified following an in-line inspection (ILI) using a magnetic flux leakage (MFL) tool. The internal anomalies identified by the ILI were concentrated along the bottom of the pipe, were relatively small (<12 mm in diameter) and exhibited wall losses of up to 3.8 mm (40% of the nominal wall thickness) with localized corrosion rates at these small pits >1 mm/year.

DOI: 10.1201/9780429355479-12

To ensure the best chance at proper identification of the corrosion mechanism, microbiological samples were collected and preserved in the field at the time the pipe was excavated and cut open for examination (Eckert and Buckingham 2017).

10.2 CORROSION MECHANISM DETERMINATION

10.2.1 Field Inspection

Figure 10.1 contains field photographs showing representative corrosion pits identified at the time of the excavation. As shown, the pits were small in diameter and located along the bottom of the pipeline. Additional discrete, isolated pits were observed along the bottom of the pipe. No evidence of significant general internal corrosion or accumulated solids was associated with the pits.

10.2.2 Solids Analyses

Solids associated with the corrosion were analyzed using qualitative spot testing for carbonates and sulfides and energy dispersive spectroscopy (EDS) in a scanning electron microscope (SEM) to determine the elemental composition of the solids. As the solids associated with the corrosion were very minimal, all testing was performed directly on the pipe surface or on 1 inch by 1 inch coupons and metallurgical cross sections that were cold cut from the pipe section, and which contained the pits. There were not enough solids to perform x-ray diffraction analysis or other laboratory analysis.

Qualitative spot tests revealed no evidence of carbonates and sulfides within the solids at or away from the corrosion pits. Elemental analyses revealed the presence of two aggressive species within the solids: chlorine and sulfur. As shown in Figure 10.2, chlorine, likely in the form of chlorides, was concentrated along the interface between the solids and the steel surface. Sulfur was also detected but tended to reside within the bulk solids inside the pits. In some cases, discrete, sulfur-rich particles, consistent with elemental sulfur, were present (Boivin and Oliphant 2011); see Figure 10.3.

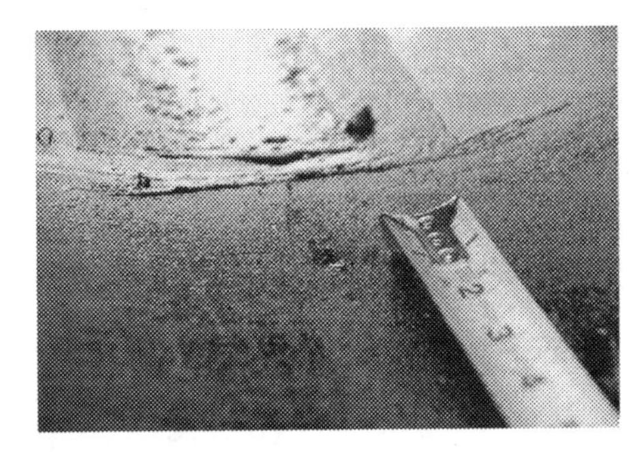

FIGURE 10.1 Field photographs showing representative pits along the bottom of the pipeline, as-found (left) and after cleaning (right).

FIGURE 10.2 SEM image and elemental map from a metallurgical cross section showing the distribution of chlorine and sulfur within deposits associated with the pit.

10.2.3 Microbiological Analyses

Culture testing using serial dilutions and molecular microbiological testing using quantitative polymerase chain reaction (qPCR) and next generation sequencing (NGS) were performed on swab samples that were removed at and away from a representative pit to determine the contribution, if any, of microorganisms to the corrosion. The swab samples were preserved in a phosphate buffered saline solution and kept cold (on ice) until they could be tested. Samples for the serial dilutions were inoculated in the field on the day that they were collected. Samples for the molecular microbiological testing were kept on ice and shipped overnight to the testing laboratory.

Table 10.1 is a summary of the serial dilution analyses performed on the samples. As shown, liquid culture media for acid-producing bacteria (APB), sulfate-reducing bacteria (SRB), aerobic bacteria (AERO), anaerobic bacteria (ANA), iron-reducing bacteria (IRB), and nitrate-reducing bacteria (NRB) were used. The results show no evidence of culturable SRB, high concentrations of AERO, IRB, and NRB, and moderate concentrations of ANA and APB in both samples. There was no evidence that specific bacteria were preferentially flourishing at the pit vs. away.

Tables 10.2 and 10.3 summarizes the results of the qPCR analyses. As shown, the total bacteria populations were similar for both samples. Similar to the culture testing, SRB, along with sulfate-reducing archaea (SRA) and acetogens, were below the minimum detection limits of the method (about 1×10^3). Methanogens and fermentative (acid producing) bacteria were also present at similar levels in both samples, while low numbers of sulfur oxidizing bacteria (SOB) were detected only in the Away sample. Sulfur oxidizers can use sulfides, elemental sulfur, thiosulfates, and other sulfur compounds in their metabolism, forming sulfuric acid as a by-product. Denitrifying bacteria (nirK and nirS genes) were also present as were iron oxidizing and iron reducing bacteria. Iron oxidizers convert Fe^{+2} in solution to solid oxides of Fe^{+3} and take advantage of available iron in solution near active corrosion. Iron and nitrate utilizing bacteria are typically found when there is some amount of oxygen in the environment to support their metabolism. *Cladosporium*, a common and ubiquitous fungus that can use organic materials as a carbon source for energy, was also found in both samples. Overall, the qPCR data showed no significant differences in the samples.

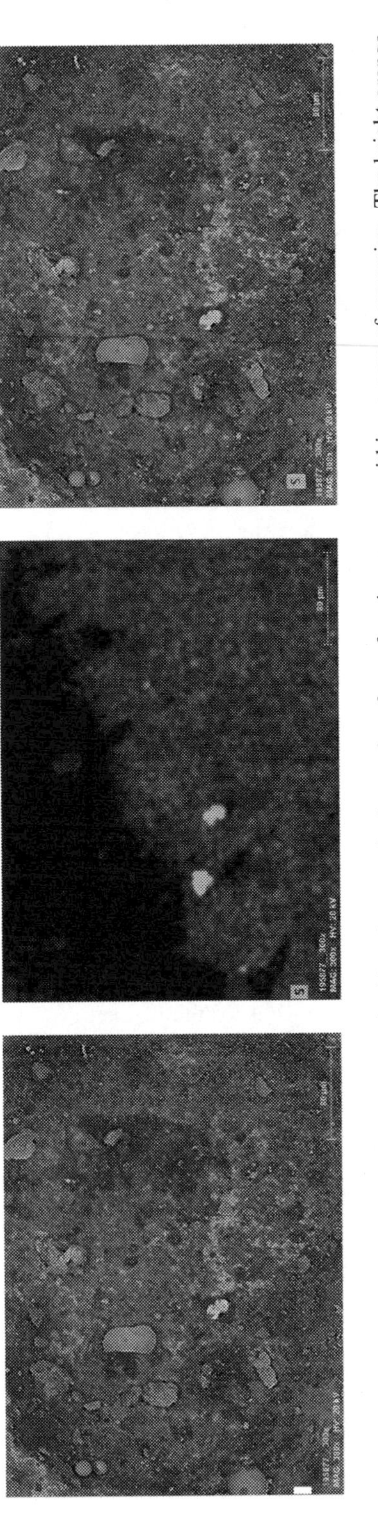

FIGURE 10.3 SEM image and elemental maps looking down at the internal surface of a pipe coupon within an area of corrosion. The bright orange areas are discrete sulfur particles.

TABLE 10.1
Summary of Culture Testing Performed Using 10 Vial Series Dilutions

Bacteria Type	Pit		Away	
	Test Result	Number of Positive Vials	Test Result	Number of Positive Vials
Aerobic	Positive	9	Positive	9
Anaerobic	Positive	3	Positive	2
Acid Producing	Positive	4	Positive	4
Sulfate-reducing	Not detected	–	Not detected	–
Iron-reducing	Positive	8	Positive	6
Nitrate-reducing	Positive	9	Positive	7

TABLE 10.2
Results of Microbiological DNA Testing Using Quantitative Polymerase Chain Reaction Method

Target Population/Functional Gene	Pit	Away
Total Bacteria	8.5×10^5	7.2×10^6
Total Archaea	7.8×10^3	8.8×10^{1} (J)
Sulfate-reducing bacteria	Not detected	Not detected
Sulfate reducing archaea	Not detected	Not detected
Methanogens	1.9×10^3 (J)	2.2×10^2 (J)
Acetogens	Not detected	Not detected
Fermenters	6.9×10^3	3.1×10^4
Cladiosporum	2.1×10^4	1.6×10^3
Iron oxidizers	4.1×10^3 (J)	5.6×10^4
IRB Anaeromyxobacter	6.5×10^4	3.1×10^5
Sulfur oxidizing bacteria	Not detected	2.1×10^2 (J)
Denitrifying bacteria (nirK)	4.71×10^5	4.92×10^5
Denitrifying bacteria (nirS)	2.35×10^6	2.27×10^6

(J) Estimate Gene Copies below practical quantitation limit, but above lower quantifiable limit

Finally, the genus classification results from the NGS analyses are depicted graphically in Figure 10.4. The top genus classification results for each sample are shown. *Sulfurimonas* and *Sulfuricurvum*, sulfur oxidizing bacteria, were both present within the Pit sample but not the Away sample. Many kinds of reduced sulfur compounds, such as sulfide, elemental sulfur, thiosulfate, and sulfite, can serve as an electron donor for the growth of *Sulfurimonas* species. *Sulfurimonas* can use

TABLE 10.3
Summary of the Corrosion Mechanism Analysis

Chemical Conditions	Microbiological Conditions
• Multiple sources for transported crude. • Sources of crude known to contain dissolved O_2 and CO_2. • Field tests for dissolved hydrogen sulfide in the water from the crude streams have historically always been near zero. • Water contained in crude is brine with high chloride content.	• No evidence of preferential flourishment of bacteria at pit vs. away. • Abundance of aerobic bacteria. • Absence of SRB and SRA. • Presence of SOB.
Corrosion and Metallurgical Information	**Design and Operation Information**
• Discrete pits with narrow openings. • Absence of general corrosion associated with pits. • Discrete particles of sulfur associated with the corrosion pits. • Concentrated layer of chlorine found at interface between corrosion deposits and pit wall. • No evidence of carbonates or sulfides detected within the deposits. • No metallurgical defects associated with corrosion. • Pipe steel meets specifications for grade of steel.	• Periods of no flow experienced. • Typical flow rates at or below calculated entrainment velocity for water.

oxygen, nitrate, and nitrite as electron acceptors and can use a variety of carbon sources including CO_2 and acetate. Sulfuric acid is produced as a by-product of their metabolism.

Pseudomonas, an aerobic biofilm former, was also detected in both samples but was more prevalent within the Pit vs. the Away solids. These microorganisms take advantage of carbon and oxygen availability. Finally, the genus *Sphingomonas* was found in both samples. *Sphingomonas* are aerobic microorganisms that can degrade toluene, naphthalene, and other aromatic compounds; they are widely found in natural environments. Their role in the corrosion of steel, if any, is not known.

Although the results from all three tests revealed no evidence that specific microorganisms were preferentially flourishing at the internal corrosion pit, the microorganisms that were identified provided key insights into the operating environment under which the corrosion occurred. Based on the presence of numerous aerobic bacteria and the atypical absence of SRB and SRA, which are strict or facultative anaerobes, it was putatively determined that some level of oxygen was likely to be present within the system, if even on an episodic basis. In general, oxygen is not a common corrosion threat for crude oil pipelines. Thus, the determination that oxygen contamination existed was key to understanding the corrosion mechanism. Further, the absence of SRA and SRB, common microorganisms

FIGURE 10.4 Pie charts showing the top genera identified, using NGS, for the microbiological samples.

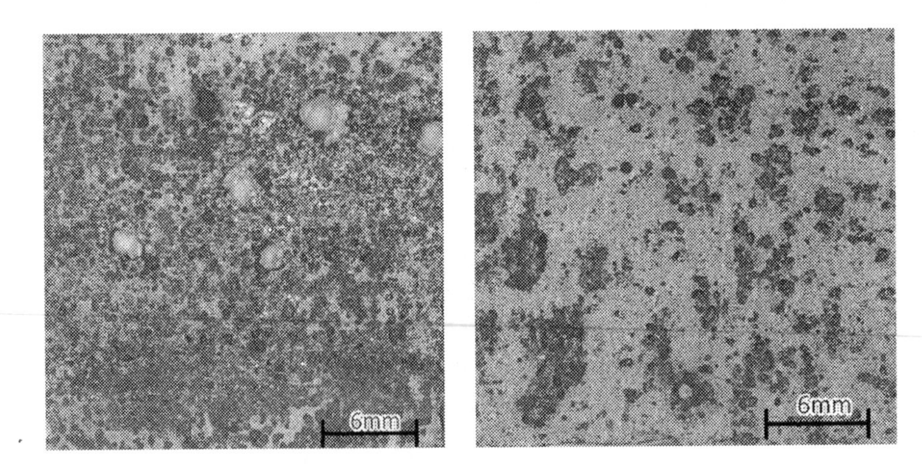

FIGURE 10.5 Light photomicrographs showing corrosion pits on the internal surface of the pipe, after acid cleaning, at the 6:00 (left) and 6:30 (right) orientations.

found in oil systems, revealed that these microorganisms were not associated with the sulfur identified within the corrosion pits.

10.2.4 Surface Profilometry

One-inch square coupons were removed from representative areas of corrosion on the pipe section for surface roughness measurements. Figure 10.5 contains light photomicrographs showing corrosion pits at the 6:00 and 6:30 orientations. As shown, the pit diameters are relatively small. The deepest and largest pits were observed along the 6:00 orientation, while the greatest pit density was observed along the bottom of the pipe, away from the 6:00 orientation.

10.2.5 Metallurgical Analyses

Metallurgical analyses were performed on cross sections that had been removed through pits to examine the morphology of the corrosion and to evaluate the steel for evidence of any metallurgical defects that may have contributed to the corrosion. Figure 10.6 is a light photomicrograph showing a representative pit in cross section after polishing and etching. As shown, the depth of corrosion attack was greater than the pit opening, and the pit was filled with deposits. No evidence of any metallurgical defects was associated with the corrosion.

10.2.6 Operational History

A review of the operational history of the line revealed that the transported crude oil comes from multiple sources. Through additional testing, the crude oil from some of these sources was found to contain dissolved oxygen and carbon dioxide. Field tests for dissolved hydrogen sulfide in the water from the crude streams have historically always been near zero.

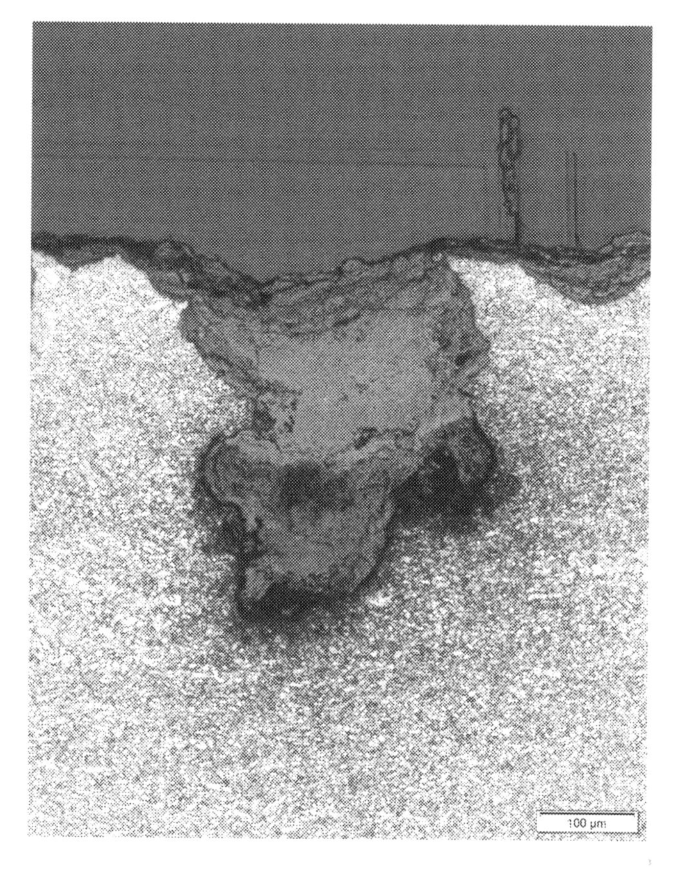

FIGURE 10.6 Light photomicrographs showing a metallurgical cross section of a pit after polishing and etching with a 2% Nital.

The potential for water accumulation within the pipeline was also confirmed as the pipeline is known to experience frequent periods of no flow. When crude is flowing within the line, the flow rates are generally at or below the calculated rate required for water entrainment. The water that is transported is typically a brine originating from upstream production that contains high levels of chlorides. Even though the transportation limit for basic sediment and water in the crude was 0.5 volume percent, episodic upsets were known to occur where large volumes of water were received.

10.2.7 Corrosion Mechanism Determination

Based on the results of the failure analysis and the operational history of the pipeline, four primary causes of the internal corrosion were evaluated: MIC, carbon dioxide corrosion, oxygen corrosion, and a sulfur-related corrosion.

MIC was eliminated as the cause of the corrosion due to the absence of preferential flourishment of bacteria at the corrosion, the absence of accumulated solids associated with the corrosion, the presence of sulfur particles within the pits, and the morphology of the pits (i.e., tight opening of the pits that is not consistent with MIC). Similarly, carbon dioxide was eliminated as the primary corrosion mechanism due to the absence of carbonates within the corrosion deposits via qualitative spot testing and the morphology of the corrosion. Carbon dioxide corrosion typically causes much larger-diameter pits that grow together to form larger pits in some cases, which was inconsistent with the observed pit morphology. Based on the morphology of the corrosion, oxygen was determined not to be the primary cause of the pitting; however, the presence of oxygen within the system likely facilitated the corrosion by promoting the oxidation of sulfides. Oxygen usually causes both general and localized corrosion; no general corrosion was observed. Elemental sulfur was identified as the most likely cause of the corrosion based on the presence of particles of sulfur within the corrosion pits, the absence of SRB and SRA that could have been associated with the identified sulfur, the presence of SOB that can use elemental sulfur as an energy source, and the morphology of the corrosion. The severity of the elemental sulfur attack was likely exacerbated by the presence of chlorides.

10.3 BASIC CAUSE ANALYSIS

10.3.1 Basic Cause Analysis Introduction

Based on the findings of the failure analysis, a basic cause analysis was performed to identify contributing factors to the corrosion experienced on the pipeline. The analysis focused on those barriers that are commonly used to prevent or mitigate internal corrosion. The barriers fell within five categories: design, operations, assessment/monitoring, mechanical cleaning (pigging), and chemical treatments. Figure 10.7 is a BowTie diagram showing the internal corrosion threat (elemental sulfur corrosion) on the left and the Top Event (unpredicted internal corrosion) on the right. Barriers commonly used to prevent internal corrosion are shown between the threat and Top Event.

10.3.2 Basic Cause Analysis Findings

The results of the basic cause analysis showed that although the operator had measures in place for each of the barriers identified, those measures were either inadequate, unreliable, or failed.

The pipeline was designed to allow for the launch and receipt of ILI tools and cleaning pigs but was not equipped for the use of flush-mounted coupons and probes to monitor corrosion in the flowing pipeline. Further, there was no capacity for online, continuous monitoring of conditions such as for dissolved oxygen, as it is not normally expected in crude oil service. Thus, the barrier for design was inadequate.

The pipeline experienced periods of no flow where water could accumulate, and the operator did not have the ability to operate the pipeline above the water entrainment velocity. Thus, operational controls could not prevent water from settling

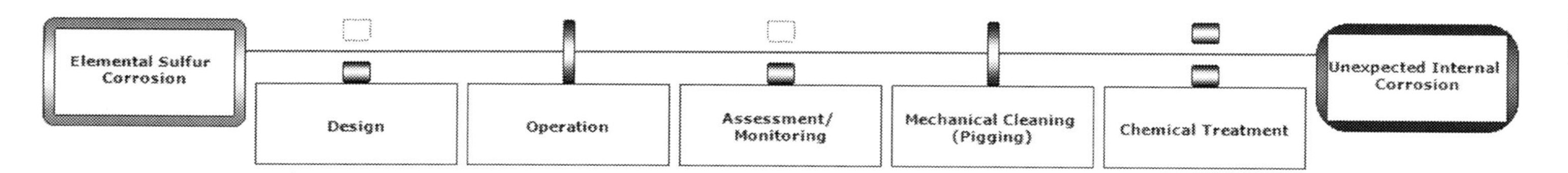

FIGURE 10.7 Preventative side of the BowTie diagram for the unexpected internal corrosion experienced on the pipeline. Barriers with one empty and one solid square were inadequate; those with a thin bar were unreliable, and barriers with two solid squares are failed.

out of the crude and accumulating on the bottom of the pipe. This means the operations barrier was unreliable.

Indirect monitoring of fluid quality was performed at the upstream and downstream ends of the pipeline. The monitoring included water testing for pH, CO_2, H_2S, Cl, Fe, and alkalinity, and liquid culture testing for planktonic and sessile bacteria (APB, SRB, and general anaerobic bacteria [GAB]). Dissolved oxygen, however, was not monitored. Corrosion rates were also assessed using weight loss coupons at sample pots at the upstream and downstream ends of the pipeline. These locations were not along the flowing pipeline and thus were not representative of the pipeline conditions where the corrosion occurred. Since the coupons were only analyzed for weight loss, they only provided general corrosion rates and did not provide any information on localized corrosion rates. Together, the water and coupon monitoring provided no insight into the potential for elemental sulfur corrosion. Thus, the assessment and monitoring barrier was inadequate.

Routine pigging for cleaning and wax removal was performed on the pipeline. While pigging is generally effective for the removal of biofilms and surface deposits, pigging can only occur when the pipeline is flowing. When pigging could be used, the narrow pit openings and undercut morphology of the corrosion present made it difficult for the pigs to remove the sulfur particles. Thus, water and solids were allowed to accumulate in the pits and promote corrosion, making the mechanical cleaning barrier unreliable.

A quaternary amine-based corrosion inhibitor was applied using continuous injection to control corrosion within the pipeline. The corrosion inhibitor, however, was not effective in the presence of oxygen and was not effective for controlling localized corrosion due to elemental sulfur. Biocide was batch applied at various points along the pipeline system and in tanks at monthly intervals. The corrosion inhibitor was not field tested for the effectiveness under actual pipeline conditions and no compatibility testing was performed with other chemicals that were used to treat the pipeline (biocide, paraffin dispersants, drag reducing agents, etc.). Thus, the chemical treatment barrier failed.

10.3.3 Basic Cause Analysis Recommendations

Based on the findings from the basic cause analysis, recommendations were made for the chemical treatment and assessment/monitoring barriers. The chemical treatment recommendations focused on improvements to the selection and evaluation process to include testing in actual crude from the pipeline and conducting compatibility and performance testing with other chemicals that were being applied, such as biocides and paraffin dispersants. The assessment/monitoring recommendations included evaluation of the installation of flush-mounted coupons and probes along the flowing pipeline to detect localized corrosion and the effectiveness of mitigation, and evaluation of technologies to monitor dissolved oxygen within the crude (or water phase) at strategic locations along the pipeline. Monitoring was also focused on locations with the highest potential for dissolved oxygen in the crude.

10.4 SUMMARY AND CONCLUSIONS

This chapter presented a case study where elemental sulfur was identified as the primary cause for aggressive internal pitting found within a crude oil pipeline. Similar to MIC, elemental sulfur can result in extremely aggressive attack of carbon steel. Although MIC was not identified as the cause of the corrosion, the results from the robust set of microbiological testing performed provided key insights into the operating environment under which the corrosion occurred. This information was essential in the determination of the actual corrosion mechanism.

REFERENCES

J. Boivin and S. Oliphant. "Sulfur corrosion due to oxygen ingress," *CORROSION* 2011, Paper 11120, March 13–17, Houston, TX.

R. Eckert and K. Buckingham. "Investigating pipeline corrosion failures," *Inspectioneering Journal*. 23(4) July/August 2017: 23–29.

H. Fang, B. Brown, D. Young, and S. Nešić. "Investigation of elemental sulfur corrosion mechanisms," *CORROSION* 2011, Paper 11398, March 13–17, Houston, TX.

M.R. Gregg, J. Slofstra, D. Thill, and W. Sudds, "Corrosion experiences and inhibition practices managing wet sour salty gas pipeline environments contaminated with elemental sulfur deposits," *CORROSION* 2003, Paper 03174, March 16–20, San Diego, CA.

M. Nagu, A. Abdulhadi, A. Huwaiji, and N. Alanazi. "Effect of element sulfur on pitting corrosion of steels," *Insights in Analytical Electrochemistry*. 14(1:6) 2018.

G. Schmitt. "Present day knowledge of the effect of elemental sulfur on corrosion in sour gas systems," *CORROSION* 1990, Paper 39, April 23–27, Las Vegas, NV.

L. Smith and B. Craig. "Practical corrosion control measures for elemental sulfur containing environments," *CORROSION* 2005, Paper 05646, April 3–7, Houston, TX.

11 MIC Investigation of Stainless Steel Seal Ring Corrosion Failure in a Floating Production Storage and Offloading (FPSO) Vessel

L.L. Machuca, T. Pojtanabuntoeng, S. Salgar-Chaparro, E. Suarez, and B. Kinsella
Curtin University

A. Darwin
Woodside Energy Ltd.

CONTENTS

DOI: 10.1201/9780429355479-13

11.1 INTRODUCTION

In the oil and gas industry, a hydrostatic pressure test is normally performed after installation of pipelines and all piping systems to verify pressure equipment does not leak and ensure they are fit for service. This quality control measure allows inspecting the strength and integrity of pipelines, vessels, and tanks before they are commissioned. The pressure of the hydrostatic test is commonly set above the design pressure (usually 125% of the Maximum Allowable Operating Pressure – MAOP), providing a safety margin against design tolerance and defect growth during the service life of assets (Machuca 2017). Besides the inspection of new systems, hydrotesting is also performed in aged systems to determine their integrity, to confirm MAOP, and to maximize their use. It has been reported that this method can reduce unnecessary replacement of aged pipelines, helping to extend their lives and avoiding repair and replacement costs (Lenhoff 1994).

Water is the most common medium used in this test since the use of gas (compressed air) represents a safety hazard in case of leak or rupture. Water can be potable, fresh, produced, or seawater (Machuca 2014). Potable water is the preferred source of water for hydrotesting, however, in offshore systems seawater is the primary source of water for hydrostatic testing due to the difficulty and costs associated with transportation of large volumes of potable water to the facilities (Machuca 2017). Seawater contains microorganisms, oxygen, and chlorides, which are known aggressive species towards metallic materials. Therefore, the quality of the hydrotest water has been discussed in several standards and code of practice (Darwin, Annadorai, and Heidersbach 2010). Nevertheless, several failures due to corrosion issues initiated by the use of inappropriate water during hydrotesting have been reported (Darwin, Annadorai, and Heidersbach 2010). Leading causes of corrosion failures of stainless steel systems soon after hydrotesting are pitting and crevice corrosion, stress corrosion cracking, and microbiologically influenced corrosion (MIC).

Prevention of corrosion due to the exposure of the system to the hydrotest water includes the use of chemically treated water and the limitation of the contact time of the water with the system. A typical chemical treatment of the hydrotest water consist of biocide, oxygen scavenger, and corrosion inhibitor (Prasad 2003). These chemicals can avoid general and localized corrosion resulting from oxygen contamination, chlorides and microbial activity. Once the system has been tested, it is recommended to drain the water and dry the system to avoid the formation of biofilms and MIC problems in the future. Hydrotesting has a duration from 8 to 10 hours; however, it is common that hydrotest fluid remains in pressure equipment for periods longer than expected (months to years) due to unexpected difficulties in commissioning (Machuca et al. 2011). During this period, stagnant conditions can favour biofilm growth, which can accelerate metal corrosion. MIC of pipelines and tanks during lay-up periods after hydrotesting has been widely reported (Borenstein and Lindsay 2002, Huang et al. 2012, Zhao et al. 2010). After sessile communities are established, the control and removal of biofilms become more difficult even using potent biocidal agents at high concentrations; hence, microorganisms can remain active and initiate corrosion, which can be accelerated after start-up (Surkein et al. 2011, Xu et al. 2013).

This study aimed to investigate the failure of piping clamp connectors in a Floating Production Storage and Offloading (FPSO) Rigid arm and RTM (riser turret mooring) that occurred during hydrostatic testing of an in-service piping system. Stainless steel (17-4 PH), nickel alloy (718), and duplex stainless steel 2205 (DSS) seal rings coupled to DSS hubs leaked due to corrosion during the testing. These rings had been exposed to production fluids during regular operation (~80°C) for over five years. After this period, the system was depressurised, flushed and filled with chemically treated seawater (biocide and oxygen scavenger). This seawater was in the pipe system for about four months before hydrostatic test. During the initial failure investigation, a crevice type corrosion was observed, with possible contributions from MIC, oxygen corrosion, galvanic corrosion, or a combination of these mechanisms. Active microorganisms were detected in all failed seal rings, which were identified using 16S rDNA sequencing as reported elsewhere (Salgar-Chaparro, Darwin, et al. 2020). Results showed that the microbial community in the failed seal rings was dominated by *Pseudomonas* and other marine microorganisms such as *Martelella*, *Marinomonas*, *Shewanella*, *Alcanivorax*, and *Halomonas*, which displayed iron-oxidising capabilities.

To establish the potential contributions from the different mechanisms, a laboratory investigation was undertaken under conditions simulating the field scenarios. Experiments were conducted in the presence and absence of a microbial consortium recovered from the corroded seal rings. Atmospheric gas in the experiments was switched from 20/80 CO_2/N_2 (1 bar) to air to simulate the transition from normal operating condition (anaerobic) to oxygen-contaminated wet lay-up exposure (aerobic). This was done to evaluate the hypothesis that corrosion initiated upon the introduction of oxygen during seawater flooding.

11.2 EXPERIMENTAL

11.2.1 Materials and Test Solutions

11.2.1.1 Test Specimens

Test specimens were machined from round bars of 2205 duplex stainless steel (DSS) (Stirlings Australia) and 17-4 PH stainless steel and nickel alloy 718 (Specialty Metals, Perth). Table 11.1 summarizes the test assemblies and alloy materials evaluated.

Crevice assembly 1 to 3 are referred to as "individual crevice" meaning the metals were coupled with an inert material (polyvinylidene difluoride, PVDF).

TABLE 11.1
Sample Arrangement and Objective of Exposure Experiments

Assembly	Material	Objectives
Crevice 1	17-4 PH/PVDF	Evaluate crevice corrosion alone
Crevice 2	DSS/PVDF	
Crevice 3	178/PVDF	
Crevice 4	17-4 PH:DSS	Evaluate synergistic effect between crevice and galvanic corrosion DSS represents the cathode
Crevice 5	17-4 PH:DSS	
Crevice 6	17-4 PH:DSS (Waterline)	
Galvanic couple 1	17-4 PH coupled to DSS	Evaluate galvanic corrosion alone
Galvanic couple 2	178 coupled to DSS	

These assemblies evaluated the susceptibility to crevice corrosion of each material. Metal samples were machined into square shape with a 7 mm center hole. The approximate exposed surface area was 50 cm^2 (two by 5 X 5 cm^2). Each assembly consisted of two electrically insulated square samples: one was soldered at a corner with an electrical wire for open circuit potential (OCP) monitoring and the second was used for mass loss and subsequent surface analysis. The soldered area was covered with epoxy resin to insulate the electrical wire from the test solution.

Metal specimens were wet ground to 600 grit finish using SiC paper, degreased with acetone and dried with nitrogen gas. Artificial crevices were formed using spring-loaded crevice assemblies (Machuca et al. 2013). Crevice formers (outside diameter 20 mm, inside diameter 7 mm, height 15 mm) were made of PVDF, ground to 1,200 grit finish, cleaned and dried before use. Four nylon coated disc springs were used to maintain a constant load corresponding to 3 Nm applied torque. Metal specimens and centralising PVC rings were sterilized by soaking in 70% ethanol for 10 minutes then exposing to UV irradiation for 15 minutes both sides. Other crevice parts were sterilized by autoclaving (121°C for 30 minutes under 15 psi of pressure).

Crevice 4 to Crevice 6 were metal-metal crevice for evaluating the combined effect of crevice and galvanic corrosion. The PVDF crevice former was replaced with a square metal sample (20 x 20 x 5 mm) with a 7 mm center hole. With this configuration, the cathode to anode ratio was estimated to be 7:1. Figure 11.1 shows an assembled metal-metal crevice sample.

For galvanic current density measurement, DSS (50 x 50 mm), 718 alloy (20 x 20 mm), and 17-4 PH (20 x 20 mm) were soldered with electrical wire and embedded in an epoxy resin. After the resin had cured, the samples were wet ground to 600 SiC paper and sterilized by immersing in 70% ethanol for 10 minutes and exposed to UV irradiation for 15 minutes.

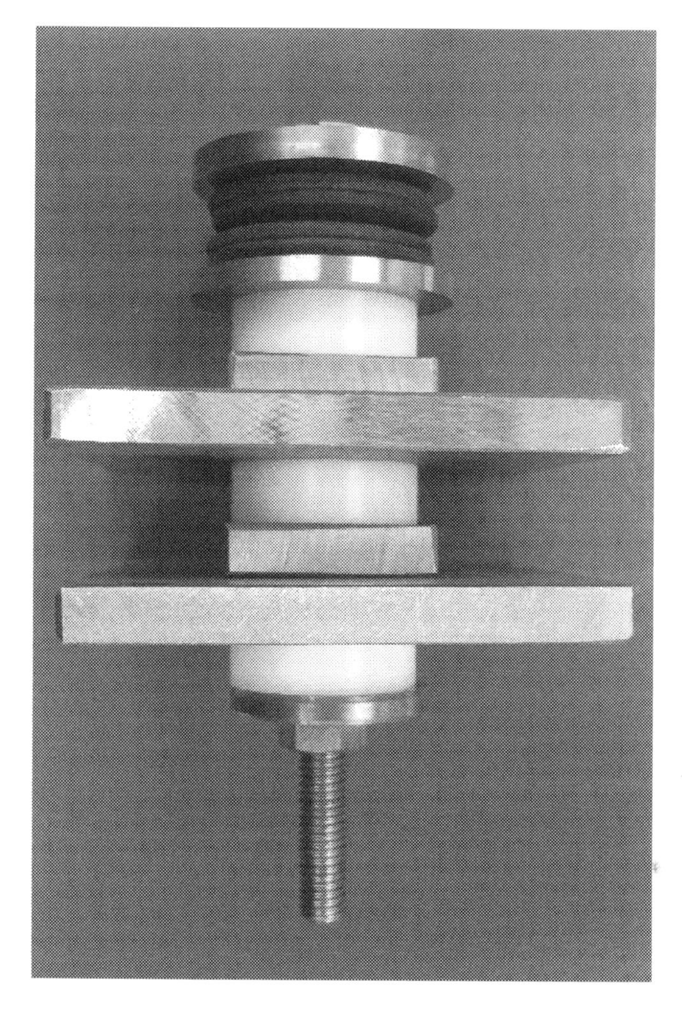

FIGURE 11.1 Photograph of a metal-metal crevice assembly. The large metal samples (2205 duplex stainless steel) represented the cathode while the small metal samples represented anode (either nickel alloy 718 or 17-4 PH precipitation hardened stainless steel).

All samples were fully immersed in the test solution, except for one sample which was maintained at the waterline (Table 11.1, crevice 6). This was done to mimic the field scenario of seal ring exposed at the waterline area.

11.2.1.2 Microbial Consortium

Tubes containing sterile culture medium for iron-oxidizing bacteria (IOB) (NACE 2014), sulphide-producing prokaryotes (SPP) (Salgar-Chaparro, Lepkova, et al. 2020), and acid-producing bacteria (APB) (NACE 2014) were inoculated with corrosion products and incubated in darkness at 40°C for 28 days. Changes in the media color and precipitation were observed after the incubation period. The incubation temperature was selected to simulate wet lay-up condition. Although results from 16S rDNA sequencing of field samples indicated that the microbial community in the failed seal rings was dominated by IOB, active APB and SPP were also detected using culture media. Therefore, the bacteria growing in the different culture media (enrichment cultures) were pooled and used as a microbial consortium to inoculate the biotic tank in this study. The diversity of the microbial consortium was expected to support microbial activity during both the anaerobic and aerobic exposure periods mimicking the field scenarios. The final inoculum contained IOB (5.7 x 10^5 cell/mL), APB (2.7 x 10^4 cell/mL), and SPP (1.6 x 10^5 cell/mL) in ratios similar to those detected in the failed seal rings using the most probable number (MPN) method (da Silva et al. 2013).

11.2.1.3 Test Solution

Filtered natural coastal seawater was used in this study. The filtration was performed using 0.22 μm membrane filters. During the anaerobic period, the test solution was sparged with 20/80 CO_2/N_2 (1 bar). The pH of the test solution was maintained at 7 ± 0.2 by the addition of sterile $NaHCO_3$ solution. For the aerobic environment used in the second half of the test, the solution was sparged with sterilized air (filtered to 0.2 μm), no pH adjustment was performed, and solution pH fluctuated around 8–8.2.

11.2.2 Procedure

11.2.2.1 Tank Setup

Experiments were conducted in two 25-L glass tanks for biotic (i.e., containing bacteria) and abiotic (no bacteria) environments. Setup and samples arrangements were identical for both tanks. Figures 11.2 and 11.3 show schematic diagrams of the tank from side view and top view, respectively. Each tank housed sterilized specimens: three individual crevice assemblies, three metal-metal crevice assemblies, and two galvanic couples. One 17-4 PH: DSS couple was suspended at the waterline (interface between gas and liquid phase). Details of each assembly was described in section 11.2.1. To elevate the assemblies from the bottom of the tank, two sterilized PTFE holders were used. After the samples were placed on the holders, wires connected to the samples were slid through the glass tubes on the acrylic lid. Parafilm was used to wrap the tube opening to avoid oxygen ingress. An immersion heater and a thermocouple (both sterilized by immersion in 70% ethanol for at least 30 minutes) were fitted into the dedicated space on the lid. Double junction reference electrodes Ag/AgCl (3 M KCl) were fitted with sterilized Luggin capillaries filled with 3% KCl/1.5% Agar. The assembled Luggin capillaries were then fitted into the dedicated space on the lid, i.e., one on each side of the tank. Two clean and sterile porous glass frits as gas inlets were placed on each side of the tank.

The glass tanks were assembled in a bio-safety cabinet equipped with a UV light, and after assembling, the tanks were UV irradiated another 20 minutes to ensure complete sterilisation.

A hotplate was placed beneath the tanks to control the rotational speed of a magnetic stirrer. To distribute the load of the tank, four specially made frames were

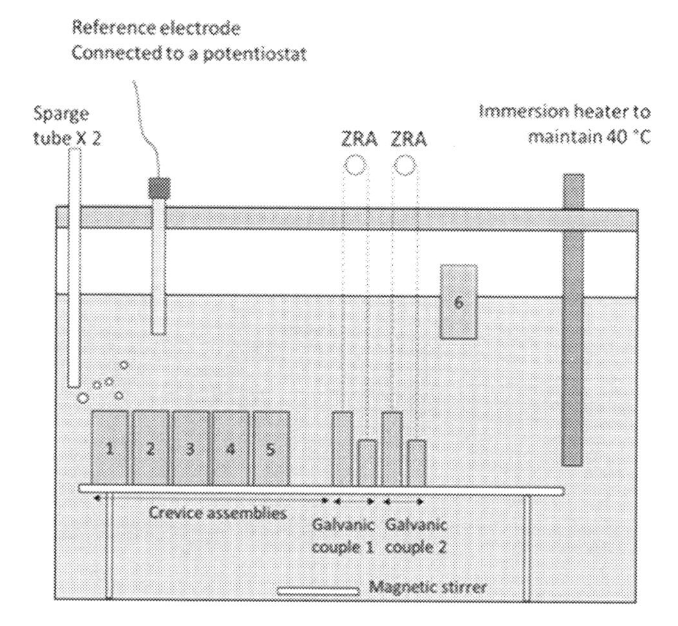

FIGURE 11.2 Schematic diagram of experimental tank setup (side view).

placed under the tub. Subsequently, 25 L of deoxygenated filtered seawater was transferred via a peristaltic pump into each tank through sterilized Silastic tubing. The test solution was mildly agitated with a magnetic stirrer bar rotated at 200 rpm and by constant gas sparging. The sparge gas was sterilized by passing it through 0.2 μm syringe filters.

11.2.2.2 Test Sequence

Experiments were conducted in the sequence depicted in Figure 11.4. The gas was sterilized through 0.2 μm syringe filters prior to entering the test tanks. The tubing was sterilized via autoclaving. The microbial consortium was inoculated at the start of the experiment only in the biotic tank. To ensure that microbes were active throughout the exposure, approx. 30% of the test solution was periodically replenished with filtered sterilized fresh seawater and inoculated with additional microbial consortium. This is illustrated in the timeline shown in Figure 11.4.

At the end of week 4, the environment was switched to aerobic by bubbling filtered air into the solutions using air pumps. Switching the gas was aimed at simulating the transition from normal operating condition (anaerobic) to oxygen contaminated wet lay-up phase (aerobic) (Salgar-Chaparro, Darwin, et al. 2020).

11.2.3 Analysis

11.2.3.1 Electrochemical Measurements

The open circuit potential (OCP) of each crevice assembly was continuously monitored using an ACM Potential 20 high impedance voltmeter. The galvanic

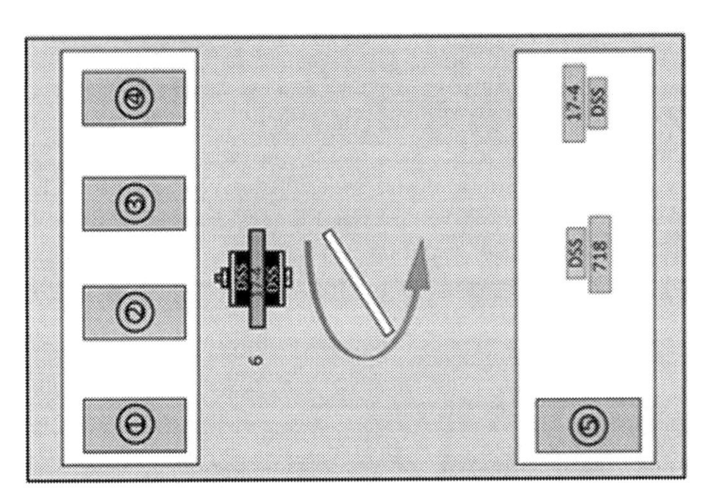

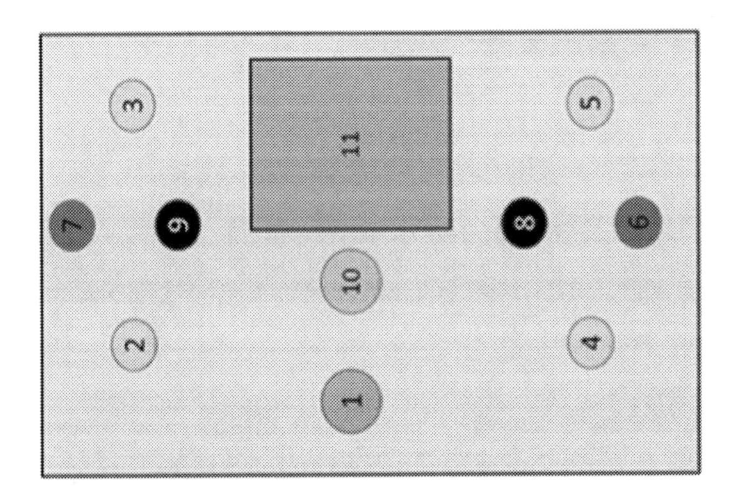

Port Number	Description
1	Immersion Heater
2,3,4,5	Test specimen wires
6,7	Sparge tubes
8,9	Reference electrodes
10	Top up port
11	Hand hole

FIGURE 11.3 Schematic diagram of the experimental tank (top view). From left to right: tank internal, tank lid, and identifications.

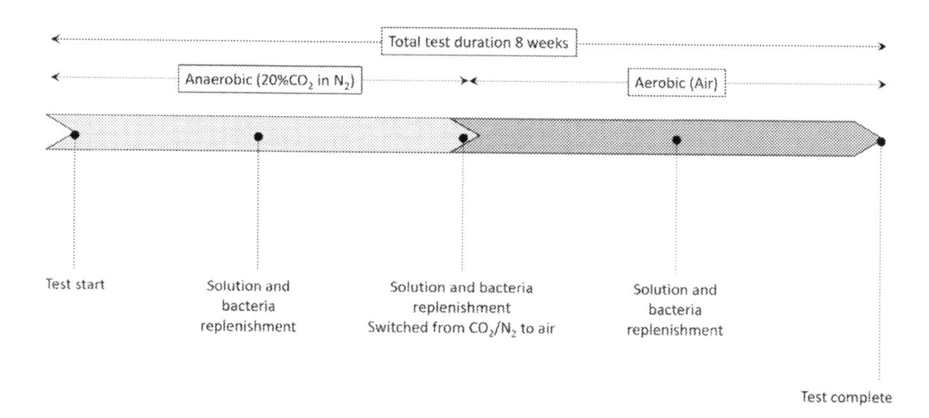

FIGURE 11.4 Experimental sequence. To maintain identical environment, solution replenishment occurred for both biotic and abiotic condition. The bacteria were only present in the biotic tank.

current density between DSS:718 and DSS:17-4 PH was monitored using a Gamry Potentiostat (Reference 600) operating in zero-resistance ammeter (ZRA) mode.

11.2.3.2 Mass Loss and Visual Inspection

At the completion of experiments, crevice assemblies were removed from the tanks, disassembled and photographed. After sessile bacteria had been removed, the samples were cleaned to remove any corrosion products and weight loss measurements performed according to ASTM G1-03 (ASTM 2017). Subsequently, the extent of localized corrosion was quantified using a 3D surface profilometer (Infinite Focus Microscope, Alicona). On samples that had undergone localized attack, maximum pit depth was recorded and converted into an equivalent localized penetration rate (PR) using the following equation (NACE 2013):

$$PR\ (mmpy) = \frac{Depth\ of\ deepest\ pit\,(mm)}{exposure\ time\,(days)} \times 365$$

The derivation of PR was based on two assumptions: 1) corrosion initiated immediately upon the introduction of oxygen and 2) corrosion propagated at a linear rate. Therefore, it should be noted that this conversion provides semi-quantitative information regarding the life of seal rings and that variations to the reported values are possible if the actual corrosion behavior deviates from one (or both) of the assumptions.

11.2.3.3 Microbiological Analysis

A culture-dependent method was employed to assess microbial activity during and at the end of the experiments. An aliquot of test solution was periodically taken (every two weeks) from the biotic tank to determine the presence of bacteria and the corresponding microbial population in the planktonic community. Further, the sessile community (attached to the metal sample) was determined at the completion of experiments using culture-dependent method. The sessile community present in 17-4 PH SS (coupled to DSS) was further characterized using 16S rRNA gene sequencing as described by Salgar-Chaparro, Lepkova, et al. (2020).

Coupons were removed and immersed in sterile phosphate-buffered saline (PBS) in sterile glass jars. The sample jars were placed into a sonicating water bath. After 20 seconds of sonication, the jars were removed and placed in an ice bath for 30–40 seconds. This process was repeated until a total two minutes of sonication was completed. Then, the coupon was removed from the suspension liquid and placed in fresh sterile PBS. The suspension was inspected under a light microscope to determine the presence of microorganisms. If bacteria were still present, the coupon was transferred to fresh liquid and the sonication process repeated until no bacteria were detected in the suspension indicating all sessile bacteria have been removed from the steel surface. Subsequently, the suspension from each round of sonication (of the same steel sample) was pooled together and centrifuged at 700 x *g* for two minutes to remove corrosion products and debris, followed by centrifugation at 5,000 x *g* for 1 hour to pellet the cells. This pellet was then resuspended in 10 mL of sterile PBS and used for subsequent analysis.

To detect and enumerate the number of microorganisms, most probable number method was carried out for assessment of the following microbial populations (NACE 2014, da Silva et al. 2013):

- Sulphide-producing prokaryotes (SPP): including detection of sulphate-reducing bacteria (SRB) and thiosulphate-reducing bacteria (TRB).
- Acid-producing bacteria (APB).
- Iron-oxidizing bacteria (IOB).

One (1) mL of each sample was inoculated into the culture media targeting the aforementioned microbial populations. Serial dilutions were conducted for each medium in duplicate to estimate the concentration of these microbes in the samples. Microbial cultures were incubated at 40°C. Following incubation, all dilution vials were examined, and bacterial numbers were estimated based on the number of vials showing a positive reaction.

11.3 RESULTS AND DISCUSSION

11.3.1 Electrochemical Measurements: Open Circuit Potential (OCP) and Galvanic Current Density

Figure 11.5a and 11.5b show the variation in OCP with time of creviced samples in the biotic and abiotic tank, respectively. The observed spikes in OCP correlate with the time at which the test solution was replenished. These spikes can be considered as an artefact of this activity and can be ignored. In the biotic tank, the OCP of all samples decreased by ~200 mV during the first days of exposure and then remained relatively stable during the anaerobic period (first 28 days). Upon switching to the aerobic condition, a marked change in OCP was observed for all samples; the OCP

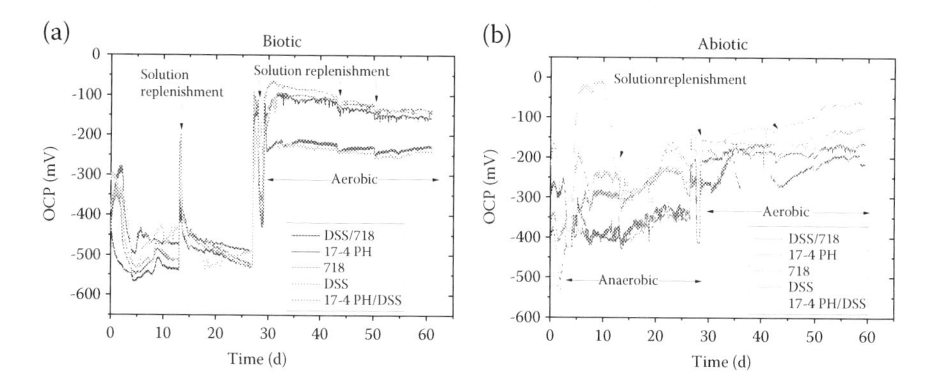

FIGURE 11.5 Open circuit potential of specimens in a) biotic and b) abiotic environments. The anaerobic condition in the first 28 days was achieved by sparging the test solutions with 20% CO_2/N_2 gas mixture while the subsequent aerobic condition was attained by sparging with air. Gas was sterilized with 0.2 μm syringe filters.

immediately increased by ~250 mV for 17-4 PH and 17-4 PH: DSS galvanic couple and by ~350 mV for DSS, 718, and 718:DSS galvanic couple.

The OCP of samples in the abiotic tank did not decrease during the first days of exposure under anaerobic conditions gas as observed in the biotic tank. Overall, the OCP for all samples gradually increased with time. The abrupt increase in OCP on the addition of oxygen observed in the biotic tank was not observed in the abiotic tank.

The more negative potential of alloys and galvanic couples under biotic/anaerobic conditions indicated that their surfaces were in a more active state compared to the abiotic/anaerobic environment where the specimens may have retained more passivity (passive film). An abrupt change in OCP is expected when transitioning from anaerobic to aerobic conditions because of the greater oxidising potential of oxygen, however, this only occurred in the biotic tank.

In Figure 11.5a (biotic condition), it is interesting to compare the OCP between 17-4 PH under anaerobic conditions (red line, left side of the graph) and DSS under aerobic conditions (pink line, right side of the graph). The OCP of 17-4 PH would approximate more the OCP of this alloy in a crevice when oxygen is depleted while the OCP of the DSS would approximate the potential outside the crevice. The difference in potential under these conditions is ~ 350 mV, which represents a significant driving potential for galvanic corrosion. The OCP of 17-4 PH in a crevice is expected to drop to even lower values due to acidification compared with pH 7, the pH maintained in the bulk fluid during biotic/anaerobic conditions. Therefore, based on the OCP values recorded at the two exposure environments, it can be assumed that galvanic corrosion is more likely to occur in 17-4 PH when coupled to DSS in biotic/aerobic conditions. It would be less likely under abiotic conditions (Figure 11.5b) because the difference in OCP is not as great. The alloy having the more positive potential behaves as the cathode driving corrosion of the more electro-negative alloy (anode). Although a difference in OCP indicates the

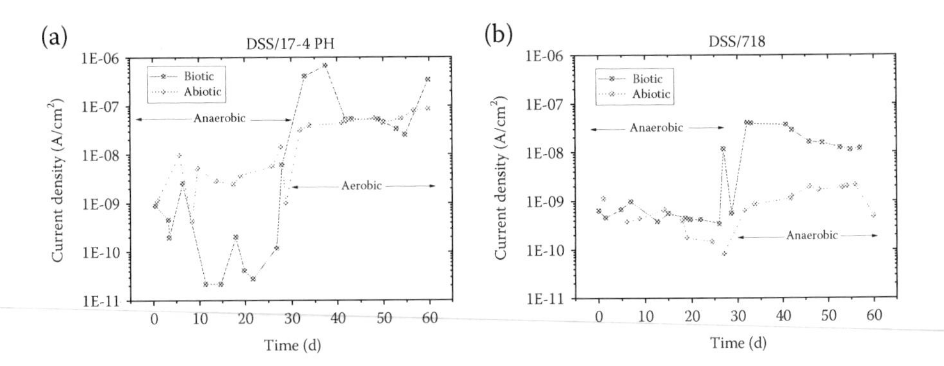

FIGURE 11.6 Galvanic current density between DSS coupled with a) 17-4 PH and b) 718 in biotic and abiotic environment. The anaerobic condition in the first 28 days was achieved by sparging the test solutions with 20% CO_2/N_2 gas mixture while the subsequent aerobic condition was attained by sparging with air. Gas was sterilized with 0.2 μm syringe filters.

possibility of galvanic corrosion, it provides no indication on the kinetics of the reaction or rate of corrosion. The kinetics of the reaction can be obtained, however, from measuring the galvanic current with a ZRA. The variation of galvanic current density between DSS specimens coupled with 17-4 PH and 718 is illustrated in Figure 11.6a and 11.6b, respectively. These measurements were performed to evaluate the influence of this form of corrosion on the overall corrosion mechanism; i.e., when 17-4 PH and 718 seal rings were electrically in contact with DSS hubs. The galvanic current density was responsive to the variation in the conditions; i.e., biotic vs abiotic, aerobic vs anaerobic. The direction of electrons flow confirmed that DSS was the more noble material and acted as the cathode. During the initial anaerobic period in the abiotic tank, the galvanic current density of the 17-4 PH vs DSS couple fluctuated between 10^{-9} and 10^{-8} A/cm^2. Switching from anaerobic to aerobic environment resulted in a slight increase in the galvanic current density, which was also observed on 718 coupled with DSS. Despite the small increase, the current density is still well below 10^{-7} A/cm^2, which is equivalent to a corrosion rate of below 10^{-3} mm/y. In the biotic tank, the increase in the galvanic current density upon the introduction of oxygen is greater compared to that in the abiotic tank. This could indicate that there is a synergistic effect between the microorganisms and dissolved oxygen. However, the galvanic current density measured is still negligible and therefore, it can be concluded that galvanic corrosion alone was not responsible for accelerated seal ring failures observed on the FPSO.

11.3.2 Visual Inspection and Corrosion Analysis

11.3.2.1 Visual Inspection

At the end of the eight weeks' exposure, crevice assemblies were removed from the test tanks and photographed as depicted in Figures 11.7 to 11.10. In the abiotic tank, individual crevice assemblies remained mostly shiny and did not exhibit severe corrosion (Figure 11.7). In 17-4 PH: DSS couples (Figure 11.8), both as a fully

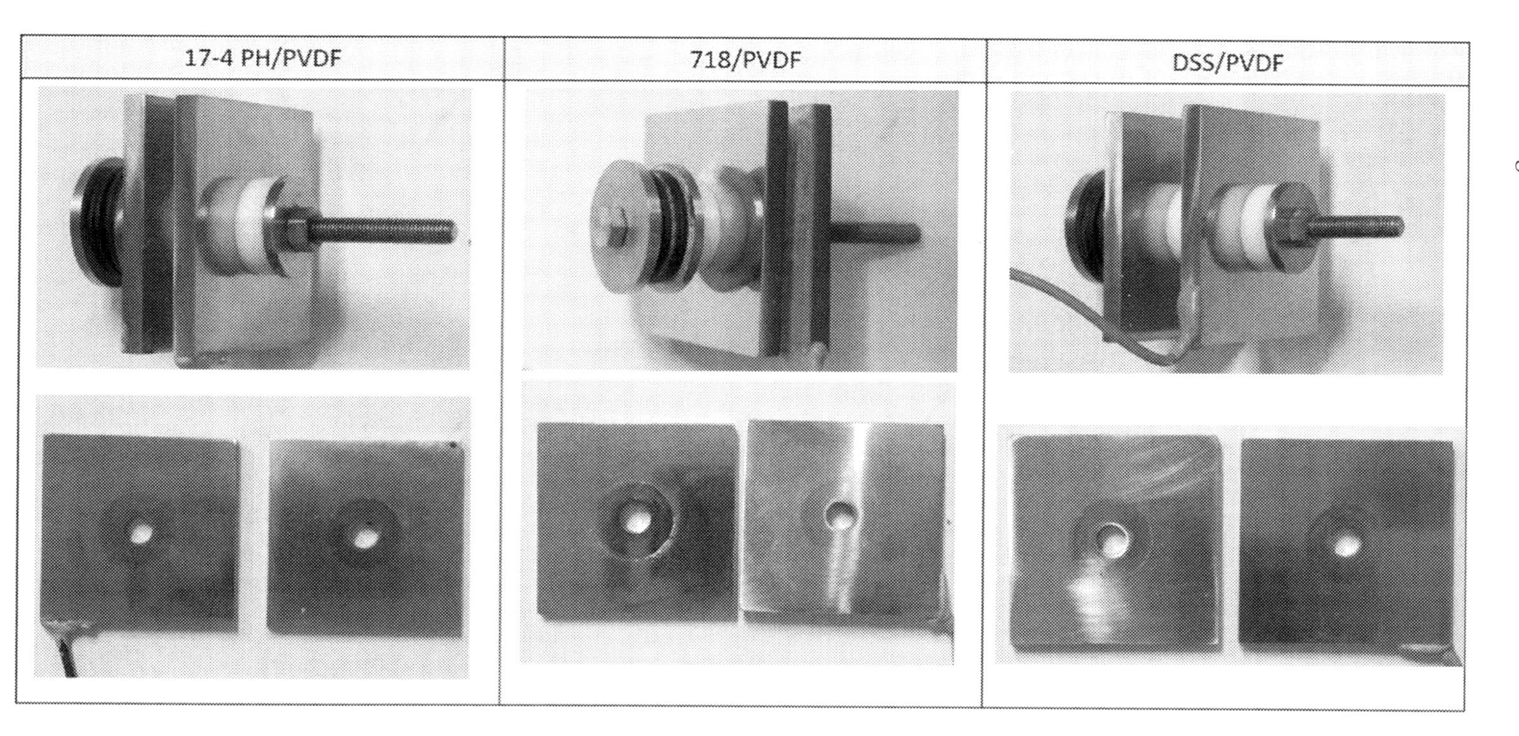

FIGURE 11.7 Photographs of individual crevice samples (PVDF crevice formers) after eight-week exposure in the abiotic environment.

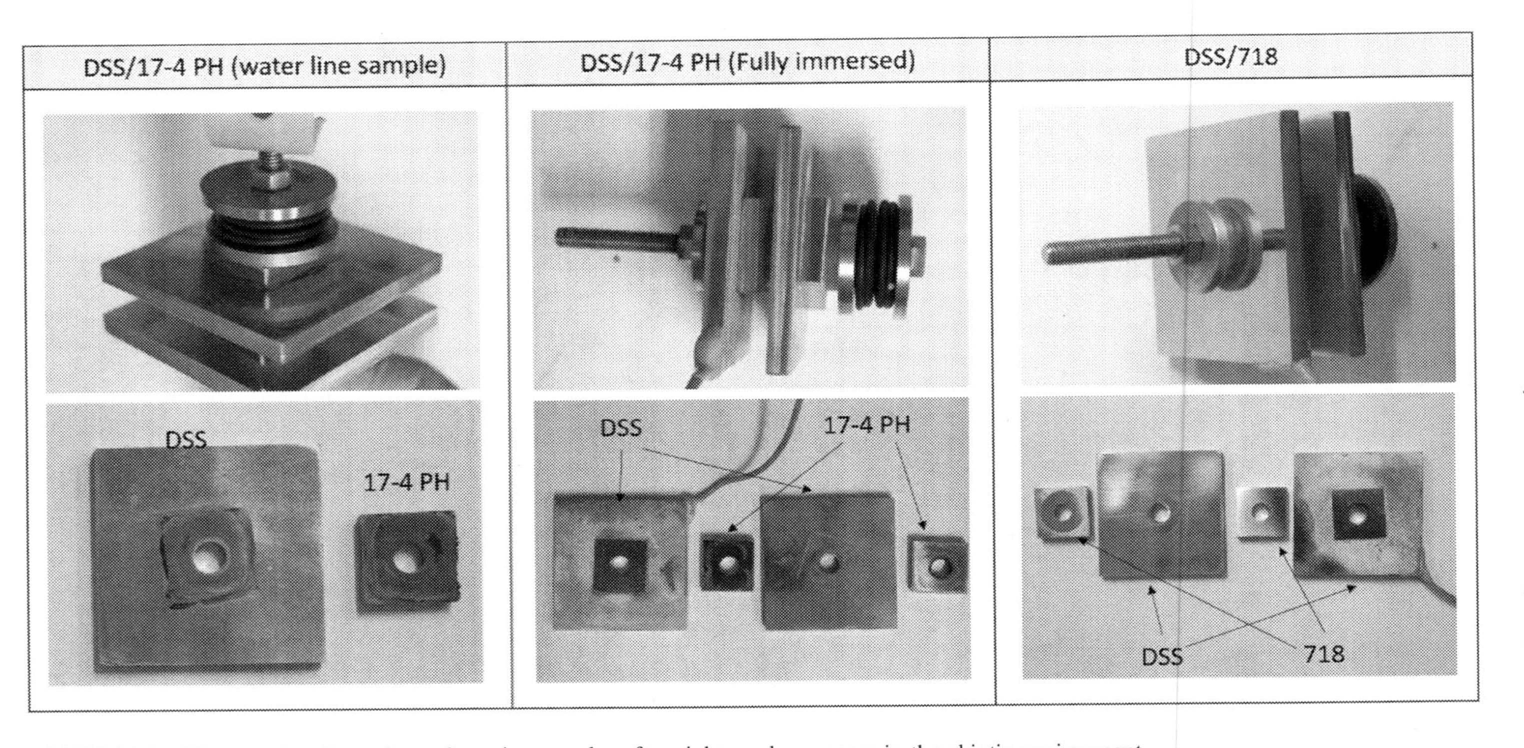

FIGURE 11.8 Photographs of metal-metal crevice samples after eight-week exposure in the abiotic environment.

immersed and as a waterline sample, corrosion product (red precipitates) is evident at the interface-crevice opening. Conversely, the 718: DSS couples did not show signs of corrosion (Figure 11.8). The white precipitates found on all samples were expected to be carbonate scale that precipitated because of the rise in solution pH from 7.0 to 8.0 after the switch from anaerobic to the aerobic environment.

In the biotic tank, samples were covered with yellowish precipitates, which may indicate that the test solution in this tank contained a higher concentration of iron ions relative to the abiotic tank. Ferrous ion released from corrosion process is oxidized to ferric ions in the presence of dissolved oxygen producing yellow iron oxides and/or hydroxide precipitates. After cleaning the samples, individual crevice samples DSS and 718 showed no sign of corrosion (Figure 11.9). Conversely, individual crevice 17-4 PH showed evidence of severe crevice corrosion (Figure 11.9). The crevice interface had corrosion product precipitates, whereas the black area inside the crevice suggests severe dissolution of the base metal. Nodules of corrosion products were also found outside crevice areas. The corrosion process was intensified when 17-4 PH was in contact with DSS, both as a fully immersed and as a waterline sample, as demonstrated by the increased amount of corrosion product and the portion of the black area within the crevice (Figure 11.10). As for the abiotic exposure, the 718: DSS couples did not show signs of corrosion in the biotic environment (Figure 11.10).

11.3.2.2 Localized Corrosion Analysis

After samples were cleaned of corrosion products and deposits, they were inspected for localized corrosion with 3D-surface profilometry. The localized penetration rates (PR) were calculated based on the maximum pit depth measured on each sample over the four-week aerobic period. It was assumed that the pit initiated with the introduction of oxygen and propagated linearly with time.

Figure 11.11 shows PR of individual crevices (alloy with PVDF crevice former). Only the results for artificially creviced 17-4 PH material are shown in the Figure since it was the only alloy that showed localized corrosion, i.e., no localized corrosion attack was found on 718 and DSS in either biotic or abiotic conditions. The maximum pit depth found at each condition is also given in the figure. It is apparent that the presence of microorganism accelerated the localized corrosion on 17-4 PH as PR increased from 0.30 mm/y in the abiotic environment to 1.6 mm/y in the biotic environment. The estimated PR for 17-4 PH may be conservative, as localized corrosion could have initiated, for example, after one-week exposure in the aerobic conditions and the pit depth resulting from only three weeks of active corrosion.

For metal-metal crevice samples, only 17-4 PH exhibited severe corrosion, particularly in the presence of microbes where crevice corrosion was accelerated by >50 times to the equivalent of abiotic condition. Corrosion was not observed on 718 and DSS in metal/metal couples (Figure 11.11). The spring-loaded artificial crevice formed in the laboratory may not be representative of actual seal ring geometry, which has a tapered seal ring-hub interface. The tapered contact edge of a seal ring creates a crevice of increasing tightness, which could have delivered the optimal crevice, and geometry for initiation and propagation of crevice corrosion in the

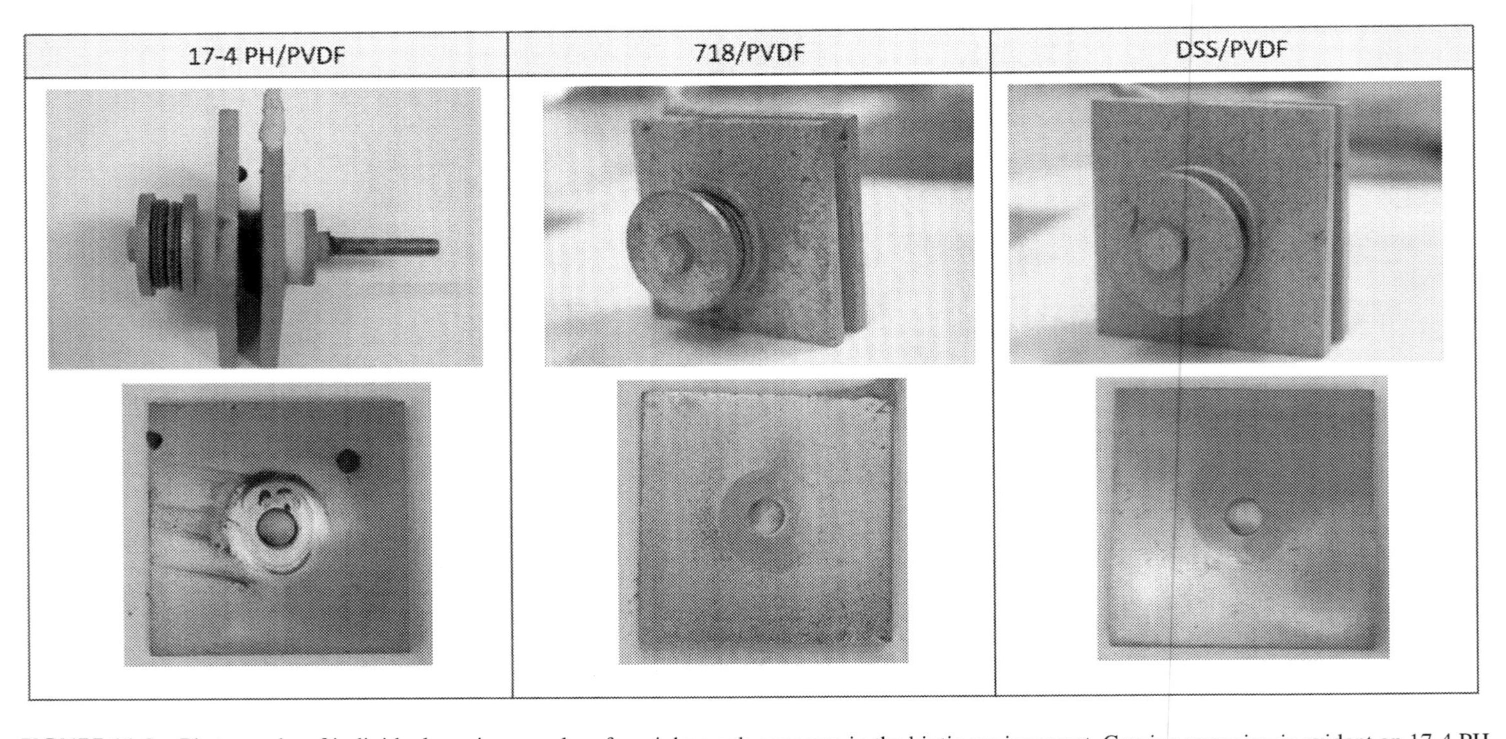

FIGURE 11.9 Photographs of individual crevice samples after eight-week exposure in the biotic environment. Crevice corrosion is evident on 17-4 PH while DSS and 718 show no corrosion.

FIGURE 11.10 Photographs of metal-metal crevice samples after eight-week exposure in the biotic environment.

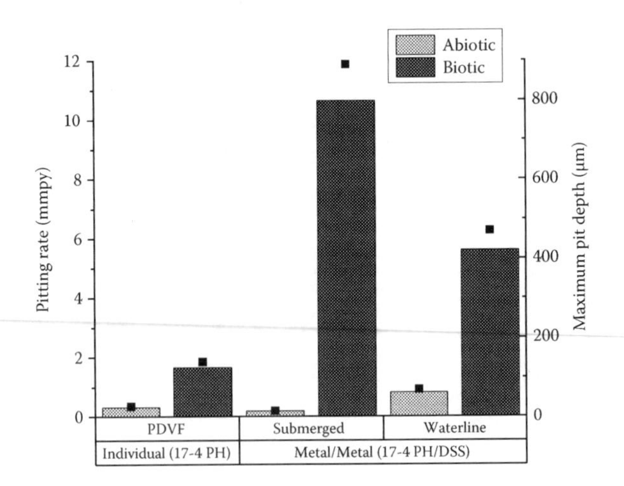

FIGURE 11.11 Pitting rate (PR) and maximum pit depth observed on 17-4 PH alloys as individual crevice sample and metal-metal crevice sample in abiotic and biotic environment. PR was estimated based on a maximum pit depth observed on a sample and assumed linear propagation. Pitting was not observed in other alloys evaluated.

oxygenated preservation fluid (treated seawater). The torque applied on the artificial crevice in this study was 3 N.m whereas much greater torque is required on actual seal rings. Greater applied torque has been shown to accelerate crevice corrosion (Thierry and Larché 2017). Likewise, a larger cathode to anode ratio in the actual system could enhance galvanic effects enabling the faster corrosion found in the field. Additionally, 718 and DSS are more resistant to corrosion compared to 17-4 PH and these alloys may require longer time for crevice corrosion to initiate and propagate to a detectable level.

11.3.2.3 Corrosion Rate by Mass Loss

Figure 11.12 compares corrosion rate by mass loss of crevice assemblies in different environments. Corrosion rates for teach alloy at the different exposure conditions are given in Table 11.2. Corrosion rates by mass loss were estimated based on the area under the crevice of each sample. This was also the case for waterline samples because the submerged area exposed to the gas and aqueous phases could not be accurately determined due to the fluctuations in the water level in the two tanks over time. It is, therefore, possible that the average corrosion rates for wetted areas of the waterline samples are greater than reported. 17-4 PH was the material most susceptible to corrosion. It exhibited the highest mass loss corrosion rate in relation to other materials. The combined effect of dissimilar metals and crevice corrosion is evident when comparing Figure 11.12a and b. In the absence of metal couples, low corrosion rates were measured.

Compared to individual crevice assemblies, under totally submerged conditions, corrosion rate values increased by approximately 50 times when 17-4 PH was coupled to DSS in the abiotic condition. Under biotic conditions, the coupling to DSS caused an additional increase in the corrosion rate in 17-4 PH from 0.16 mm/y

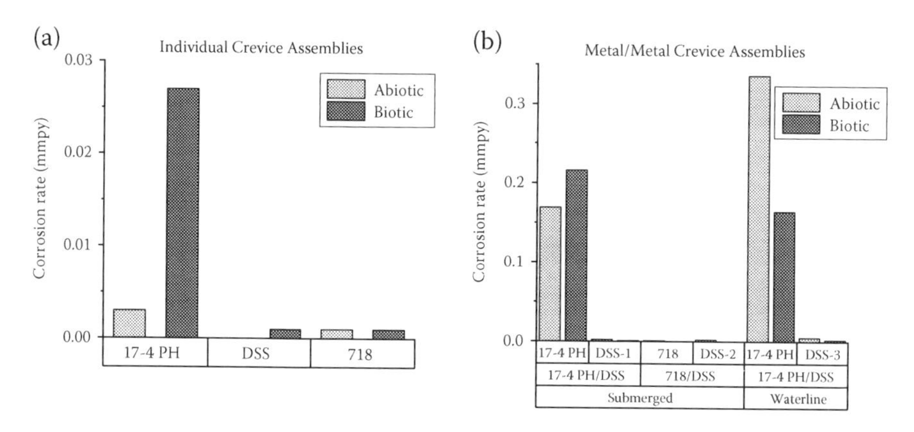

FIGURE 11.12 Comparison of weight loss corrosion rate observed in abiotic and biotic environment. a) individual crevice assemblies and b) metal-metal crevice assemblies.

TABLE 11.2
Corrosion Rates by Mass Loss (mmpy)

Sample	Abiotic	Biotic
17-4 PH	<0.01*	0.03
DSS	<0.01	<0.01
718	<0.01	<0.01
17-4 PH/DSS	0.17/<0.01	0.22/<0.01
718/DSS	<0.01/<0.01	<0.01/<0.01
17-4 PH/DSS (Waterline)	0.34/<0.01	0.16/<0.01

Note
* Limit of Reporting

to 0.22 mm/y. The presence of microbes aggravated corrosion of 17-4 PH specimens in the fully submerged condition but it did not have the same effect at the waterline samples.

Corrosion results highlight the importance of localized corrosion measurements (Figure 11.11), particularly as stainless steels and high alloy steels do not usually undergo general or uniform corrosion.

11.3.3 Microbiological Analysis

Planktonic cells were collected every two weeks and MPN measurements were carried out. Results are shown in Figure 11.13. IOB had the largest amount of growth and were active throughout the eight-week period. APB were in high numbers during the anaerobic phase but decreased significantly after the

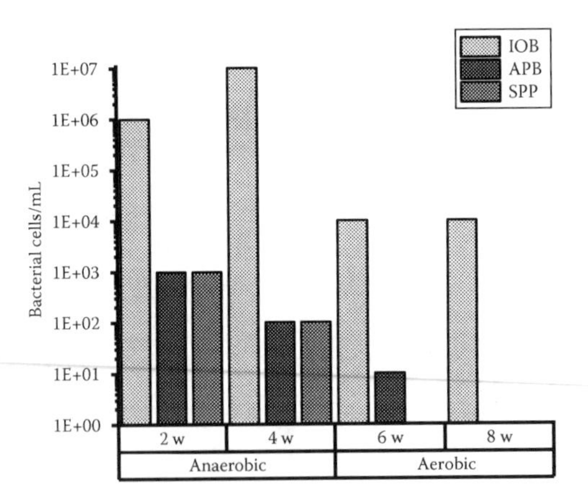

FIGURE 11.13 Enumeration of microbial populations IOB, APB, and SPP in planktonic state in the biotic tank.

introduction of oxygen; after eight weeks no APB were detected in the tank. SPP were only present in low numbers during the anaerobic phase and were not detected at six or eight weeks during the aerobic phase. The inoculum was not counted at week 8 because there was no replenishment at that time.

Figure 11.14 displays the results of microbial groups of sessile communities. The dominant group of sessile microbes were IOB, followed by APB (as observed in the planktonic community). No active SPP were detected at eight weeks, i.e., after four weeks of aerobic condition. SPP are generally strict anaerobes. Therefore, there is no detectable growth in both planktonic and sessile growth tests. Many IOB and some APB can grow with (or tolerate) oxygen, this is reflected in the growth tests, where IOB and APB were detected at end of the experiment. The survival of IOB and APB and the abundance of IOB over APB after the aerobic period is in agreement with the microbial analysis of failed seal rings (Salgar-Chaparro, Darwin, et al. 2020).

16S rRNA gene sequencing showed that the sessile microbial community found in 17-4 PH coupled to DSS was dominated by microorganisms associated with the following families: Rhodobacteraceae (29%), Alteromonadaceae (18%), Pseudomonadaceae (11%), unclassified Caldithrixales (7%), Phyllobacteriaceae (4%), unclassified Phycisphaerales (3%), Rhodospirillaceae (2%), Flammeovirgaceae (2%), Hyphomicrobiaceae (2%), and Planctomycetaceae (2%). Other families were detected in lower abundance.

Even though the exact mechanism cannot be concluded from the scope of this work, it can be hypothesized that IOB were the main microorganisms involved in the failure.

11.3.4 Proposed Synergistic Effects of Contributing Factors on Seal Rings Failure

The results from this investigation show that the presence of oxygen, a galvanic couple, a crevice and bacteria had synergistic effects on corrosion rate. The

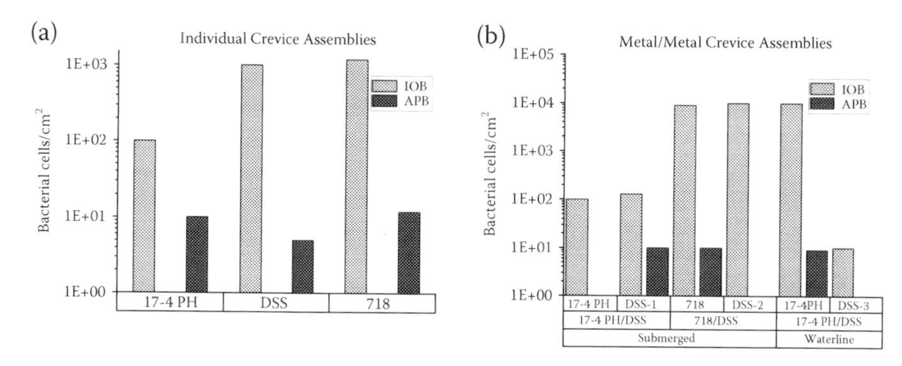

FIGURE 11.14 Enumeration of microbial populations IOB, APB and SPP in sessile state on 17-4 PH, 718, and DSS materials.

presence of oxygen appears to be necessary for corrosion to initiate in the occluded areas within crevices, where the dissolved oxygen concentration gradually depletes. It is known that differential aeration cells eventually develop and metal dissolution is accelerated inside crevices via an autocatalytic process (White, Weir, and Laycock 2000). Migration of chloride ions into occluded areas inside crevices to maintain electro-neutrality causes accumulation of chloride ion and acidification through hydrolysis of metals. The high concentration of chloride ions and low pH prevents alloys repassivating. This process is then aggravated by galvanic corrosion (coupling with a more noble alloy, DSS) and even more so by MIC (the presence of microbes, IOB in particular). The relative contribution of each parameter was qualitatively demonstrated in Figure 11.15.

IOB appears to be responsible for passivity breakdown and localized corrosion in creviced 17-4 PH coupled to DSS. It can be hypothesized that IOB couples iron oxidation (Fe to Fe^{2+} and/or Fe^{2+} to Fe^{3+}) to reduction of oxygen resulting in the precipitation and accumulation of corrosion products. It is known that IOB can produce dense layers of deposit causing oxygen exclusion immediately under the occluded area (crevice). This differential aeration cell enhances the formation of an anodic site in the more susceptible steel, i.e., 17-4 PH SS in this study. The propagation is then similar to that of the autocatalytic process of classical pitting and crevice corrosion. The presence of IOB facilitates and drives the conversion of Fe^{2+} to Fe^{3+}, thereby accelerating overall metal dissolution. Furthermore, APB was also detected in the sessile community. MPN results suggest that APB survived aerobic condition despite some reduction in cell numbers. The presence of both groups indicates that they may work synergistically. It is known that the fermentative bacteria such as APB release metabolites that can acidify the environment. The acidic condition then promotes depassivation of steel and induces metal dissolution. For iron-based alloys, Fe^{2+} will be generated which can subsequently be oxidised by IOB forming Fe^{3+}. A recurring question in MIC investigations is whether IOB actually cause corrosion or are opportunists attracted by the release of ferrous ions from abiotic corrosion processes. The high localized corrosion rates of 17-4 PH in the biotic experiment (under the predominant activity of IOB) indicates an active

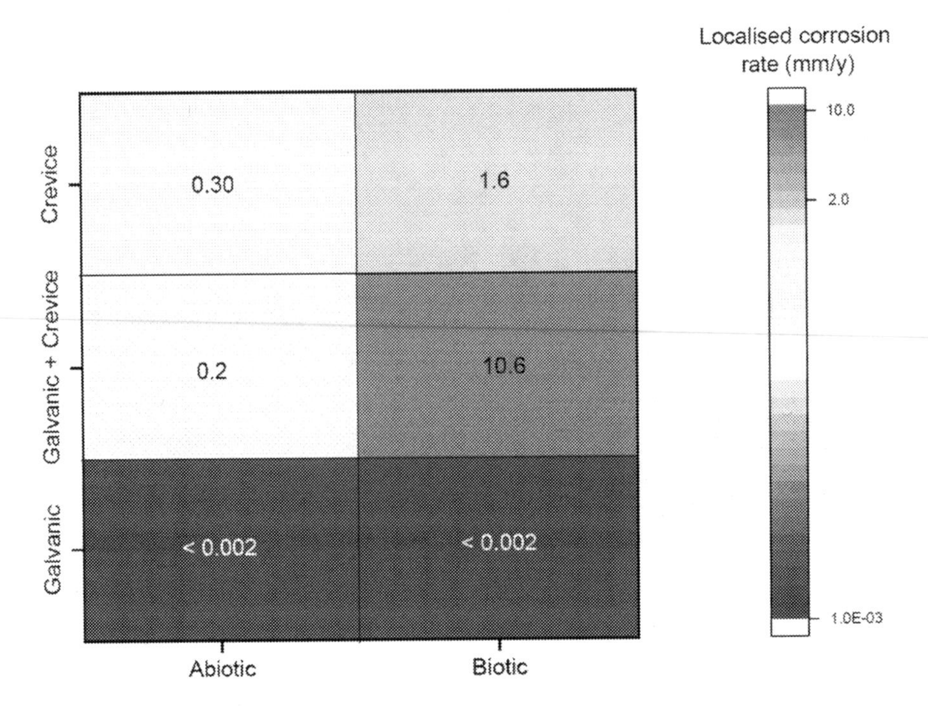

FIGURE 11.15 Diagram demonstrating relative and mutual influence of bacteria, crevice, and galvanic effects on pitting rate of 17-4 PH after eight-week exposure to test solution. PR were extrapolated from maximum pit depth and four-week exposure to aerobic condition.

involvement of IOB in the corrosion process and supports the hypothesis that microbes played a key role in the severe corrosion failure of 17-4 PH seal rings on the FPSO.

11.4 CONCLUSIONS

The following outcomes were derived from the experimental results of this investigation:

- Crevice corrosion was detected only on 17-4 PH and was found to be accelerated by the presence of microbes (>50 times). The most severe localized penetration rate of 10.6 mm/y was obtained when 17-4 PH was coupled with DSS and in the presence of microbes. The instant increase in galvanic current density upon introduction of air also suggests that corrosion was triggered by the introduction of oxygen (air). Overall, results indicate that the corrosion failure of seal rings was the result of a combination of mechanisms including oxygen ingress, galvanic corrosion, crevice corrosion, and MIC.
- 17-4 PH should not be used as a seal ring material for processes where seawater flooding and potential oxygen ingress are anticipated due to its

susceptibility to crevice corrosion, particularly when coupled with a higher-grade alloy.

- Iron oxidising bacteria (IOB) were found to be the dominant sessile bacteria population on steel samples suggesting that they were responsible for the accelerated corrosion of 17-4 PH. APB were also detected and co-existed with IOB on most samples, albeit at lower numbers.
- No localized corrosion was observed on 718 and DSS both as individual crevice and as metal/metal crevice couples during the test period (eight weeks). However, caution must be taken in interpreting data considering that crevice geometries can be more aggressive in actual seal rings compared to artificial crevices used in the experiments. The discrepancy in corrosion between field observation and experimental results can be attributed to the difference in the geometry of seal rings and the artificial crevice used in this study. Spring-loaded artificial crevice formed in the laboratory may not be representative of actual seal ring geometry. Using the actual seal rings and hubs in this study was considered to be impractical within the investigation time and cost constraints. Failures on 718 and DSS seal rings occurred during three to four months of exposure to treated seawater. However, the total experimental time for conducting these experiments was eight weeks which may not be sufficient to initiate and propagate crevice corrosion of 718. Because of the above reasons, it is not possible to confirm the corrosion failure mechanisms of the 718 and DSS seal rings and to determine the duration at which seal rings may be exposed to oxygenate seawater before significant corrosion occurs.
- It is important to note that this work did not fully address the risk of corrosion during the operating condition (anaerobic condition and high temperature) because the samples were not analysed after the anaerobic period, but only after additional exposure to aerobic environments. There is a possibility that exposure during normal operating conditions could make the alloys more sensitive to crevice, galvanic, and MIC when exposed to oxygen environments. It also remains unknown whether bacteria alone can initiate corrosion during normal operating conditions. However, because oxygen and bacteria are identified as two key parameters determining corrosion of the FPSO's seal ring materials, this indicates that corrosion during regular service condition is unlikely if the system remains oxygen-free.
- Further from the above, ensuring the efficiency of chemical treatment and avoiding oxygen ingress is crucial to avoid similar corrosion failures. The preservation plan on the FPSO rigid arm and RTM consisted of addition of an oxygen scavenger and biocide to the hydrotest water. However, results indicated that these chemicals did not perform as expected, i.e., active microbes were recovered from corrosion products covering the failed rings. It can be speculated that oxygen ingress had a negative effect on the efficiency of the chemicals, i.e., depleting oxygen scavenger and reducing the concentration of active biocide chemical.
- This MIC investigation demonstrated a direct comparison between the damage observed in the laboratory (location, nature, corrosion rate) to that observed in the field.

REFERENCES

ASTM. 2017. "ASTM G1 Standard Practice for Preparing, Cleaning, and Evaluating Corrosion Test Specimens." West Conshohocken, PA: ASTM International.

Borenstein, Susan, and Philip Lindsay. 2002. "MIC Failure of 304L Stainless Steel Piping Left Stagnant After Hydrotesting." *Materials Performance* 41 (6):70.

da Silva, Neusely, Marta Hiromi Taniwaki, Valéria Christina Amstalden Junqueira, Neliane Ferraz de Arruda Silveira, Margarete Midori Okazaki, and Renato Abeilar Romeiro Gomes. 2013. *Microbiological Examination Methods of Food and Water: A Laboratory Manual / Neusely da Silva... [et al.].* Leiden: CRC Press/Balkema.

Darwin, Adam, Karthik Annadorai, and Krista Heidersbach. 2010. "Prevention of Corrosion in Carbon Steel Pipelines Containing Hydrotest Water - An Overview." *CORROSION 2010*, San Antonio, Texas, 2010/1/1/.

Huang, Weiji, Dake Xu, Greg Ruschau, Jie Wen, Jennifer Hornemann, and Tingyue Gu. 2012. "Laboratory Investigation of MIC Due to Hydrotest Using Seawater and Subsequent Exposure to Pipeline Fluids With and Without SRB Spiking." *CORROSION 2012*, Salt Lake City, Utah, 2012/1/1/.

Lenhoff, D. A. 1994. "Determining the Integrity of a Pipeline by Hydrotesting." *SPE Mid-Continent Gas Symposium*, Amarillo, Texas, 1994/1/1/.

Machuca, Laura L. 2014. "Microbiologically Influenced Corrosion: A Review Focused On Hydrotest Fluids In Subsea Pipelines." *ACA Corrosion and Prevention*, Darwin, Australia.

Machuca, L. L. 2017. "Microbiologically Induced Corrosion Associated with the Wet Storage of Subsea Pipelines (Wet Parking)." Edited by Dr. Torben Lund Skovhus, Dr. Dennis Enning and Dr. Jason S. Lee. 1st ed, *Microbiologically Influenced Corrosion in the Upstream Oil and Gas Industry*. Boca Raton, FL: CRC Press.

Machuca, Laura L., Stuart I. Bailey, Rolf Gubner, Elizabeth L. J. Watkin, Maneesha P. Ginige, Anna H. Kaksonen, and Krista Heidersbach. 2013. "Effect of Oxygen and Biofilms on Crevice Corrosion of UNS S31803 and UNS N08825 in Natural Seawater." *Corrosion Science* 67:242–255. doi: 10.1016/j.corsci.2012.10.023.

Machuca, L. L., S. Bailey, R. Gubner, E. Watkin, and A. Kaksonen. 2011. "Microbiologically Influenced Corrosion of High Resistance Alloys in Seawater." *CORROSION 2011*, Houston, Texas.

NACE. 2013. "SP0775 Preparation, Installation, Analysis, and Interpretation of Corrosion Coupons in Oilfield Operations." In *Standard Practice*. Houston, Texas: NACE International.

NACE. 2014. "NACE TM0194 Field Monitoring of Bacterial Growth in Oil and Gas Systems." In *Standard Test Method*. Houston, TX: NACE International.

Prasad, Rupi. 2003. "Chemical Treatment Options for Hydrotest Water to Control Corrosion and Bacterial Growth." *CORROSION 2003*, San Diego, California.

Salgar-Chaparro, Silvia J., Adam Darwin, Anna H. Kaksonen, and Laura L. Machuca. 2020. "Carbon Steel Corrosion by Bacteria from Failed Seal Rings at an Offshore Facility." *Scientific Reports* 10 (1):12287. doi: 10.1038/s41598-020-69292-5.

Salgar-Chaparro, S. J., K. Lepkova, Thunyaluk Pojtanabuntoeng, A. Darwin, and L. L. Machuca. 2020. "Nutrient Level Determines Biofilm Characteristics and the Subsequent Impact on Microbial Corrosion and Biocide Effectiveness." *Applied and Environmental Microbiology* 86. doi: 10.1128/AEM.02885-19.

Surkein, Michael, Weiji Huang, Erin Sullivan, and Gregory Ruschau. 2011. "Hydrotesting of LNG Tanks Using Untreated Brackish Water." *CORROSION 2011*, Houston, Texas.

Thierry, Dominique, and Nicolas Larché. 2017. "Crevice Corrosion of Stainless Steel in Tropical Seas." *CORROSION 2017*, New Orleans, Louisiana.

White, Stephen P., Graham J. Weir, and N. J. Laycock. 2000. "Calculating Chemical Concentrations During the Initiation of Crevice Corrosion." *Corrosion Science* 42 (4):605–629. doi: 10.1016/S0010-938X(99)00097-9.

Xu, Dake, Yingchao Li, Fengmei Song, and Tingyue Gu. 2013. "Laboratory Investigation of Microbiologically Influenced Corrosion of C1018 Carbon Steel by Nitrate Reducing Bacterium Bacillus licheniformis." *Corrosion Science* 77:385–390. doi: 10.1016/j.corsci.2013.07.044.

Zhao, Kaili, Tingyue Gu, Ivan Cruz, and Ardjan Kopliku. 2010. "Laboratory Investigation Of Mic in Hydrotesting Using Seawater." *CORROSION 2010*, San Antonio, Texas, 2010/1/1/.

12 Failure Analysis of Tubing in an Electrical Submersible Pump Well

A. Harmon, K. Crippen, and S. Leleika
GTI Energy

CONTENTS

12.1 INTRODUCTION AND BACKGROUND

Failure analysis was conducted on two pipe samples installed in an electrical submersible pump (ESP) well. The well was about 2,100 meters deep and produced 70,000 liters of water per day with an average water cut of approximately 96%. The well was continually treated with scale inhibitor via backside flush and had not been treated with corrosion inhibitor. Brand new or reused tubing that contained less than 15% metal loss was used for the installation. The well was running for 48 days before being pulled for low production. The pipe submitted for failure analysis was from the first instance of a severe corrosion attack.

During the removal operation for an ESP well the tubing string is scanned and pulled. Each joint of tubing is "stood up" or stacked vertically within the pulling rig. Tubing that contains more than 30% wall loss is laid down on the ground and hauled off for scrap. Once the replacement tubing arrives, the remaining tubing is re-run with the brand-new tubing. This process usually happens relatively quickly

DOI: 10.1201/9780429355479-14

only taking one to two days to complete, so the "stood up" tubing should not be out in the open air for more than a few days. If the well is scheduled to be down for a longer period, then the good tubing is stored in a pipe yard for future use in the same well.

12.2 SAMPLE DESCRIPTION

The first sample seen in Figure 12.1, Joint 580 meters, contained one through wall hole and several pits. The second sample seen in Figures 12.2 and 12.3, Burst Point, contained a through wall burst point and several pits and was located approximately 15 meters below the Joint 580 meters sample. Table 12.1 lists the samples analyzed for this failure investigation.

To investigate the root cause of failure, the following tests were performed:

FIGURE 12.1 Photo of interior pipe failure joint at 580 meters.

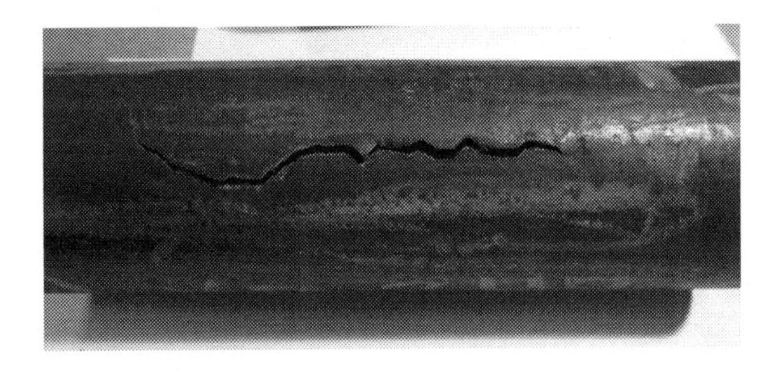

FIGURE 12.2 Photo of outer diameter pipe failure sample burst point.

FIGURE 12.3 Photo of inner diameter pipe failure sample burst point.

TABLE 12.1
Samples Analyzed for This Failure Analysis

Description	Received Date
Joint at 580 meters	May 7, 2015
Burst Point	May 7, 2015

1. Photo documentation and optical microscopy.
2. Chemistry analysis on pipe debris samples.
3. Metallurgical examination of the inner and outer surfaces.
4. Microbiological analysis using qPCR.

12.3 EXPERIMENTAL

12.3.1 Initial Sample Examination

All sample examination was performed on laboratory bench tops that are washed and allowed to dry. All sample handling is performed wearing gloves. The initial examination was conducted by a metallurgist, chemist, and microbiologist by their unaided eyes to characterize the sample and determine the necessary subsamples for testing.

12.3.2 Examination of the Pipe Surface Conditions

Visual examination was the first tool used to inspect the failure site. Surfaces are illuminated via both annular and remote light sources. The interior and exterior

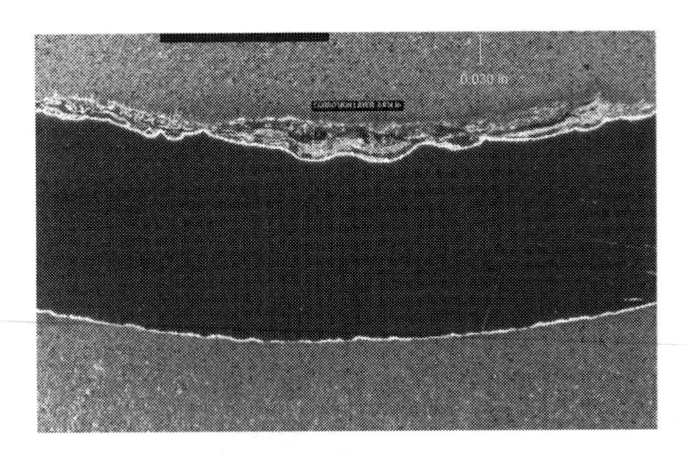

FIGURE 12.4 Microphotograph of corrosion layer in sample joint at 580 meters.

surfaces of the samples were observed and photographed for surface defects/damage, corrosion pits, and cracks. Micro-samples were cut at locations with corrosion pits. These specimens were examined by stereo optical microscopy.

12.3.3 Microbiological Analysis

Debris was collected from the interior pits and surrounding areas of each pipe sample and were analyzed using quantitative Polymerase Chain Reaction (qPCR) techniques for microbes commonly associated with Microbiologically Influenced Corrosion (MIC). The results were reported as the number of copies per gram of sample.

The specific microbes targeted for MIC utilizing qPCR were sulfate-reducing bacteria (SRB), sulfate-reducing archaea (SRA), acid-producing bacteria (APB), iron-oxidizing bacteria (IOB), denitrifying bacteria (DNB), and methanogens. SRB have been isolated from a wide range of environmental samples and are considered an aggressive corrosion-causing bacterium (Enning 2014). It is reasonable to assume that SRA will cause the same microbial corrosion as SRB based on their functional mechanisms of dissimilatory sulfate reduction (Thauer and Kunow 1995). In addition to SRB and SRA, APB are also considered aggressive corrosion-causing bacteria. The APB number reported here includes acetic acid–producing bacteria and butyric acid-producing bacteria. The IOB number reported here includes *Leptothrix*, *Sphaerotilus*, and *Gallionella*. Lastly, DNB and methanogens are frequently retrieved from pipeline samples, and also cause corrosion (Al-Nabulsi et al. 2015, Uchiyama et al. 2010).

12.3.4 Chemical Analysis

Ion chromatography was performed to test for the presence of anions. The debris was extracted with water, filtered, and injected onto an anion column using a carbonate/bicarbonate eluent and conductivity detection with ion suppression. An

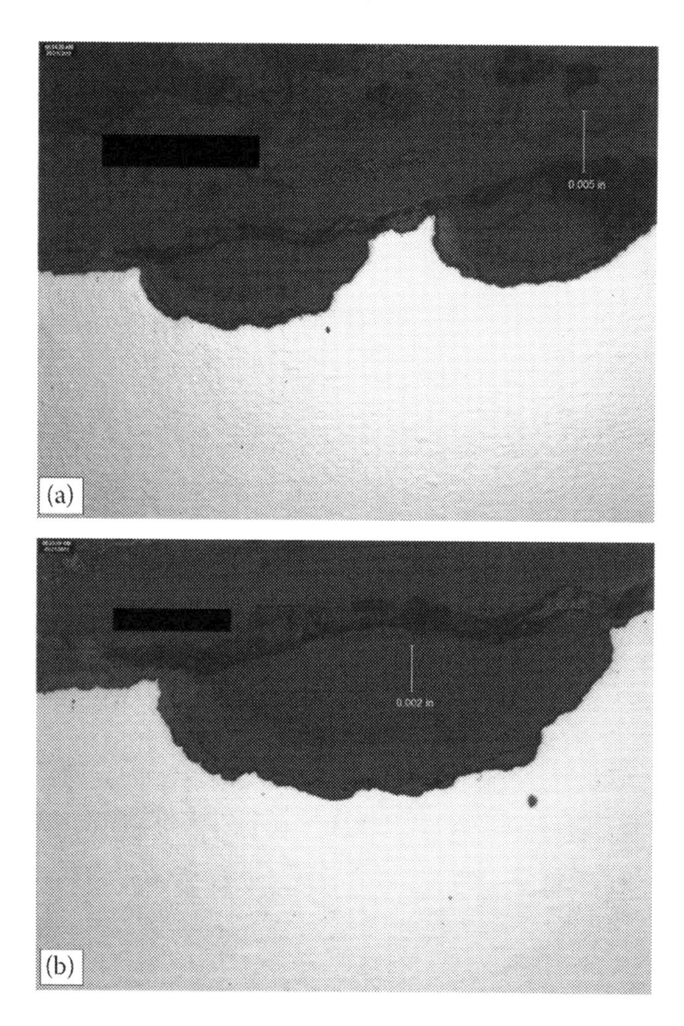

FIGURE 12.5 a and b Microphotographs of corrosion pits in sample joint at 580 meters.

additional representative specimen was cut from the pipe substrates. The chemical contents were analyzed by Glow Discharge Spectroscopy (GDS) except carbon and sulfur elements which were analyzed per ASTM E1019 (2018): *Standard Test Methods for Determination of Carbon, Sulfur, Nitrogen, and Oxygen in Steel, Iron, Nickel, and Cobalt Alloys by Various Combustion and Fusion Techniques*.

12.4 RESULTS AND DISCUSSION

12.4.1 Results of Visual and Metallurgical Examination

Joint at 580 meters sample contained one through wall hole and several pits. Corrosion occurred from the inner diameter to the outer diameter. Figure 12.4 microphotograph shows the corrosion layer of 0.864 millimeter (0.034 inch) that formed

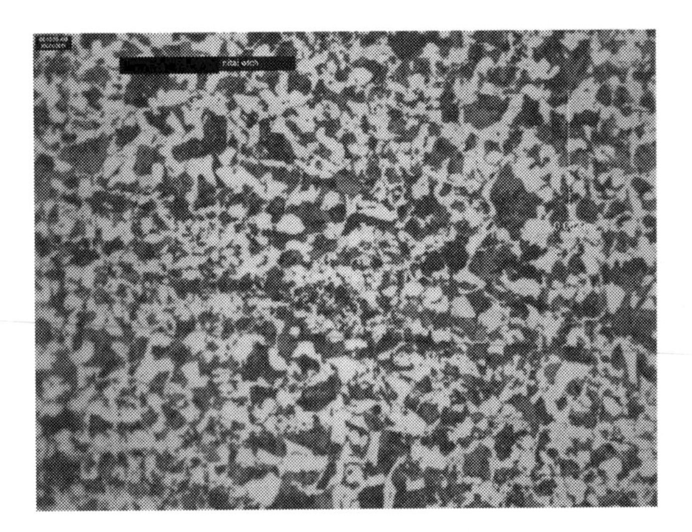

FIGURE 12.6 Microstructure photograph of sample joint at 580 meters.

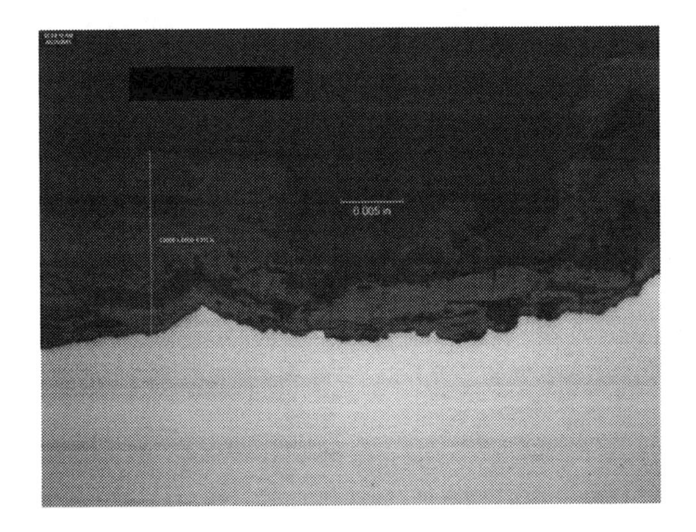

FIGURE 12.7 Microphotograph of corrosion layer in sample burst point.

on the interior pipe surface. There were several internal corrosion pits along the entire length of the pipe sample. Figures 12.5a and 12.5b are microphotographs of a few of the pits to illustrate the depth of the corrosion and their elliptical morphology. In addition, Figure 12.6 shows the microstructure of the pipe sample. The white areas are ferrite (normalized), and the dark areas are pearlite (normalized).

Burst Point sample contained several inner diameter pits along with the burst point failure. In Figure 12.7 a 0.381 millimeter (0.015-inch) oxide layer is observed that formed on the inner diameter of the sample.

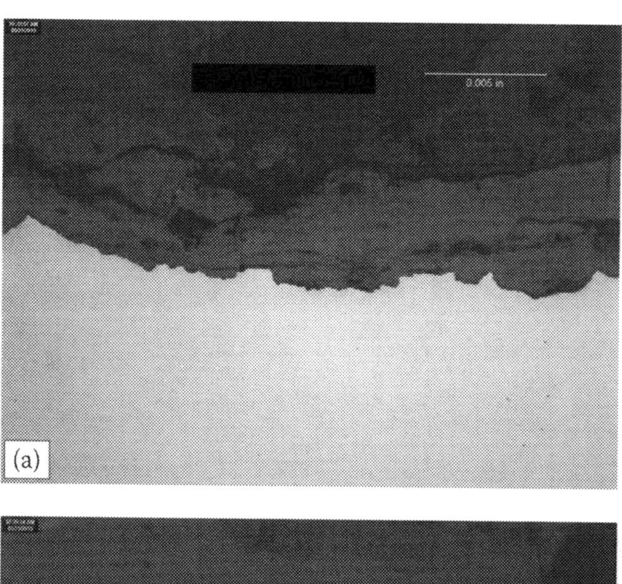

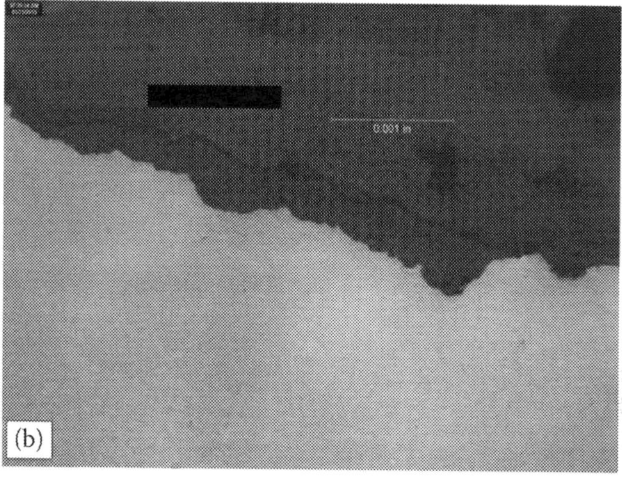

FIGURE 12.8 a and b Microphotographs of corrosion pits in sample burst point.

Figure 12.8a and b are microphotographs of various cross sections of the Burst Point sample. The morphology of the pits is shallow and parabolic. The microstructure of the pipe sample, Figure 12.9, is observed to be quenched and tempered martensite. This differs from the structure of Joint at 580 meters sample which was normalized ferrite/pearlite. Both types of structure are allowed by the N80 specification (API). The tempered martensitic structure is usually considered slightly better from the standpoint of corrosion resistance (Kadowaki et al. 2017).

12.4.2 Results of Microbiological Analysis Using qPCR

Table 12.2 shows the microbe (DNA) analysis results for the debris collected from the samples. 0.02338 gram of debris from sample Joint at 580 meters was used for

FIGURE 12.9 Microstructure photograph of sample burst point.

qPCR analysis. The through wall pit contained the highest concentration in bacteria, archaea, and methanogens that ranged in the 10^7 to 10^8 copies per gram. The sample from the through wall hole also detected sulfate-reducing bacteria, sulfate-reducing archaea, acid-producing bacteria, and iron-oxidizing bacteria in the 10^4, 10^5, and 10^6 copies per gram concentrations. An additional sample (0.01749 gram) of debris was collected directly across from the through wall hole. Similarly, the sample contained the highest concentrations bacteria (10^8 copies per gram), archaea (10^7 copies per gram), and methanogens (10^7 copies per gram). Sulfate-reducing bacteria, sulfate-reducing archaea, acid-producing bacteria, and *Sphaerotilus* and *Leptothrix* species of iron-oxidizing bacteria was detected in concentrations of 10^4 to 10^5 copies per gram.

Two debris subsamples (0.06274 gram of subsample "a" and 0.10149 gram of subsample "b" were used for qPCR analysis) were collected from the Burst Point sample and used for qPCR analysis. Each subsample contained high concentrations of bacteria, archaea, and methanogens in the 10^7 to 10^8 copies per gram range. Concentrations of sulfate-reducing bacteria (10^4 copies per gram), sulfate-reducing archaea (10^5 copies per gram), denitrifying bacteria (10^4 copies per gram), acid-producing bacteria (10^5 copies per gram), and *Leptothrix* and *Sphaerotilus* species of iron-oxidizing bacteria (10^5 copies per gram) were also detected.

12.4.3 Results of Chemical Analysis of the Pipe Samples

The API 5CT chemistry specification for N80 pipe has no limits beyond 0.030 maximum phosphorus and sulfur (API). The steel chemistry results in Table 12.3 indicate that the material of the pipe used at Joint at 580 meters is consistent with N80 grade steel. The steel chemistry results in Table 12.4 indicate that the material of the pipe used at Burst Point is consistent with N80 grade steel. Analysis of each pipe debris subsample from both Joint at 580 meters and Burst Point by ion chromatography show high levels of chloride, sulfate, and thiosulfate as seen in Table 12.5.

TABLE 12.2
Microbe (qPCR) Analysis Results for the Pipe Samples

qPCR Assay (results in copy # per gram)		Joint 580 meters Through Wall Pit	Joint 580 meters Opposite of Through Wall Pit	Burst point (a)	Bust point (b)
Total Bacteria		**2.42E + 08**	**2.75E + 08**	**1.83E + 07**	**2.28E + 07**
Sulfate-Reducing Bacteria (SRB)		**4.53E + 06**	**4.37E + 06**	**3.16E + 04**	**4.45E + 04**
Denitrifying Bacteria (DNB)		**BDL**	**BDL**	**2.82E + 04**	**2.78E + 04**
Acid Producing Bacteria (APB)	**Total**	**3.12E + 04**	**2.07E + 04**	**1.94E + 05**	**1.44E + 05**
	Acetic Acid-Producing Bacteria	2.78E + 04	1.78E + 04	9.57E + 04	6.51E + 04
	Butyric Acid-Producing Bacteria	3.38E + 03	2.92E + 03	9.80E + 04	7.93E + 04
Iron-Oxidizing Bacteria	**Total**	**5.80E + 05**	**1.84E + 06**	**1.57E + 05**	**2.99E + 05**
	***Leptothrix* and *Sphaerotilus* Species**	5.80E + 05	1.84E + 06	1.57E + 05	2.99E + 05
	***Gallionella* Species (Gall)**	ND	ND	ND	ND
Total Archaea		**9.84E + 07**	**8.06E + 07**	**4.89E + 08**	**3.28E + 08**
Sulfate-Reducing Archaea (SRA)		**1.89E + 06**	**1.07E + 06**	**2.25E + 05**	**1.27E + 05**
Methanogens		**3.12E + 07**	**2.15E + 07**	**1.19E + 08**	**7.88E + 07**

BDL = Below Detection Limit; ND = Not Detected

12.5 CONCLUSION OF ROOT CAUSE FAILURE ANALYSIS

12.5.1 Primary Failure Mechanisms

Based on the available data presented in the results section, the through wall hole and pitting in both pipe samples is due to the presence of microorganisms that are known to cause corrosion. The bacteria and archaea populations detected indicate the presence of a very active prokaryote community that was able to coexist together and corrode the steel substrate by both electrical or Type I MIC (steel substrate serves as an electron donor for metabolism) and chemical or Type II MIC (metabolic byproduct, like acetic acid, corrode the steel substrate). The Burst Point failed due to the loss of integrity sustained by MIC.

TABLE 12.3
Joint at 580 Meters Pipe Steel Substrate Chemistry Results

Element	Sample Result	Element	Sample Result
Aluminum, wt%	0.008	Phosphorus, wt%	0.016
Cobalt, wt%	<0.01	Silicon, wt%	0.32
Copper, wt%	0.02	Titanium, wt%	<0.005
Chromium, wt%	0.02	Vanadium, wt%	0.12
Manganese, wt%	1.51	Carbon, wt%	0.330
Molybdenum, wt%	<0.01	Sulfur, wt%	0.00328
Nickel, wt%	0.01	Niobium, wt%	<0.005

TABLE 12.4
Burst Point Pipe Steel Substrate Chemistry Results

Element	Sample Result	Element	Sample Result
Aluminum, wt%	0.027	Phosphorus, wt%	0.010
Cobalt, wt%	0.02	Silicon, wt%	0.28
Copper, wt%	0.18	Titanium, wt%	<0.005
Chromium, wt%	0.09	Vanadium, wt%	<0.005
Manganese, wt%	1.41	Carbon, wt%	0.261
Molybdenum, wt%	0.01	Sulfur, wt%	0.0047
Nickel, wt%	0.12	Niobium, wt%	<0.005

TABLE 12.5
Ion Chromatography Results for Pipe Samples

	Joint at 580 Meters Subsample A through Wall Pit	Joint at 580 Meters Subsample B Opposite of through Wall Pit	Burst Point Subsample A	Burst Point Subsample B
Chloride, mg/L	3,400	2,800	2,160	2,230
Nitrate, mg/L	28	36	6	3
Bromide, mg/L	42	37	17	17
Sulfate, mg/L	2,250	2,560	1,670	1,430
Thiosulfate, mg/L	2,290	2,300	1,890	1,730

12.5.2 High Concentrations of Sulfate and Thiosulfate

The nutritious environment sustained in the ESP well is the source for the high levels of corrosion-associated microorganisms and thus rapid pitting and subsequent failure. For example, the levels of sulfate and thiosulfate present in the debris from both samples serves as a food source for sulfate-reducing bacteria and sulfate-reducing archaea, which are known corrosion-causing microorganisms.

Chemical Conditions	***Microbiological Conditions***
Sulfate, thiosulfate, and chloride (corrosive compounds) detected in debris from pipe samples.	Microbial concentrations in 10^7–10^8 copies per gram and a diverse community of MIC-associated microbes. The steel substrate is an electron donor, and the metabolic byproducts can corrode the steel substrate.
Corrosion and Metallurgical Conditions	***Design and Operation Conditions***
Pipe materials consistent with N80 standards. Corrosion pits parabolic in nature.	ESP well, 2,100 meters deep, 440 barrels of water per day production rate, water cut ~96%, treated with scale inhibitor, tubing installed had less 15% wall loss.

REFERENCES

Al-Nabulsi K.M., Al-Abbas F.M., Rizk T.Y., Salameh A.M. (2015) Microbiologically Assisted Stress Corrosion Cracking in the Presence of Nitrate Reducing Bacteria. *Engineering Failure Analysis*, 58 (1): 165–172. 10.1016/j.engfailanal.2015.08.003.

API Specification 5CT Casing and Tubing. American Petroleum Institute, Washington, DC, www.api.org

ASTM E1019-18. (2018) Standard Test Methods for Determination of Carbon, Sulfur, Nitrogen, and Oxygen in Steel, Iron, Nickel, and Cobalt Alloys by Various Combustion and Inert Gas Fusion Techniques. ASTM International, www.astm.org West Conshohocken, PA.

Enning D. (2014) Corrosion of Iron by Sulfate-Reducing Bacteria: New Views of an Old Problem. *Applied and Environmental Microbiology*, 1226–1236.

Gu T. (2014) Theoretical Modeling of the Possibility of Acid Producing Bacteria Causing Fast Pitting Biocorrosion. *Journal Microbial Biochemical Technology*, 068–074.

Kadowaki M., Muto I., Sugawara Y., Doi T., Kawano K., Hara N. (2017) Pitting Corrosion Resistance of Martensite of AISI 1045 Steel and the Beneficial Role of Interstitial Carbon. *Journal of The Electrochemical Society*, 164:C962.

NACE Standard TM-0212-2018 Standard Test Method: Detection, Testing, and Evaluation of Microbiologically Influenced Corrosion on Internal Surfaces of Pipelines. NACE International, Houston TX.

Thauer R.K., Kunow J. (1995) Sulfate-Reducing Archaea. In: Barton L.L. (eds), *Sulfate-Reducing Bacteria. Biotechnology Handbooks*, Vol 8. Springer, Boston, MA. 10.1007/978-1-4899-1582-5_2.

Uchiyama T., Ito K., Mori K., Tsurumaru H., Harayama S. (2010) Iron-Corroding Methanogen Isolated from a Crude-Oil Storage Tank. *Applied and Environmental Microbiology*, 76 (6): 1783–1788.

13 Pitting Biocorrosion in Internal Pipeline Welds

Vitor Liduino and João Payão-Filho
Federal University of Rio de Janeiro

Márcia Lutterbach
National Institute of Technology

Eliana Sérvulo
Federal University of Rio de Janeiro

CONTENTS

13.1 INTRODUCTION

The relationship between the presence of microorganisms on metallic surfaces and corrosion process has been extensively studied over a long period, since they are widespread both in nature and in industrial environments (Coetser and Cloete, 2005; Little and Lee, 2014). The biofilm-induced corrosion process, known as microbiologically influenced corrosion (MIC) or biocorrosion, involves interactions between microbes, metals, and the environmental conditions that is prevailing. Biofilms formed in natural and industrial environments are heterogeneous and composed of more than one group of microorganisms (algae, fungi, archaea, bacteria) that coexist in a consortium (Beech and Sunner, 2004). Corrosion is generally estimated to account for about 3% of global gross domestic product (GDP) through

DOI: 10.1201/9780429355479-15

the destruction and replacement costs of steel infrastructure. More precise estimates of costs specific to MIC compared to corrosion in general are challenging. Nonetheless, it is reasonable to say that on a global basis MIC runs up to 40% of all degradation of metallic structures in the oil and gas industry (Walch, 1992).

In general, MIC is associated with pitting by sulfate-reducing bacteria (SRB) and is the main cause of fast and unexpected pipeline failure. For instance, Kuwait researches found indications of severe internal corrosion on an oil pipeline from carbon-steel with corrosion depth of more than half of pipeline wall thickness. In spite of its 30-year design life, the pipeline was corroded within 7 years of operation. Based on morphology of localized corroded sites like large terraced pits and detection of high concentrations of SRB and other anaerobic bacteria on the coupons immersed in a wet tank with same medium of pipeline, they concluded that MIC is the reason of severe localized corrosion on inner wall of crude oil pipeline. Moreover, a great amount of sulfide in the downstream fluid was another signal for MIC corrosion of inside pipeline (Al-Sulaiman et al., 2010). This corrosion scenario is often reported by many oil and gas facilities all over the world.

As a matter of fact, pipeline accidents by MIC have been reduced due to the use of steels recommended by the American Petroleum Institute (API), but nevertheless are still reported. Commonly, some of these reports are addressed to corrosion of welded joint (Shabani et al., 2018). The corrosion behavior of the different regions of a welded joint is expected to be subject to the local microstructure changes that occur with welding thermal cycles. These local microstructure changes in the heat-affected zone (HAZ) deserve special attention for its role in the reduced corrosion resistance. Corrosion of welded joints of oil pipelines is a complex phenomenon known to depend on several factors, such as produced water chemistry, type of steel, galvanic interactions, and fluid flow velocity (Askari et al., 2019; Liduino et al., 2019). Additionally, petroleum product pipelines contain large numbers of different groups of microorganisms that can directly or indirectly increase the speed of electrochemical corrosion reactions, especially in sections of low elevation along the pipelines where the presence of small amounts of produced water enhance the microbial activity.

This chapter discusses the pitting corrosion in internal weld zones of steel pipeline circumferential joints welded with two different processes usually applied by the oil and gas industry in the construction of long-distance pipelines. Both welds were exposed to natural seawater in a pipe flow simulator. The objective was to develop correlations among flow velocity, microstructure of weld zones and pit density in the base metal (BM), heat-affected zone (HAZ), and weld metal (WM). Also, microbiological characteristics of the biofilms formed during the biocorrosion experiments were elucidated.

13.2 STUDY DESIGN

13.2.1 Welded Working Coupons

Welded joints were made by shielded metal arc welding (SMAW) and gas tungsten arc welding (GTAW) using rectangular flat pieces of an API 5L X65 low alloy steel. Prior to the welding process, rectangular flat pieces of the BM (API 5L X65) were blasted with 200 μm diameter glass spheres to provide an uneven surface similar to the inner wall of oil pipelines. Each welded joint was made of two rectangular flat pieces joined by a weld seam. The filler metals employed for welding were an AWS ER70S-6 rod and an AWS E7018 stick electrode for GTAW and SMAW, respectively. Analyses of the chemical elements of the BM and the filler metals were performed using inductively coupled plasma-atomic emission spectroscopy and are given in the Table 13.1.

The welding procedures adopted are detailed in the Table 13.2 and were carried out in accordance with good welding practices. Post-weld heat treatment was not performed because the thickness of the BM (5 mm) does not require it. Also, non-welded coupons were used as controls. After welding, the joints were degreased in acetone, sanitized by exposure to germicidal ultraviolet (UV) light for 30 min, settled in sterile Petri dishes, and kept in a desiccator until the beginning of the experiments.

Figure 13.1 shows the welded joints before the biocorrosion experiments. Note that the figure demonstrates just one sample out of a triplicate, showing each examined zone (BM, HAZ, and WM). Differences in the filler metal deposition, which was peculiar to each welding process, resulted in a lower surface roughness (Ra) for WM-GTAW (6.8 ± 0.5 μm) than for WM-SMAW (8.7 ± 1.4 μm).

13.2.2 Corrosive Fluid

Seawater collected from Guanabara Bay, Rio de Janeiro, Brazil (220 24’ S and 430 33’ W), was used. This fluid has been used due to its high microbial diversity,

TABLE 13.1
Chemical Compositions of the API 5L X65 Base Metal and the Filler Materials for GTAW and SMAW

Element	Base Metal API 5L X65	Filler Material	
		GTAW AWS ER70-S6	SMAW AWS E7018
C	0.04	0.07	0.1
Mn	1.4	0.1	0.55
Si	0.25	0.61	0.2
P	0.01	0.02	0.02
S	0.002	0.02	0.03
Ni	0.5	0.02	0.15
Cr	0.02	0.1	0.15
Nb	0.04	–	–
Mo	0.01	0.01	–
Al	0.01	–	–
V	0.06	0.05	0.03
Mg	0.06	–	–
Cu	0.07	–	0.5
Fe	Bal.	Bal.	Bal.

TABLE 13.2
Parameters Used for Welding the API 5L X65 Steel by GTAW and SMAW Processes

Parameter	Welding Process	
	GTAW	SMAW
Arc voltage, U [V]	11.5	22
Welding current, I [A]	85	76
Welding speed, WS [mm/s]	2	4
Electrode diameter, ϕ [mm]	3.2	2.4
Shielding gas	Argon (99.9%)	N. A.

N. A. means not applicable.

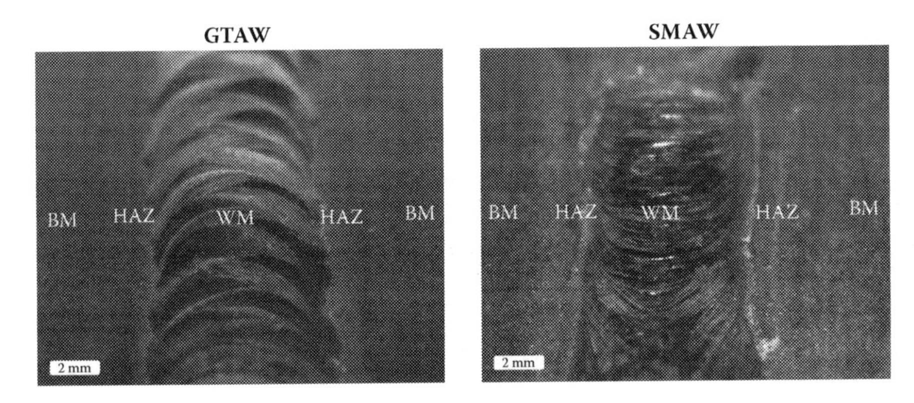

FIGURE 13.1 Joints welded by gas tungsten arc (GTAW) and shielded metal arc (SMAW) before biocorrosion experiments. BM, base metal; HAZ, heat-affected zone; WM, weld metal.

including microorganisms that are usually implicated in biocorrosion, caused by constant and undue disposal of sewage as well as industrial effluents. This justifies the need not to select, prepare and add a microbial inoculum. Before testing, the seawater was supplemented with a nutritional solution of short chain organic acids (4.25 g/L lactic acid, 3.12 g/L butyric acid, 3.5 g/L propionic acid, and 3.12 g/L acetic acid) as a carbon source to maintain microbial activity. Table 13.3 shows the main chemical, physicochemical, and microbiological parameters of the used seawater sample, according to the American Public Health Association (2020) methodologies.

13.2.3 Pipeline Simulator

The test program consisted of lab-scale exposure of specimens taken by both SMAW and GTAW and nonwelded coupons to laminar and turbulent seawater

TABLE 13.3
Chemical, Physicochemical, and Microbiological Properties of the Seawater from Guanabara Bay

Property	Value
Salinity	33.2 ± 0.02 g/L
Conductivity	56.8 ± 0.05 mS/cm
Total sulfide	2.6 ± 0.4 mg/L
pH	6.8 ± 0.2
Temperature	$22.8 \pm 0.01°C$
Dissolved oxygen	7.9 ± 0.3 mg/L
Biochemical oxygen demand	10.4 ± 1.7 mg/L
Turbidity	1.21 ± 0.1 NTU
Total suspended solids	320 ± 52 mg/L
Heterotrophic aerobic bacteria	$7.5 \times 10^5 \pm 2.5 \times 10^4$ MPN/mL
Iron-oxidizing bacteria	$4.5 \times 10^2 \pm 9.5 \times 10$ MPN/mL
Acid-producing anaerobic bacteria	N. D.
Sulfate-reducing bacteria	$3.0 \times 10^2 \pm 7.5 \times 10$ MPN/mL

N. D. means not detected.

flows, comparatively. Each laminar and turbulent flow experiments were performed in single with the flow in uninterrupted recirculation of the same fluid for 28 days at $23 \pm 2°C$. All specimens were placed horizontally into dynamic system, so that the weld bead of the welded specimens was transversal to the flow direction in order to mimic the pipeline girth welding. Weld biocorrosion experiments were performed using a closed pipeline system filled of 25 L seawater, as illustrated in Figure 13.2.

13.2.4 Bacterial Quantification

Triplicate biofilm samples were scraped from three independent specimens for each type of welded joints (GTAW and SMAW) and nonwelded coupon collected at 7, 14, and 28 days of experiments. Quantitative determinations of different bacteria populations in the scraped cell suspension were performed using the MPN method in appropriate culture medium (Harrison, 1978). The populations of heterotrophic aerobic bacteria, acid-producing anaerobic bacteria, iron-oxidizing bacteria and sulfate-reducing bacteria were cultured in nutrient broth, phenol red broth (supplemented with 1% (w/v) glucose), ferric ammonium citrate broth and Postgate E medium (supplemented with 1.2% (w/v) sodium thioglycolate), respectively. All culture media were amended with 35 g/L NaCl to mimic the salinity of used seawater. The conditions and incubation time of each culture medium inoculated with samples were ensured according to the requirements of each bacterial population.

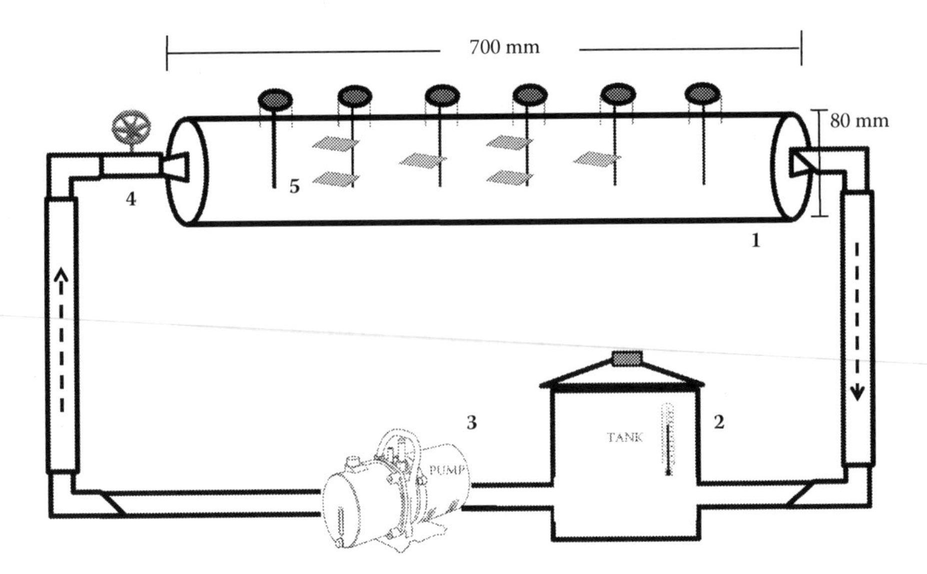

FIGURE 13.2 Illustration of the pipeline simulator (loop system): (1) acrylic pipe where the specimens were fixed, (2) seawater reservoir, (3) pump, (4) fluid flow controller, and (5) welded specimens with the weld metal transversal to the fluid flow direction (arrows).

13.2.5 Examination of Weld Zone Surfaces

The surface roughness of the weld beads was characterized in three different positions of each coupon to calculate the average value using a profilometer (Bruker Contour GT-K1). The welded joints were sectioned by wet abrasion for microstructural analysis of the BM, HAZ, and WM. The surfaces underwent abrasion using sandpaper with different granulometries (100, 220, 320, 400, 600, and 1,200 grit) and then were polished with diamond paste (sequence 6 μm, 3 μm, and 1 μm). To reveal the microstructure, nital 2% (2% nitric acid in ethyl alcohol) was used according to standard metallography procedures. The microstructure and non-metallic inclusions were observed using a Zeiss Axio Imager M2m optical microscope coupled to an AxioCam 503 digital camera. Pitting corrosion was investigated according to ASTM G-46/94 (2005) standard employing a 3D measurement system (Zeiss Smartzoom 5 microscope). The evolution of pitting was assessed by quantifying pitting parameters including the: (i) density (the total number of pits per m^2) and (ii) average depth of five random pits present in each weld zone (BM, HAZ, and WM).

13.3 RESULTS AND DISCUSSION

13.3.1 Microstructure of the Welded Joints

The microstructures of the BM, as well of the WM and HAZ of the welded joints, are shown in Figure 13.3. The microstructure of the API 5L X65 low alloy steel consisted of approximately 90% fine equiaxed grains of ferrite and 10% fine-

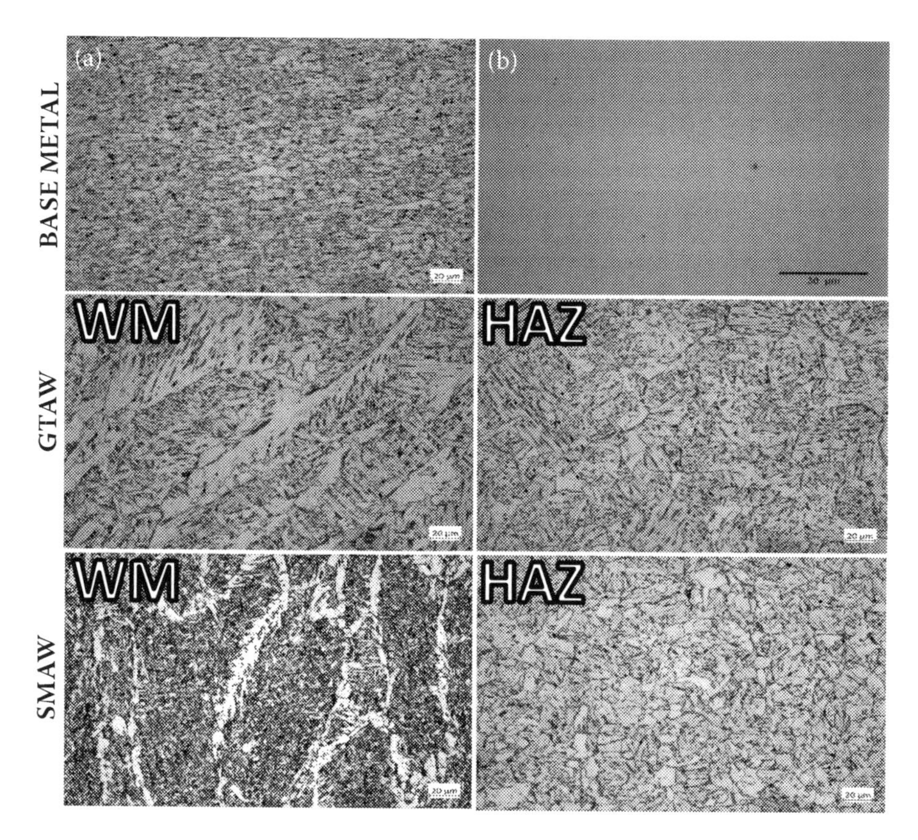

FIGURE 13.3 Optical micrographs of the API 5L X65 base metal, (A) microstructure and (B) nonmetallic inclusions; and weld metals (WM) and heat-affected zones (HAZ) for gas tungsten arc welding (GTAW) and shielded metal arc welding (SMAW).

grained perlite with fine lamels. The HAZ microstructure of the GTAW joint was coarser than those observed in the HAZ of SMAW. As the BM was the same for both welded joints, this difference in the HAZ microstructure was due to the following: (i) the heat input or (ii) the root opening was higher during GTAW than SMAW, or (iii) both factors concomitantly. According to Marques et al. (2009), a high heat input or large root opening reduces the cooling rate of the welded joint, which promotes a coarse grain size and the formation of intragranular Widmanstätten ferrite (IWF). Our study also showed that IWF was slightly greater in the HAZ-GTAW joint than in the SMAW one, which agrees with Marques and collaborators (2009). Most of the studies report that changes in the grain size of a steel affect its resistance to corrosion (Park et al., 2017; Rai et al., 2018). However, the existing literature is often contradictory, even within the same alloy class, and a coherent understanding of how grain size influences pitting corrosion response is largely lacking.

13.3.2 Microbial Adhesion to Welds

The microbial enumeration of the biofilms formed on corroded joints (SMAW and GTAW) and nonwelded coupons under laminar and turbulent flows are shown in the Figure 13.4. In 7 days of laminar flow experiment, the number of heterotrophic aerobic bacteria (HAB) was similar on GTAW and SMAW joints and slightly smaller on nonwelded coupons. Otherwise, SRB and acid-producing anaerobic bacteria (APAB) were numerically higher in the GTAW joint than in the SMAW joint and even higher compared to nonwelded coupons. Xu et al. (2016) reported that biofilms formed by acid-producing bacteria and SRB can reduce the pH to 2 on the surface of the metal, supplying locally high concentrations of oxidants, such as protons, that are directly corrosive. Also, the synthesis/release of organic acids by APAB may have raised the carbon source within biofilms, contributing to high SRB growth. Other than supporting SRB, biogenic acids are widely known by their capacity to accelerate iron corrosion, principally in oil pipelines. So, it is possible to consider that surfaces with weld metals favored a rapid and strong adhesion of microorganisms known to be associated to MIC.

From 14 days of laminar flow experiment, both welded joints showed rather similar microbial populations profiles. Note that there was a quantitative dominance of SRB in those biofilms overtime. Nevertheless, biofilms formed on nonwelded coupons exhibited another behavior, with slightly dominance of HAB. Those facts may be related to the high roughness (rather waviness profile) formed during welding, which may enhance the conditions so that the anaerobes are better accommodated on welded joint surfaces. Also, it must be pointed out that iron oxidizing bacteria (IOB) were not detected in all biofilm samples from the laminar flow. The low dissolved oxygen, which already decreased in the early days of experiment (<2.0 mg/L), and/or the increasing of total sulfide content (>97 mg/L) in the laminar fluid may have inhibited IOB in this system. Data of total sulfide and dissolved oxygen in the fluids during weld biocorrosion experiments are shown in the Figure 13.5.

Under turbulent flow, 7-day-old biofilm experiments showed rather different biofilm profiles compared to laminar flow, in particular for IOB population probably due to the availability of dissolved oxygen (>6.5 mg/L) and small accumulation of sulfide (<7 mg/L) (Figure 13.5). Both aerobic bacteria, i.e., HAB and IOB, were the dominant populations in welded and nonwelded coupons. For 14 days of experiments, SRB number was similar to that of HAB, suggesting that oxygen level on biofilm surface was proper for the HAB EPS-producers, which lead to a biofilm thickness triggering oxygen depletion within biofilm enough to improve anaerobic cellular activity. However, it must be highlighted that under laminar flow, SRB and HAB abundances were equal since the 7th day of experiment, suggesting that the laminar flow fast promoted sulfate-reducers growth. In 28 days, biofilms on the welded joints, mainly in the GTAW one, showed a reduction of bacteria populations probably due to detachment of biofilm pieces (sloughing) to bulk water, which is much more intense in turbulent fluid flow. Liu et al. (2017) showed that the increase of flow velocity to 1.0 m/s reduced or inhibited the biofilm formation on API 5L X70 pipe steel exposed to oilfield produced water. In contrast, Song et al. (2016)

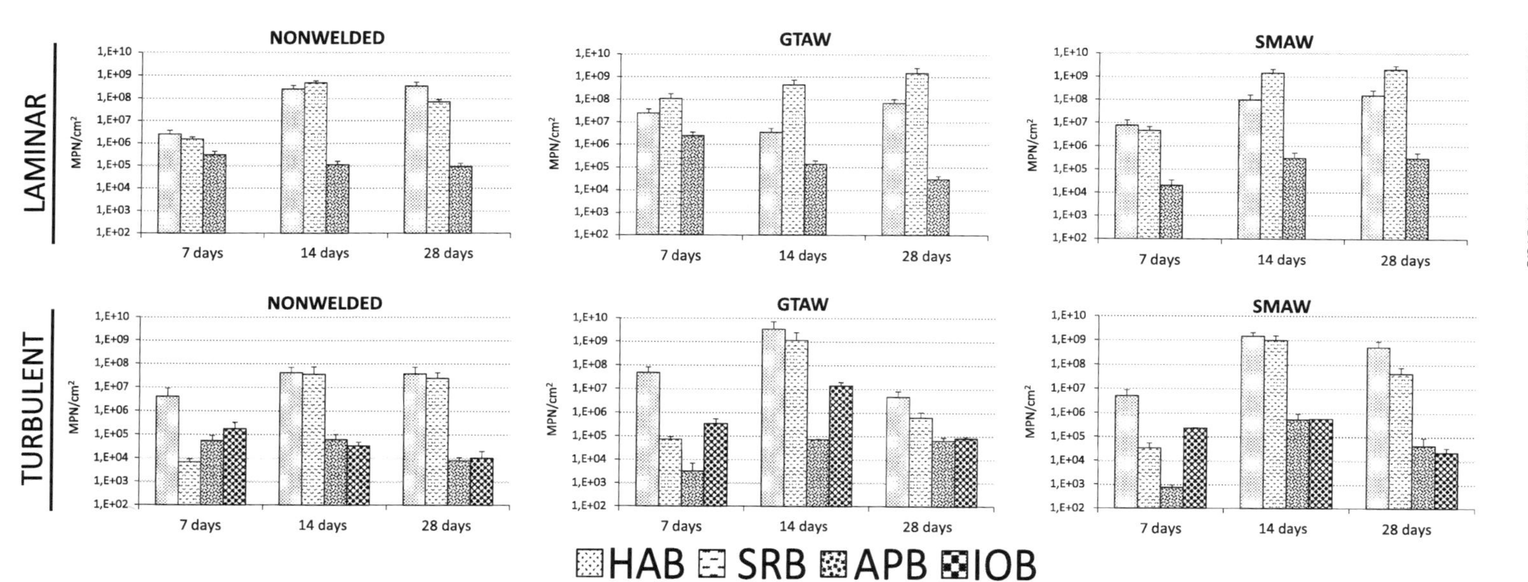

FIGURE 13.4 Microbiological monitoring of biofilms developed on welded joints (GTAW and SMAW) exposed to different flow regimes. HAB, heterotrophic aerobic bacteria; SRB, sulfate-reducing bacteria; APAB, acid-producing anaerobic bacteria; IOB, iron-oxidizing bacteria.

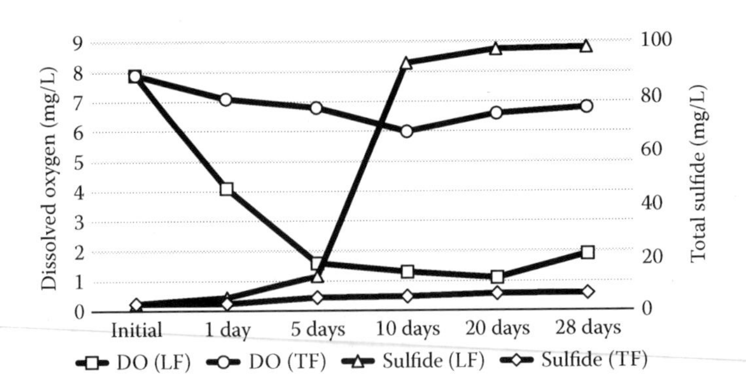

FIGURE 13.5 Dissolved oxygen (DO) and total sulfide (TS) monitoring in the fluids under laminar and turbulent flows during weld biocorrosion experiments. LF, laminar flow; TF, turbulent flow.

discussed that the adhesion of HAB, SRB and IOB to inner wall of pipelines was not reduced even at flow velocity of 3.0 m/s. In general, considering the microbial enumeration on welded joints, the findings of this study were similar to those of Song et al. (2016). Furthermore, the results showed that under turbulent flow there was higher microbial adhesion in welded joints than in nonwelded coupons, clearly emphasizing that microbial adhesion is tight influenced by surface roughness. Thus, literature data may be contradictory and some caution is required since only increasing the flow to high velocities is not a safe strategy to control and prevent the biocorrosion of oil pipelines.!

13.3.3 Pitting Distribution on Weld Zones

A detailed pitting corrosion analysis of representative welded joints recovered from the experimental system is shown in Table 13.4. Both welded joints from the laminar flow case showed a high pit incidence on the entire surface, and the density values were rather similar in the BM, HAZ, and WM. However, the pit depth in the HAZ was higher than that for in other regions regardless of the welding process used. The HAZ in the GTAW showed pits 20% deeper than pits formed in the BM and WM. For the SMAW joint, the pits in the HAZ were 25% deeper than those in other regions.

These findings suggest that the microstructural changes of API 5L X65 low alloy steel caused susceptibility of the HAZ to pitting corrosion, which may have been aggravated by the presence of microorganisms. Changes in the steel microstructure were peculiar to each welding process (Fig. 13.3), and the coarse-grained HAZ in the GTAW joints showed pits 12% deeper compared to the fine-grained HAZ in the SMAW joints. However, the GTAW weld metal (AWS ER70S-6 rod) was more resistant to corrosion than the SMAW one (E7018 stick electrode), which also exhibits different microstructures (grain size and orientation). Other studies in the literature have also attributed deep pits in the HAZ to changes in the microstructure as well as residual stress, but detailed justifications have not been proposed yet

TABLE 13.4
Pit Data on Zones of the Welded Joints*

	Laminar Flow				Turbulent Flow			
	GTAW		SMAW		GTAW		SMAW	
	Density (pit/m^2)	Average Depth (μm)	Density (pit/m^2)	Average Depth (μm)	Density (pit/m^2)	Average Depth (μm)	Density (pit/m^2)	Average Depth (μm)
BM left side	4.6×10^5	31.8	3.1×10^5	24.3	9.7×10^4	15.2	6.8×10^4	13.7
HAZ left side	8.5×10^5	40.5	4.2×10^5	35.0	2.8×10^4	17.5	3.6×10^4	19.1
WM	3.5×10^5	19.1	1.5×10^5	23.5	4.2×10^5	20.8	8.5×10^5	23.4
HAZ right side	7.9×10^5	38.6	3.6×10^5	34.8	3.1×10^6	33.5	2.7×10^6	32.0
BM right side	6.0×10^5	28.5	1.2×10^5	23.9	1.6×10^5	16.9	7.2×10^5	15.1

Notes
* Zones of the welded joints were defined as illustrated in Fig. 13.1.

(Chaves and Melchers, 2011; Ramkumar et al., 2017). Chaves and Melchers (2011) verified the highest severity of pits in the HAZ compared to that in the WM and BM, but the pit depth in these regions only differed after one year of in situ exposure to Pacific Ocean water. However, these authors did not consider the participation of seawater microorganisms in the biocorrosion process of the welded joints. On the other hand, Ramkumar et al. (2017) inferred that the combination of both *Bacillus* and *Pseudomonas* species have caused extensive damage to HAZ of stainless-steel welds by causing pitting corrosion and entering the cavities and widening the pits when the exposure time in synthetic marine medium was extended to 30 days.

Welded joints from turbulent flow showed shallow pits compared to the pits on welded joints exposed to laminar flow; likewise, the HAZ was the region with a high severity of pitting corrosion. However, the pit density was different for regions on the right side compared to those on the left side of the coupons. The density and depth of pits in the HAZ-GTAW on the right side were higher than the pits in the HAZ on the left side. The same pitting profile was observed for the SMAW joint. In general, the HAZ on the right side of the WM showed an increasing pit density by two orders of magnitude compared to the HAZ on left. The BM and WM of the GTAW and SMAW joints also showed a different pit severity on each side that was analyzed, where it was always higher in the regions on the right side, but the average pit density for the WM was calculated considering the total area. Briefly, the turbulent flow reduced pit depth but increased their number on the welded joint surface. This behavior can result in a greater pitting severity than that with laminar flow, mainly in long-term oil pipeline operation, since the pits would deepen, causing accidental damage to the material/structure.

The higher deterioration in the right side of the welded joints may be due to their intentional position in the experimental system to reproduce the pipeline girth welding. The left regions of the coupons were subjected to direct exposure of the turbulent fluid, which, when colliding with the weld metals, may have generated a strong local swirling, reducing the adhesion of microorganisms and consequent biocorrosion, as illustrated in the Figure 13.6. The idea of lower microbial activity in the regions on the left is justified by the lower density and depth of pits than those on the right side. In contrast, the right-side regions of the coupons may have been "protected" from the turbulent flow due to the weld metal reinforcement, which worked like a barrier/wall, generating an amenable environment where the microorganisms could easily adhere, form a biofilm and initiate or accelerate biocorrosion. In these "protected regions", the density and depth of the pits were the highest, especially in the HAZ, indicating an intense activity of microorganisms metabolically associated with metallic corrosion. It is emphasized that the investigation of the formation and evolution of pits is an arduous work, especially in short-term biocorrosion experiments, as in this study. The small size of pits makes their detection difficult, especially in the early stages, when pits may have small diameters and depths of only a few tens of nanometers. Additionally, the pits may be covered by corrosion products even after the metal surfaces are cleaned with chemical solution. Therefore, the pit numbers presented here may be under estimated, thus the biocorrosion severity in welded joints may be much higher than that considered by our findings.

To estimate biocorrosion severity away from the welds, all nonwelded coupons recovered from laminar flow showed a pit density of 1.2×10^4 pits per m^2 and an average pit depth of 21.7 μm. Increased fluid velocity reduced the pit density to 8.5×10^3 pits per m^2 and average pit depth to 11.4 μm. These findings indicate that for a base metal not affected by heat, turbulent flow can be an important strategy to avoid pitting corrosion as a result of microbial activity; however, the same

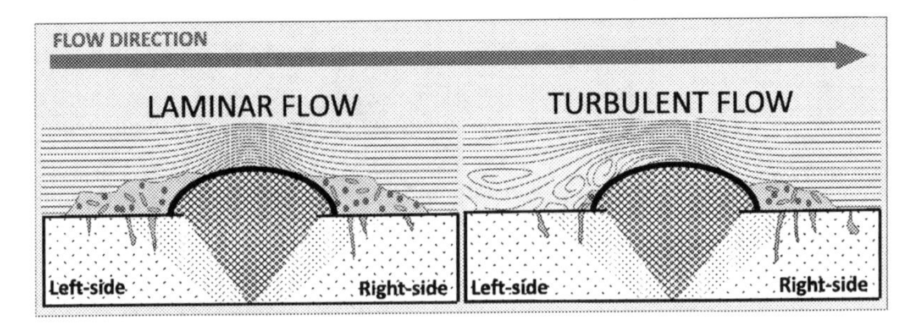

FIGURE 13.6 Comparative illustration of biofilm morphology grown under laminar and turbulent flows.

conclusion cannot be drawn for pitting corrosion mitigation in welded joints. Although nonwelded coupons were used as control, it is difficult to compare their results with those obtained from the corrosion of welded joints because even in the absence of microorganisms and without microstructural changes, weld corrosion tends to be higher due to the formation of a galvanic cell between the WM and the BM. These regions are metals with different chemical compositions (Table 13.1), resulting in differentiated corrosion potentials, even though both are classified as carbon steel. It is also important to note that the presence of chloride ions in the seawater may have intensified the pitting corrosion. Several studies have reported that chloride has the power to cause localized oxidative dissolution, which can generate a pit density propagation rate up to four times faster than demineralized water (Liu et al., 2018; Guo et al., 2018).

13.4 FINAL REMARKS

The type of flow regime directly affected the sessile microbial community since laminar flow quickly provided biofilms with a large anaerobic microbial community (7 days), while turbulent flow supported a community with anaerobic and aerobic groups until the end of the experiment (28 days), regardless of welded joint analyzed (GTAW and SMAW). However, it is highlighted that both laminar and turbulent flows showed a sufficient microbial activity to induce weld biocorrosion. Microstructural differences among the base metal, weld metal and heat affected zone resulted in a greater localized corrosion attack on the heated zone, but there is no simple correlation between grain size and pit depth. In addition, pitting profile on HAZ was rather influenced by the type of flow regime and, consequently, attached microbial community. Under laminar flow, the two HAZs around both welded joints showed high severity of pits with similar density and depth. However, in turbulent flow case, only one HAZ was affected by very high pit density; while the other HAZ in the same joint showed few pits, regardless of GTAW or SMAW. The knowledge acquired in this study indicates that the turbulent flow investigated here may trigger a higher biocorrosion severity in welded joints in the long term compared to that of laminar flow.

REFERENCES

Al-Sulaiman, S., Al-Mithin, A., Al-Shamari, A., Islam, M., Prakash, S. S. 2010. Microbiologically influenced corrosion of a crude oil pipeline. Proceedings of the NACE International Conference and Expo 2010, San Antonio, Texas.

American Public Health Association. 2012. *Standard methods for the examination of water and wastewater*, 22nd edition, American Public Health Association, Washington.

Askari, M., Aliofkhazraei, M., Afroukhteh, S. 2019. A comprehensive review on internal corrosion and cracking of oil and gas pipelines. *Journal of Natural Gas Science and Engineering* Vol. 71, 102971. doi:10.1016/j.jngse.2019.102971

ASTM G46-94. 2018. Standard guide for examination and evaluation of pitting corrosion, West Conshohocken, PA:ASTM International. www.astm.org

Beech, I. B., Sunner, J. 2004. Biocorrosion: towards understanding interactions between biofilms and metals. *Current Opinion in Biotechnology* 15: 181–186. doi:10.1016/j.copbio.2004.05.001

Chaves, I. A., Melchers, R. E. 2011. Pitting corrosion in pipeline steel weld zones. *Corrosion Science* 53: 4026–4032. doi:10.1016/j.corsci.2011.08.005

Coetser, S. E., Cloete, T. E. 2005. Biofouling and biocorrosion in industrial water systems. *Critical Reviews in Microbiology* 31: 213–232. doi:10.1080/10408410500304074

Guo, P., La Plante, E. C., Wang, B., Chen, X., Balonis, M., Bauchy, M., Sant, G. 2018. Direct observation of pitting corrosion evolutions on carbon steel surfaces at the nano-to-micro- scales. *Scientific Reports* 8: 7990. doi:10.1038/s41598-018-26340-5

Harrison, A. P. 1978. Microbial succession and mineral leaching in an artificial coal spoil. *Applied and Environmental Microbiology* 36: 861–869.

Liduino, V. S., Cravo-Laureau, C., Noel, C., Carbon, A., Duran, R., Lutterbach, M. T., Camporese Sérvulo, E. F. 2019. Comparison of flow regimes on biocorrosion of steel pipe weldments: Community composition and diversity of biofilms. *International Biodeterioration & Biodegradation* 143: 104717. doi:10.1016/j.ibiod.2019.104717

Little, B. J., Lee, J. S. 2014. Microbiologically influenced corrosion: An update. *International Materials Reviews* 59: 384–393. doi:10.1179/1743280414y.0000000035

Liu, T., Cheng, Y. F., Sharma, M., Voordouw, G. 2017. Effect of fluid flow on biofilm formation and microbiologically influenced corrosion of pipelines in oilfield produced water. *Journal of Petroleum Science and Engineering* 156: 451–459. doi:10.1016/j.petrol.2017.06.026

Liu, H., Sharma, M., Wang, J., Cheng, Y. F., Liu, H. 2018. Microbiologically influenced corrosion of 316L stainless steel in the presence of *Chlorella vulgaris*. *International Biodeterioration & Biodegradation* 129: 209–216. doi:10.1016/j.ibiod.2018.03.001

Marques, P. V., Modenesi, P. J., Bracarense, A. Q. 2009. *Welding – fundamentals and technology*, 3rd edition, UFMG, Belo Horizonte.

Park, C., Kang, N., Liu, S. 2017. Effect of grain size on the resistance to hydrogen embrittlement of API 2W grade 60 steels using in situ slow-strain-rate testing. *Corrosion Science* 128: 33–41. doi:10.1016/j.corsci.2017.08.032

Pickering, H. W., Frankenthal, R. P. 1972. On the mechanism of localized corrosion of iron and stainless-steel I: Electrochemical Studies. *Journal of the Electrochemical Society* 119: 1297–1304. doi: 10.1149/1.2403982

Rai, P. K., Shekhar, S., Mondal, K. 2018. Development of gradient microstructure in mild steel and grain size dependence of its electrochemical response. *Corrosion Science* 138: 85–95. doi:10.1016/j.corsci.2018.04.009

Ramkumar, D. K., Dagur, A. H., Kartha, Ashwin, A. N., Subodh, M. A., Vishnu, C., Arun, D., Kumar, M. G. V., Abraham, W. S., Chatterjee, A., Abraham, J., Abraham, J. 2017. Microstructure, mechanical properties and biocorrosion behavior of dissimilar welds of AISI 904L and UNS S32750. *Journal of Manufacturing Processes* 30: 27–40. doi:10.1016/j.jmapro.2017.09.001

Reynolds, O. 1883. An experimental investigation of the circumstances which determine whether the motion of water shall be director sinuous, and of the law of resistance in parallel channels. *Philosophical Transactions of the Royal Society* 174: 935–982.

Shabani, H., Goudarzi, N., Shabani, M. 2018. Failure analysis of a natural gas pipeline. *Engineering Failure Analysis* 84: 167–184. doi:10.1016/j.engfailanal.2017.11.003

Song, X., Yang, Y., Yu, D., Lan, G., Wang, Z., Mou, X. 2016. Studies on the impact of fluid flow on the microbial corrosion behavior of product oil pipelines. *Journal of Petroleum Science and Engineering* 146: 803–812. doi:10.1016/j.petrol.2016.07.035

Walch, M. 1992. Microbial corrosion. In *Encyclopedia of microbiology 1*, J. Lederberg, Ed. New York: Academic, 585–591.

Xu, D., Li, Y., Gu, T. 2016. Mechanistic modeling of biocorrosion caused by biofilms of sulfate reducing bacteria and acid producing bacteria. *Bioelectrochemistry* 110: 52–58. doi:10.1016/j.bioelechem.2016.03.003

14 Impact of Metallurgical Properties on Pitting Corrosion in a High-Pressure Seawater Injection Pipeline

Recep Avci
Montana State University and University of Oklahoma

Joseph M. Suflita
University of Oklahoma

Gary Jenneman
GJ Microbial Consulting, LLC and University of Oklahoma

David Hampton
ConocoPhillips Company

CONTENTS

DOI: 10.1201/9780429355479-16

14.1 INTRODUCTION

Two connecting sections of high-pressure pipeline from a seawater injection platform were found to be differentially corroded on their inner surfaces. The pipe was located in a minimum-flow line downstream a high-pressure seawater injection pump. Photos of the submitted sample are shown in Figure 14.1, and the subsections are labeled as base metal A, base metal B, and weld. This pipeline section was a dead leg most of the time. Corrosion was particularly evident on the section of the pipe labeled as base metal A (Figure 14.1a). The metallurgical reasons for the more pronounced localized corrosion of base metal A relative to base metal B were investigated. Since the two base metals were subject to identical physical, chemical, and biological environments over a 20-year period, it was reasoned the underlying material properties and metallurgical history of the base metals must differ in a fundamental manner.

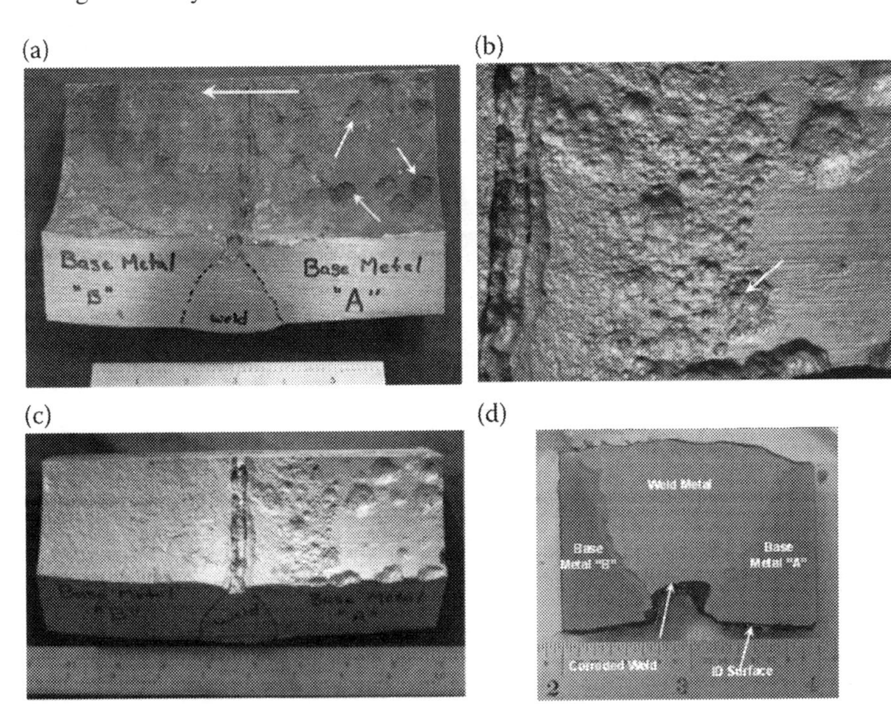

FIGURE 14.1 (a) Photograph of the as-received pipe section showing differences in the degree of surface pitting in base metals A and B and a connecting weld area. The flow through the pipe was from the right to the left (yellow arrow). The degree of surface pitting was more pronounced on base metal A (white arrows) and particularly evident in (b) a close-up photograph. A photograph of the pipe section following bead-blasting to more clearly image the surface pitting morphology is seen in (c). (d) Photo of pipe cross section taken through the weld after it was mounted, polished, and etched with Nital to highlight microstructural features. The ruler in (a) and (d) is in inches.

In addition, model corrosion experiments were conducted on the respective base metals and verified that base metal A exhibited a greater propensity for localized corrosion than base metal B. Although evidence is provided that the main cause of this pitting corrosion was the result of MIC, it was the purpose of this case study to explain why the pitting corrosion was more evident on base metal A and not as prominent on base metal B.

14.2 BACKGROUND

14.2.1 Chemical and Microbiological Analyses

An unpreserved pipe section was originally submitted to the ConocoPhillips laboratories to determine whether the surface pitting was due to microbiologically influenced corrosion (MIC). The pipe section initially had a diameter of 203 mm (8 inches), was constructed from API 5L Grade X52 material, and had been in service for more than 20 years. Historically, the seawater system was batch treated daily for 1 hour with 1,500 ppmv biocide, as supplied (50% glutaraldehyde, as active), but is currently treated 3 days a week. The seawater in the area of the corrosion was also deoxygenated using vacuum degassers to a specification of 20 ppb. A chemical analysis was performed on the corrosion deposits in the corrosion pits and pipe surface on base metal A, as well as the corrosion adjacent to the weld. According to XRD analyses, the corrosion products in the pits (Figure 14.1b), but not on the pipe surface or near the welds, contained 1 wt. % mackinawite, (FeS) suggesting the pitting corrosion on base metal A (Figure 14.1c) was the result of the metabolic activity of sulfate reducing bacteria (SRB). No attempts were made to culture SRB or conduct molecular analyses due to the non-preserved state of the deposits and the lack of available sample. However, it had been demonstrated through extensive culture-based monitoring over the 20+ years of service that the pipeline seawater contained viable SRB.

14.2.2 Metallographic Examination

A metallographic specimen was removed from the pipe in the area of the weld. A cross section of the weld and associated base metals was cut out, mounted, polished, and etched (Figure 14.1d). The cross section revealed that the weld was preferentially corroded. The cross section of the inner diameter surface also exposed the microstructural features of the two base metals. Base metal A was a "banded" pearlitic microstructure in a matrix of ferrite (Figure 14.2a), while no banding structure was evident in base metal B even though pearlite structures were clearly present (Figure 14.2b).

The microstructural differences between the base metals were most likely due to the original thermal processing temperatures the two metals experienced during manufacture. The "banded" microstructure in base metal A is typically formed at finishing temperatures less than 760°C, right around or slightly below the ferrite-austenite transition temperature. The microstructural features in base metal B are more consistent with finishing temperatures above 871°C, in the gamma (austenitic) phase of iron. This temperature difference in the final processing of the steel created

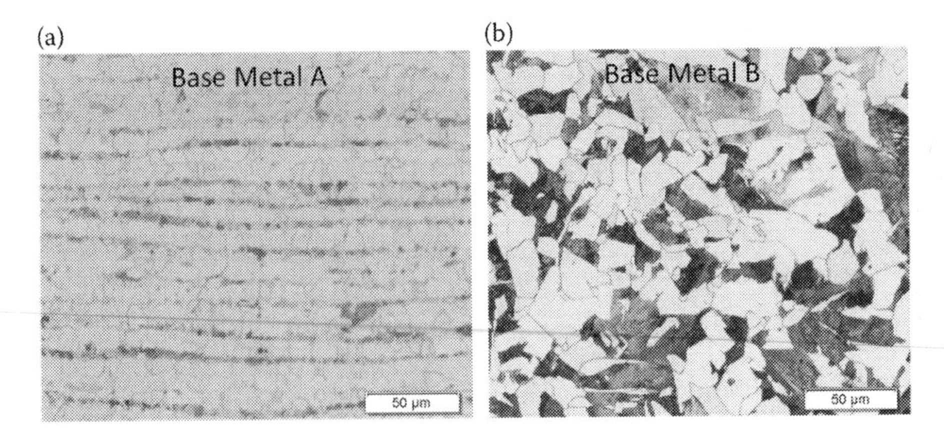

FIGURE 14.2 Photomicrographs of the microstructure of base metals A and B (a and b, respectively). The microstructure of base metal A consists of "banded" pearlite in a matrix of ferrite, while the pearlite and ferrite features of base metal B show no obvious banding. The samples were etched with Nital.

very different microstructural features that should only slightly affect steel mechanical properties such as hardness and impact strength. However, different temperature treatments and subsequent annealing of the two steels were certainly consistent with the difference in the susceptibility of the base metals to corrosion.

14.2.3 Chemical Composition

Chemical analyses of the pipe sections and weld were performed with an optical emission spectrometer (FoundryMaster™). The results are shown in Table 14.1 along with the requirements of API 5L Grade X52 material for comparison purposes.

The analyses revealed that both base metal sections met the chemical requirements of API 5L for Grade X52. Note in particular that the carbon content of base metal B was about twice that of base metal A. The differences in the rest of the trace elements were regarded as of little significance for this analysis. For example, the excess Mn in base metals A and B will simply occupy the body-centered cubic (bcc) lattice sites of the ferrite iron.

14.2.4 Hardness Testing

Hardness testing to assess the mechanical properties of the pipe sections was conducted on a cross section of the piping and included both base metals, their heat affected zones (HAZ), and the weld metal (Table 14.2).

Conversion of the average hardness readings of the carbon steel material revealed that base metals A and B had approximate tensile strength values of about 496 and 545 MPa, respectively. Both of these parameters are above the minimum tensile strength requirements for API 5L X52 material (460 MPa). The hardness of

TABLE 14.1
Chemical Composition of the Pipe and Weld Samples (wt.%)

Chemical Element	Base Metal A	Base Metal B	Weld Metal	API 5L Grade X52 Requirements
Carbon	0.085	0.156	0.023	0.28 max.
Silicon	0.215	0.207	0.51	N/A
Manganese	1.65	1.83	1.15	1.4 max.
Sulfur	0.002	0.032	0.004	0.03 max.
Phosphorus	0.016	0.027	0.017	0.03 max.
Chromium	0.171	0.127	0.033	N/A
Molybdenum	0.043	0.018	0.015	N/A
Nickel	0.079	0.085	0.024	N/A
Copper	0.159	0.268	0.038	N/A
Iron	Balance	Balance	Balance	Balance

TABLE 14.2
Hardness Testing of the Piping Sample (Rockwell B Scale)

Sample	Base Metal B	HAZ	Weld	HAZ	Base Metal A
1	83.6	93.4	79.6	90.2	80.9
2	85.5	94.7	78.6	88.4	80.3
3	86.2	95.8	78.4	95.4	80.5
Average Reading	**85.1**	**94.6**	**78.9**	**91.3**	**80.6**

base metal B was slightly greater than that of base metal A, a feature that is consistent with the increased carbon content of the former (i.e., more carbide is expected to produce harder steel). However, the relatively small differences in hardness between the two base metals suggests that this feature was unlikely to account for the large discrepancies in corrosion behavior.

On the other hand, the weld likely corroded preferentially because of the minor differences in chemical properties between the base metals and the weld metal (Mahajanam and Joosten 2011, Vander Woort n.d.). These differences likely created a local anode (the weld) and cathode (the base metals), which led to the preferential [galvanic] corrosion of the weld as the predominant loss mechanism.

14.2.5 Salient Aspects of Carbon Steel Metallurgy

The API 5L Grade X52 carbon steel base metals exhibited the carbon and sulfur content differences detailed in Table 14.1. Sulfur typically imparts undesirable characteristics that compromise the mechanical integrity of steel and give rise to sulfur embrittlement, also known as red-short or hot-short. This is due to the sulfur forming iron sulfide/iron mixtures at the metal grain boundaries that have a lower melting point and can cause the steel to separate when heated. Manganese is typically added in large excess (many orders of magnitude) of the sulfur content to low-carbon steels to trap the residual sulfur in less-soluble inclusions such as MnS (Lehmann and Nadif 2011).

During the final stages of the hot rolling and the annealing process, the metal is transformed from the austenitic (γ-Fe) to the ferrite (α-Fe) phase at ~727 °C. In γ-Fe the carbon occupies the interstitial positions in the relatively large face-centered cubic (fcc) lattice, while in α-Fe the carbon atoms segregate into pearlite phase lamellar structures (Porter and Easterling 1992) locked into the Fe_3C cementite structure. Pearlite lamellar structures are repeated submicron-thick orthorhombic cementite and bcc Fe ferrite structures and are expected to be highly strained because of the lattice mismatch. In fact, we predict that as a consequence of the large surface area of cementite/ferrite interfaces in the pearlite phases, the strained ferrite Fe layers would be highly susceptible to the initiation and growth of localized corrosion (Davis 2013, Avci et al. 2018). An example of this is shown in Figure 14.3.

As noted, pearlite bands are often formed in low-carbon steel during the rolling process, as the steel transitions from the austenite to the ferrite phase. Under the high pressure of the rolling process, pearlite bands are stretched along the rolling axis, as observed in base metal A. At even higher temperatures, MnS inclusions are also stretched along the rolling direction during hot rolling, giving rise to stringers (Krauss 2003).

When hot rolled at 870 °C, iron is in the γ-Fe phase and the highly malleable MnS inclusions are stretched into stringers along the rolling axis. This is because the MnS, with a melting temperature of about 1,617°C, has already solidified. On the other hand, carbon atoms are randomly distributed interstitially in the austenite fcc lattice. The phase diagram of Fe-C (not shown) at 0.1–0.2 wt.% carbon concentrations suggests that as the steel slowly cools, a quasi-thermodynamic equilibrium is maintained and islands of ferrite iron (α-Fe) form below ~870 °C. This is a bcc lattice with a smaller lattice size than that of fcc γ-Fe. The α-Fe structure cannot accept much carbon interstitially because of the smaller lattice size. Hence, in the temperature range between 870 and 727°C, the carbon steel is a mixture of α-Fe and γ-Fe phases whose composition is determined by the lever rule (used to determine the mole or mass fraction of each phase in a binary phase diagram). While some of

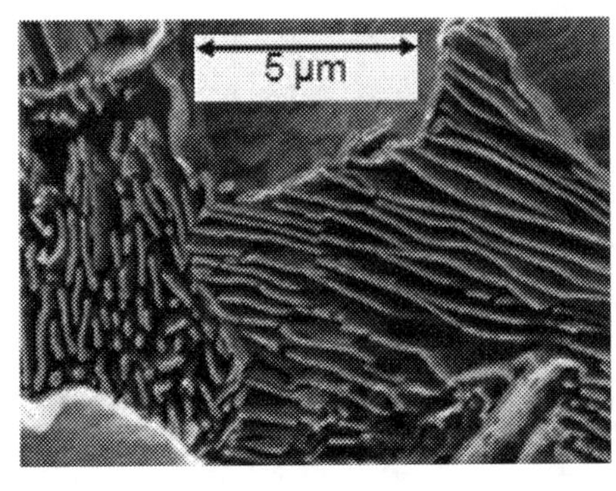

FIGURE 14.3 Corrosion of ferrite lamellae of pearlite structures.

the interstitial carbon is randomly distributed in the γ-Fe phase, in the fcc lattice, the remaining carbon starts forming the pearlite phase mixed into the α-Fe. When the temperature is reduced slowly below 727°C, there is no γ-Fe left in the mixture and all the Fe grains are bcc α-Fe. This effectively means that most of the carbon atoms are locked into the pearlite phase, a mixture of Fe_3C (iron carbide or cementite) layers and α-Fe iron phases with submicron-thick lamellar structures. Cementite is a very hard carbide, and in alternating layers with strained ferrite (Avci et al. 2013, 2015, Martin 2014) produces the well-known pearlite lamellar structure. It is important to recognize the distinction between the two ferrite phases in carbon steel: the thin layers of ferrite phase squeezed between two Fe_3C cementite layers are expected to be highly strained because of the lattice mismatch between the bcc α-Fe and the orthorhombic cementite structure. Atoms residing at grain boundaries are also expected to be strained because of the lattice mismatch. The ferrite crystals making up the majority of the steel will be relatively less strained. These two ferrite structures have been shown experimentally to corrode at different rates (Avci et al. 2013, 2018).

14.2.6 Importance of Inclusions

The inclusions typically found in carbon steel deserve particular attention. Previous work has shown that the immediate surroundings of MnS inclusions are primary locations of pit initiation and growth (Avci et al. 2013, 2015, 2018). As molten steel cools and solidifies, MnS inclusions are formed (Krauss 2003, Behrens and Webster 2011, Choudhary and Ghosh 2008, Choudhary 2012, Ito, Masumitsu, and Matsubara 1981, Kimura et al. 2002, Yu et al. 2006, Sun, Militzer, and Jonas 1992, Liu et al. 2006, Oikawa et al. 1995, Avci et al. 2018). The precipitation behavior of manganese sulfide in mild carbon steel is thoroughly considered by Ito and colleagues (Ito, Masumitsu, and Matsubara 1981). Various forms of MnS inclusions precipitate at different stages during the metallurgical process, nucleate at specific sites, and exhibit characteristic physical traits in terms of size and shape. Earlier work on MnS inclusions in carbon steel suggests that the phase of some of these inclusions is fcc cubic symmetry belonging to space group 255 with m3m symmetry. Here, the term "*MnS inclusions*" is adopted loosely to describe secondary phases whose elemental composition is dominated by Mn and S. Investigations of the surface and bulk composition, as well as considerations of crystalline structure and the identification of phases, have shown that these inclusions are complex and often heterogeneous. The bulk composition of these inclusions has been found to include Cu, Fe, Ca, Ti, and Si. Surface-sensitive techniques have shown that some inclusions are covered with a thin (5–20 nm) layer of copper sulfide, Mn oxide and FeS, which could result in drastically different electrochemical behavior. It has become clear that the term "MnS inclusion" is a broad generalization and is addressed more detail in the previous publication (Rieders 2021).

The literature is replete with articles on the size, distribution and types of MnS inclusions (Baker and Castle 1993, Spitzig 1983a, 1983b, Spitzig and Sober 1981, Gainer and Wallwork 1979). The nucleation and formation of inclusions occur at austenitic Fe temperatures and their geometrical shapes—long stringers extending hundreds of microns along the rolling axis—are formed during the hot rolling process (Krauss 2003, Cyril, Fatemi, and Cryderman 2008). The lattice mismatch between the MnS inclusions and ferrite iron creates lines of enhanced strain areas permeating deep into the material alongside and around the MnS inclusions. Other inclusions, such as silicon oxide, are less malleable, retain their spherical shape, and do not become elongated along the rolling direction. We routinely observe that MnS inclusions and pearlite phases are co-deposited, an observation that is consistent with the bulk of the scientific literature. For instance, Krauss argues that pearlite preferentially forms near elongated MnS inclusions in the iron matrix (Krauss 2003). Our observations on other carbon steel samples suggest more than the randomly expected number of cases in which MnS inclusions and pearlite bands are superimposed along the rolling direction of the steel. This was also the case with the microstructure of base metal A (Figures 14.2a and 14.4). This was an important observation to help account for the corrosion properties of the two base metals (A and B) considered here. The MnS stringers will form in γ-Fe and persist in α-Fe, but pearlite bands can only form during rolling at relatively low temperatures (they start forming below 870°C and are completed at ~725°C). During hot rolling at or above 870°C, no pearlite structures should be formed. The carbon tends to be more randomly distributed interstitially in the fcc lattice of γ-Fe. When annealed and cooled to room temperature, the pearlite structures are randomly deposited (without an ordered pattern), as observed with base metal B (Figure 14.2b) and confirmed with further analysis (below). This distinction becomes an important property when considering the relative susceptibility to corrosion of the two base metals (Clover et al. 2005, Igwemezie and Ovri 2013).

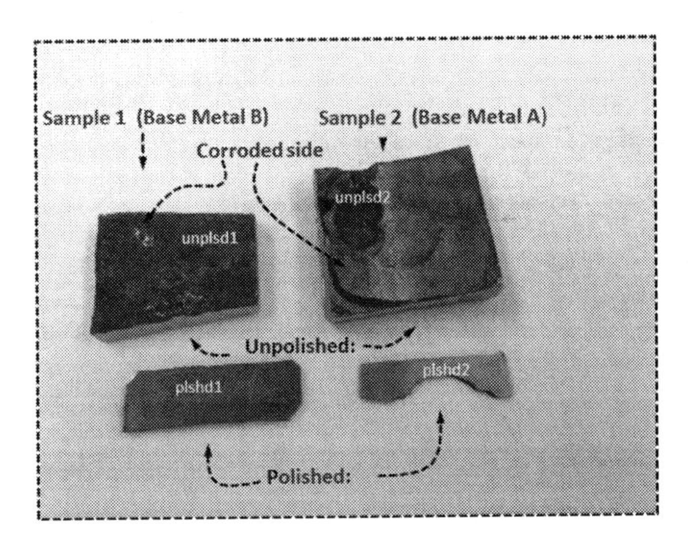

FIGURE 14.4 Corroded pipe samples received from the field (top). Cross sections were cut from corroded pipe and polished to a 50-nm finish (bottom). The coupons were used for a series of abiotic corrosion experiments. Abbreviated names were used for labeling SEM images.

There are many literature reports of carbon steel deterioration in which pits, cracks, or blisters contribute to accelerated localized corrosion. This deterioration tends to initiate at three types of microstructure sites: i) the immediate vicinity of MnS inclusions, ii) pearlite grains, and iii) grain boundaries (Avci et al. 2018). For example, Luu and Wu (1995) used electrochemical permeation measurements and hydrogen microprint techniques to determine that the main diffusion path of hydrogen in mild carbon steel is along the MnS stringers, at the interfaces between the Fe-lattice and in the immediate surroundings of MnS stringers. Ju and Rigsbee in their experiments involving cathodically hydrogen-charged low-carbon steels found that deterioration occurred preferentially at grain boundaries, at the interphase boundaries between ferrite and carbide in pearlite, and at the interfaces of ferrite/MnS inclusions (Ju, Don, and Rigsbee 1986, Ju and Rigsbee 1985, 1988). The damage was manifest as blisters and cracks. Hydrogen-induced dislocations were found to stack up at the same sites, causing deterioration of the steel and accelerated corrosion.

Although MnS inclusions have long been known to be sites of corrosion pitting in carbon steels, the main mechanism of this deterioration has not been satisfactorily addressed. For example, Gainer (Gainer and Wallwork 1979) suggested carbon steel pitting in neutral chloride solutions initiates at MnS inclusions that are anodic to the surrounding matrix. Regions of relatively high MnS inclusion density form macropits through the coalescence of multiple micropits. A similar inference can be seen in the work of both Sephton (Sephton and Pistorius 2000) and Krauss (Krauss 2003), wherein more inclusions per unit area was associated with more severe steel corrosion. Reformatskaya (Reformatskaya et al. 2004) claims that the severity of corrosion at an inclusion is dictated by the phase structure of the inclusion and how coherent it is with the surrounding matrix, rather than by the chemical composition of the inclusion. Wranglen (Wranglen 1974), too, suggests that "active" inclusions corrode more readily than "inactive" ones. He suggests that pitting is initiated in the matrix immediately surrounding inclusions by a fine dispersion of sulfide nanoparticles that infects the iron matrix, making it less noble than the inclusion itself or the bulk steel.

Herein we contend that localized corrosion in carbon steel is driven by residual strain at dislocations due to plastic deformation of crystalline structure and at lattice mismatches (i.e., pearlite and grain boundaries). Whether these strained areas are localized around MnS inclusions or at phase boundaries in pearlite lamella, these are the sites that are micro-anodes relative to the less strained grains of the carbon steel. In effect, micro-galvanic cells distributed throughout the steel surface determine the propensity for corrosion of these areas provided the depositional reactions involving Fe^{2+} ions in solution are not disrupted. In fact, the presence and activity of microorganisms function essentially to change the rate of corrosion but do not determine the locations where the deterioration will initiate and grow. The latter are determined strictly by the metallurgical properties of the steel.

14.3 EXPERIMENTAL METHODS AND RESULTS

14.3.1 Characterization of Strained Areas

To further explore the micrometallurgical characteristics of the metal and to examine its susceptibility to abiotic corrosion, cross sections of the base metals were obtained and polished after adherent deposits were removed with an acidic (Clarke) solution (Lizama and Wade 2015, ASTM 2017, Davis 2013) (Figure 14.4). The resulting rinsed and dried coupons were renamed Sample 1 and Sample 2 and originated from base metals B and A, respectively.

The coupons were microscopically characterized before being used in a series of comparative corrosion experiments. In each experiment, the newly cleaned surface of the coupons served as the reference point for subsequent weight loss determinations. Figure 14.5 shows the SEM images and corresponding orientation images for Samples 1 and 2 obtained by electron backscatter diffraction (a feature of the field emission microscope). The significance of these images is that they show that the annealing process (applied after the rolling process was completed) produced proper ferrite domains without traces of austenitic features. All the orientations

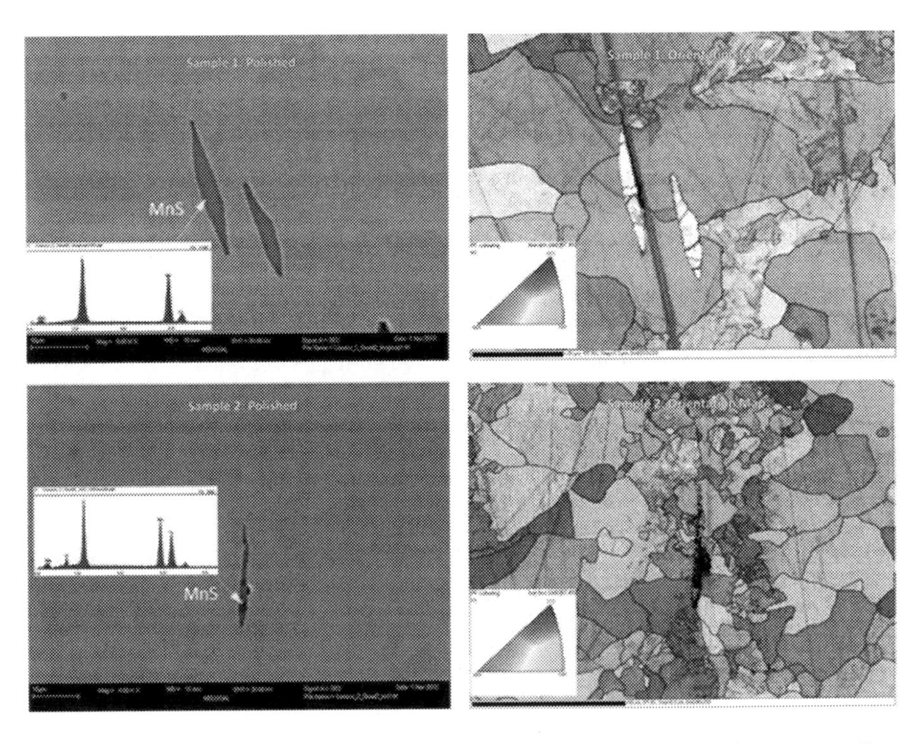

FIGURE 14.5 SEM images (left frames) showing MnS inclusions and the corresponding orientation maps (right frames) obtained from Sample 1 (top row) and Sample 2 (bottom row) prior to experimentation. Insets in the SEM images (left) shows EDX spectra with Mn, S, and Fe signatures. Insets in the orientation images (right) show the color code for the crystal domain orientations: 111 = blue, 101 = green, and 001 = red.

associated with the domains are part of the bcc structure of ferrite lattice. However, the quality of the orientations varies in regions where pearlite structures dominate.

These regions correspond to a highly strained lattice wherein two-phased pearlite lamellar structures composed of alternating layers of highly strained ferrite and cementite structures dominate. Such structures were not visible in the SEM images (Figure 14.5, left frames) because the surfaces were not treated with Nital to emphasize these features (as in Figure 14.2). However, such structures are very apparent in the corresponding orientation maps (Figure 14.5, right frames). The pearlite structures are highly strained because of the lattice mismatch between the orthorhombic cementite and the cubic bcc ferrite lattice structures. The orientation maps (Figure 14.5) can be used to derive strain maps (sometimes called misorientation maps) (Figure 14.6 right frames). It is important to note that such strain maps can be collected before metal samples are exposed to a corrosive environment.

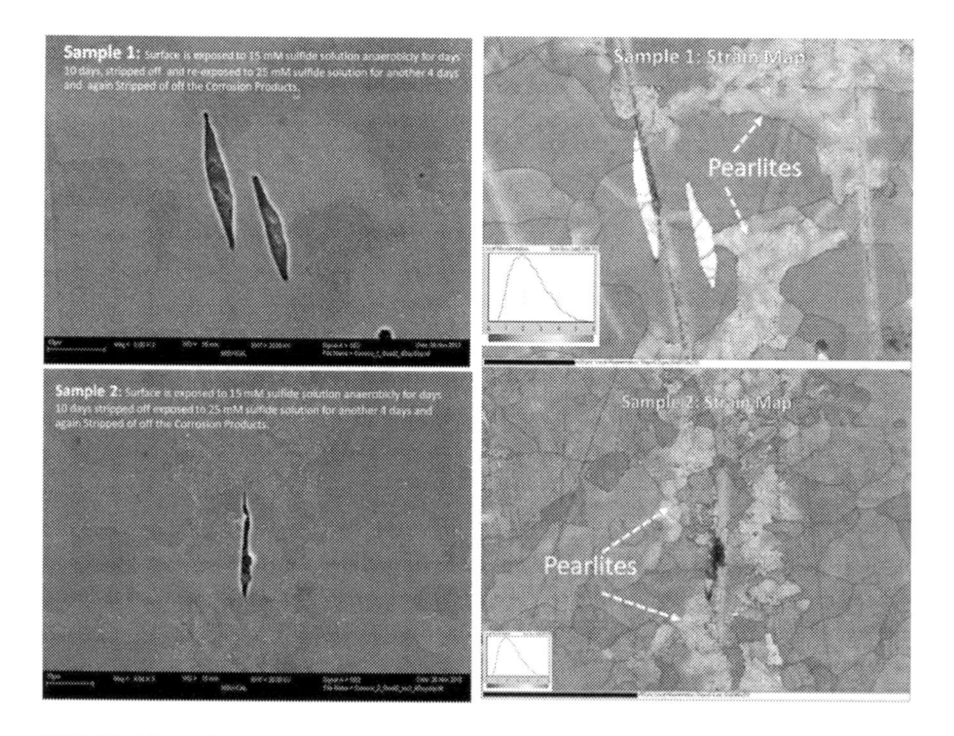

FIGURE 14.6 SEM images (left frames) showing inclusions and corresponding derived strain maps (right frames) obtained from Sample 1 (top row) and Sample 2 (bottom row). The strain maps were obtained before the metal was exposed to a series of corrosion experiments. The SEM images were taken following a 2-week anoxic incubation of the coupons in a high-strength sulfide solutions (experiments 1 and 2, see text) and surface cleaning (Clarke solution) to remove adherent corrosion products. This treatment revealed the highly strained pearlite band locations, which were well predicted by the strain maps made prior to the experimentation. Insets in the misorientation images (right) show the color code of relative concentrations of local misorientation ranging from 0 to 6 degrees of crystal planes from the ideal orientations.

The determination of strain effectively allows for the prediction of where localized corrosion will initiate and propagate. A comparison of predicted strain areas and the micrometallurgical features revealed after a series of abiotic corrosion experiments on Samples 1 and 2 is shown in Figure 14.6. This comparison revealed that the more highly strained metallurgical areas are indeed areas that tend to corrode differentially.

14.3.2 Abiotic Corrosion Experiments

We tried to experimentally accelerate and reproduce the field deterioration of the metal by immersing the polished coupons (Samples 1 and 2) in a series of corrosive environments and then assessing corrosion product formation, degree of pitting and weight loss. Given the micrometallurgical differences detailed previously, we predicted an increase in corrosion in Sample 2 (base metal A) relative to Sample 1 (base metal B).

We conducted five comparative experiments (1–5), the first four of which were strictly anoxic incubations; the fifth incorporated a low level of oxygen. Experiments 1 and 2 involved immersing the coupons in filter-sterilized seawater that was amended with 15 mM sulfide solution for the first 10 days and 25 mM sulfide for four more days thereafter. At the end of each experiment, the coupons were recovered, analyzed for corrosion products, stripped with Clarke solution, analyzed for evidence of pitting, and reweighed to relate weight loss to the experimental treatment.

Figure 14.6 (left frames) illustrates that there was minimal corrosion of the metal after exposure of the coupons to an anoxic sulfide-rich incubation for 2 weeks. Similarly, the coupons, once recovered, were incubated in 100 mM sulfide solution made with nanopure water for 48 days (experiment 3). SEM and EDX analysis of adherent material showed no appreciable corrosion of the coupons for this treatment either (data not shown). The lack of corrosion is most likely due to the formation of a protective (passivating) sulfide layer precluding further deterioration of the metal surfaces.

The coupons were again recovered and processed as indicated above for experiment 4. In this experiment, the freshly cleaned coupon surfaces were anoxically exposed to NS_4 (Xu and Cheng 2013) solution for 28 days in an attempt to stimulate CO_2 corrosion. This treatment also failed to produce appreciable levels of corrosion, but calcium oxide deposits were noted on the metal surface of both coupons in SEM images (not shown).

A fifth experiment was conducted by again immersing the recovered and cleaned coupons into sulfide-amended (25 mM) filter-sterilized seawater, but no effort was made to remove oxygen from the incubation. This low-oxygen experiment was conducted in an effort to disrupt a presumed passivating layer by allowing oxygen to react abiotically with sulfide. As in the other experiments, this incubation was conducted in 27 mL bottles containing 20 mL of sulfide-amended filter-sterilized seawater. Each bottle had about 6.7 mL of headspace filled with ambient air and was sealed from the atmosphere. A simple calculation shows that if all the O_2 in the headspace had reacted with H_2S, this would have minimally reduced the sulfide concentration, from 25 mM to 22 mM. The dissolved O_2 in the seawater would react with sulfides relatively rapidly (within an hour). However, the diffusion rate of O_2 from air to water is fairly slow, calculated at about 3.44 x 10^{13} O_2 molecules per second per cm^2. This slow rate would help provide a low level of oxyen diffusing to the coupon surface over the period of incubation. The experiments were conducted for 48 days under these conditions and dense corrosion product adhered to the coupon surfaces (Figure 14.7). The exposure to sulfide and low levels of oxygen produced more obvious evidence of abiotic corrosion than was seen in the other experiments.

The corrosion products generated in experiment 5 tended to be sulfides and some oxides of Fe and/or Mn (see zoomed-in images in Figure 14.7). The deposition of MnS particles is somewhat surprising, since there was less than 2 wt.% Mn in the base metals (Table 14.1). While a very small fraction of the Mn is in MnS stringers, the majority is scattered throughout the ferrite lattice, occupying some of the bcc

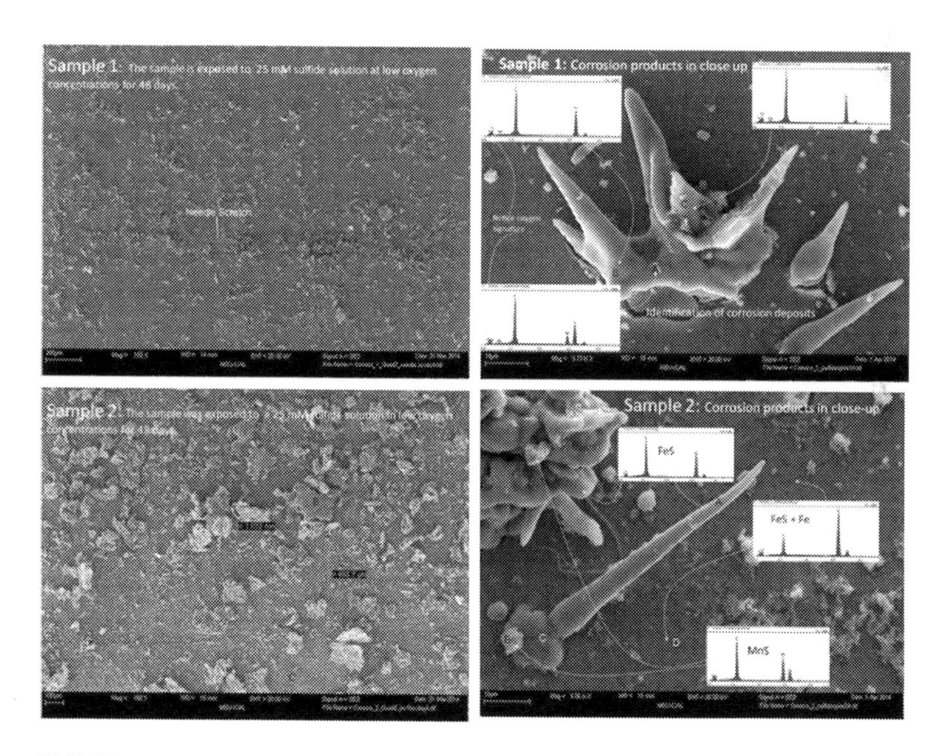

FIGURE 14.7 SEM images of corrosion products on Sample 1 (top row) and Sample 2 (bottom row). The samples were exposed to a 25 mM sulfide solution under low oxygen concentrations in seawater for a period of 48 days. The column on the right shows close-up images of some of the corrosion products. Also shown in the close-up images are examples of EDX spectra from selected points on the corrosion products. These identify the elemental composition of these corrosion deposits. Insets in the right panels show mostly Fe-K_α, K_β (~5.4 & 7 KeV), Fe-L (~0.7 eV), S-K_α (~2.3 KeV), Mn-K_α (~5.9 KeV), and low concentrations of O-K_α (~0.5 KeV) as seen in the images.

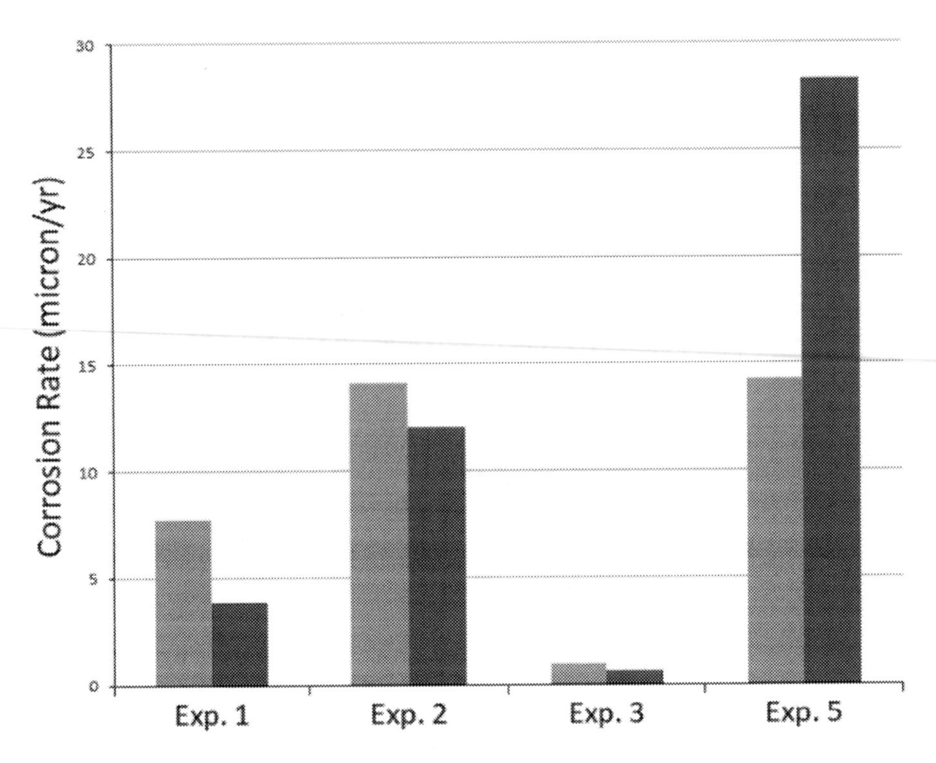

FIGURE 14.8 Comparison of metal corrosion rates (Sample 1—blue bars; Sample 2—red bars) in sulfide-amended experiments. There was minimal or no differential corrosion in Experiments 1–3. These surfaces were likely passivated by high sulfide concentrations (see text for details) that slowed or prevented anodic and cathodic processes. When the coupons were exposed to sulfide and a low level of oxygen (experiment 5), differential corrosion was evident, presumably reflecting the reaction of oxygen with the passivating layer to reveal the base metals.

sites. The observation of small particles of MnS-like particles on FeS spires in Figure 14.7 is surprising given the far greater mass of iron in the system.

The abiotic corrosion rate estimations based on the weight loss determinations for the various experiments are shown in Figure 14.8. It was evident from this comparison that minimal or no differential corrosion occured in experiments 1–3. However, Sample 2 (from base metal A) corroded much faster than Sample 1 (from base metal B) when traces of oxygen were present in sulfide-replete incubations (experiment 5). The imposed incubation conditions involving traces of oxygen in sulfide-replete seawater likely better reflected the differential corrosion noted in the field.

14.3.3 Laboratory to Field Extrapolation

We presumed that the high sulfide environment in the field would largely be a function of the metabolic activity of sulfate reducing bacteria. We also presumed

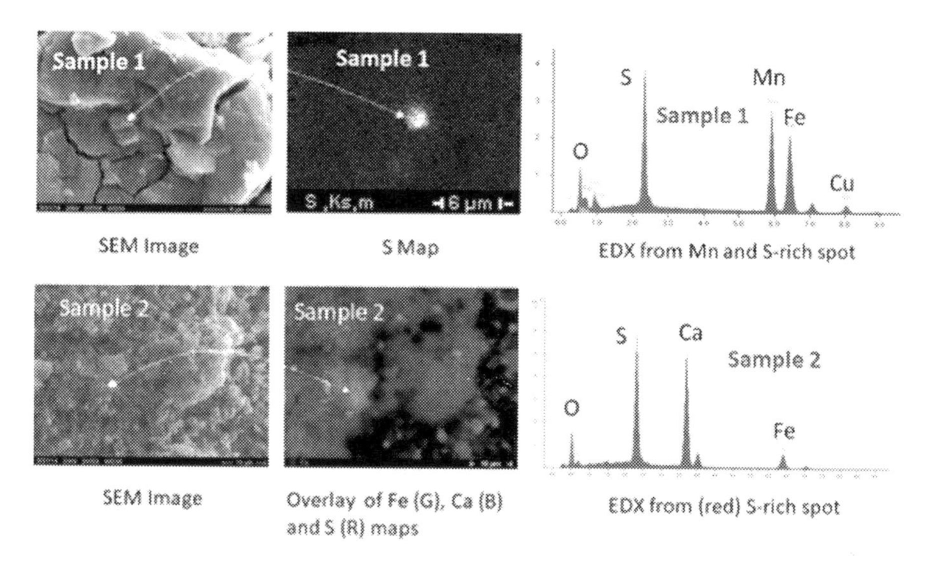

FIGURE 14.9 EDX spectra obtained from the two as-received unpolished steel coupons (Figure 14.4) showing evidence of sulfide as in MnS and FeS, indicating SRB activity in the vicinity of the pipe. Elemental maps of Fe, Ca, and S are also shown.

that traces of oxygen would be fairly common in the field, even in a system where oxygen was vigorously scavenged. If this is true, we would expect to detect abiotic corrosion products that were similar in composition to the type observed in experiment 5. We searched for evidence of Mn and S signatures on the adherent materials on the unpolished base metals (Figure 14.1). These findings are depicted in Figure 14.9.

Morphologically, the corrosion deposits from the field were quite different from those produced in experiment 5. However, compositionally, there were similarities such as FeS and MnS minerals in common on both samples (compare Figures 14.9 and 14.7).

When the coupons from experiment 5 were recovered, cleaned with Clarke solution and imaged again (Figure 14.10), the formerly near pristine coupon surfaces were severely scored by localized corrosion, particularly in and around pearlite and MnS features. Pearlites dominate the steel structure, and where there was pearlite, severe localized corrosion was observed. It is not difficult to imagine that in prolonged field exposure of the steel with trace oxygen levels, corrosion would be similarly localized, giving rise to pits where there were high concentrations of pearlite, MnS inclusions, and related micro-imperfections.

A prior publication (Avci et al. 2018) poses that all localized corrosion can be traced to the concept of residual strain and dislocations generated in the ferrite matrix of carbon steel, particularly at banded and random pearlite grains, MnS/ferrite interfaces and MnS/pearlite rich areas. As indicated, when steel is rolled near the transition temperature (~730 °C) the bands of pearlite and some MnS stringers align as in Figure 14.10 (Sample 2 close-up view). This is a disastrous combination

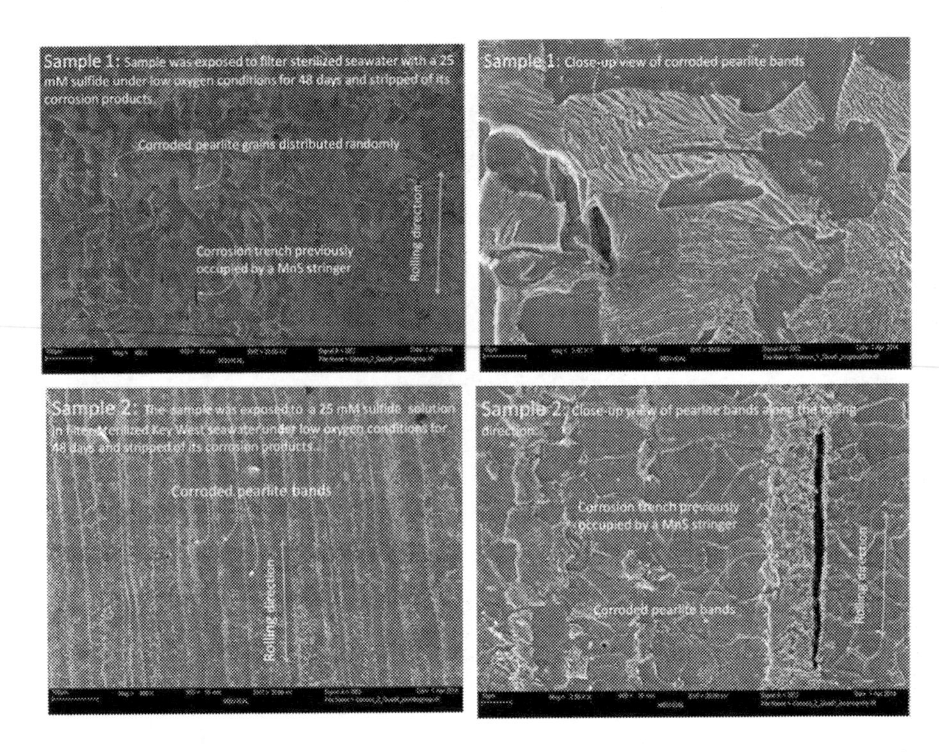

FIGURE 14.10 Comparison of the corrosion damage to coupons recovered from experiment 5. The images show the microstructural features revealed after the corrosion products were removed. Sample 1 (top row) and Sample 2 (bottom row).

for corrosion (Avci et al. 2015, 2018) since the strain around MnS/ferrite interfaces and in the pearlite ferrite/cementite interfaces represent preferential locations for corrosion, making Sample 2, originating from base metal A, far more susceptible to localized corrosion than Sample 1.

We also wanted to determine how deep the localized corrosion penetrated into the steel so we would have a qualitative idea about the rate of localized corrosion over 48 days of exposure. Hence, we determined the depth of pitting using techniques developed previously (Avci et al. 2015). An example of typical depths inside the MnS stringers is shown in Figure 14.11. The model determines the path-lengths of Fe-Kα X rays generated at the bottom of the pit as they travel through the ferrite matrix towards the X-ray detector at 55 degrees from the surface normal. The pit depths are found to be typically 3–4 microns, which is fairly deep compared to the starting surface that was polished to a nearly flat submicron finish.

We also sought to better account for the fraction of O_2 that was incorporated into corrosion products in the experiment 5 incubations. However, EDX spectra (as in Figure 14.9) are not reliable since the low-energy X-rays of the low-Z elements are known to be less than entirely reliable in the absence of appropriate reference spectra. That is, EDX data for low-Z elements can only be interpreted as a qualitative guide. X-ray photoelectron spectroscopy (XPS) is a more reliable technique

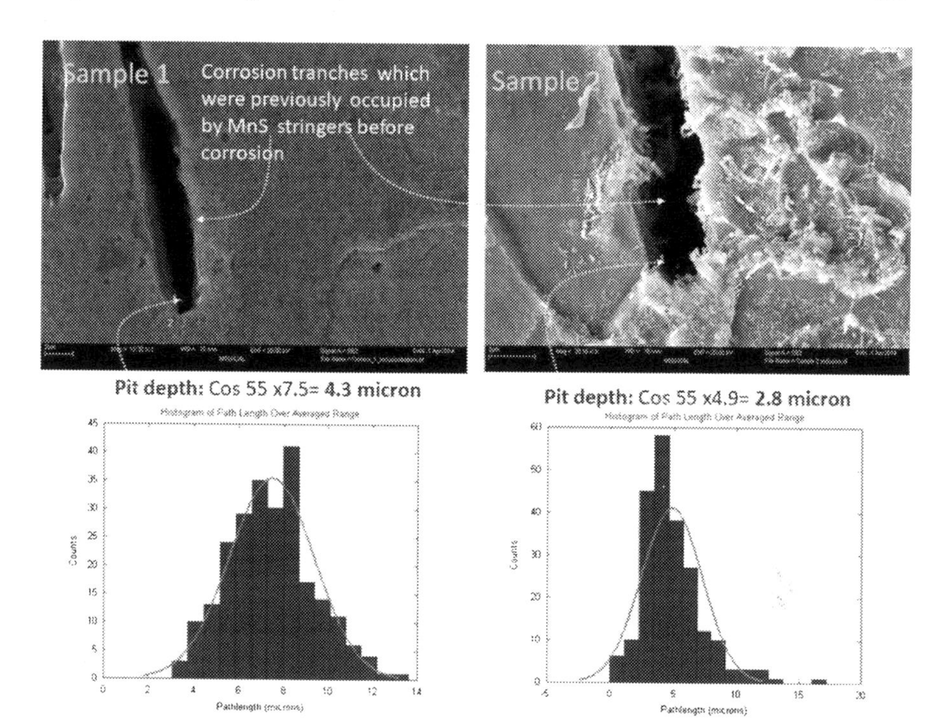

FIGURE 14.11 Typical pit depths determined using the X-ray attenuation method.

for quantitative analysis of the low-Z elements on material surfaces (Figure 14.12). The surface concentrations of various elements in the abiotic corrosion products on Sample 2 are presented as an example (Figure 14.12 upper frame). Presumably, some of the oxygen in the system reacted with sulfide and Fe^{2+} ions to generate a number of composite minerals containing both sulfides and oxides.

However, a comparative EDX survey (Figure 14.12, lower frame) showed relatively little oxygen incorporated into corrosion deposits that were otherwise dominated by sulfides. This suggests that the trace levels of oxygen in the system reacted with the Fe^{2+} ions in solution, disrupting the formation of a protective sulfide layer on the steel surfaces. If there had been more oxygen in the system, the majority of the corrosion deposits would presumably have been a series of Fe-oxides, as observed in previous carbon steel corrosion experiments. Furthermore, we have seen Fe/S/O-rich deposits in previous biotic and abiotic corrosion experiments in which trace concentrations of oxygen were made available in sulfide-rich incubations.

14.4 DISCUSSION

A model of localized corrosion in carbon steel is conceptually depicted in Figure 14.13 as it relates to strain, but it does not differentiate whether the deterioration is taking place near inclusions, in pearlite grains, or elsewhere. The model

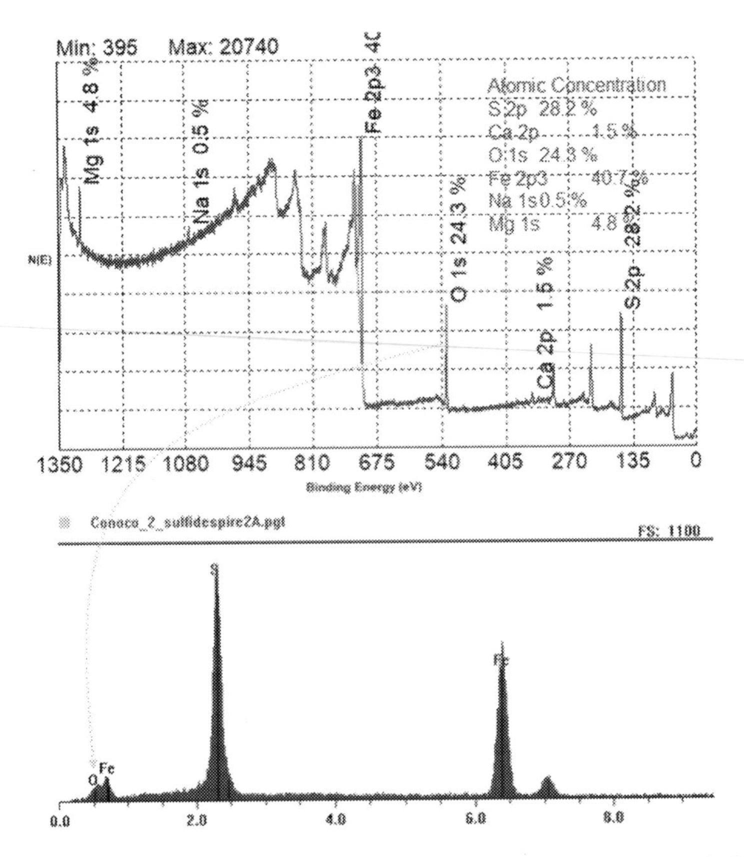

FIGURE 14.12 Confirmation of oxygen in the abiotic corrosion product deposited during experiment 5. The top frame shows an XPS spectrum taken from a wide (~1 mm^2) area of Sample 2. The bottom frame is EDX data taken from the sulfide spires (see Figure 14.7), revealing the presence of oxygen. The sulfide spires are yet to be identified minerals and are likely a combination of Fe-sulfide and Fe-oxide features. The XPS spectrum of Sample 2 was taken before the corrosion products were removed, but the surface was sputtered with an Ar+ ion beam (by about 20–30 nm) to clean off the environmental residues.

will be the same wherever a region of the ferrite matrix is strained as a result of a stress—be it plastic deformation, lattice mismatch, or the presence of a foreign inclusion. Such disturbances in a localized area increase the Gibbs free energy of the ferrite matrix. The increase in the thermodynamic potential of the iron matrix causes dissolution of Fe^{2+} ions into the medium, creating an excess electron density in the metal at the strained location. The excess electron density creates a micro-electrochemical potential difference ($\Delta\phi$, Figure 14.13). The galvanic potential is then more positive at the unstrained site than at the strained location. The magnitude of the difference will depend on the degree of strain and the rate of depositional reactions. If dissolved Fe^{2+} ions are cleared from the strained area (e.g., by deposition reactions), more opportunity for fresh dissolution of Fe^{2+} ions is then

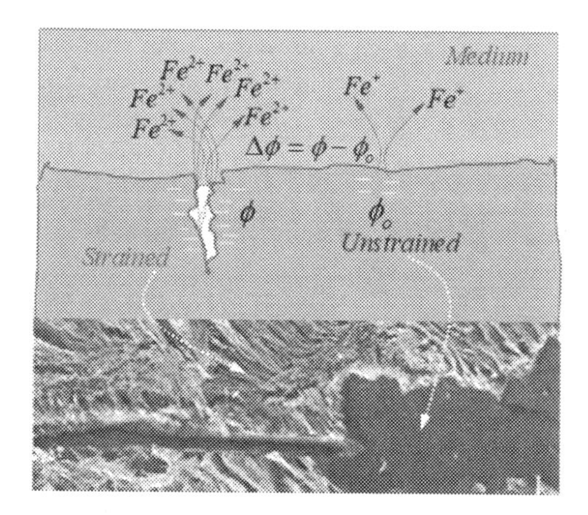

FIGURE 14.13 Proposed model for the localized corrosion of carbon steel. The corrosion is driven by the *mechanochemical effect*: Regions of the metal surface that are strained during the metallurgical process become more anodic relative to unstrained regions. The galvanic potential differences, $\Delta\phi$'s, are established throughout the surface. The yellow features represent excess electrons left after ferrous iron leaves the local area. These cause a negative increase in the electrochemical potential of the strained regions relative to the unstrained regions throughout the metal. As long as the depositional chemistry in the liquid phase does not interfere (i.e., formation of oxides or sulfides), these galvanic processes at the micron scale effectively continue localized corrosion and pitting of the metal surface at the microscale.

available. Localized corrosion will persist as long as quasi-static thermodynamically stable conditions persist that effectively etch the iron matrix.

It is instructive to review the underlying electrochemical processes that give rise to excess dissolution of Fe^{2+} ions from the strained localized areas. Because of the annealing and hot and cold rolling treatments, the carbon steel surface contains a high density of areas with plastic strain, dislocations, MnS stringers, pearlite phases and lattice defects. As discussed previously, there is a strong correlation between the propensity for corrosion and the degree of strain in these areas. The dissolution of MnS in acidic environments ($pH < 4$) generates abiotically highly corrosive products such as elemental S and H_2S, which will participate in cathodic reactions on the Fe sites away from the surface, i.e., along the length of the MnS stringers. The hydrolysis of Fe ions in the immediate surroundings of MnS inclusions or localized corroding sites via, for example, $Fe^{2+} + 2H_2O \rightarrow Fe(OH)_2 + 2H^+$ will acidify the local area. Under low-pH conditions, MnS starts dissolving via $MnS + 2H^+ \rightarrow H_2S + Mn^{2+}$, generating abiotic H_2S along the MnS inclusions (stringers), which extend hundreds of microns into the carbon steel matrix along the rolling direction (although the inclusion width is < 1 μm). H_2S species react with the carbon steel surface and contribute to its dissolution through the reaction $2H_2S + Fe^0 \rightarrow 2HS^- + H_2 + Fe^{2+}$, generating H_2 gas in the process. Corrosion along a MnS inclusion causes micro-pits, which then coalesce with other micro-pits, giving rise to larger pits, and so on (Avci et al. 2013, 2018).

Coupled with the MnS inclusions, now consider the pearlite, which may or may not be ordered into bands. If, however, the pearlite is in bands because of the rolling process while most of the Fe matrix is in the ferrite phase, for reasons yet to be understood the MnS inclusions and pearlite bands are inclined to occupy the same physical space (more than would be expected from random coincidence) (Krauss 2003), as shown in Figure 14.10 for Sample 2 (base metal A). Both pearlite bands and the MnS/α-Fe interfaces are highly concentrated regions of strained ferrite. The resulting mechano-electrochemical properties give rise to excess corrosion sites relative to the randomly distributed pearlite bands and MnS distributions in hot rolled carbon steel (Sample 1; base metal B). The difference between the two metals gives rise to the excess corrosion in Sample 2.

An understanding of why strain introduces an excess local chemical potential stems from the pioneering work of E.M. Gutman (1968, 1998). This under-appreciated theoretical treatise predicted that the corrosion potentials of metal alloys would change to more active potentials as a result of elasto-plastic deformations. Gutman's predictions for changes in the corrosion potentials for elastic and plastic deformations, respectively, are given as:

$$\Delta\phi_e = -\frac{\Delta P V_m}{zF} \text{ and } \Delta\phi = -\frac{TR}{zF}\ln\left(\frac{\nu\alpha}{N_o}\varepsilon + 1\right) \tag{14.1}$$

where z is typically 2, the number of electrons left in metal as a result of release of one Fe^{2+} ion into the solution, T is temperature and $\Delta\phi_e$ does not contribute substantively to the local potential difference and will be ignored in this consideration. In contrast, $\Delta\phi$ is particularly important. The parameters listed in the $\Delta\phi$ formulation are shown in Figure 14.14.

One of us (RA) has conducted model calculations with 1018 carbon steel using the parameters in Figure 14.14 in an attempt to assess the values of $\Delta\phi$ due to elastic deformation. Plastic strain ε, introduced during the metallurgical working of the

- m = local misorientation
- d = 0.25 μm (EBSD step size)
- ν = 0.45 (orientation factor)
- $\alpha = 10^{10}$ cm^{-2} (dislocation modulus)
- $N_o = 10^{10}$ cm^{-2} (dislocation density in the strained iron)
- $\Delta\phi$ = galvanic potential difference (V)
- ε = plastic strain
- b = 0.025 (average Tafel slope)
- i_o = 1.6 10^{-6} A/cm^2 (corrosion current density)
- $\Delta\tau$ = excess corrosion rate (mm/yr)
- Δi = excess corrosion current density (A/cm^2)
- R = Gas constant
- F = Faraday constant
- n = 2 (number of electrons exchanged)

FIGURE 14.14 Parameters used to calculate the mechano-electrochemical potential differences in carbon steel.

steel, can be determined using local misorientation m mapping (in degrees) as in Figure 14.5. Misorientation m is then translated into plastic strain ε through Equation 14.2:

$$\varepsilon = \frac{(m - 0.1)/100}{-0.0027d^2 + 0.041d} \tag{14.2}$$

where the value of d is obtained from the literature (Kamaya 2009). From this we can quantify the mechano-electrochemical potential differences using $\Delta\phi$ from Equation 14.1:

$$\Delta\phi = -\frac{TR}{zF}\ln\left(\frac{\nu\alpha}{N_o}\varepsilon + 1\right).$$

The values listed in Figure 14.14 can then be used to construct strain and local mechano-electrochemical potential maps of the strained surface, as in Figure 14.15. We conducted corrosion experiments on the 1018 carbon steel surface shown in Figure 14.15, measured pit depths generated by the excess corrosion, and compared them to the model predictions. There were generally good (within 10–15%) agreements with predictions (Martin 2014). This finding helped illustrate how electrochemical potential differences relate to the excess corrosion current and the excess corrosion propensity. It is well-known from the thermodynamics of the corrosion process that the excess corrosion current density is given by

$$\Delta i = i_o e^{\frac{|\Delta\phi|}{b}} \text{ in A}/m^2 \tag{14.3}$$

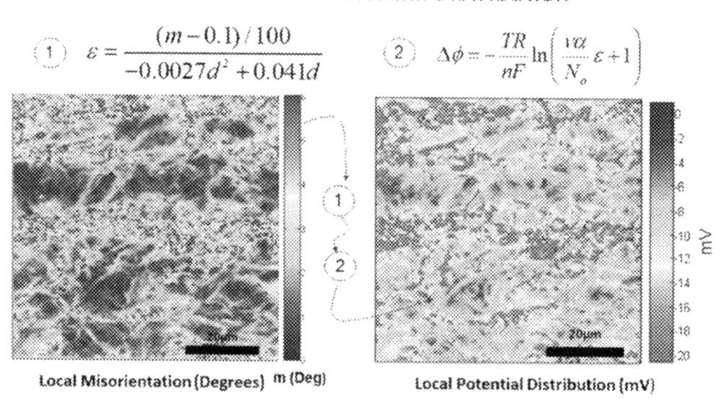

FIGURE 14.15 Strain map and local mechano-electrochemical potential differences predicted by the Gutman formulation as it applies to freshly polished 1018 carbon steel before it is subjected to corrosion. The yellow and red areas represent the pearlite bands running along the horizontal direction.

and parameter b is listed in Figure 14.14. Similarly, the excess corrosion rate can be calculated using a method described in a thesis (Martin 2014) and proved in Annex 1 as

$$\Delta\tau = 1.16 \times 10^4 \Delta i \text{ in mm/yr.} \quad (14.4)$$

The important point is that the mechano-electrochemical distributions shown in Figure 14.15 and local potential differences were the reason the surfaces of the test steels were etched differentially and one metal corroded more obviously than the other.

14.5 CONCLUSIONS

It is important to emphasize a number of fundamental features that help account for the differential corrosion in the pipe and the role of the resident microorganisms. The presence and activity of microorganisms likely had a substantive role in overall process. Though they likely colonized the surfaces of metal more or less randomly, when coupled with the metallurgical history of the pipe metal, their role is amplified. Their most important contribution to corrosion is to generate H_2S and related corrosive cathodic reactants. The pipe locations that corrode are entirely determined by the metallurgical history of the metal. Pipes will corrode at predetermined locations regardless of whether the system has active microorganisms in it or not. The activity of microorganisms will affect the rate of corrosion but not where the localized corrosion takes place. The latter locations are determined by the metallurgy of the pipe material.

For a practical application, the banding patterns of pearlite are a good indication of which pipe or pipe sections might be more resilient against localized corrosion. These banding patterns can be determined by conducting standard microstructure analyses from cross sections of pipe exhibiting different corrosion patterns (Figure 14.2). This information is, in turn, useful for determining the direct cause of the corrosion and for making recommendations for repair or replacement. In the case reported here it was important to realize that microorganisms were a contributing factor and thus MIC was a reasonable assessment. However, the microstructural history of the pipe sections during manufacturing accounted for the differential corrosion patterns.

Whether a particular pipe has banded pearlite features is a secondary issue as all carbon steel pipes have strained areas that produce localized micro-galvanic cells that are like an underlying cancer throughout the carbon steel. Minimizing these mechano-electrochemical galvanic couplings is predicted to produce the most rewarding steel for oil and gas production and other marine applications.

ACKNOWLEDGMENTS

Partial support for this work by ConocoPhillips through the University of Oklahoma Biocorrosion Center is gratefully acknowledged. Federal grants from NSF/NNCI (ECCS¬1542210) and the Office of Naval Research (N00014-10-1-0946) also helped finance the work. The effort of Montana State University to underpin the ongoing efforts of the Imaging and Chemical Analysis Laboratory (ICAL) is also appreciated. The authors would like to thank Mr. J. Martin, Dr. M. Wolfenden, Ms. N. Equall, and Ms. L. Kellerman for their acquisition of some of the spectroscopic and microscopic data, Prof. I. Beech and Prof. D. Mogk of Montana State University for valuable discussions regarding the localized corrosion of carbon steel, Mr. B.H. Davis, Mr. N. Rieders and Mrs. E. Roehm for some of the SEM-related work and Mrs. L. Avci for copy editing.

ANNEX 1: DERIVATION OF CORROSION RATE

At the steady state equilibrim the polarization curve represented by current density i vs. ϕ (the excess voltage above the open circuit potential E_{ocp}) plot can be represented by means of Volmer-Butler Equation as $i \simeq i_0(e^{\phi/\alpha} - e^{-\phi/\beta})$ where i_o is the corrosion current density, $\alpha = RT/\gamma nF$ and $\beta = RT/(1-\lambda)nF$, where γ and λ are electron transfer coefficients for anodic and cathodic processes, respectively and F = 9.65 × 10^5 Coulomb/mol is the Faraday's constant, n = 2 on account of Fe^{+2} ions, T = 298 K and R = 8.31 J $K^{-1}mol^{-1}$. Here we make the approximation that $\gamma \approx \lambda \approx 0.5$, which is a reasonable approximation introduces at most 10–20% uncertainty. From this we can calculate that $\alpha \approx \beta \approx 0.026\,V$. Now let us expand $i \simeq i_o(e^{\phi/\alpha - e^{-\phi/\beta}})$ into Taylor's series around $\phi/\alpha = \phi/\beta \approx 0$ which immediately reduces to $i \approx i_o(1/\alpha + 1/\beta)\phi = \phi/R$, where R = corrosion resistivity. Considering $\alpha \approx \beta \approx 0.026\,V$, it is easy to note that $\frac{1}{R} = \frac{i_o}{1.3 \times 10^{-2}}$ or $i_o = \frac{1.3 \times 10^{-2}}{R}$ where R is in $ohm - cm^2$ and i_o is in A/cm^2. It is then very easy to find the R (resistivity) from slope of a straight line associated with ϕ vs. i polarization plot, from there one can determine the corrosion current density i_o using $i_o = \frac{1.3 \times 10^{-2}}{R}$. Taking into account that i_0 is in the units of coulombs per second per square cm, and each dissolved Fe^{+2} ion removes 2 × 1.6 × 10^{-19} coulombs from the Fe and taking into the account the density of carbon steel one can easily determine the rate of Fe eroded from the surface by the Corrosion rate $\tau \approx 1.2 \times 10^4 \times i_o\,(mm/year)$.

REFERENCES

ASTM. 2017. Standard practice for preparing, cleaning, and evaluating corrosion test specimens. In *G1*. West Conshohocken, PA.

Avci, Recep, Bret H. Davis, Nathaniel Rieders, Kilean Lucas, Manjula Nandasiri, and Dave Mogk. 2018. Role of metallurgy in the localized corrosion of carbon steel. *Journal of Minerals and Materials Characterization and Engineering* 6(6):618–646. doi: 10.4236/jmmce.2018.66044.

Avci, R., B. H. Davis, M. L. Wolfenden, I. B. Beech, K. Lucas, and D. Paul. 2013. Mechanism of MnS-mediated pit initiation and propagation in carbon steel in an anaerobic sulfidogenic media. *Corrosion Science* 76:267–274. doi: 10.1016/j.corsci.2013.06.049.

Avci, Recep, Bret H. Davis, Mark L. Wolfenden, Laura R. Kellerman, Kilean Lucas, Joshua Martin, and Muhammedin Deliorman. 2015. A practical method for determining pit depths using X-ray attenuation in EDX spectra. *Corrosion Science* 93:9–18.

Baker, M. A., and J. E. Castle. 1993. The initiation of pitting corrosion at MnS inclusions. *Corrosion Science* 34 (4):667–682.

Behrens, Harald, and James D. Webster. 2011. Studies of sulfur in melts–motivations and overview. *Reviews in Mineralogy and Geochemistry* 73 (1):1–8.

Choudhary, Shiv Kumar. 2012. Thermodynamic evaluation of inclusion formation during cooling and solidification of low carbon Si-Mn killed steel. *Materials and Manufacturing Processes* 27 (9):925–929.

Choudhary, S. K., and A. Ghosh. 2008. Thermodynamic evaluation of formation of oxide-sulfide duplex inclusions in steel. *ISIJ international* 48 (11):1552–1559.

Clover, D., B. Kinsella, B. Pejcic, and R. De Marco. 2005. The influence of microstructure on the corrosion rate of various carbon steels. *Journal of Applied Electrochemistry* 35 (2):139–149. doi: 10.1007/s10800-004-6207-7.

Cyril, Nisha, Ali Fatemi, and Bob Cryderman. 2008. Effects of sulfur level and anisotropy of sulfide inclusions on tensile, impact, and fatigue properties of SAE 4140 steel. SAE Technical Paper.

Davis, Bret H. 2013. Anaerobic pitting corrosion of carbon steel in marine sulfidogenic environments. MS, Physics, Montana State University.

Gainer, L. J., and G. R. Wallwork. 1979. The effect of nonmetallic inclusions on the pitting of mild steel. *Corrosion* 35 (10):435–443.

Gutman, E. M. 1968. Thermodynamics of the mechanico-chemical effect. *Soviet Materials Science* 3 (3):190–196.

Gutman, E. M. 1998. *Mechanochemistry of Materials.* Cambridge, UK: Cambridge International Science Publishing.

Igwemezie, V. C., and J. E. O. Ovri. 2013. Investigation into the effects of microstructure on the corrosion susceptibility of medium carbon steel. *The International Journal of Engineering and Science (IJES)* 2 (6):16.

Ito, Yoichi, Noriyuki Masumitsu, and Kaichi Matsubara. 1981. Formation of manganese sulfide in steel. *Transactions of the Iron and Steel Institute of Japan* 21 (7):477–484.

Ju, C. P., J. Don, and J. M. Rigsbee. 1986. A high voltage electron microscopy study of hydrogen-induced damage in a low alloy, medium carbon steel. *Materials Science and Engineering* 77:115–123.

Ju, C. P., and J. M. Rigsbee. 1985. The role of microstructure for hydrogen-induced blistering and stepwise cracking in a plain medium carbon steel. *Materials Science and Engineering* 74 (1):47–53.

Ju, C. P., and J. M. Rigsbee. 1988. Interfacial coherency and hydrogen damage in plain carbon steel. *Materials Science and Engineering: A* 102 (2):281–288.

Kamaya, Masayuki. 2009. Measurement of local plastic strain distribution of stainless steel by electron backscatter diffraction. *Materials Characterization* 60 (2):125–132. doi: 10.1016/j.matchar.2008.07.010.

Kimura, Sei, K. Nakajima, S. Mizoguchi, and H. Hasegawa. 2002. In-situ observation of the precipitation of manganese sulfide in low-carbon magnesium-killed steel. *Metallurgical and Materials Transactions A* 33 (2):427–436.

Krauss, George. 2003. Solidification, segregation, and banding in carbon and alloy steels. *Metallurgical and Materials Transactions B* 34 (6):781–792.

Lehmann, Jean, and Michèle Nadif. 2011. Interactions between metal and slag melts: steel desulfurization. *Reviews in Mineralogy and Geochemistry* 73 (1):493–511.

Liu, Zhongzhu, Yoshinao Kobayashi, Kotobu Nagai, Jian Yang, and Mamoru Kuwabara. 2006. Morphology control of copper sulfide in strip casting of low carbon steel. *ISIJ international* 46 (5):744–753.

Lizama, S. A., and Y. Wade. 2015. Clarke's solution cleaning used for corrosion product removal: effects on carbon steel substrate. Corrosion and Prevention 2015, Adelaide, Australia.

Luu, W. C., and J. K. Wu. 1995. Effects of sulfide inclusion on hydrogen transport in steels. *Materials Letters* 24 (1):175–179.

Mahajanam, S., and M. Joosten. 2011. "Guidelines for filler-material selection to minimize preferential weld corrosion in pipeline steels." Proceedings of SPE International Conference on Oilfield Corrosion, Aberdeen, UK, May 24–25 2010; Paper SPE 130513.

Martin, Joshua Daniel. 2014. "Biocorrosion of 1018 steel in sulfide rich marine environments; A correlation between strain and corrosion using electron backscatter diffraction." Masters of Science MS, Chemical Engineering, Montana State University.

Oikawa, Katsunari, Hiroshi Ohtani, Kiyohito Ishida, and Taiji Nishizawa. 1995. The control of the morphology of MnS inclusions in steel during solidification. *ISIJ international* 35 (4):402–408.

Porter, D. A., and K. E. Easterling. 1992. *Phase Transformation in Metals and Alloys (Revised Reprint)*. 3rd ed. CRC Press. https://doi.org/10.1201/9781439883570.

Reformatskaya, I. I., I. G. Rodionova, Yu A. Beilin, L. A. Nisel'son, and A. N. Podobaev. 2004. The effect of nonmetal inclusions and microstructure on local corrosion of carbon and low-alloyed steels. *Protection of Metals* 4 (5):447–452.

Rieders, N., M. Nandasiri, D. Mogk, and R. Avci. 2021. New insights into sulfide inclusions in 1018 carbon steels. *Metals* 11 (3):428–444.

Sephton, Michelle, and P. C. Pistorius. 2000. Localized corrosion of carbon steel weldments. *Corrosion* 56 (12):1272–1279.

Spitzig, W. A. 1983a. Effect of sulfide inclusion morphology and pearlite banding on anisotropy of mechanical properties in normalized C-Mn steels. *Metallurgical Transactions A* 14 (1):271–283.

Spitzig, W. A. 1983b. Effect of sulfides and sulfide morphology on anisotropy of tensile ductility and toughness of hot-rolled C-Mn steels. *Metallurgical Transactions A* 14 (2):471–484.

Spitzig, W. A., and R. J. Sober. 1981. Influence of sulfide inclusions and pearlite content on the mechanical properties of hot-rolled carbon steels. *Metallurgical Transactions A* 12 (2):281–291.

Sun, W. P., M. Militzer, and J. J. Jonas. 1992. Strain-induced nucleation of MnS in electrical steels. *Metallurgical Transactions A* 23 (3):821–830.

Vander Woort, G.F. n.d. Microstructures of Hot and Cold Worked Metals and Alloys. http://www.georgevandervoort.com/mic_met_pdf/Hot_and_Cold_Deformation.pdf."

Wranglen, G. 1974. Pitting and sulphide inclusions in steel. *Corrosion Science* 14 (5):331–349. doi: 10.1016/S0010-938X(74)80047-8.

Xu, L. Y., and Y. F. Cheng. 2013. Development of a finite element model for simulation and prediction of mechanoelectrochemical effect of pipeline corrosion. *Corrosion Science* 73:11.

Yu, Hao, Yong-Lin Kang, Zheng-Zhi Zhao, and Hao Sun. 2006. Morphology and precipitation kinetics of MnS in low-carbon steel during thin slab continuous casting process. *Journal of Iron and Steel Research International* 13 (5):30–36.

15 Appearance of MIC in Well-Flowlines Producing from a Sour Reservoir

Amer Jarragh, Sandip Anantrao Kuthe, Akhil Jaithlya, Farah Al-Tabbakh, and Israa Mohammad
Kuwait Oil Company, Kuwait

CONTENTS

15.1 INTRODUCTION

Microbiological influenced corrosion (MIC) has been reported to cause approximately 40% of all internal corrosion events in crude oil pipelines (Santillan et al. 2015, Maruthamuthu et al. 2011). In oil and gas handling industries, microbial acceleration of the localized corrosion of iron and steel have been observed particularly in water-flooded systems and in those environments that are soured by sulfate-reducing bacteria (SRB)–produced sulfides (Skovhus and Whitby 2019). Worldwide, about 70% of water flooded reservoirs have experienced biogenic reservoir souring, irrespective of whether they are stand-alone or comingled with seawater. The main reason for souring is the lack of appropriate water treatment and

DOI: 10.1201/9780429355479-17

mitigation to control the proliferation and activity of microorganisms in the reservoir, which is required from the start of water flooding. Seawater flooding in reservoirs, due to water breakthrough, introduce SRBs [mesophilic (m-SRB) and thermophilic (t-SRB)] into the well fluid posing a risk for MIC in the wet crude pipelines (Eden et al. 1993).

By definition, MIC is corrosion affected by the presence or activity, or both, of organisms (NACE Standard TM0212-2018-SG 2018). Evaluating MIC failures requires a multidisciplinary approach, which includes chemical and microbial analysis of the fluid, analysis of the corrosion deposits, physical factors such as fluid flow conditions, pressure, temperature, and damage morphology (Little et al. 2006). To investigate MIC in the oil and gas industry, Molecular Microbiological Methods (MMM) have been increasingly used by several researchers (Eckert and Skovhus 2011).

Although it has been extensively studied, the MIC mechanism is still open for discussion. It has been established that MIC is linked to the presence of complex stratified biofilms formed of a complex microbial consortia that may include metal-oxidizing bacteria, sulfate-reducing bacteria, sulfate-reducing prokaryotes, acid-producing bacteria, metal-reducing bacteria, and methanogens that interact in complex ways within the biofilms at the metal surface (Gerard and Alfons, 2008, Little et al. 1991, 1996). Formation of biofilms is a complex process and depends on several factors. In pipelines, the flowrate of the fluid is one of the critical factors affecting the formation of biofilms and its structure. Though MIC was previously considered to be taking place under stagnant conditions, reports are available showing that microbial corrosion can occur under fluid flow (Song et al. 2016, Dunsmore et al. 2002). Rapid corrosion of metal takes place in the presence of SRB with recorded corrosion rates of steel that are much faster than with normal oxygen-promoted corrosion (Little et al. 1991, 1996, Brozel and Cloete 1989). The corroded metal is pitted rather than evenly corroded. The rate of pitting increases with time, indicating that the process is autocatalytic in character. Corroded pipes therefore perforate rather than disintegrate (Barton 1997). Based on visual examinations, MIC does not form a unique form or morphology of corrosion; however, MIC damage characteristics such as cup-shaped pits within pits on carbon steel in cooling water service have been documented (API RP571 2003). In crude oil service, the presence of acid gases can simultaneously corrode metal surfaces, thereby changing the MIC pit morphology (Skovhus et al. 2017, Eckert, Skovhus, and Sylvie 2014).

In the present case study, corrosion failures in wet crude flowlines producing from a sour reservoir have been discussed. Several failures due to internal corrosion were reported in the old and new flowlines transporting wet crude from oil wells to the Gathering Centre (GC). Some flowlines failed more than once within the last couple of years and leaks were located mostly aboveground while a few underground sections have also failed. The flowlines are connected to single oil wells producing from a reservoir, which recently turned sour. Failure investigations were conducted for selected flowline leaks. CO_2 and CO_2/H_2S and MIC corrosion rate modelling was carried out using available corrosion models for various operating periods where substantial changes in the fluids and operating parameters were observed. The flowlines with high probability for CO_2 or CO_2/H_2S and/or MIC corrosion were prioritized for inspection in the field to validate the predicted corrosion rates. Guided Wave Ultrasonic Testing or Long-Range Ultrasonic Testing (LRUT) measurements were carried out on the flowlines along their entire lengths and defect distribution charts were prepared.

The findings based on failure investigations, field data, operational data, and their impact on the internal corrosion of flowlines is presented here.

15.2 DATA COLLECTION

15.2.1 Fluid Sample Analysis

An in-house database for the wells was used to gather data and for trending the variations in the water cut, gas phase CO_2 and H_2S analyzed by using the Portable Gas to Oil Ratio (GOR) Multi-rate testing carried out by specialized contractor at the wellsite. This data is highlighted against the flowline connected to it since each well is connected to a single flowline unless otherwise mentioned (Table 15.1). During secondary recovery, seawater and effluent water was injected in this reservoir to enhance the performance of the oil wells. Information on the oil wells that are associated with either seawater injectors and/or effluent water injectors was received from field development group. Fluid samples were collected by specialized contractor and separated water was tested for dissolved sulfates and chlorides and analyzed accordingly; e.g., the level of sulfate ions from the separated water of the oil well associated with a seawater injector were found to be different than that from the well associated with an effluent water injector.

TABLE 15.1
Wet Chemical Analysis of Separated Water from Crude Oil Samples of Various Wellheads

Flowline No.	CO_2 (mol %)	H_2S (ppm)(v/v)	Water cut (%)	Sulfate SO_4^{2-} (ppm)
FL-1***	2	180	98	2,500
FL-2*	2	0	96	1,900
FL-3*	1.5	25	98	n.a.
FL-4*	1.5	200	94	2,075
FL-5*	2	5,000	98	2,275
FL-6*	1	0	97	1,891
FL-7**	1	25	98	1,250

Notes
* Wells associated with seawater
** Wells associated with effluent water
*** acetate up to 90 ppm was detected
n.a. means not available

15.2.2 Corrosion Rate Prediction

Mixed phase flow velocity and predicted CO_2 or mixed CO_2/H_2S corrosion rates were calculated using the Hydrocorr† corrosion model and tabulated below (Table 15.2). A MIC assessment model (Wang and Jain 2016) was used to predict the MIC propensity for one of flowlines based on data from the last five years (Table 15.3).

15.2.3 qPCR Analysis of Injected Water (Seawater Flooding)

Water samples from a seawater injection facility were analyzed by applying a suite of quantitative polymerase chain reaction (qPCR) assays for enumeration of total bacteria and archaea, and key indicator organisms (sulfate-reducing bacteria (SRB), sulfate-reducing archaea, and methane-producing archaea). In addition, full population structure fingerprinting was carried out for a few of the selected samples by using next generation sequencing (NGS) technologies that target bacteria and archaea. All of the previous work was conducted under the operating company contract by a third-party laboratory† and the data is compiled in Table 15.4.

15.2.4 Detailed Visual Inspection

Site visits were conducted and location details such as the orientation of the pipe, clock position of the corrosion, and external deposition on the pipe were recorded and photographed. Leak samples were cut open in the fabrication shop and analyzed. Samples were cut across the section and sweep blasted. The bottom of

TABLE 15.2
Predicted General Corrosion Rates and MIC Corrosion Rates for Flowlines

Flowline No.	Predicted Corrosion Rate CO_2 /CO_2/H_2S (mpy) (Average for Last 10 Years)	MIC Propensity Corrosion Rate (mpy) (Based on 2019 Data)	Mixed Velocities (m/s) Hydrocorr†	Length of Required Pipe Peplacement as per LRUT
FL-1	10.0	117	0.20	500 m
FL-2	8.0	47	0.40	275 m
FL-3	10.0	116	0.43	2.89 km
FL-4	7.0	116	0.65	700 m
FL-5	15.0	117	0.30	6.5 km
FL-6	5.0	114	0.95	1.3 km
FL-7	9.5	46	0.70	540 m

Note
† Trade Name

TABLE 15.3
Predicted General Corrosion Rates and MIC Corrosion Rates for FL-1 for Last Four Years

Date	CO_2 (mol %)	H_2S (ppm v/v)	Water cut (%)	Shutdown Period (days)	Mixed Velocity (m/s)	MIC Propensity (mpy)	Bottom Corrosion (mpy)
26 March 2019	1.2	180	85.78	210	0.2	60	13
1 July 2018	2	12	80.79	54	0.2	50	60
12 May 2017	1	0	26.12	Not available	0.3	47	3
26 September 2016*	1	0	31.24	47	0.2	50	3
1 November 2015	0.1	0	8.69	96	0.3	53	0.4

Note
* The sudden increase in water cut has not been considered; instead, average water cut data has been taken for prediction; mpy, mils/year.

TABLE 15.4
Bacteria Testing of Injector Wells Water Samples by the MMM Method

Sample Reference	Unit	Date	Bacteria	Archaea	SRB	SRA
Injector well	cells/ml	25 March 2017	100,000	1,400	3,400	BDL*
Injector well	cells/ml	30 March 2017	2,400	BDL	100	BDL
Injector well	cells/ml	04 April 2017	84,000	4,600	850	BDL

Note
* BDL—Below Detection Limit

line corrosion (5–7 o'clock) and pitting near the perforation was observed in most of the flowline samples. Loose deposits were observed in some cases while thin black film was observed on the other cut sections of flowline FL-1 (Figures 15.1 and 15.2).

After removing deposits, the general corrosion and corrosion pits were visible on FL-1 (Figures 15.3 and 15.4).

Several deep pits surrounding the perforation were observed on one section while isolated perforation was observed on the other section of FL-1 (Figures 15.5 and 15.6).

FIGURE 15.1 Brown-orange deposit near perforation and thick, black, loose deposits away from perforation (5 to 7 o'clock) observed on FL-1 sample 1. SRB were detected in this deposit.

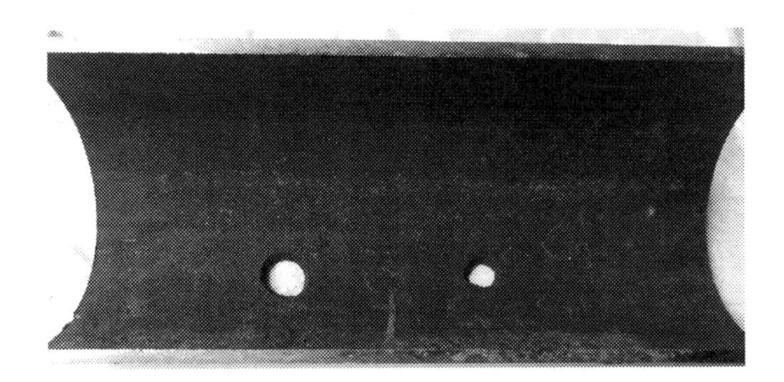

FIGURE 15.2 Thin, black deposit near perforation and away from perforation (5 to 7 o'clock) observed on FL-1 sample 2. No SRB were detected in this deposit.

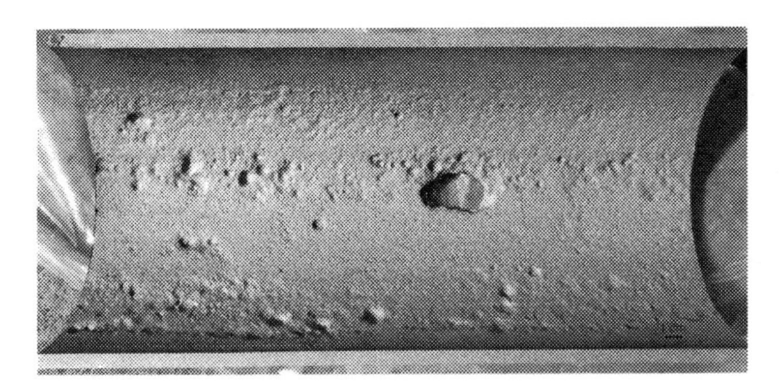

FIGURE 15.3 Perforation at 6 o' clock, deep pitting with thinning on bottom section (5 to 7 o'clock) observed on sweep blasted section of FL-1 sample 1.

Thick, heavy deposition was observed on the cut sections of flowline FL-4 (Figure 15.7). After prolonged exposure, the deposits turn orange-brown due to oxidation of iron sulfides (Figure 15.8).

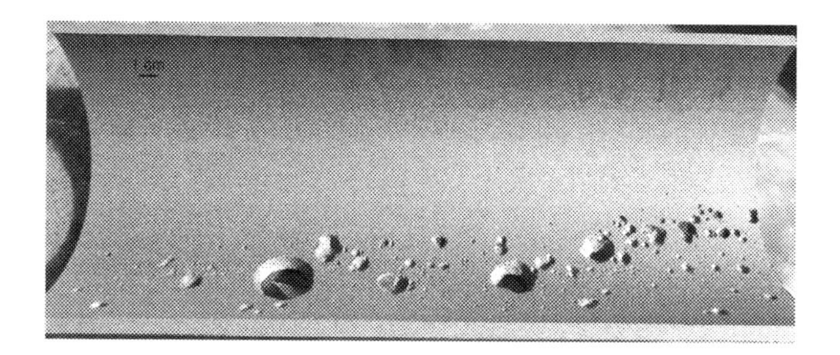

FIGURE 15.4 Perforation, deep pitting without thinning on bottom section (5 to 7 o'clock) observed on sweep-blasted section of FL-1 sample 2.

FIGURE 15.5 Close-up photograph showing several deep pits surrounding perforation on FL-1 sample 1. Pit boundary diffused and sloped.

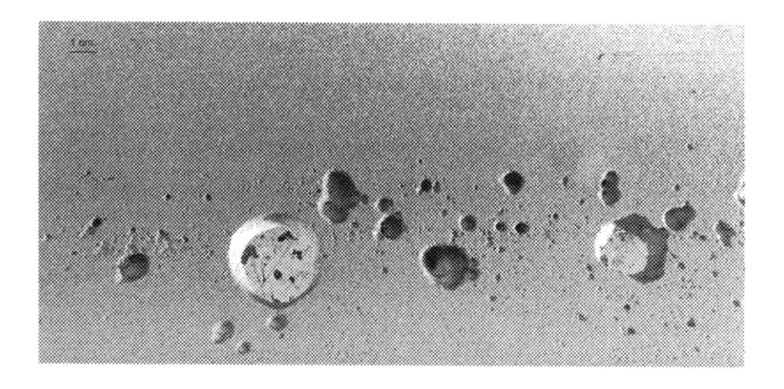

FIGURE 15.6 Close-up photograph showing perforation on FL-1 sample 2. Perforation pit boundary is sharp and not surrounded by pits.

Continuous pitting and perforation was observed on the sweep-blasted sections of flowline FL-4 sample 1 (Figure 15.9) while pits within pits arrangement were visible in FL-4 sample 2 (Figure 15.10).

FIGURE 15.7 Blackish deposit near and away from the perforation (5 to 7 o'clock) observed on FL-4 sample 1 section taken from underground portion of the flowline.

FIGURE 15.8 Blackish and orange-brown deposit near the perforation (5 to 7 o'clock) observed on FL-4 sample 2 section taken from aboveground portion of the flowline.

A closer look of Figure 15.9 indicated pits growing together along the axis (Figure 15.11). Smooth, silky surface was observed on the FL-6, sample 2 (Figure 15.12).

Physical measurements were taken at and around the leak location to establish the severity of the corrosion on the flowline section. Typical length and width of the pit and average pit diameter were measured and both the maximum pit depth to

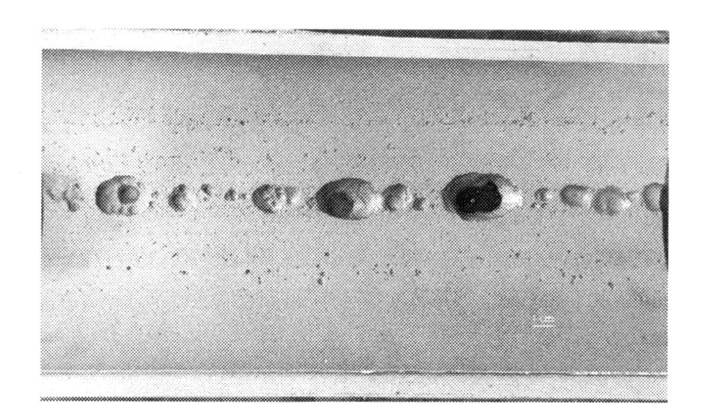

FIGURE 15.9 Pitting and channels between two perforations along the bottom of FL-4 sample 1.

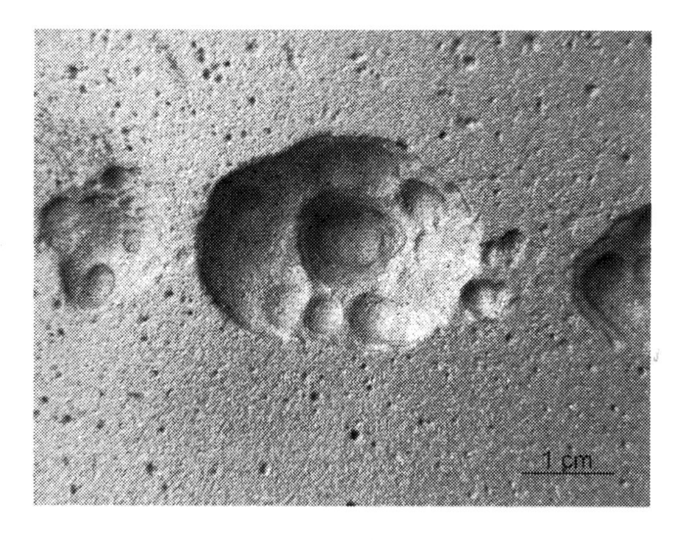

FIGURE 15.10 Pits within pits arrangement (5 to 7 o'clock) on FL-4 sample 2. Diffused boundary between large and small circular pits with sloped edges can be seen.

maximum pit diameter and average pit depth to average pit diameter ratio were calculated (Table 15.5).

15.2.5 Analysis of Deposits

As per the maintenance procedures, the flowline was depressurized, drained, and washed by wash water prior to inspection. The sections were cut and transported to a nearby fabrication workshop. It took approximately two working days from cutting the pipe section and making it available for deposits collection from the internal surface of the flowline. Deposit (mostly oil wet) samples were collected

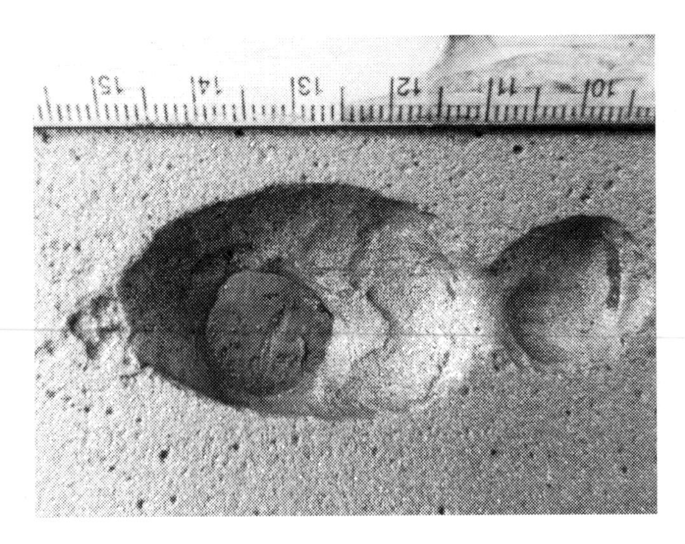

FIGURE 15.11 Typical association of a large pit with a small pit on FL-4 sample 2.

FIGURE 15.12 Pits surrounding perforation and sloped pattern on bottom section (5 to 7 o'clock) of FL-6, sample 2.

from the flowlines and a part of it was used for X-Ray Diffraction (XRD) analysis after washing the deposit with xylene to remove the oily layer. The dry deposit was then grounded to fine powder and subjected to XRD analysis. The diffraction patterns show predominantly iron sulfides (Figure 15.13).

Another portion of the deposit samples was immediately inoculated at the site in two different media, representing produced water (high TDS) and seawater (low TDS), which were then incubated at 35°C for 28 days. The bacteria counts

TABLE 15.5
Visual Inspection and Corrosion Pit Measurement Data (mm)

Flow Line #		Max Pit Depth	Avg Pit Depth	Max Pit Diameter	Avg Pit Diameter	Pit Length/ Pit Width	Max Pit Depth/ Max Dia.	Avg Pit Depth/ Avg Dia.
FL-1		7.00	16.16	40.00	27.00	0.88	0.17	0.6
FL-4	1	10.00	6.72	55.00	33.75	1.67	0.18	0.2
	2	10.00	4.57	40.00	21.67	1.03	0.25	0.21
FL-6		11.00	10.52	80.00	48.33	1.00	0.14	0.22

were evaluated by using the MPN method (NACE TM0194 2014) and the assessment is provided in Table 15.6.

15.2.6 Chemical Composition of a Metal Sample Piece from a Leaking Flowline

Chemical analysis of a cut section of the flowline was performed by spectroscopy and the results confirmed to the API 5L Gr. B pipe, as shown in Table 15.7.

15.2.7 Guided Wave (Long-Range) Ultrasonic Testing Measurement of Flowlines

Guided Wave/Long-Range Ultrasonic Testing (LRUT) measurements were carried out on all the flowlines along the entire length. Guided Wave Ultrasonic Testing (GWUT) or LRUT is a non-destructive technique which was deployed for the flowlines to assess the change in cross-sectional area of pipe wall due to corrosion. This technique was deployed for rapid screening and full inspection coverage of corroded sections with target inspection of the suspected area by closer examination using localized ultrasonic testing methods. The LRUT method was implemented for aboveground sections of the flowlines with some coverage at the underground section near road crossing. LRUT can detect metal loss indications from corrosion/erosion inside the pipe or corrosion on the outside of the pipe. The detection of additional mode converted signals from defects aids discrimination between pipe features and metal loss (NACE SP0313 2013). The main benefit of using LRUT is that flowline may be inspected in a relatively faster manner to determine the area where corrosion is occurring and the area which are inaccessible for direct inspection such as road crossings and pipe casings, etc.

As the resolution of defect sizing measured by this technique depends on the pipe sizes, a confirmatory manual ultrasonic testing was conducted to do verification of defect depth. To understand the relationship between the LRUT and the observed corrosion morphology a defect distribution graph was prepared based on

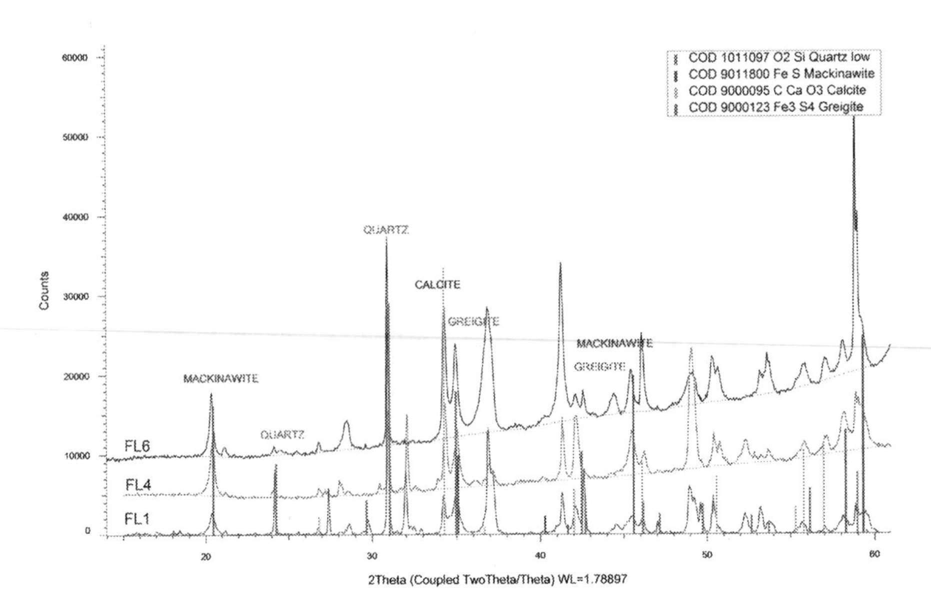

FIGURE 15.13 XRD of deposits collected from FL-1, FL-4, and FL-6 showing iron sulfide corrosion products with varying concentrations.

the data gathered from the LRUT and failure investigation of the cut sample. The average diameter and approximate distance between the pits (edge to edge) observed on the cut sample of typical FL-1 was measured and used in preparing the defects (pits) distribution having a depth <50% and >50% (Figure 15.14).

15.3 RESULTS AND DISCUSSION

Data was plotted for flowlines mentioned in Table 15.1 for its age (years of service), stagnation experienced for last 10 years due to well workovers (cumulative shutdown period), and the percentage of flowline length replaced due to internal corrosion (Figure 15.15).

It was observed that flowlines as old as 32 years in service and as recent as 9 years in service have experienced severe internal corrosion and several hundred meters of pipeline length have been recommended for replacement (Table 15.2). FL-3 and FL-5 required replacement of entire flowline length due to loss of metal thickness in excess of 50% of its original value. The possible explanation for severe internal corrosion losses observed in the flowlines under study is depicted below.

It was observed from Table 15.1 that all the wells produced significantly high levels of water cut reaching more than 90%. At the same time significant decrease in the salinity levels (measured in terms of dissolved chlorides) was recorded with the increase in water cut. Figures 15.16 and 15.17 show variations of salinity with water cut in FL-1.

The variation in the production chemistry of flowline FL-1 indicated a decrease in the salinity with the increase in the water cut over the last 10-year period with

TABLE 15.6
Sessile Bacteria Proliferation Data of the Corrosion Deposits by MPN

Flowline Number	Sample Number	Sample Location	Media	(Count/gm) SRB	GAB	GAnB
FL-1	1	Deposit from cut section	*High TDS	10,000	1,000	10,000
	2		High TDS	0	100	10
			High TDS	0	10,000	100
FL-4	1	Deposit from cut section	Low TDS	0	10,000	10,000
		Under deposit		0	100	10
		Scale deposit in cut section		0	10	0
		Under deposit	High TDS	0	0	10
	2	Deposit from cut section	Low TDS	0	100	100
		Under Deposit		0	1,000	10
FL-6	1	Deposit from cut section	High TDS	0	0	0
	2	Near Failure	High TDS	0	0	0
		Under Deposit		0	0	0
		Near Failure	Low TDS	0	1,000	10
		Under Deposit		0	100	0

*High Total Dissolved Solids (TDS) =150,000 ppm, **Low TDS = 30,000 ppm

TABLE 15.7
Material Composition of Flowlines Obtained Using Portable Spectroscope (Arc/Spark Optical Emission Spectrometry (OES))

Description of Material	Fe %	C %	Si %	Mn %	P %	S %	Cr %	Mo %	Ni %
API 5L Gr. B Specification (3)	Rest	0.28	-	1.2	0.03	0.03	≤0.5	≤0.15	≤0.5
FL-1	98.1	0.14	0.30	0.65	0.009	0.005	0.13	0.059	0.15
FL-4	98.9	0.15	0.10	0.63	0.017	<0.002	0.015	0.015	0.037
FL-6	98.7	0.034	0.22	0.73	0.015	0.009	0.030	<0.003	0.017

maximum rate of change of salinity with water cut observed in June 2016 (Figure 15.16). Appearance of H_2S was recorded in July 2018, which increased ten-fold in the subsequent analysis carried out in 2019 (Figure 15.17). This decrease in

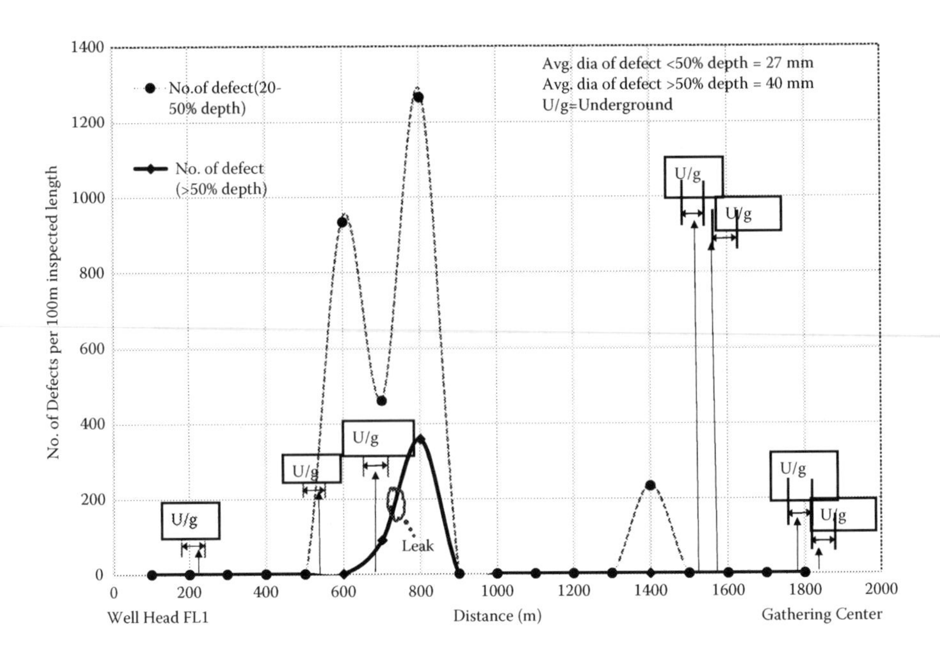

FIGURE 15.14 Defect distribution on FL-1 evaluated based on the LRUT data.

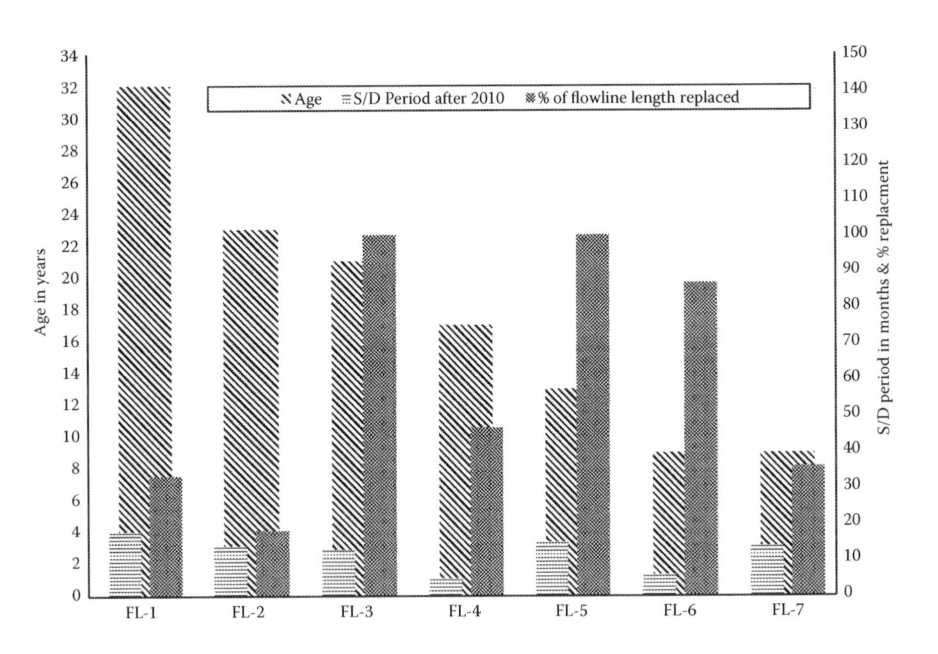

FIGURE 15.15 Correlation between the age of the flowlines, shutdown period experienced by the flowlines and the percentage of flowline length replaced.

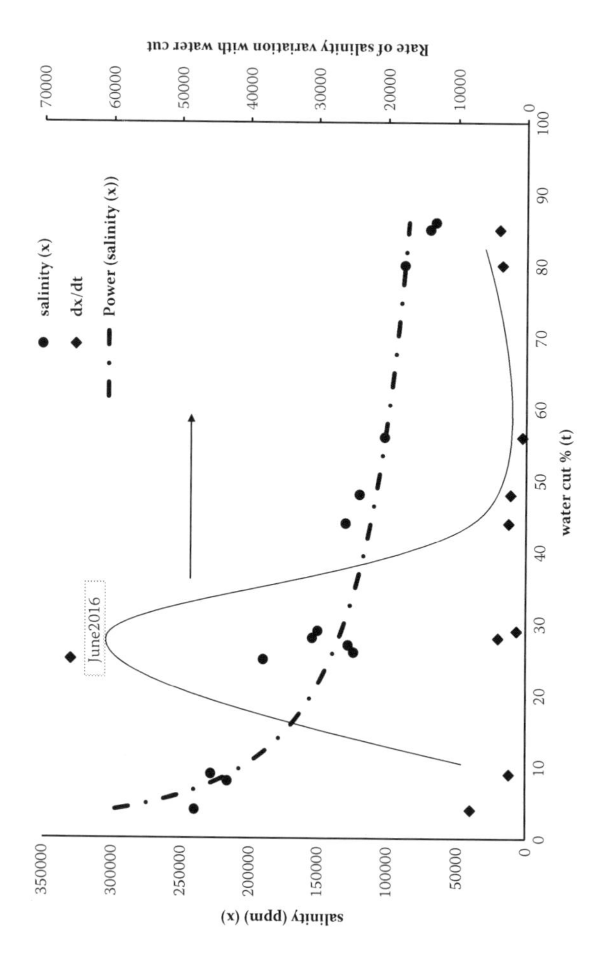

FIGURE 15.16 Rate of change of salinity with produced water cut from FL-1; peak appears in June 2016, indicating major breakthrough of injected seawater.

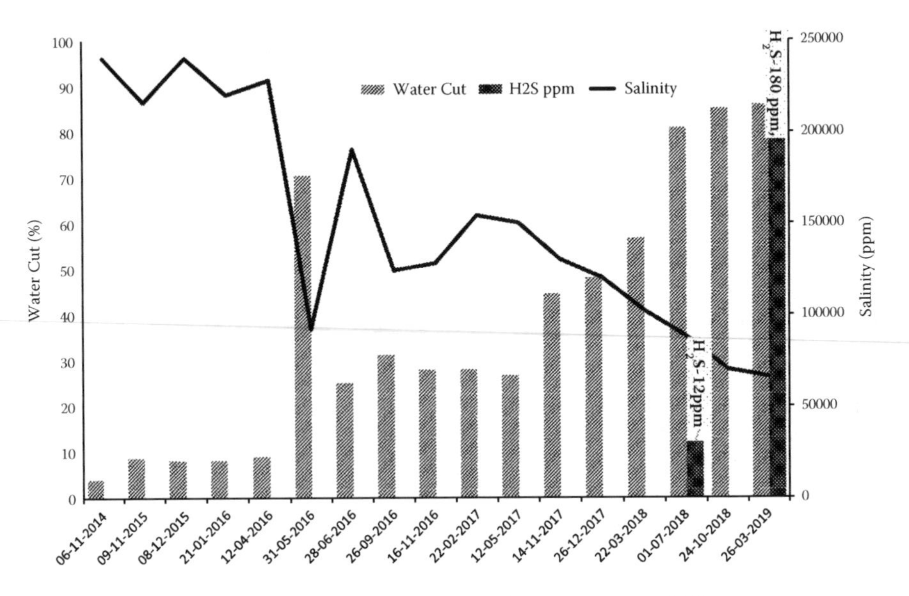

FIGURE 15.17 Increase in water cut in FL-1 with appearance of H_2S.

salinity of the produced water with the increase in water cut was attributed to the dilution effect due to the mixing of injected seawater in the production zone. The increased levels of dissolved sulphate ions in the separated water from the crude wells-flowlines (Table 15.1) support the findings that seawater breakthrough and mixing in the production zone has occurred at multiple well locations.

The predicted general corrosion rates (CO_2 or CO_2/H_2S mixed) based on the most recent fluid analysis data were below 10 mpy (mils/year) for most of the well-flowlines (Table 15.2). However, episodes of high and low corrosion rates were expected for the flowlines for various operating periods where substantial changes in the fluids and operating parameters were observed. Predicted corrosion rate data for flowline FL-1 for the last four years was tabulated (Table 15.3) and predicted loss in flowline thickness due to CO_2 or CO_2/H_2S corrosion is explained below.

Based on the available data, corrosion rate due to CO_2 or CO_2/H_2S was predicted to be in the range of min 0.4 mpy to a max 60 mpy (Table 15.2). Though this well (connected to FL-1) was producing for the last 32 years, water cut (in excess of 25%) started in the last 10 years and a sudden increase was recorded in the last 4 years. Considering 10 mpy as the average corrosion rate the maximum wall loss expected along the flowline for the last decade would be 2.5 mm (against the nominal thickness of the schedule 40 pipe which is aprrox. 7 mm) without considering the pitting probability. In worst-case scenario corrosion rate of 60 mpy for one year can be included to predict a maximum wall loss of 3.5 mm, which is half the thickness of the pipe. This indicates that corrosion due to CO_2 or CO_2/H_2S alone cannot explain the severe corrosion and leaks in several of the new flowlines.

The other possible explanation considering MIC as the predominant corrosion mechanism and predicted wall loss is explained below.

The prediction of MIC was highlighted in most of the flowlines (Table 15.2). This was expected since the lines were flowing with a very low flow velocity of the order of 0.2 to 0.3 m/sec and a flow regime that was stratified smooth. High MIC propensity and higher pitting rates were predicted and considered as the dominant corrosion mechanism when H_2S presence was detected (in most flowlines first H_2S appearance was recorded in year 2012 and onwards). It can be seen from Figure 15.15 that these flowlines experienced some stagnation period during the well workover and the untreated fluid remained in the flowline for an extended period of time. Stagnant conditions can result in lower MIC rates because the supply of nutrients by diffusion is rate limited (Nyborg 2002). This downtime factor that was evaluated based on the empirical data has been considered in the MIC prediction model and the data in Table 15.3, which shows that the MIC rates were always predicted above 50 mpy since 2015. A predicted average MIC rate calculated over these five years approximates to 52 mpy (Table 15.3).

Based on the above predictions, the MIC rate at 52 mpy (1.3 mm/year) can alone consume the entire thickness of the flowlines (at localized sections) in six years (considering the nominal thickness of the schedule 40 pipe which is aprrox. 7 mm); moreover, the MIC corrosion rate on the pre-corroded (due to CO_2 or CO_2/H_2S) sections can be even more severe.

It can be seen from Table 15.4 that SRBs were present in the injected seawater in significant quantities. Based on the sequencing data, it was revealed that sulfate-reducing *Desulfovibrio sp.* and *Desulfobacter sp.* were present in these injected water samples. Species within *Desulfovibrio* are known to directly scavenge electrons from carbon steel, and thus leads to direct and severe corrosion. The deposits from the corrosion coupons installed on the injected seawater distribution header also revealed the presence of sessile SRB of the order of 10^5 counts/g. Similarly, cleaning pig deposits from the seawater injection distribution header was analyzed to find the presence of sessile SRB of the order of 1.7×10^5 counts/g. Thus, it was quite possible that the continuous and copious amounts of SRBs in the seawater injection system introduced bacteria in the crude oil production zone. This fact was substantiated with the findings from the deposits from the leaked spool pieces of the flowlines FL-1 which revealed the presence of sessile SRB within a range of 10 to 10^5 counts/g (Table 15.6). The fluid analysis data from these well-flowlines showed that there was a significant drop in the salinity of the produced water and increases in the dissolved sulphate content in recent years. Presence of fatty acid was also found in the produced water from FL-1 (Table 15.1). The temperature of all these well fluid sample collected at the well head was found out to be around 50°C. Further, the low-flow rates and stagnant conditions especially during workovers created a most conducive atmosphere for the bacteria to grow rapidly in the flowlines. Considering the limitations on the availability of failed pipe pieces over time, availability of fresh deposit samples, precise sampling of bacteria in the field, and the limitation on the detection capability of the serial dilution method itself; it is important that due consideration is given to any positive identification of live SRB in the leaked flowline corrosion deposit.

Table 15.5 shows the ratio of both the maximum pit depth to the maximum pit diameter and average pit depth to average pit diameter. It was observed for most of the corrosion pits that the ratio of maximum pit depth to maximum pit diameter was < 1 indicating slower penetration rates and more surface coverage. The higher pit length over pit width indicates that corrosion pits are elongated axially indicating sudden change in the flow conditions (Skovhus et al. 2017), probably when the flowline starts up after a long stagnation period.

Figures 15.5 to 15.12 show the pitting morphology on all of the leaked flowline sections. The pits appeared to be elongated with etched sides, slopy edges, and satiny bottoms. A typical oblong-shaped corrosion pit with two or more interconnected pits with the perforation in the center of the bigger pit can be seen in Figure 15.11. Smooth, silky surface can be seen in Figure 15.12, while pits within pits arrangement is visible in Figure 15.10. Similar features such as cup-shaped pits within pits on carbon-steel in cooling water service have been characterized and documented as MIC (API RP571 2003). However, in crude oil service, the presence of acid gases can simultaneously corrode metal surfaces, thereby changing the MIC pit morphology (Skovhus et al. 2017).

As mentioned previously during seawater flooding in reservoirs, due to water breakthrough, SRBs [mesophilic (m-SRB) and thermophilic (t-SRB)] possibly was introduced into the well fluid and proliferated in the flowlines during stagnant condition at low spots, forming biofilms. Bacterial action reduces the sulfate present in the water to sulfide, using the hydrogen to form hydrogen sulfide. The hydrogen sulfide reacts with the dissolved ferrous iron, which is released at the anode to form an iron sulfide precipitate (Von Holy 1987). Formation of such black-colored corrosion product on failed sections of all the three flowlines can be seen from Figures 15.1, 15.7, and 15.8 and thin FeS film in Figure 15.2 and identified as iron sulfide by XRD in Figure 15.13.

The FeS film initially protects the surface, decreasing the corrosion rate. However, the film is dependent on the free ferrous ion concentration. At low ferrous ion concentration (2 ppm), it transforms to the more corrosive greigite (Fe_3S_4) (Figure 15.13) and crumbles at higher (10 ppm) concentrations (Brozel and Cloete 1989). These iron sulfide films on metal surface did not offer protection from anaerobic MIC. Moreover, the formation of the biofilms on the iron sulfide films could not avoid the risk of localized corrosion in the presence of chlorides (Lee and Characklis 1990). Thus, it can be said that corrosion could have possibly initiated due to breaking of mackinawite film in the presence of chlorides. At a later stage, biofilms formation took place on the iron sulfide surface or on the fresh surface, aggravating the localized corrosion under the iron sulfide deposits.

The field observations are captured for FL-1 using LRUT data analysis, as shown in Figure 15.14. It can be observed that the corrosion defects were highly concentrated on aboveground sections. As expected, the leak was observed within the region where defect concentrations were higher. Based on the observed type of attack on the cut sections of this leaked flowline it was assumed that the bacterial colonies might have formed in this region and spread further. However, the growth rate depended upon the local conditions existing on the flowline sections (low spots, pre-corroded sections) and so the severity of internal corrosion varied accordingly, causing leaks at the most vulnerable locations.

In conclusion, the studied fluid conditions, physical conditions in the flowlines, the presence of SRB on the corroded component of the flowline, corrosion pit morphology, presence of iron sulfides in the corrosion deposit, and the corrosion prediction support MIC and CO_2 and CO_2/H_2S corrosion are the most possible causes of the premature failures of old and new flowlines.

ACKNOWLEDGMENTS

The authors would like to thank Kuwait Oil Company management for support in publishing this chapter. We would like to thank Mr. Carlos Caicedo-Martinez (M/s Shell) for providing corrosion prediction data. We would like to thank Mr. Priya Ranjan Kumar (field development SA team) for providing wells data. We would like to thank Mr. Mohammed Shalwan Al Hajri (pipeline inspection team) for providing LRUT data. We would like to thank Mr. Meshal Mohammed Al-Doub (I&C workshop team) for arranging sectioning, sand blasting, and composition analysis of cut pieces. We would also like to thank Mr. Subramaniyam, Mr. Shibu John, and Mr. Al Asgar from M/S DNV for performing the chemical analysis of fluids, deposits, and bacteria culturing.

REFERENCE

API RP571. 2003. Damage mechanism affecting fixed equipment in the refining industry. The Woodlands, TX: API.

Brozel, V. S., Cloete, T.E. 1989. The role of sulfate reducing bacteria in microbial induced corrosion. *Paper SA*. 11.12.89:30–36.

Barton, L.L. 1997. Sulfate-reducing bacteria. International Workshop on Industrial Biofouling and Biocorrosion. Mulheim, Germany, September.

Dunsmore, B., Jacobsen, A., Hall-Stoodley, L., et al. 2002. The influence of fluid shear on the structure and material properties of sulfate-reducing bacterial biofilms. *J. Ind. Microbial. Biotechnol.* 29:347–353.

Eckert Richard, Skovhus, Torben Lund. 2011. Using molecular microbiological methods to investigate MIC in the oil and gas industry. *Mater. Perform.* 50(8):2–6.

Eckert, Richard, Skovhus, Torben Lund, le Borgne, Sylvie. 2014. A Closer Look at Microbiologically Influenced Corrosion. *Material Performance*. 53(1):32–40.

Eden, B., Laycock, P.J., Fielder, M. 1993. Oilfield reservoir Souring, OTH 92 385 Offshore Technology Report prepared by Capsis Ltd, UMIST and BP exploration for the Health and Safety Executive.

Gerard, M., Alfons, J.M. 2008. The ecology and biotechnology of sulfate-reducing bacteria. *Nat. Rev.* 6:441–454.

Lee, W., Characklis, W.G. 1990. Corrosion of mild steel under anaerobic biofilm. *Corros.* 49(3):186–199.

Little B.J. 1991. Electrochemical behavior of 304 stainless steel in natural seawater. *Biofouling* 3:43–59.

Little, B.J., Wagner, P.A., Hart, K.R., et al. 1996. Spatial relationships between bacteria and localized corrosion. *Corros.* 96:278–284.

Little, B.J., Wagner, P.A. 1997. Myths related to microbiologically influenced corrosion. *Mater. Perform.* 36(6):40–44.

Little, B.J., Lee, J.S., Ray, R.I. 2006. Dignosing microbiologically influenced corrosion: a state-of-the-art review. *Corros.* 62(11):1006–1017

Maruthamuthu, S., Kumar, B.D., Ramachandran, S., et al. 2011. Microbial corrosion in petroleum product transporting pipelines. *Ind. Eng. Chem. Res.* 50:8006–8015.

NACE Standard TM0212-2018-SG. 2018. Detection, testing, and evaluation of microbiologically influenced corrosion on internal surfaces of pipelines. Houston, TX: NACE.

NACE SP0313. 2013. Standard practice for guided wave technology for pipeline application. Houston, TX: NACE.

NACE TM0194. 2014. Field monitoring of bacterial growth in oil and gas systems. Houston, TX: NACE.

Nyborg, Rolf. 2002. Overview of CO_2 corrosion models for wells and pipelines. NACE International Corrosion 2002 Conference & Expo, Denver, Colorado, April 7–11, 2002. Paper 02233.

Santillan, E.F.U., Choi, W., Bennett, P.C., et al. 2015. The effects of biocide use on the microbiology and geochemistry of produced water in the Eagle Ford formation, Texas, U.S.A. *J. Petro. Sci. Eng.* 135:1–9.

Skovhus Torben Lund, Whitby, Corinne. 2019. Oilfield microbiology. Boca Raton: CRC Press/Taylor & Francis.

Skovhus, Torben Lund, Enning, Dennis, Lee, Jason S. 2017. Microbiologically influenced corrosion in the upstream oil and gas industry. Boca Raton: CRC Press/Taylor & Francis.

Song, X., Yang, Y., Yu, D., et al. 2016. Studies on the impact of fluid flow on the microbial corrosion behavior of product oil pipelines. *J. Petro. Sci. Eng.* 146:803–812.

Von Holy, A. 1987. New approaches to micorbiologically induced corrosion. *Water Sewage and Effluent* 39–44.

Wang, Y., Jain, L. 2016. MIC assessment model for upstream production and transport facilities. NACE International Corrosion 2016 Conference & Expo, Vancouver, British Columbia, Canada, March 6–10, 2016. Paper 7769.

Part III

MIC in Other Engineered Systems

16 Failure Analysis of Pipe in a Fire Suppression System

A. Harmon, K. Crippen, and S. Leleika
GTI Energy

CONTENTS

16.1 INTRODUCTION AND BACKGROUND

An overhead fire suppression system located in an unheated building developed pin hole leaks after three years of service. The pipe was hot-dipped galvanized with a nominal 6.4 centimeters (2.5 inches) diameter. During normal operation, the system is charged with a nominal 412 kilopascals (45 PSIG) air. The water is periodically tested with potable water, drained, and reset with air. During testing, a leak was identified and subsequently a 4 millimeters (0.158 inch) hole was found in a section of pipe.

Figures 16.1, 16.2, and 16.3 are photographs of the portion of failed pipe received for determining the root cause of failure analysis. Five water samples from various distances away from the failure were collected including water from the fire hydrant that is approximately 45 meters from the pipe failure and the main drain that is less than 30 meters upstream of the pipe failure. The other water samples are

DOI: 10.1201/9780429355479-19

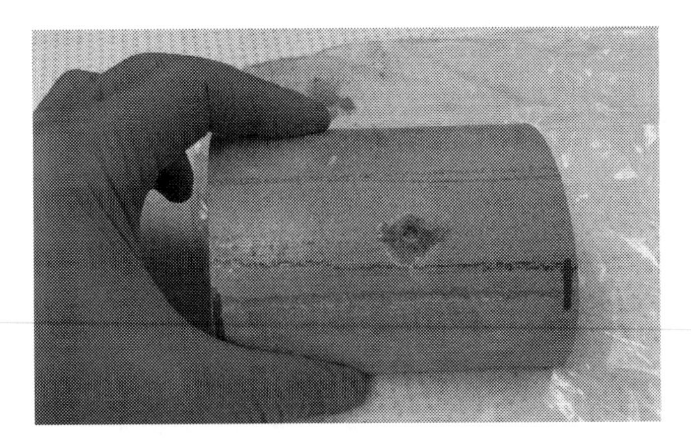

FIGURE 16.1 Macro photo of pipe failure sample as received.

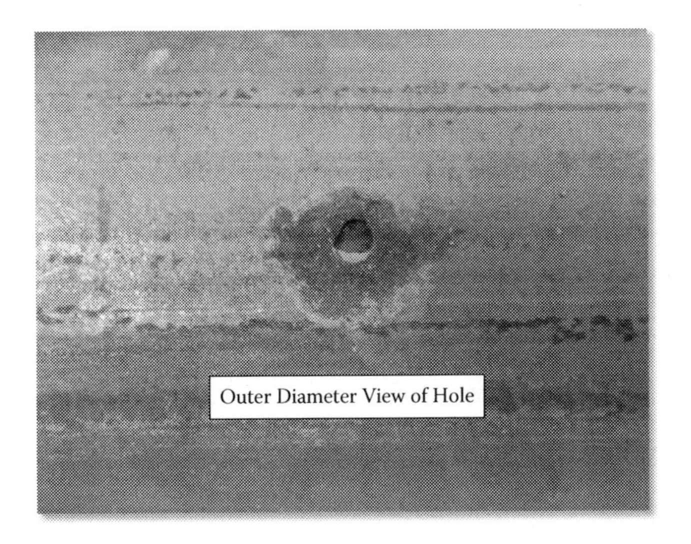

FIGURE 16.2 Macro photo of pipe failure sample as received.

from around the site and several tens of meters away from the leak. Table 16.1 lists the samples analyzed for this failure investigation.

To investigate the root cause of failure, the following tests were performed:

1. Chemistry analysis on water and pipe debris samples.
2. X-ray diffraction on pipe debris sample.
3. Metallurgical examination of the inner and outer surfaces.
4. Microbiological analysis using qPCR.

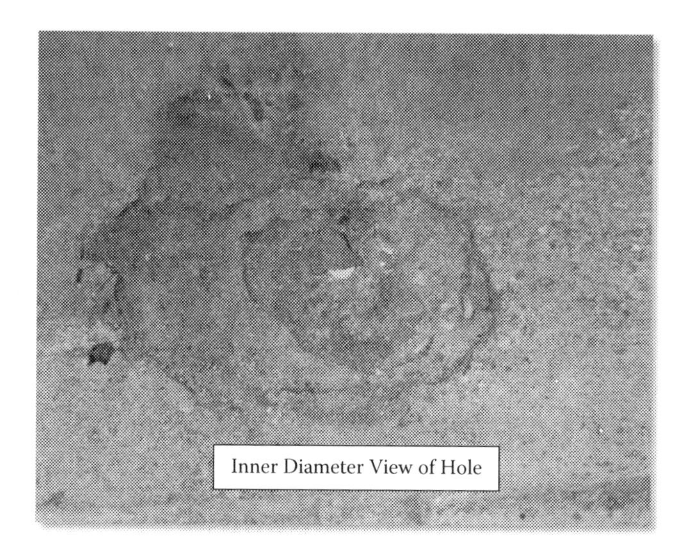

FIGURE 16.3 Macro Photo of Pipe Failure Sample as Received

TABLE 16.1
Samples Analyzed for This Failure Analysis

Description	Received Date
Fire Hydrant	February 5, 2015
Building #1-Main Drain	February 5, 2015
Building #2-Main Drain	February 5, 2015
Building #3-Main Drain	February 5, 2015
Building #3-Close to pipe failure	February 5, 2015
Pipe	February 5, 2015

16.2 EXPERIMENTAL

16.2.1 Initial Sample Examination

All sample examination is performed on laboratory bench tops that are washed and allowed to dry. All sample handling is performed wearing gloves. The initial examination is by the unaided eye to locate the failure site.

16.2.2 Examination of the Pipe Surface Conditions

Visual examination is the first tool used to inspect the failure site. Surfaces are illuminated via both annular and remote light sources. The interior and exterior surfaces of the sample were observed and photographed for surface defects/damage, corrosion pits, and cracks.

16.2.3 Microbiological Analysis Using qPCR

One hundred milliliters of each liquid sample and 0.26512 gram of debris collected from the interior hole and surrounding area of the pipe sample were analyzed using quantitative Polymerase Chain Reaction (qPCR) techniques for microbes commonly related to Microbiologically Influenced Corrosion (MIC). The results were reported as the number of copies per milliliter or gram of sample.

The MIC-associated microbes targeted in the qPCR analysis were sulfate-reducing bacteria (SRB), acid-producing bacteria (APB), iron-oxidizing bacteria (IOB), and denitrifying bacteria (DNB). SRB have been isolated from a wide range of environmental samples and are widely considered an aggressive corrosion-causing bacterium in various environments (Enning and Garrelfs 2014). In addition to SRB, APB are also considered aggressive corrosion-causing bacteria. The total APB number reported here includes acetic acid–producing bacteria and butyric acid–producing bacteria. The IOB number reported here includes *Leptothrix*, *Sphaerotilus*, and *Gallionella*. Lastly, denitrifying bacteria have been found in MIC-associated biofilms and may play an important role in microbe communities due to their nitrogen metabolism. Research also indicates that nitrate (the metabolism byproduct of denitrifying bacteria) can result in increased metal corrosion rates (Xu, Li, Song, and Gu 2013).

16.2.4 Chemical Analysis and X-Ray Diffraction

A representative specimen was cut from the pipe substrate. The chemical contents were analyzed by Glow Discharge Spectroscopy (GDS), except carbon and sulfur elements which were analyzed per ASTM E1019: *Standard Test Methods for Determination of Carbon, Sulfur, Nitrogen, and Oxygen in Steel, Iron, Nickel, and Cobalt Alloys by Various Combustion and Fusion Techniques* (2018).

X-ray diffraction (XRD) analysis determines the molecular structure of a crystal, in which the crystalline lattice structure causes a beam of incident X-rays to diffract into many specific directions. Each crystalline solid has its unique characteristic X-ray powder diffraction pattern which may be used as a "fingerprint" for its identification (Young 1996). It was used to identify the debris material. Additionally, qualitative chemical analysis by Energy Dispersive X-ray Fluorescence (EDXRF) was conducted on the pipeline debris.

The water and debris samplers were analyzed by ion chromatography to determine if anions related to acids or MIC were present. The water samples were also examined for the presence of iron.

16.3 RESULTS AND DISCUSSION

16.3.1 Results of Visual and Metallurgical Examination

Figure 16.4 shows a 4 millimeter (0.158 inch) diameter interior hole in the pipe sample received. The dried-out debris collected from the interior side of the hole, and used for subsequent testing, was yellow/orange, red, and brown in color and was removed easily from the pit. There were no visible biological accumulations in the corroded areas. Figure 16.5 is the image of the hole after ultrasonic cleaning.

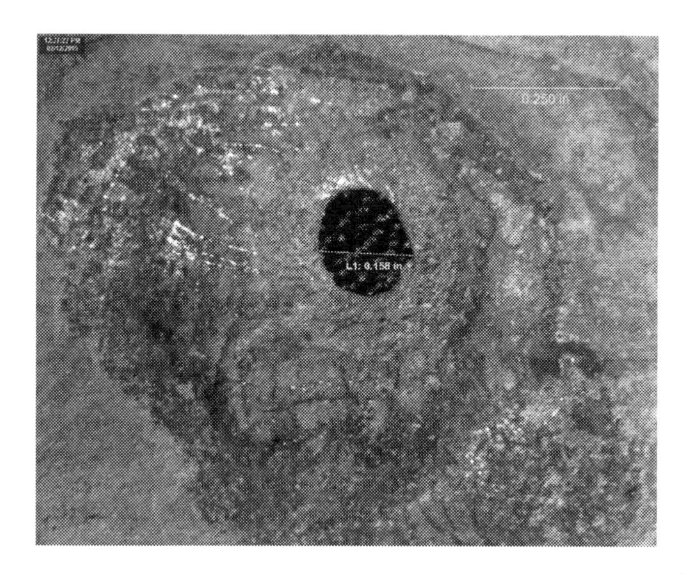

FIGURE 16.4 Interior micro photo of pipe hole

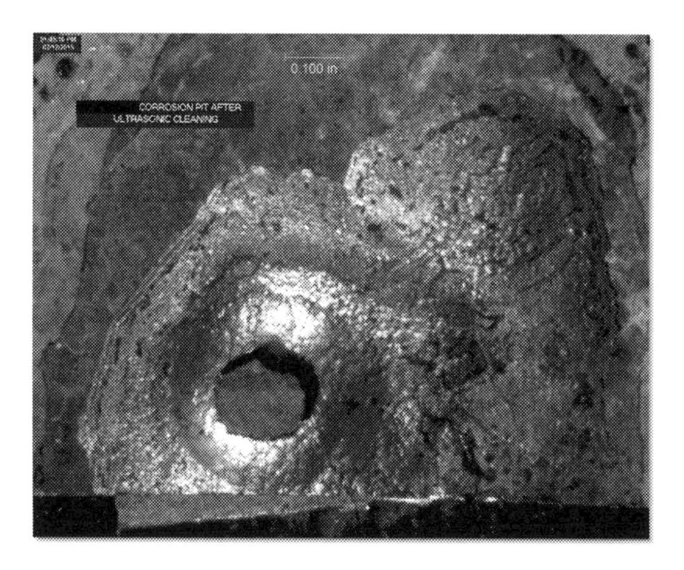

FIGURE 16.5 Micro photo of interior pipe hole after ultrasonic cleaning

The morphology of the surrounding pitting of the hole is elliptical in nature (not uniform).

Figures 16.6 and 16.7 show the remnants of the zinc layer on the steel substrate with a 0.3 millimeter (0.012 inch) of corrosion layer. The intact zinc layer in the non-corroded areas (not shown) indicates that the galvanization process was sufficient to form an appropriate protective zinc layer. Where the corrosion occurred, the zinc layer is cracked and depleted, leaving the steel substrate exposed. Cracking of the zinc layer

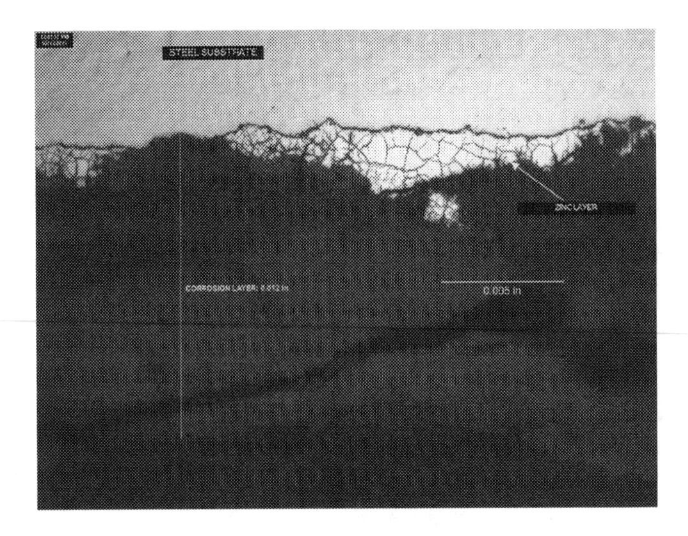

FIGURE 16.6 Cross section of pipe with corrosion layer.

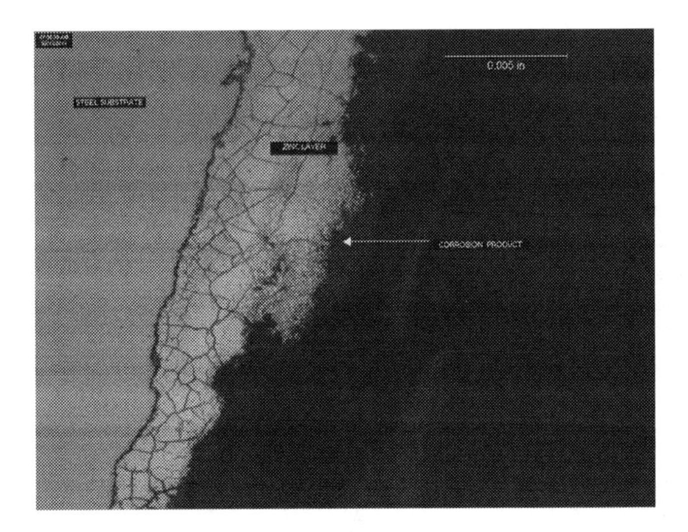

FIGURE 16.7 Cross section of pipe with corrosion product.

is often the result of a chemical attack against the zinc. Figure 16.8 is a cross section of the pipe with some minor pitting after the degradation of the zinc layer.

16.3.2 Results of Microbiological Analysis Using qPCR

Table 16.2 shows the microbe (qPCR) analysis results for the liquid samples. The sample Building #3 Close to Pipe Failure had a concentration of bacteria at 10^7 copies per milliliter, denitrifying bacteria at 10^4 copies per milliliter, and *Leptothrix*

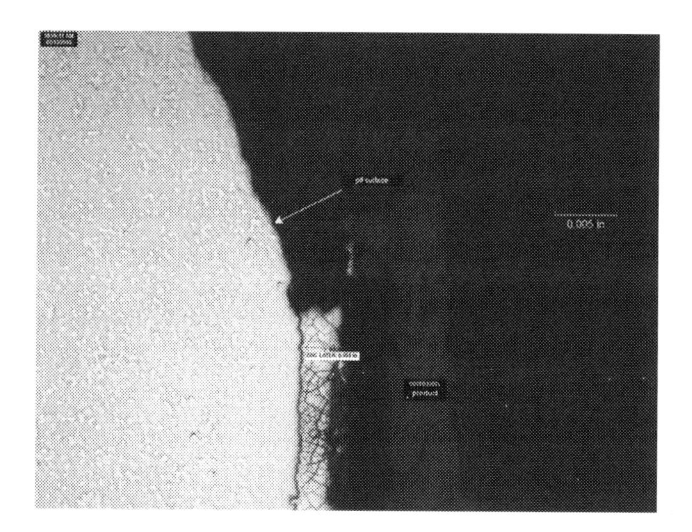

FIGURE 16.8 Cross section of pipe pit surface.

and *Sphaerotilus* species of iron-oxidizing bacteria at 10^4 copies per milliliter. The qPCR analysis also detected sulfate-reducing (10^2 copies per gram) and acetic acid–producing bacteria (10^2 copies per gram). The Fire Hydrant and Building #3 Main Drain samples contained concentrations of bacteria at 10^4 copies per milliliter. The Building #3 Main Drain sample also contained concentrations of denitrifying bacteria at 10^4 copies per milliliter and sulfate-reducing bacteria at 10^2 copies per milliliter.

Table 16.3 shows the microbiological (qPCR) analysis results from debris collected from the pit on the pipe. The debris contained a concentration of Bacteria at 10^7 copies per gram. The sample also contained concentrations of *Leptothrix* and *Sphaerotilus* species of iron-oxidizing bacteria and denitrifying bacteria at 10^5 copies per gram. Lastly, qPCR analysis detected acetic acid–producing bacteria at 10^3 copies per gram.

16.3.3 Results of Chemical Analysis of Liquid and Pipe Samples

The steel chemistry results in Table 16.4 indicate that the chemistry of the pipe used for the fire sprinkler system meets specifications for steel C1008 grade (ASTM A519 2017). Analysis of the pipe debris by X-ray diffraction spectroscopy, seen in Figure 16.9, indicates that the debris is primarily zinc oxide. The red lines in the same figure are the stick patterns for a reference zinc oxide analysis. Iron is also present, from a qualitative analysis by Energy Dispersive X-ray Fluorescence as seen in Table 16.5. Iron is not seen in the XRD analysis, likely because it is present as amorphous iron oxides and not in a crystalline state. Hydrated amorphous iron oxides are often yellow and orange in color. The results of the water samples from Table 16.6 are consistent with potable water results (U.S. Safe Drinking Water Act;

TABLE 16.2
Microbiological (qPCR) Analysis Results for the Liquid Samples

qPCR Analysis (Results in Copy # per Milliliter)		Fire Hydrant	Building #1 Main Drain	Building #2 Main Drain	Building #3 Main Drain	Building #3 Close to Pipe Failure
Total Bacteria		**6.50E + 04**	**3.82E + 03**	**2.56E + 03**	**6.45E + 04**	**1.33E + 07**
Sulfate-Reducing Bacteria (SRB)		**BDL**	**BDL**	**BDL**	**5.25E + 00**	**2.01E + 02**
Denitrifying Bacteria (DNB)		**2.39E + 02**	**1.85E + 02**	**2.87E + 02**	**1.09E + 04**	**3.94E + 04**
Acid Producing Bacteria (APB)	**Total**	**5.75E + 00**	**2.05E-01**	**BDL**	**8.95E-01**	**1.26E + 02**
	Acetic Acid-Producing Bacteria	5.75E + 00	2.05E-01	ND	8.95E-01	1.26E + 02
	Butyric Acid-Producing Bacteria	BDL	BDL	BDL	BDL	BDL
Iron-Oxidizing Bacteria	**Total**	**2.84E + 02**	**7.60E + 02**	**1.56E + 02**	**3.68E + 03**	**4.70E + 04**
	***Leptothrix* and and Sphaerotilus *Sphaerotilus* Species**	2.84E + 02	7.60E + 02	1.56E + 02	3.68E + 03	4.70E + 04
	***Gallionella* Species (Gall)**	ND	ND	ND	ND	ND

BDL = Below Detection Limit; ND = Not Detected

United States Government 1986). The analytical chemistry results in Table 16.7 indicate that there is a high concentration of acetate (520 μg/g) in the solid debris collected from the pipe sample.

16.4 CONCLUSIONS

16.4.1 Primary Failure Mechanisms

Based on the available data presented in the Results and Discussion section, the hole in the pipe specimen is due to the 1) breakdown of the protective zinc layer from a chemical attack from metabolites secreted from bacteria exposing the metal substrate, and 2) creation of a galvanic cell between the bare carbon steel and the surrounding environment (biofilm, water, pressurized air), eventually leading to a through wall hole. Both processes can be attributed to chemical and electrical microbial influenced corrosion. This is reinforced by the shape of the corrosion pitting which is representative of MIC pitting morphologies.

TABLE 16.3
Microbiological (qPCR) Analysis Results for the Pipe Sample Debris

qPCR Analysis (Results in Copy # per Gram)		Pipe
Total Bacteria		**1.67E + 07**
Sulfate-Reducing Bacteria (SRB)		**ND**
Denitrifying Bacteria (DNB)		**1.68E + 05**
Acid Producing Bacteria (APB)	**Total**	**1.80E + 03**
	Acetic Acid-Producing Bacteria	1.80E + 03
	Butyric Acid-Producing Bacteria	BDL
Iron-Oxidizing Bacteria	**Total**	**2.04E + 05**
	***Leptothrix* and and Sphaerotilus *Sphaerotilus* Species**	2.04E + 05
	***Gallionella* Species (Gall)**	ND

BDL = Below Detection Limit; ND = Not Detected

TABLE 16.4
Fire Sprinkler Pipe Steel Substrate Chemistry Results

Element	Pipe	Element	Pipe
Aluminum, wt%	0.01	Phosphorus, wt%	0.008
Cobalt, wt%	<0.01	Silicon, wt%	0.02
Copper, wt%	0.15	Titanium, wt%	<0.005
Chromium, wt%	0.07	Vanadium, wt%	<0.005
Manganese, wt%	0.34	Carbon, wt%	0.085
Molybdenum, wt%	0.02	Sulfur, wt%	0.0092
Nickel, wt%	0.06		

16.4.2 Protective Zinc Layer/Acetic Acid

The breakdown of the protective zinc layer was caused by a chemical attack, possibly acetic acid, which is a by-product of the metabolism of acetic acid–producing bacteria. The high concentration of acetate found in the debris from the pipe sample is common in many biological systems. Acetic acid is a weak acid; however, it is corrosive because acetic acid in free acid form and coupled with iron-oxidation is thermodynamically favorable and kinetically not retarded (Gu 2014). The debris collected from the hole of the pipe sample contained iron-oxidizing bacteria and acetic acid–producing bacteria.

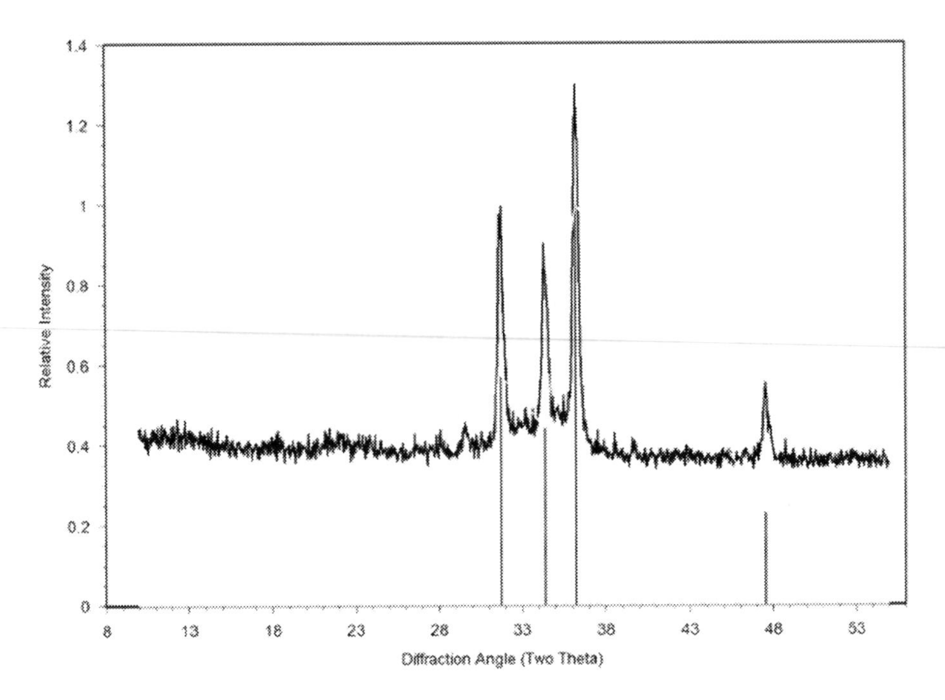

FIGURE 16.9 XRD pattern for debris sample

TABLE 16.5
EDXRF Qualitative Analysis

Element	Pipe Debris	Element	Pipe Debris
Carbon	8	Sulfur	0.3
Oxygen	10	Chlorine	0.8
Sodium	0.2	Calcium	0.2
Magnesium	0.2	Manganese	0.3
Aluminum	0.2	Iron	9
Silicon	0.2	Zinc	71

16.4.3 Denitrifying Bacteria Levels

Although there was not a large concentration of nitrate found in the pipe debris sample, denitrifying bacteria was detected in the sample. Denitrifying bacteria are ubiquitous in the environment and previous lab research indicates these bacteria pit more aggressively than sulfate-reducing bacteria due to a large thermodynamic driving force (Xu, Li, Song, and Gu 2013). Denitrifying bacteria reduce nitrate to nitrogen gas and water and require a carbon source to act as an electron donor (nitrate is the electron acceptor).

TABLE 16.6
Chemistry Results for Water Samples

	Fire Hydrant	Building #1 Main Drain	Building #2 Main Drain	Building #3 Main Drain	Building #3 Close to Pipe Failure
Formate, mg/L	<0.2	<0.2	<0.2	<0.2	<0.2
Acetate, mg/L	<1	<1	1	<1	<1
Fluoride, mg/L	0.4	0.4	0.4	0.4	0.4
Chloride, mg/L	16	13	14	19	13
Nitrate, mg/L	1	<0.3	<0.3	<0.3	<0.3
Sulfate, mg/L	23	17	18	20	3
Iron, mg/L	<0.2	<0.2	<0.2	<0.2	<0.2
pH, standard units	6.10	7.33	7.21	7.30	7.30

TABLE 16.7
Chemistry Results for Solid Residue

	Pipe
Formate, μg/g	170
Acetate, μg/g	520
Fluoride, μg/g	<10
Chloride, μg/g	86
Nitrate, μg/g	<25
Sulfate, μg/g	250

16.4.4 Microbial Influenced Corrosion in the Fire Suppression System

Microbiologically influenced corrosion in fire protection systems is defined as an "electrochemical corrosion process that is concentrated and accelerated by the activity of specific bacteria within a fire sprinkler system resulting in the premature failure of metallic system components" (Clark and Aguilera 2007). Failures due to both a concentrated and accelerated corrosion process similar to the one experienced at the facility have been observed in wet, dry, and pre-action fire protection systems. Although the pipe was hot dip galvanized (providing a zinc oxide layer for corrosion protection), bacteria that are known to cause corrosion can also initiate the corrosion process, as observed in the chemical microbial-influenced corrosion degrading preferentially the zinc oxide layer. Galvanization does protect the underlying steel pipe; however, the protection mechanism is only effective if the zinc coating is continuous and free of localized defects (Van Der Schijff 2008). Lastly, pressurized air and any residual water left from hydro-testing can sustain the existing corrosion process.

TABLE 16.7
Chemistry Results for Solid Residue

Chemical Conditions	*Microbiological Conditions*
Corrosive compounds: acetate, sulfate, formate, and chloride found in debris from pipe sample	Microbiological concentrations (16S rRNA) at 10^7 copies per gram for pipe debris sample (10^5 copies per gram of denitrifying bacteria and iron-oxidizing bacteria 10^3 copies per gram of acid-producing bacteria). Exposed pipe substrate is an electron donor and metabolic by-product from bacteria corrodes the steel substrate. Metabolic by-product corrodes zinc oxide layer.
Corrosion and Metallurgical Condition	***Design and Operation Condition***
Pipe materials consistent with C1008 grade	Three years in operation
Zinc oxide detected in XRD analysis	System charged with 412 kilopascals of air
Iron from EDXRF	Tested with potable water

REFERENCES

ASTM A519/A519M-17 (2017) Standard Specification for Seamless Carbon and Alloy Steel Mechanical Tubing. ASTM International, West Conshohocken, PA. www.astm.org.

ASTM E1019-18 (2018) Standard Test Methods for Determination of Carbon, Sulfur, Nitrogen, and Oxygen in Steel, Iron, Nickel, and Cobalt Alloys by Various Combustion and Inert Gas Fusion Techniques. ASTM International, West Conshohocken, PA. www.astm.org.

Clark B, Aguilera A (2007) Microbiologically Influenced Corrosion in Fire Sprinkler Systems. *Automatic Sprinkler Systems Handbook Supplement* 3: 955–964.

Enning D, Garrelfs G (2014) Corrosion of Iron by Sulfate-Reducing Bacteria: New Views of an Old Problem. *American Society of Microbiology* 80: 1226–1236.

Gu T (2014) Theoretical Modeling of the Possibility of Acid Producing Bacteria Causing Fast Pitting Biocorrosion. *Journal Microbial Biochemical Technology* 6: 068–074.

NACE Standard TM0212, Standard Test Method: Detection, Testing, and Evaluation of Microbiologically Influenced Corrosion on Internal Surfaces of Pipelines. 2018 NACE International, Houston, TX.

United States Government (1986) Pub.L. 99–359; 100 Stat. 642. Safe Drinking Water Act Amendments of 1986.

Van Der Schijff O (2008) MIC in Fire Sprinkler Systems Field Observations and Data. NACE Corrosion Conference & Expo.

Xu D, Li Y, Song Fengmei, Gu T (2013) Laboratory Investigation of Microbiologically Influenced Corrosion of C1018 Carbon Steel by Nitrate Reducing Bacterium *Bacillus licheniformis*. *Corrosion Science* 77: 385–390.

Young RY (1996) The Rietveld Method. Oxford University Press, Oxford.

17 Analysis of Field Observations of Severe MIC of FPSO Mooring Chains

Robert E. Melchers
University of Newcastle

Tim Lee
AMOG Pty. Ltd.

CONTENTS

17.1 INTRODUCTION

Floating production storage and offloading (FPSO) vessels are floating steel platforms, similar to oil tankers, or are converted oil tankers, used in the oil and gas industry for offshore operations. Typically, they have extensive topside equipment that includes, for production platforms, pipework, pumps, tanks, etc. and equipment for handling the considerable length of pipe for production pipelines (risers) that extend to the seafloor, as well as associated equipment.

FPSOs are finding increasing application in deep waters, currently up to about 3 km deep, often located many kilometers offshore. For safety, environmental and economic reasons it is important for the vessels to be kept "on-station" to quite close tolerances, lessening the risk of riser failure and hence crude oil spills. Typically, this is achieved by a system of 8–12 mooring lines. Typically, the mooring lines consist of steel chain at the top and through the wave zone into the fully immersed zone, followed by steel mooring cables or synthetic hawsers and near the seafloor back into steel chain. Usually all the chains are stud-less to

DOI: 10.1201/9780429355479-20

save weight and for improved fatigue performance. Mooring line tension due to self-weight of the mooring lines can be an issue, and in some cases underwater buoys are employed to add buoyancy.

The anchorage point(s) for the mooring lines often is at a turret mounted on the bow of the vessel, although there also are systems where the turret is within the vessel hull, or the mooring lines connect directly to the vessel hull. Typically, chain is used at the top of the mooring lines to connect to the vessel platform. This permits easier logistics for attaching the mooring lines to the platform and a more mechanically robust connection. While alternative systems have been considered and used in some instances, the use of chains is still the industry standard. Chain link sizes currently are often 76 mm diameter but sizes up to 150 mm and more are also employed. The steel cables are around 100 mm in diameter. Although non-metallic cables are being trialed, steel is still the predominant material. Typically, the cables are galvanized. Together with the grease used in their construction and remaining inside the multiple strands making up a cable, they are not particularly prone to corrosion, noting also that they are entirely in the immersion zone. Similarly, the lower chains, while usually bare steel, normally are not prone to significant corrosion, at least no more than would be expected at depth. The most severe corrosion problem is for the upper chains and mainly in the intertidal/wave action zone (Lee et al., 2015). Particularly in the tropics, some chains have been found to show severe, usually quite localized corrosion, even after only some 8–10 years of continuous exposure (Figure 17.1). This is of considerable practical interest as increasingly FPSOs are being deployed in tropical waters.

Discovery of the localized corrosion shown in Figure 17.1 during routine inspections as part of a major research study (SCORCH-JIP) to ascertain the effect of operating in the tropics (Lee et al., 2015) caused considerable alarm because of the potential implications for FPSO operational safety. This led the research project to investigate the potential reasons for such corrosion, noting that the FPSOs involved were many kilometers offshore and thus in what were assumed to be unpolluted ocean waters. However, water quality testing showed that this was not the case and that there were high concentrations of Dissolved Inorganic Nitrogen (DIN) in the seawaters, indicating that microbiologically influenced corrosion (MIC) was likely to be involved (Melchers, 2014a). The next section reviews the background for the link between MIC and DIN in seawater, drawing on earlier work for accelerated low water corrosion (ALWC) and the correlation between DIN and MIC for

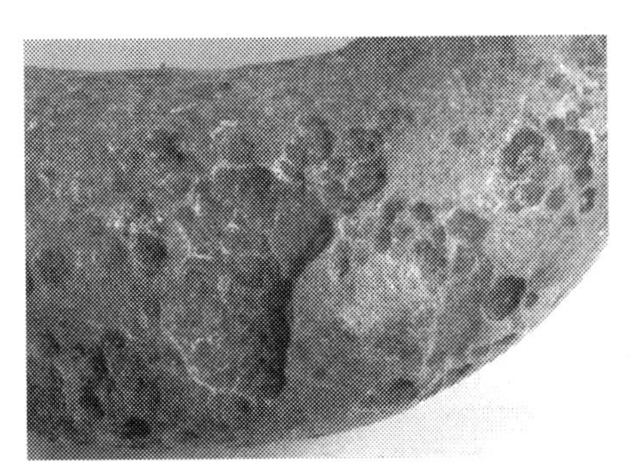

FIGURE 17.1 Part of corroded steel chain link (76 mm diam.) showing evidence of highly localized corrosion after removal of rust layers.

corrosion loss. It also considers the potential role of other nutrients for microbiological metabolism and their potentially limiting function. The corrosion mechanism(s) likely to be involved are considered, including the most likely reason for the rather large but usually isolated areas of localized corrosion as seen in Figure 17.1. Brief consideration is given to possible ameliorating strategies and also why MIC at depth appears to be unlikely.

17.2 BACKGROUND

It has been established for some time that for field exposure conditions the concentration of critical nutrients in sea (or other) waters is likely to influence the activity of microorganisms and thus, potentially corrosion loss and pitting, provided other aspects such as shelter and energy sources are available (Melchers, 2005). The presence of bacteria and archaea alone is not sufficient. Microorganisms can multiply at very fast rates relative to the rate of marine corrosion (Little and Lee, 2007). It follows that the rate-limiting step for microorganism activity, and thus potentially for their effect on the rate of corrosion, is a function of the rate of supply, and thus of the availability, of critical nutrients. For corrosion of steel in seawater, the critical nutrient has been shown to be dissolved inorganic nitrogen (DIN) (Melchers, 2005; Melchers and Jeffrey, 2012, 2013). Other nutrients critical for microorganism activity and metabolism typically are available to excess in seawater, including sulphates, phosphorous, and organic carbon. For the corrosion of steels (and alloys containing some Fe component) the critical micro-nutrient, ferrous iron, is supplied by the corrosion process. That process also supplies the energy source in terms of electron flow, while biofilms and rusts can provide the necessary shelter.

For steel corrosion coupons exposed over several years in various seawaters, the effect of DIN concentration in the seawater local to unprotected steel structures has been quantified using corrosion loss data obtained from a wide variety of field exposure sites and correlated with data about nutrient concentrations (Melchers, 2014a). In terms of the bi-modal model for the corrosion loss of steels in seawater over extended periods of exposure (years), the effect of bacterial activity stimulated by nutrient availability is most prominent at the beginning of the second mode (Figure 17.2). The effect may be simplified for longer-term corrosion using the tangent line AC. It represents the long-term trend without nutrient availability. The effect of the present of nutrients is to raise the tangent AC as shown, roughly parallel to AC. The tangent is parameterized by the intercept c_S and the slope r_S.

In what is essentially an (environmental) input—(corrosion severity) output relationship, the relative increase in corrosion loss for waters with elevated concentration of DIN compared to corrosion in low-DIN seawaters is approximately a linear function of nutrient concentration (Melchers and Jeffrey, 2013). The change in the parameter c_s with increased DIN is as shown in Figure 17.3 (Melchers, 2014a) for moderate changes in DIN and for different average water temperatures. Similarly, Figure 17.4 shows the changes with DIN and average water temperature in the long-term slope of the corrosion losses, defined by parameter r_s (Melchers, 2014a).

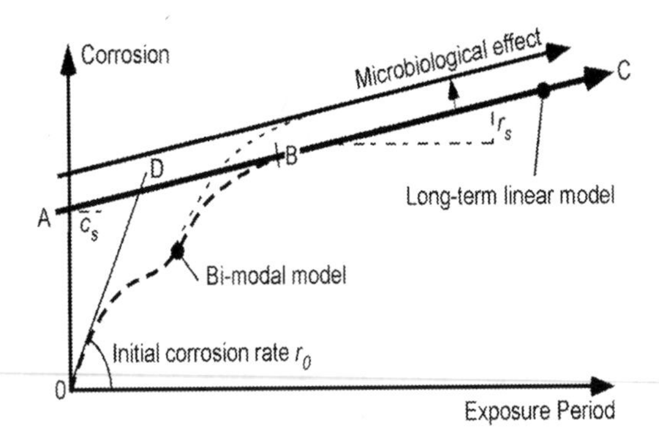

FIGURE 17.2 Bi-modal corrosion model, long-term linear corrosion, and effect of microbiologically influenced corrosion (Melchers, 2014a).

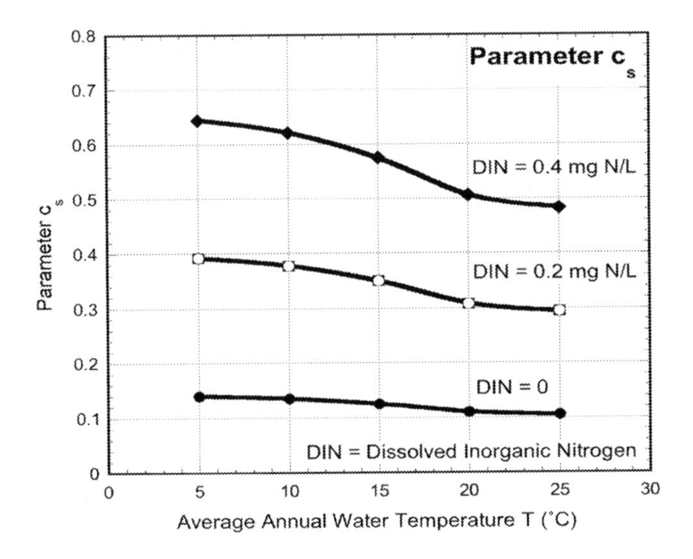

FIGURE 17.3 Relationship between parameter c_s, seawater temperature, and DIN (Melchers, 2014a).

These two figures may be used to construct, for a given average annual seawater temperature, the long-term trend AC but now with the inclusion of the possible microbiological corrosion effect (i.e., with MIC), defined through the concentration of the critical nutrient DIN. Figure 17.5 shows the result for seawater at 25°C and various DIN concentrations. It is an extrapolation to DIN = 3.0 mgN/l of the results in Figures 17.3 and 17.4 applied to the long-term trend AC in Figure 17.2. The validity of the extrapolation has not been proven but is plausible. Note that Figures 17.2, 17.3, and 17.4 deal with corrosion loss defined as "uniform" corrosion, as conventionally obtained from mass loss observations. It does not deal specifically with localized corrosion or pitting corrosion.

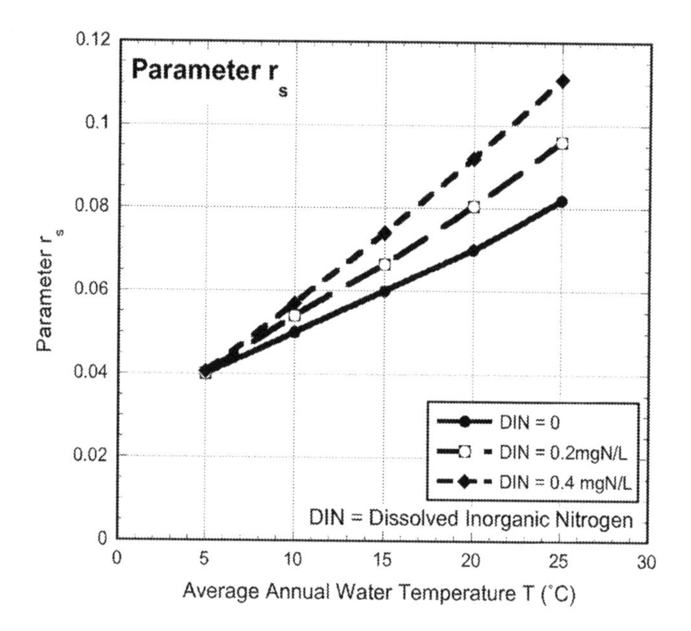

FIGURE 17.4 Relationship between parameter r_s, seawater temperature, and DIN (Melchers, 2014a).

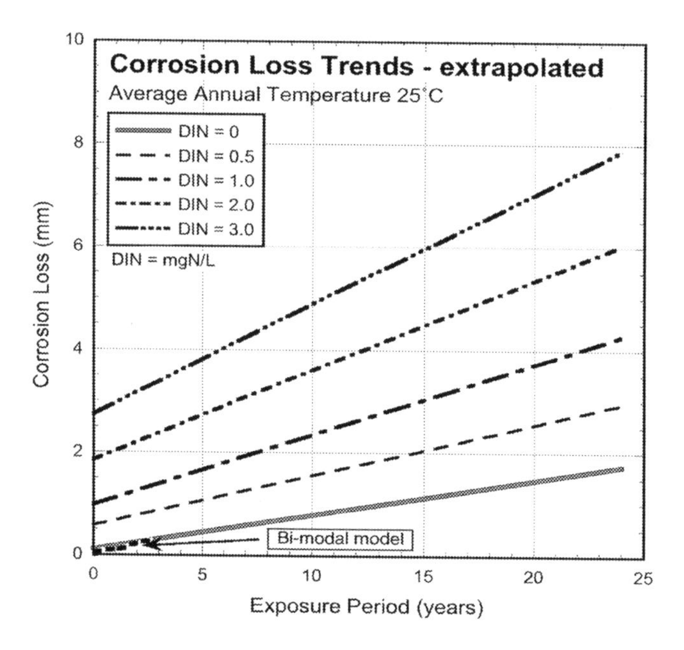

FIGURE 17.5 Extrapolation of trends for longer-term corrosion with MIC, based on Figure 17.2.

The fact that MIC usually is associated with significant localized corrosion and deeper pitting corrosion is well-known, even for laboratory observations under rather artificial conditions and for short-term exposure periods (Little and Lee, 2007). For longer-term exposures, the effect of MIC for localized corrosion can be estimated from test results for steel strips suspended vertically in seawater through the immersion, tidal, plash, and atmospheric zones. This experimental arrangement is similar to sheet piling in harbors. Under elevated DIN conditions these strips (and sheet piling) show accelerated low water corrosion (ALWC). From many experimental observations for steel strips (Melchers and Jeffrey, 2013), and from data recovered for sheet piling in various U.S. harbors (Melchers, 2013), the trend shown in Figure 17.6 was obtained. It shows the localized corrosion in the low water zone (measured by *a*—see inset) and the amount of corrosion in the immersion zone (shown as *b*). It also shows the ratio *R* (= *a*/*b*) as a function of the total dissolved inorganic nitrogen (DIN) measured for the corresponding exposure sites. For most of these the data were reasonably precise, but for some, shown circled, there is considerable degree of uncertainty, mainly because of difficulty in obtaining reliable data for average DIN at or close to the sites where the corrosion losses were reported (Lee et al., 2015). The U.S. data alone gives the linear trend shown, with

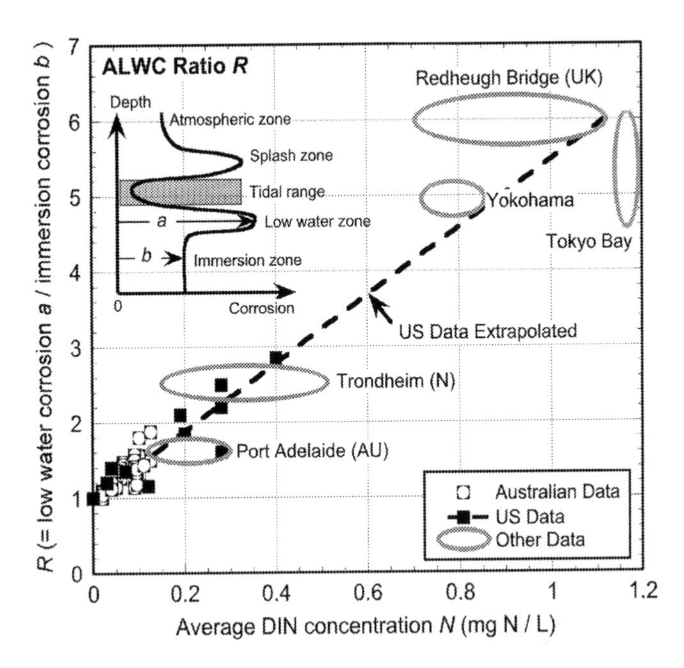

FIGURE 17.6 Relationship between corrosion ratio *R* and DIN, with *R* defined as $R = a/b$, where *a* is the local severity of corrosion for the lower water zone and *b* the local severity of corrosion in the immersion zone, as defined in the inset. It shows, schematically, the different corrosion zones for seawater tidal conditions defined along the vertical axis, against the degree of corrosion along the horizontal axis. $R = 1$ indicates no accelerated low water corrosion and corresponds to $N = 0$.

$R = 1.055 + 4.5\ N$ [r = 0.715]. For the Australian data, covering a lower range of N, the trend line is very similar, with function $R = 0.98 + 4.49\ N$ [r = 0.982].

The information in Figure 17.6 is of interest because the corrosion in the low-water zone is quite localized, noted already many years ago by Evans (1960) for corrosion at the water-air interface for still water conditions. For tidal conditions that zone moves up and own, but it was quite evident in the experimental samples from actual seawater exposures over several years (Melchers and Jeffrey, 2012, 2013). Whereas the data for the U.S. sites was opportunistic, extracted many years after the original observations were made, the Australian data are from controlled experimental work in which the local corrosion loss in the low-water zone, and in fact the corrosion losses in all the zones were captured by cutting the long strips into short (100 mm long) segments and determining the mass loss for each of these. This produced profiles of localized corrosion loss similar to that shown schematically in the inset in Figure 17.6. The importance of this in the present context is that the size of the segments so produced is of the same order of magnitude as the extent of the most severe localized corrosion observed for the chain links most affected by localized corrosion. This is now described.

17.3 OBSERVATIONS

The chain links considered herein were recovered from operational FPSO sites off West Africa, and in Indonesian and East Timor waters. There also is some information about corrosion of FPSO mooring chains in the Gulf of Mexico and in the North Sea. The (few) cases that were available for examination in the SCORCH-JIP project (Lee et al., 2015) are summarized in Table 17.1. In each case, the mooring line became available as a result of field inspection and the decision by the platform operator to take the mooring line out of service. This decision is governed by a balance between the risk of mooring line failure and the considerable cost of a new mooring line and the cost of downtime. There was also the issue of whether the operator was willing to let the chains be examined in what was essentially a process that permitted access by competitors, consultants, and academics. A number of the participants in the SCORCH project actively participated and provided in-kind support as well as the data summarized in Table 17.1.

The mooring lines available for the project were recovered at sea, usually high-pressure water blasted to remove marine growth and loose rusts, transported to land, the chain length of the upper part of the mooring line separated and then grit blasted to attempt to remove as much as possible of the rusts including as much as possible inside the pits inside the localized corrosion regions and elsewhere and thus to expose the remaining surface of the chain links (cf. Figure 17.1). The links were then scanned using an infra-red scanner and from the data so produced 3D computer models could be developed (Lee et al., 2015; Asadi and Melchers, 2017). These were used together with the known nominal sizes of the chains to estimate the general or uniform corrosion losses over the surface of the links. In addition, the maximum depth of localized corrosion penetration was determined. This was also checked manually wherever possible.

TABLE 17.1
General Corrosion Losses (mm) and Maximum Corrosion Penetration (all 76 mm Diam. Chains)

Location	Exposure Period (years)	Estimated DIN (mgN/l)	Average Corrosion Loss *b* (mm)	Maximum Penetration *a* (mm)	*R* = *a*/*b*	Remarks*
East Timor	5		0.68			Unit 16
Gulf of Mexico	8	0.3–0.5	1.65	4.3 ± 0.9	≈ 3–4	Unit 8
Brazil	11	?	1.7	?		Unit 9
Indonesia A	5	0.4–1.0 [a]	1.13 ± 0.6			Unit 14
	7		1.5–2.0	10–20	4–5	Unit 14
Indonesia B	15	2–2.3 [b]	11.5 ± 0.45	15–20	≈ 2	Unit 15
West Africa	7	12.08 [c] (2.71)	5.5–7.8	14–28	3–5.1	Unit 16
	8	12.08 [c] (2.71)	7	?		Unit 4
	15	12.08 [c] (2.71)	8.48	20–30	2–4.5	Unit 3
	15	12.08 [c] (2.71)	11.0	20–30	2–3	Unit 5
North Sea	?	0.2–0.3			≈ 2 [d]	–

Notes
a Source: Chevron IndoAsia TS Lab. reported as NO_3^- and converted to N.
b Source: Duri Lab., Jakarta, at 5 m depth.
c Source: Chevron Nigeria Ltd Escravos Laboratory. Reported as 12.08 mg/l as N - more likely it is NO_3^- as obtained from standard ion chromatography. This is equivalent to 2.71 mgN/l (see text).
d Estimated from photographs.
* Unit refers to the notation in AMOG Consulting (2014).
Note: sulphates equivalent to typical seawater, phosphorous detected.
Based on data in the Final Report for the SCORCH project (AMOG Consulting, 2014).

Typically, the regions of severe localized corrosion were observed to occur relatively isolated from one another at various locations on the chain links, typically surrounded by larger areas of apparently uniform corrosion. Not all chains in all tropical seawaters showed severe localized regions of corrosion. Typically, the regions of localized corrosion have a complex topography (Figure 17.7).

The topographies shown in Figure 17.7 may be interpreted as showing pits formed on plateaus within the corroded regions. There is clear evidence of pitting on the lowest plateaus, that is, in the deepest overall part of the local corrosion region. Evidently the size of the localized region is comparable to or greater than the chain link diameter and the overall pit depths 20–30 mm (Table 17.1). Because these links were recovered from field operations the original link diameters were not

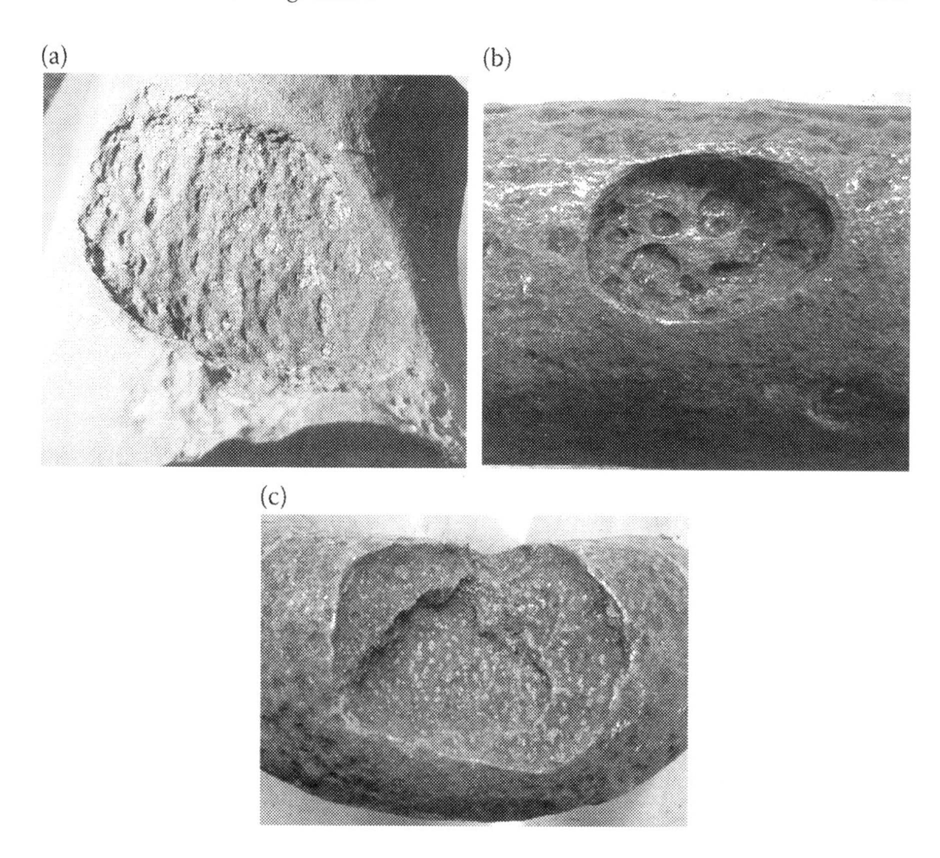

FIGURE 17.7 Close-up views of regions of severe localized corrosion of 76 mm diameter chain links after 8–15 years exposure in near-surface immersion conditions and subsequent water jet cleaning and grit-blasting to remove the corrosion products. Figure 17.7(a) shows that the grit-blasted surface still has some remnants of corrosion products in the pits. Figure 17.7(b) shows evidence of pits within the localized corroded zones and Figure 17.7(c) shows plateaus of corroded surfaces within the larger localized regions of corrosion, with total depth of penetration some 20–30 mm.

known precisely at all points along the links, noting that their manufacture results in the actual original link diameters varying by several millimeters from the nominal.

The water quality (WQ) of the seawater local to the mooring lines was determined at each of the sites from which mooring lines were recovered. This was done in the standard manner of taking a sizeable amount of seawater from the site, immediately taking it (often by helicopter) to a certified water quality testing laboratory on-shore for testing. The water quality testing included salinity, pH, conductivity, temperature, dissolved oxygen, nitrites, nitrates, ammonia, sulphates, phosphate, total phosphorous, and calcium. In Table 17.1, only the DIN values are reported (i.e., the inorganic N content derived from nitrites, nitrates, and ammonia). These values are quite high when compared to typical coastal and harbor conditions.

In one case, initial WQ readings showed very low (background) values of DIN and when these were queried it became apparent that the samples had been taken some distance away from the FPSO vessel, and had measured, essentially, open sea levels of nutrients. The values reported in Table 17.1 are those for water samples taken in the immediate vicinity of the mooring line, noting that discharges from FPSOs with poor housekeeping may be the critical agents for elevated DIN and other nutrients. A parallel situation has been reported for offshore fixed production platforms (Odom, 1993). For the North Sea chains, the DIN values shown in Table 17.1 were drawn from the OSPAR study of the North Atlantic and North Sea (OSPAR, 2000).

17.4 ANALYSIS

The chain links in Figures 17.1 and 17.7 show that the most severe corrosion is highly localized and that large areas of the chain links have experienced much less corrosion, mainly in a form that could be classified as 'uniform' or 'general' corrosion. A parallel situation was found for the corrosion of vertical steel piling (Melchers and Jeffrey, 2013). Particularly in seawaters with elevated DIN concentrations usually there is a contrast between the depth of highly localized corrosion, in the so-called accelerated low water corrosion (ALWC) immersion zone, compared with the corrosion for the remainder of the pile in the immersion zone (Figure 17.6). For the present analysis it should be noted that the surface area extent of the localized corroded region for ALWC is comparable in size to the more severe cases of localized corrosion of the 76 mm diameter chain links. In both exposure situations the concentration of DIN in the seawater appears to be an indicator of the severity of localized corrosion (Figure 17.6 and Table 17.1).

To make progress in the analysis of the data for the chain links (Table 17.1), consider now the parameters a and b employed in Figure 17.6. Let these be applied to the data in Table 17.1 for generalized corrosion and for the deepest penetrations respectively for the chain links (Figure 17.8).

As a first step, consider the overall "uniform" corrosion losses (Table 17.1) and now described by parameter b. Using the corrosion loss trends for increasing values of DIN shown in Figure 17.5, and extrapolated to higher DIN values, the data for b

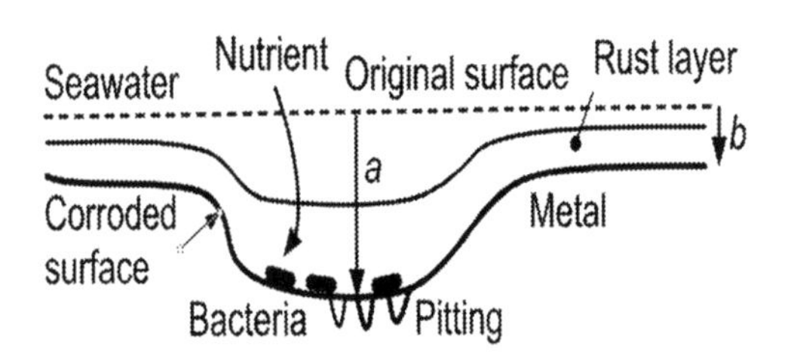

FIGURE 17.8 Parameters a and b applied to the corrosion of the surfaces of chain links. A schematic of the rust layer and the diffusion paths of inorganic nitrogenous nutrients also are shown. The schematic rust layer thickness shown is based on the observation in Figure 17.1(a) that the depressions caused by localized corrosion are visible on the outer rust layers.

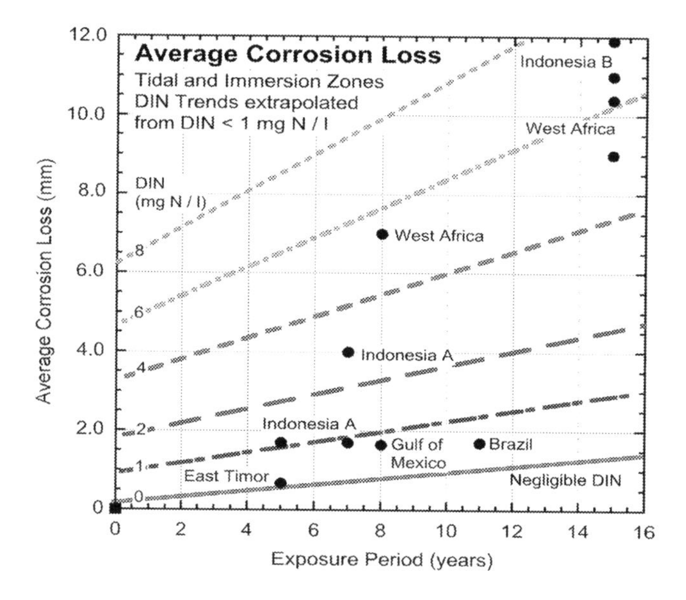

FIGURE 17.9 Average corrosion loss b of chain links as a function of exposure period plotted on an extrapolated version of Figure 17.5. Comparison to Table 17.1 shows some discrepancies between the DIN values recorded in the field (shown as •) and the expectations shown by the linear extrapolations of the trends in Figure 17.5.

are as shown in Figure 17.9. Comparing to the DIN values in Table 17.1 for the cases of Indonesia B and West Africa at 15 years' exposure it is apparent that the corrosion losses shown are considerably greater than would be expected, at least compared to the DIN values for the trends shown. For all the other data there is reasonable correspondence between the DIN values in Table 17.1 and the corrosion losses for the trends for the different DIN values. The potential reasons for the apparent discrepancy for the higher DIN values is considered further below.

The second comparison can be made for the ratio $R = a/b$ between the deepest localized corrosion values a and the general corrosion losses b using the ratios given in Table 17.1. The comparison can be made by adding the R and N values to Figure 17.6. This produces the plot shown in Figure 17.10. It is clear that the linear trend fitted to the data in Figure 17.6 cannot be extrapolated *ad infinitum* for the chain link corrosion data. There is a clear limit on the value for R. The potential reasons for this are explored in the next section.

17.5 DISCUSSION

As noted, the data for general corrosion of the chain links as plotted on Figure 17.9 show reasonable consistency with the extrapolated trends derived from Figure 17.6 for the cases with lower concentrations of DIN. The cases with somewhat higher DIN have values for corrosion loss that align, on Figure 17.9, with DIN values that are much higher than the actual field values for DIN. In part, this may be because the general corrosion losses reported in Table 17.1 relate to b as measured by the

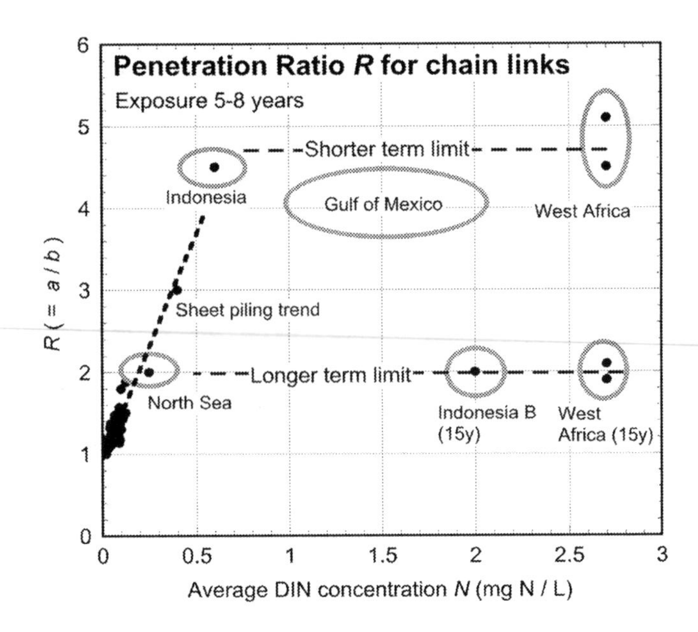

FIGURE 17.10 Plot of ratio R against DIN N for both the ALWC cases and the trend from Figure 17.6, with extreme localized corrosion data from chain links added. The effect of increased exposure period from 'shorter term' to 'longer term' is shown as producing a reduction of R, consistent with the field data.

observed diameters compared with the original, but nominal diameters and these are not usually the actual localized diameters. For example, the corrosion losses for Indonesian chains at five years' exposure were mostly calculated from observed diameters and nominal diameters to be negative, clearly indicating that the original, nominal diameters were not sound bases for comparisons. The other aspects is that the change in diameter is unlikely to be a sufficient measure of overall corrosion since that would not include the, in some cases significant, loss of metal occasioned by the presence of localized corrosion and pitting, such as seen, for example, in Figure 17.1b. Overall, however, these aspects do not appear to offer an explanation for the very considerable differences seen for the 15-year Indonesia B and the 15-year West Africa data.

The most likely reason for the differences between the trend lines for different DIN concentrations and the concentrations of DIN recorded in Table 17.1 for the 15-year Indonesia B and the 15-year West Africa corrosion losses lies with the DIN values. Longer-term corrosion can be considered to be related to average values of DIN. This is important because it is known that coastal region DIN values can vary considerably throughout the year. For example, for temperate zones such as the North Sea the DIN values are much higher in winter compared with summer (OSPAR, 2000). This is owing to the variations in freshwater inflow and freshwater catchment fertilizer application between these seasons, with the winter months having the highest river waters inflow, corresponding to the higher DIN values in the offshore zones.

In the tropics, such as would apply to Indonesia and West Africa, a parallel scenario might be expected, with the wet season (April–October) supplying both more river flow into the coastal zone and also higher DIN, although information about the DIN aspects appears to be poor. The values recorded in Table 17.1 were obtained in the dry season, and thus are likely to be underestimates of DIN for the Indonesia and West Africa offshore sites. It follows that the average DIN values could be considerably higher, not inconsistent with the DIN values corresponding to the corrosion loss values in Figure 17.9. However, it is clear that this aspect requires further investigation. This is an area for further research.

Turning now to the severity of localized corrosion, it is clear from Figures 17.7(a–c) that pitting is very much involved in the development of the regions of localized corrosion. The mechanism for this has been described (Jeffrey and Melchers, 2007) and discussed previously (Melchers, 2018). In brief pit depth is constrained by potential limitations and eventually any further corrosion within a pit can occur only sideways, permitting, for pits in close proximity (which is not uncommon, see Figure 17.7b), the development of corrosion plateaus. On these, new pitting can form (Figure 17.7b) and the process can then repeat, a process that has been referred to as clustering of pit depths (Asadi and Melchers, 2018). It follows that the processes inside corrosion pits (Galvele, 1976) are relevant to the pits themselves but not to the overall localized corrosion cavity—for these overall conditions almost certainly will prevail, not unlike those for other areas of the corroding surface not affected by the localized corrosion scenario.

The most likely constraint on the depth of localized corrosion as measured by R is diffusion through the rust layers (Figure 17.8). Unlike the rusts that form over pits (the rust "cap") the rusts that form over the inner surfaces of the localized corrosion region are more likely to follow the contours of the depression (see Figure 17.8). Similar to the recognized concept that the buildup of rust layers in oxygen-rich environments limits inward diffusion of oxygen, they also will inhibit the inward diffusion of nitrates, nitrites, and ammonium, and will do so more effectively than inhibiting oxygen diffusion because of their larger molecular size, assuming potential differences are comparable. Interestingly, because of the small molecular size of hydrogen, the thickness and permeability of the rusts are unlikely to have any significant effect on outward hydrogen diffusion, and thus on the rate of the hydrogen evolution cathodic reaction relevant for longer-term exposures (Melchers, 2014b).

In the early stages of exposure both a and b are small and comparable, with a driven by the same considerations that cause pitting to be selective in location or by crevice corrosion initiation conditions (Wranglen, 1974; Butler et al., 1972). Once such differential corrosion has set in, the process is similar to that for other localized corrosion by differential aeration effects, except that there could be the contributing effect of microbiological metabolites contributing to lowering the pH inside the pits. Fundamentally this is no different to the processes involved in MIC, including its often localized nature, induced by local in-homogeneities in the steel or caused by deposits or localized microbiological, bacterial or algal marine growth (Little and Lee, 2007). For the present analysis the precise mechanisms involved in the initiation and the earlier stages are not of particular interest—what is important is the longer-term development of the resulting localized corrosion. This has now been shown to be, like MIC more generally, related directly to the DIN concentration in the seawater immediately to the corroding surfaces, although with an upper limit of about $R = 4–5$, as shown in Figure 17.10. The dependence on N is less clear in view of the comments relating to Figure 17.9 that the values for DIN in Table 17.1 may be underestimates for Indonesia and West Africa owing to the time of the year when the DIN in the seawater was measured. However, it should be clear that this does not affect the reasons for the limitation on R.

Figure 17.10 shows that the limit on R depends on the period of exposure, with a lower value for 15 years' exposure compared with 7 or 8 years' exposure. This effect can be attributed to the gradual build-up of rusts, with those at 15 years being thicker and likely more dense than those for shorter-term exposures. In the localized corrosion region the rate of buildup will be faster as, inside the pits, more metal is lost, and thus the rate of growth of the rusts in this region will be faster. As a result, inward diffusion of nutrients will be slowed. The difference will tend to level out with time, as a and b also come closer in value, with consequent reduction in $R \geq 1$. In addition, the thickness of the rust layer is controlled, in the classical model, by the balance between the inward buildup of rust layers utilizing the ferrous ions released at the anodic parts within pits and the external removal of the rusts to the external environment, through oxidation and perhaps erosion (Evans and Taylor, 1972). Erosion is less likely for depressed region of corrosion product, as can be seen for example already after a few years' exposure under immersion conditions (e.g., Figure 17.1a).

The cases considered herein are all for chain links exposed to near-surface immersion conditions, or in the splash zone. In both zones seawater soluble DIN can be available, even some distance from shore if environmental and hydrodynamic conditions permit the transportation of the necessary nutrients. Since particularly the oxidized form of DIN, i.e., nitrate, undergoes degradation with time, and is highly soluble in water, it is unlikely to reach great depths. At depth, the temperature is also lower, which suppresses the rate of corrosion relative to the warmer surface waters, and therefore the rate of electron flow to support MIC also is reduced. This suggests that MIC of steel chain at depth, sometimes suspected (Miller et al., 2012), is unlikely, unless the presence of nutrients, particularly DIN, can be verified at that location. At depth there is also the issue of the ultimate electron acceptor – in surface waters this is still O_2 but at depth the concentration of O_2 is almost always extremely low (Svedrup et al., 1942).

Finally, the possibilities for avoiding MIC in nutrient-polluted seawaters appear to be limited, largely through practical considerations. The most obvious is to select operational sites with low or negligible nutrient concentrations. But this is seldom an option. Impressed current cathodic protection (CP) would require the links to be electrically connected, either to each other or to a common point, and this is considered to be both expensive and infeasible under the usually exposed conditions to which most FPSOs are exposed. The present understanding is that direct electrical connectivity through the wearing surfaces between links is low to very low. For this reason, too, sacrificial anodes for CP would need to be applied to each link. This has the disadvantage of adding bulk to the chain links and hence increasing drag on the chains. The anodes also would need to be replaced at regular intervals and this would be difficult below the water surface. Of course, CP is effective only for surfaces continuously immersed—other surfaces would need protective coatings of some type. Coatings composed of, or incorporating, zinc or aluminium sometimes are used, but these have little wear resistance, important at the inter-link region, and also may cause hydrogen embrittlement issues. The use of biocides or micro-organisms to oppose the effects of MIC has obvious practical and environmental limitations in what is often a hostile offshore environment. Clearly the avoidance, amelioration, or elimination of MIC for offshore systems such as the mooring lines used for FPSOs and similar structures is an area ripe for innovative research.

17.6 CONCLUSION

The investigations described herein led to the following conclusions:

1. Field observations have shown that localized corrosion can be severe, reaching depths of 20 mm within 8 years and 30 mm in 15 years. The available evidence suggests that this highly localized corrosion is instigated by marine organisms, marine biofilm, and deposits.
2. The effect of DIN on general corrosion of chain links is generally consistent with that found earlier for steel coupons in seawaters with DIN concentrations less than about 0.5 mgN/l. The data indicate that the trend will continue for higher DIN values, provided these are average annual values.
3. The relatively greater depth of localized corrosion compared to more uniform corrosion increases with the concentration of DIN in the local seawater up to a relative ratio of depth of corrosion R of 4–5 for shorter-term exposures (say <8 years) and thereafter gradually becomes more comparable with uniform corrosion. This is largely independent of DIN concentrations greater than about 0.5 mgN/l, indicating that localized corrosion is more critical for earlier exposures.

ACKNOWLEDGMENTS

This chapter was presented in an earlier form at the 2019 Eurocorr conference in Sevilla, Spain. The authors acknowledge the funding and very considerable in-kind support provided by the industry partners in the SCORCH-Joint Industry Project (https://amog.consulting/joint-industry-projects-jips/scorch-jip). That project was managed, coordinated, and investigated by AMOG Consulting, Melbourne, Australia and Houston, TX, under the umbrella of the FPSO Research Forum (www.fpsoforum.com). The corrosion research group in the Centre for Infrastructure Performance and Reliability at The University of Newcastle provided both practical and theoretical input to the project. Both authors acknowledge the collegiality, incredible enthusiasm and driving force for all things marine and offshore of their friend and colleague, and CEO of AMOG, the late Dr. Andrew Potts.

REFERENCES

AMOG Consulting. (2014) Corrosion and wear of mooring chain links, final report. Melbourne, Australia:AMOG Consulting.

Asadi, Zohreh Soltani and Melchers, R.E. (2017) Pitting corrosion of older underground cast iron pipe. *Corros. Eng. Sci. Technol.*, 52(6): 459–469.

Asadi, Zohreh Soltani and Melchers, R.E. (2018) Clustering of corrosion pit depths for buried cast iron pipes. *Corros. Sci.*, 140: 92–98.

Butler G., Stretton, P. and Beynon, J.G. (1972) Initiation and growth of pits on high-purity iron and its alloys with chromium and copper in neutral chloride solutions. *Br. Corros. J.*, **7**(7): 168–173.

Evans, U.R. (1960) *The corrosion and oxidation of metals: scientific principles and practical applications*. London: Edward Arnold (Publishers).

Evans, U.R. and Taylor, C.A.J. (1972) Mechanism of atmospheric corrosion. *Corros. Sci.*, 12: 227–246.

Galvele, J.R. (1976) Transport processes and the mechanism of pitting of metals. *J. Electrochemical Soc.*, 123: 464–474.

Jeffrey, R. and Melchers, R.E. (2007) The changing topography of corroding mild steel surfaces in seawater. *Corros. Sci.*, 49: 2270–2288.

Lee, T.M., Melchers, R.E., Beech, I.B., Potts, A.E. and Kilner, A.A. (2015) Microbiologically influenced corrosion (MIC) of mooring systems: diagnostic techniques to improve mooring integrity. Proceedings. of the 20th Offshore Symposium, Feb. 2015, Houston, TX, Society of Naval Architects and Marine Engineers (SNAME).

Little, B.J. and Lee, J.S. (2007) Microbiologically influenced corrosion. Hoboken: Wiley.

Melchers, R.E. (2005) Effect of nutrient-based water pollution on the corrosion of mild steel in marine immersion conditions. *Corrosion*, 61: 237–245.

Melchers, R.E. (2013) Influence of dissolved inorganic nitrogen on accelerated low water corrosion of marine steel piling. *Corrosion*, 69(1): 95–103.

Melchers, R.E. (2014a) Long-term immersion corrosion of steels in seawaters with elevated nutrient concentration. *Corros. Sci.*, 81: 110–116.

Melchers, R.E. (2014b) Microbiological and abiotic processes in modelling longer-term marine corrosion of steel. *Bioelectrochemistry*, 97: 89–96.

Melchers, R.E. (2018) A review of trends for corrosion loss and pit depth in longer-term exposures. *Corros. Mater. Degrad.*, 1: 4. DOI:10.3390/cmd1010004.

Melchers, R.E. and Jeffrey, R. (2012) Corrosion of long vertical steel strips in the marine tidal zone and implications for ALWC. *Corros. Sci.*, 65: 26–36.

Melchers, R.E. and Jeffrey, R.J. (2013) Accelerated low water corrosion (ALWC) of steel piling in harbours. *Corros. Eng. Sci. Technol.*, 48(7): 496–505.

Miller, J., Warren, B. and Chabot, L. (2012) Microbiologically influenced corrosion of Gulf of Mexico mooring chain and 6,000 feet depths. Proceedings of the ASME Intl. Conf. on Ocean, Offshore and Arctic Engineering (OMAE), Rio de Janeiro, Brazil, paper 84067.

Odom, J.M. (1993) Industrial and environmental activities of sulfate-reducing bacteria. J.M. Odom and Rivers Singleton Jr. (Eds.). *The sulfate-reducing bacteria: contemporary perspectives*. Berlin: Springer Verlag, 189–210.

OSPAR. (2000) Regional quality status report II for the Greater North Sea. OSPAR Commission for the Protection of the Marine Environment of the North-East Atlantic. http://www.ospar.org/eng/html/qsr2000, posted on 29 June 2004.

Sharland, S.M. and Tasker, P.W. (1988) A mathematical model of crevice and pitting corrosion - I. The physical model. *Corros. Sci.*, 28(6): 603–620.

Svedrup, H.U., Johnson, M.W. and Fleming, R.H. (1942) The oceans: their physics, chemistry and general biology. New York: McGraw-Hill.

Wranglen, G. (1974) Pitting and sulphide inclusions in steel. *Corros. Sci.*, 14: 331–349.

18 MIC in the Fire Water Sprinkler System at St. Olavs Hospital, Trondheim, Norway

H. Parow and R. Johnsen
Norwegian University of Science and Technology

Torben Lund Skovhus
VIA University College

CONTENTS

DOI: 10.1201/9780429355479-21

18.1 INTRODUCTION

For water-based fire protection systems (FPS) made from carbon steel, clogging and leaks caused by corrosion are two of the main problems for maintenance and operation. Repair of corrosion related damage leading to mechanical failure causes a significant increase in life cycle costs of a fire sprinkler system (Su and Fuller, 2014). St. Olavs Hospital uses drinking water in the FPS, and the inlet is taken from the water utility services operated by the Trondheim municipality. No water treatment is indicated in the system, which implies that the water is not chemically treated while in operation in the FPS, and the water is stagnant until a fire occurs. The pipework is mainly made of carbon steel, although some parts have been replaced by AISI 304/AISI 316 due to corrosion (Parow, 2018). During an inspection in the fall of 2017 the pipes in the 1902 building and Women's and Children's Center were removed, visually inspected, and replaced due to suspicion of internal corrosion. Upon closer inspection, the pipes showed severe internal corrosion attacks. Through visual inspection large quantities of corrosion product and tubercles were discovered on the pipe surface, which indicated that corrosion caused by bacterial activity (MIC – Microbiologically Influenced Corrosion) was a contributing factor to the corrosion attacks.

Detailed experimental work was conducted, with focus on the determination of MIC. Microbial Community Analysis (MCA) of the corrosion product from the pipes was conducted to determine the total number of bacteria, as well as different types of bacteria present in the system. Water samples were collected from the public drinking water supply in Trondheim and from the fire sprinkler system, which had been stagnant without replacement for over a period of five years, to determine its chemical parameters (corrosivity) and bacterial content.

18.2 CORROSION IN FIRE SPRINKLER SYSTEMS

18.2.1 THE FIRE SPRINKLER SYSTEM AT ST. OLAVS HOSPITAL

Automatic fire protection systems consist of water supply, pipe network, valves, and heat-activated fire sprinklers. Wet sprinkler systems, where the water is held constantly in the pipe network under pressure, is normally used. If the sprinklers are activated, the water is immediately spread onto the fire. The FPS consists of a network of pipes gradually decreasing in diameter from the water supply out to each sprinkler head, see Figure 18.1. Check valves isolate the water flow from the water supply into the fire system, regulating the water flow in the sprinkler system. The check valve closes when the pressure on the sprinklers exceeds the water supply pressure. If the pressure equalizes or falls below the water supply pressure, the valve opens, enabling the water to flow into the FPS. The sprinkler head controls the water supply when a fire occurs. Each individual sprinkler head activates when exposed to a specific temperature over a defined period (Cote, 2008).

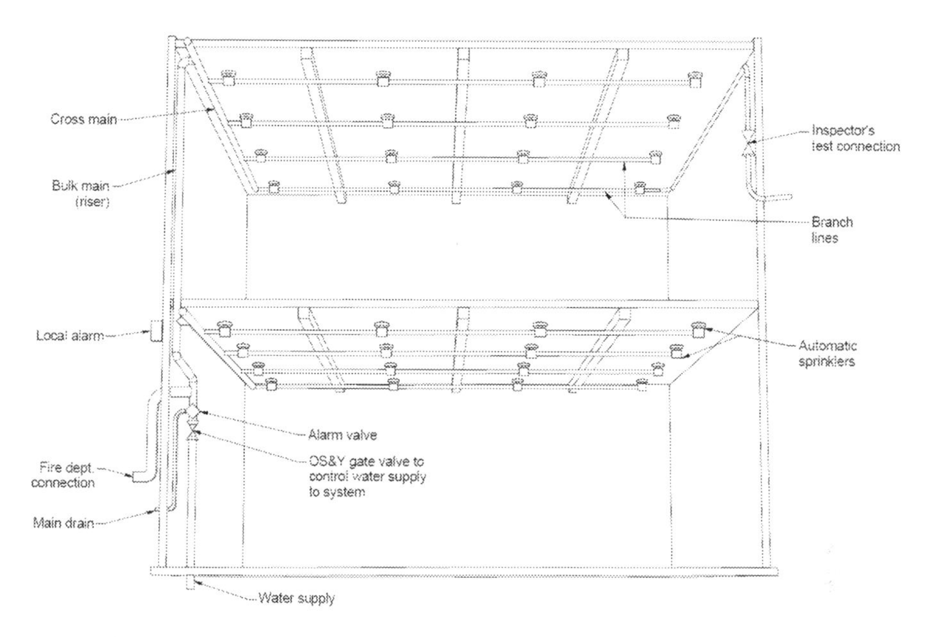

FIGURE 18.1 Overview of a standard fire protection system.

The fire protection system at St. Olavs Hospital is connected through a large underground pipe network, and the network extends to each section and building of the hospital. The drinking water supply comes from Vikelvdalen water treatment facility. At the facility, CO_2 gas is added to the water, and the water is filtered through a limestone filter of 1–2.5 mm pore size to prevent gravel and sediments from entering the water pipeline network. The water is then chlorinated and put through UV radiation before being sent out to consumers. This treatment process leads to increased calcium content, alkalinity and pH value to prevent corrosion in the pipe network (Tronhus, 2020). The water is untreated in the FPS and is stagnant until a fire occurs. The pipes are commonly placed in ventilated and heated rooms, where the pressure is held constant at around 7–8 bar, and the temperature varies from 20 to 30°C. The sprinkler heads will activate in the affected area of the hospital and open for water flow when the temperature reaches 67°C. Due to the size of the hospital, and consequently the FPS, service, and maintenance of the system happens at different times and in different parts of the system. Therefore, no data was available on how often inspections were conducted on the filter system, water quality, valves and corrosion damages in the system. Subsequently, there were no information about how often new fresh water, oxygen and air was supplied to the system. However, it is assumed that this happened relatively infrequently (Parow, 2018).

Inlet pipes from three different parts of the hospital have been replaced due to corrosion. The pipes varied in size, length, and number of years in operation before being replaced, see Table 18.1 and Figures 18.2–18.4. These three pipe spools were the subject for the experimental work in determining the main cause of the corrosion problems.

TABLE 18.1
Overview over Retrieved Pipe Spools Examined to Document Corrosion and MIC in the FPS

Location	Type	Dimension		Years in operation
		Length (m)	Diameter (mm)	
Knowledge Center	Pipe	24	168.3	5
Women's and Children's Center	U-bend	1.5	168.3	10
1902 building	Pipe	2	114.3	8

FIGURE 18.2 Inlet pipe in DN150 standard from the Knowledge Center. The bottom pipe was replaced April 2018 due to corrosion. The pipe had been in operation for five years.

FIGURE 18.3 Inlet pipe in DN150 standard from the Women's and Children's Center. The pipe was a 1.5-meter U-bend and was replaced in 2017 due to corrosion. The pipe had been in operation for 10 years.

18.2.2 Corrosion in FPS

Corrosion occurs in all sprinkler systems made from carbon steel and is one of the main problems for maintenance and operation of a FPS. Corrosion products and mineralization occurrences can limit the flow of water through the system and impair the mechanical integrity of the firefighting equipment. Corrosion can lead to blockage of pipes and sprinklers, as well as leakage. Corrosion in wet sprinkler

FIGURE 18.4 Inlet pipe in DN100 standard from the 1902 building. The bottom pipe was 2 meters long and was replaced in 2017. The pipe had been in operation for eight years.

systems occurs as a result of several influencing factors; trapped air, corrosive water chemistry, periodic access to oxygenated water, stagnant conditions, and microbiological activity (Su and Fuller, 2014).

The presence of oxygen is the primary cause of corrosion-related problems in a wet sprinkler system. Oxygen trapped in air pockets at the top of the pipe is usually the cause of the corrosion, and the corrosion will be most aggressive in the air/water interface. Periodic testing and replenishment of drinking water leads to a supply of fresh oxygen which accelerates the corrosion process. The cumulative effect of frequent exposure to fresh oxygen is the formation of oxygen-depleted areas under the deposit of corrosion product, which can result in formation of pits and, ultimately, leaks (Kochelek, 2009).

MIC is a localized form of corrosion which is caused by microorganisms. Microorganisms are present in all water sources and fire protection systems (Kochelek, 2009). The activity of microorganisms and corrosion in carbon steel pipe systems can lead to the formation of tubercles and rust mounds on the pipe surface, as shown in Figure 18.5. The consequence may be obstruction of the water flow and blockage of valves when the fire-water system is activated (Little et al. 2011; Su and Fuller, 2014).

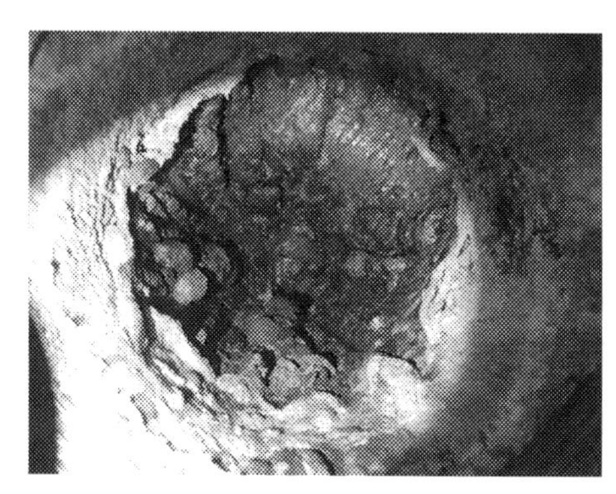

FIGURE 18.5 Example of tubercles and rust mounds in a carbon steel pipe from the Women's and Children's Center at St. Olavs Hospital.

18.3 EXPERIMENTAL WORK

The determination of the corrosion mechanism was conducted through experimental work, involving water analysis, Microbial Community Analysis (MCA), visual inspection of corrosion damages, as well as Infinite Focus Microscopy (IFM) for topographical imaging.

18.3.1 Sample Preparation and Visual Inspection

The pipes were cut in 50 mm long sections and split, as shown in Figure 18.6. The sections were visually inspected, and maximum height of the rust mounds on the pipe surface were measured using a caliper. An odor test was performed by applying 10% diluted hydrochloric acid to the corrosion product to stimulate the formation of H_2S if FeS was present.

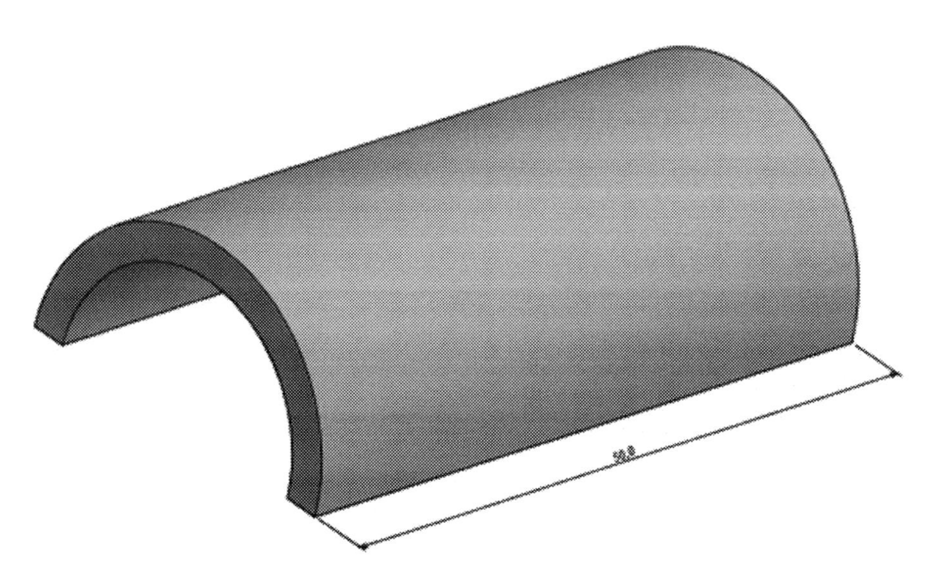

FIGURE 18.6 Sections of pipe cut to 50 mm width.

18.3.2 Preparation of Samples for Infinite Focus Microscopy (IFM)

A 10 × 10 mm piece of each pipe was etched in an ultrasound bath to remove any corrosion products from the surface. The etchant consisted of 20% hydrochloric acid, 80% distilled water, and 1.75 grams of hexamethylenetetramine mixed in a 500 ml container. The etching was conducted in three rounds of 20 minutes in the ultrasonic bath, and the samples were washed with ethanol and dried between each round.

18.3.3 Water Analysis

Water from the fire system was sampled when the pipe from the Knowledge Center was replaced in April 2018, shown in Figure 18.7. Water samples were obtained from the pipe, where the water had been stagnant for five years, and from the drinking water supply in Trondheim, at the inlet of the FPS. The water samples were analyzed for bacterial content and water chemistry. Water samples were

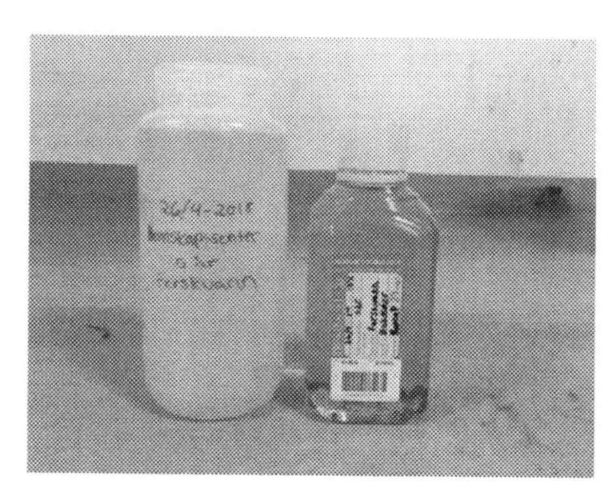

FIGURE 18.7 Container to the left for mineralogical analyses. Container to the right for bacterial content.

analyzed for number of coliform bacteria, *E. coli,* and intestinal enterococci, in addition to Heterotrophic Plate Counts at 22°C (HPC-22). The HPC-22 is a measure for all culturable bacteria and microorganisms in the water sample grown at 22°C (Eurofins, 2018). Coliform bacteria, *E. coli,* and intestinal enterococci come from pollution and fecal contamination, and according to the drinking water regulations, the maximum allowed concentration is 0 cfu per 100 ml. The limit value for HPC-22 is 100 cfu per ml (Helse- og omsorgsdepartementet HOD, 2016).

The analysis for the water's chemical parameters tested for dry solutes, turbidity, conductivity, alkalinity, hardness, pH, color number, and concentration of chlorides, sulfates, magnesium, calcium, nitrates, and iron. High content of magnesium and calcium cause hard water, and the hardness should be less than 3.5 dH° (Eurofins, 2018). The presence of chlorides and sulfates affect the corrosivity of the water, and the content should not exceed 200 mg/l and 100 mg/l, respectively. A high content of iron can indicate ongoing corrosion of the pipe material, and the value should be less than 200 µg/l. pH indicates the acidity/alkalinity of the water and should be between 6.5 and 9.5. High conductivity of the water can indicate a high content of dissolved minerals and metals, and the value should be below 250 mS/m. Turbidity is a measure of the amount of suspended particles in the water (cloudiness), and can indicate a high iron content. The turbidity should be lower than 4 NTU (Nephelometric Turbidity Units). The color unit represents the color of the water, where a high value can indicate a high iron content or humic acids. The color number should be less than 20. Nitrates present in high concentrations can indicate contamination and should not exceed 10 mg/l (Eurofins, 2018). Alkalinity is a measure of the concentration of the salts of carboxylic acids (CO_3^{2-}, HCO_3^-). The measure of dry solutes indicates the content of dissolved solutes in the water and the value should not exceed 500 mg/l (Eurofins, 2018; Helse- og omsorgsdepartementet, 2016; Folkehelseinstituttet, 2016).

18.3.4 Analysis of Biological Samples

When the pipe was replaced from the Knowledge Center in April 2018, test samples were collected from the corrosion product and tubercles located inside the pipe. The samples were collected using a scalpel which was washed in pure alcohol between each sampling. Figure 18.8 shows where the samples were collected from opposites ends of the pipe. A total of eight samples, four from the bottom of the pipe and four from the top of the pipe, were analyzed.

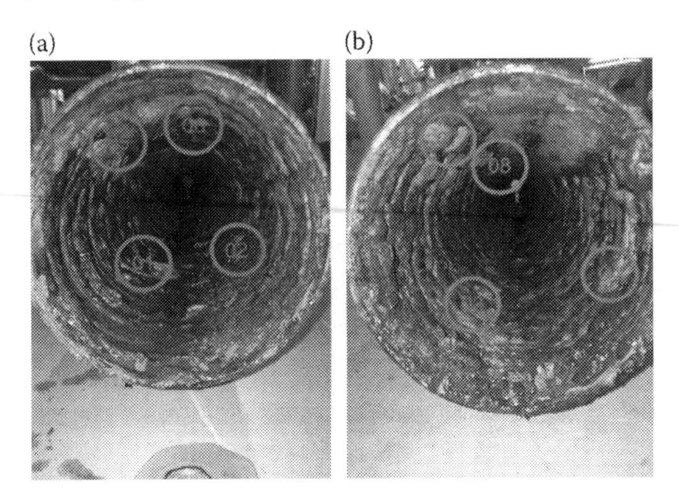

FIGURE 18.8 Sampling points for collecting corrosion product from the replaced pipe in the Knowledge Center.

The sample holders consisted of 50 ml tubes which were filled with 5–10 ml of corrosion product, shown in Figure 18.9. The samples were then placed in a cooler with dry ice so that the temperature would stay below −20°C.

FIGURE 18.9 Sample holders consisting of 50 ml tubes filled with 5–10 ml of corrosion product each.

Microbial Community Analysis (MCA) was conducted to measure microbial activity and abundance. Molecular Microbiological Methods (MMM) such as qPCR and Amplicon Sequencing of the 16S rRNA gene were applied. Amplicon sequencing is a semi quantitative molecular technique based on the 16S rRNA gene present in the bacteria. The nucleotide sequence of the 16S rRNA gene is unique to each bacterial species, and acts as a "fingerprint" for the bacteria. The method gives a list of the most abundant species/genera of bacteria in the sample, quoted as relative abundance (Skovhus et al., 2017b). In addition to 16S rRNA amplicon sequencing, qPCR for the total number of bacteria was performed, which is a quantitative method also based on the 16S rRNA gene (Parow, 2018). Finally, qPCR-SRB for the total number of sulfate-reducing bacteria, which is a quantitative method based on the DSR gene, unique for SRB, was performed (Parow, 2018). qPCR measures both viable and inactive cells in the sample, which must be taken into account during data interpretation (Cangelosi and Meschke, 2014).

18.4 RESULTS

18.4.1 Visual Inspection

18.4.1.1 1902 Building

The samples from the 1902 building contained several rust mounds in red and orange color. The largest tubercle was measured to be 7.18 mm in maximum height, as shown in Figure 18.10. After testing with diluted hydrochloric acid, no significant smell of H_2S was detected.

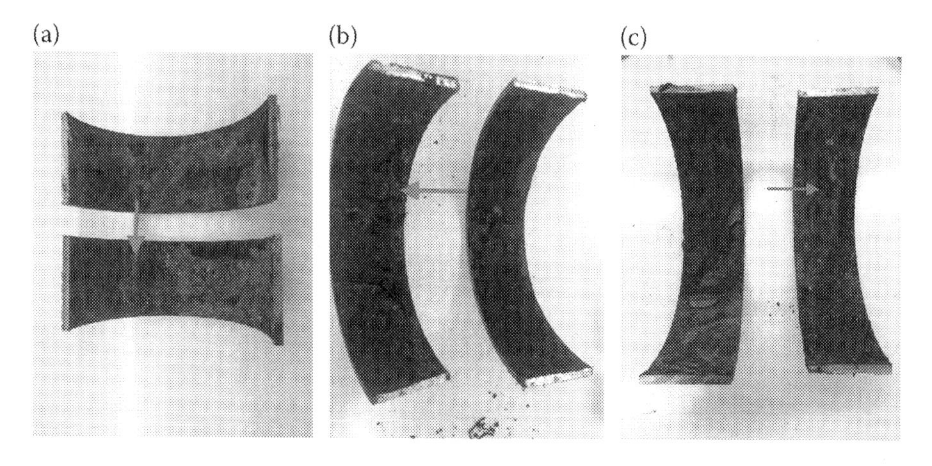

FIGURE 18.10 Metal samples in carbon steel from the pipe in the 1902 building, the Women's and Children's Center, and the Knowledge Center. The arrows indicate the highest measured tubercle in the pipes.

All the samples had a strong odor of rust and metal. The odor combined with many deposits appearing orange/red in color suggested a high amount of dissolved metal (iron ions) on the pipe surface.

18.4.1.2 Women's and Children's Center

The samples from the Women's and Children's Center contained rust mounds and deposits ranging from dark orange to almost black in color. The largest tubercle was measured to be 11,20 mm in maximum height, as shown in Figure 18.10. After testing with diluted hydrochloric acid, a significant smell of H_2S (rotten eggs) was detected.

18.4.1.3 Knowledge Center

The samples from the Knowledge Center contained several rust mounds and deposits in bright and dark orange colors. The largest tubercle was measured to be 10.13 mm in maximum height, as shown in Figure 18.10. After testing with diluted hydrochloric acid, a significant smell of H_2S was detected.

18.4.2 Infinite Focus Microscopy (IFM)

18.4.2.1 1902 Building

The result from the IFM measurement performed on the sample from the 1902 building is shown in Figure 18.11. The darkest area in the figure to the right represents the deepest point in the sample.

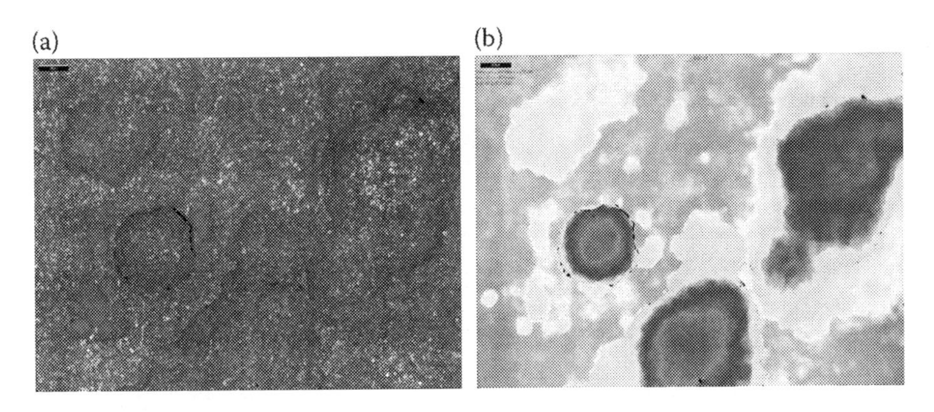

FIGURE 18.11 Topography imaging of the sample from the 1902 building.

A depth profile was measured, as shown in Figure 18.12. The deepest pit on the sample from the 1902 building was measured to be 306 μm deep.

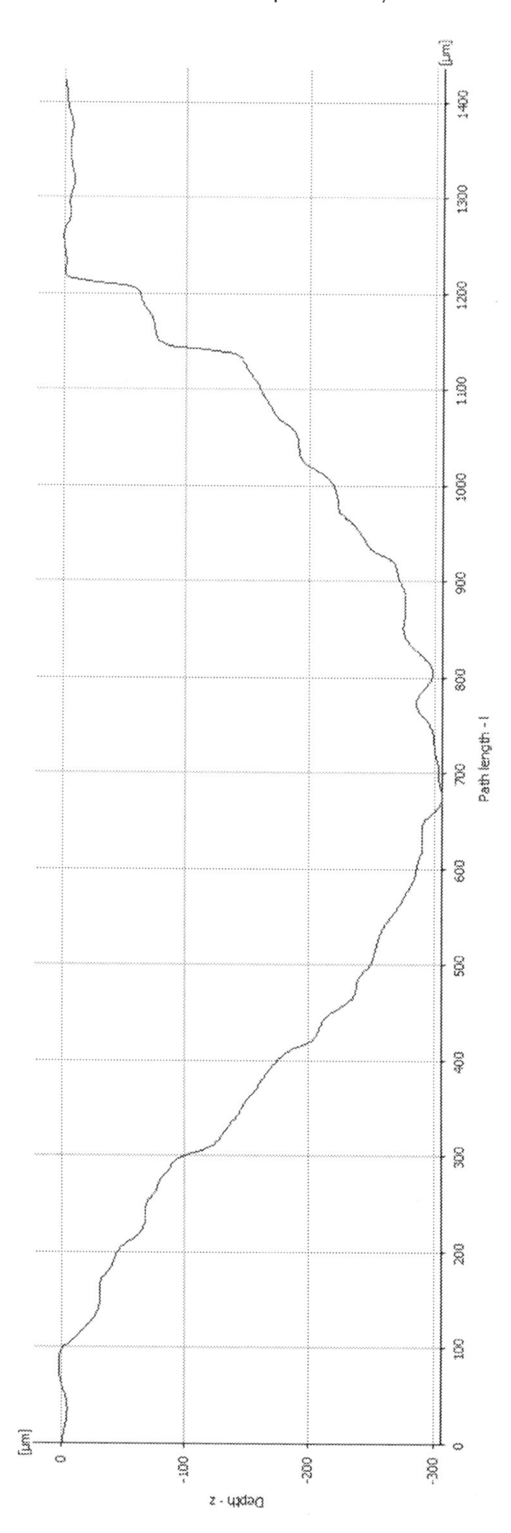

FIGURE 18.12 Depth profile measurement of the deepest pit in the sample from the 1902 building.

18.4.2.2 Women's and Children's Center

The result from the IFM measurement performed on the sample from the Women's and Children's Center is shown in Figure 18.13. The darkest area in the figure to the right represents the deepest point in the sample.

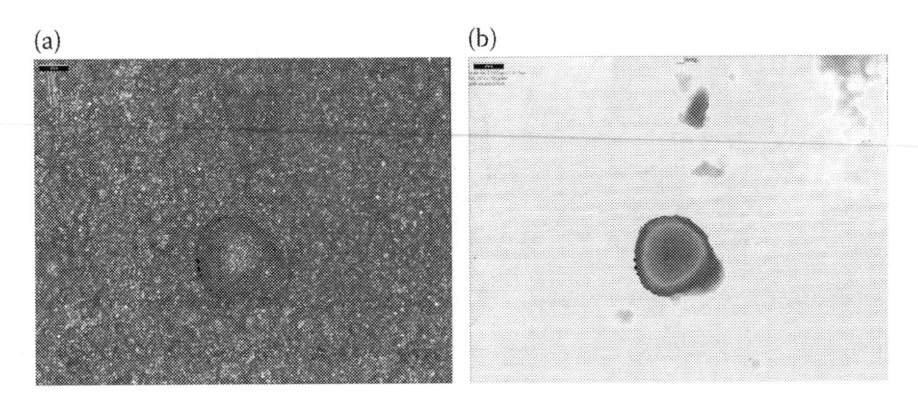

FIGURE 18.13 Topography imaging of the sample from the Women's and Children's Center.

A depth profile was measured, as shown in Figure 18.14. The deepest pit on the sample from the Knowledge Center was measured to be 363 μm deep.

18.4.2.3 Knowledge Center

The result from the IFM measurement performed on the sample from the Knowledge Center is shown in Figure 18.15. The darkest area in the figure to the right represents the deepest point in the sample.

A depth profile was measured, as shown in Figure 18.16. The deepest pit on the sample from the Knowledge Center was measured to be 318 μm deep.

The results from the three parts of the hospital show that the pipes have been subjected to localized corrosion in the form of pits. The most severe pits were discovered in the sample from the Women's and Children's center, which had been in operation for 10 years.

18.4.3 Water Analysis

Bacterial and mineralogical analysis were carried out on the water from the fire protection system at St. Olavs Hospital, as well as on the water from the municipal drinking water supply in Trondheim. Table 18.2 show the results from the bacterial analysis. The table shows no manifestation of intestinal bacteria (coliform bacteria, *E. coli,* and intestinal enterococci), which should be expected of samples from the drinking water supply. The results for HPC-22 showed a value of 550 cfu/ml for the water from the Knowledge Center that had been stagnant for five

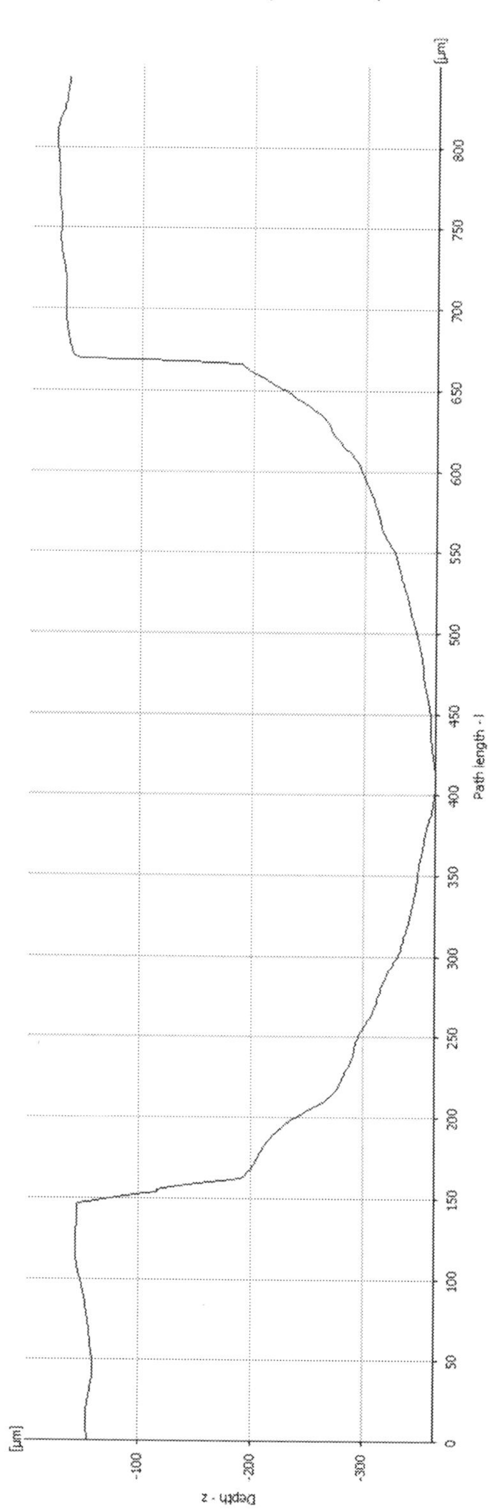

FIGURE 18.14 Depth profile measurement of the deepest pit in the sample from the Women's and Children's Center.

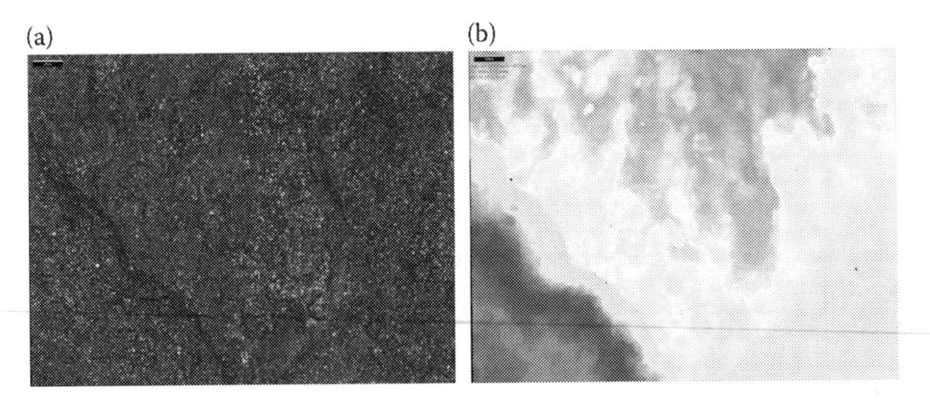

FIGURE 18.15 Topography imaging of the sample from the Knowledge Center.

years, and a value of 1,200 cfu/ml for the water from the drinking water supply, which is a 12× higher value than what is allowed for the drinking water in Trondheim (100 cfu/ml).

Table 18.3 show the results from the chemical and mineralogical analysis of the water samples from the hospital and from the municipal network in Trondheim. The table shows that the result for dry solutes, pH, conductivity, color number, and hardness are within the acceptable values as described in chapter 18.3.3. In addition, the values for chlorides, sulfates, and nitrates are low, which suggests low corrosivity and contamination of the water. Turbidity and iron content are shown to have a higher value than the permissible limit value (4 NTU and 200 µg/l, respectively), and give an indication that the water contains iron ions due to corrosion of the pipe material.

18.4.4 Bacterial Analysis of Corrosion Product

Figure 18.17 shows the results from the qPCR for the total number of bacteria in the corrosion product from the pipe in the Knowledge Center. Sample numbers 1–4 were taken from the bottom of the pipe, and samples 5–8 were taken from the top of the pipe. The figure shows a higher number of bacteria in the samples from the bottom of the pipe compared to the top. This corresponds to the larger amounts of tubercles and deposits in the bottom of the pipe as compared to the top. The number of bacteria registered in the samples is high (between 2.2×10^6 and 3.7×10^8 cells/ml) (Parow, 2018).

Figure 18.18 shows the total number of SRB present in the corrosion product. qPCR SRB was carried out on selected samples from the bottom and top of the pipe at each pipe end, shown as sample numbers 1, 3, 6, and 8 in Figure 18.18. The figure shows that there is a significant number of SRB (> 10^5 cells/ml) even though the SRB only constitutes a small portion of the total number of bacteria. %SRB per total number of bacteria constitutes 0.3%, 0.2%, 0.8%, and 0.4% for samples 1, 3, 6, and 8, respectively. This shows that sulfate-reducing bacteria only constitutes a

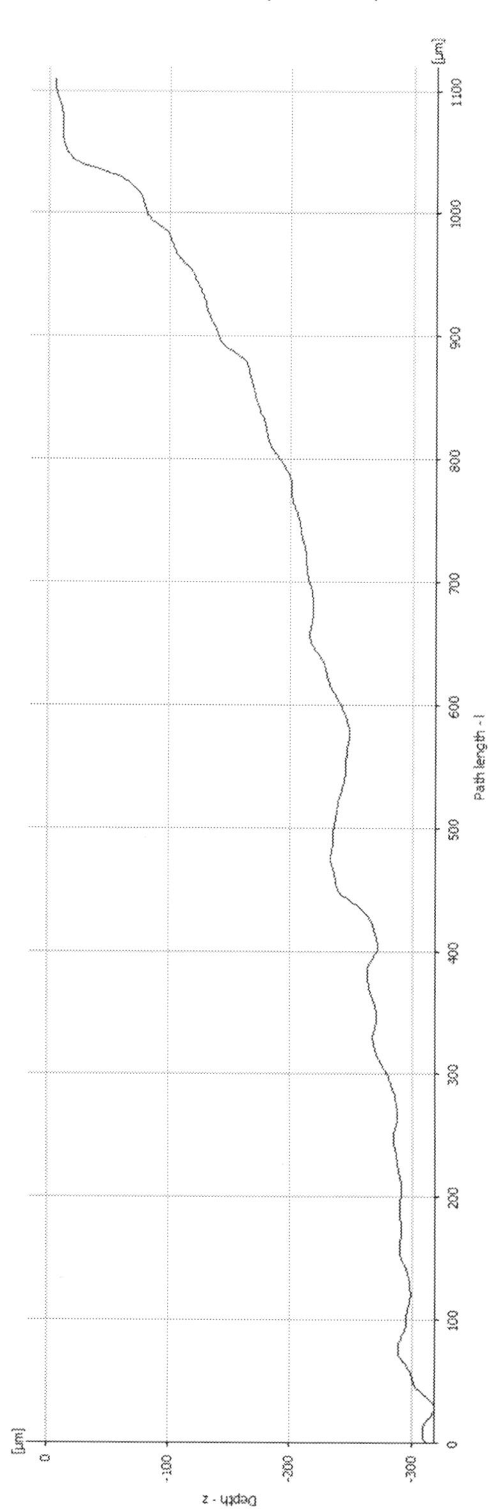

FIGURE 18.16 Depth profile measurement of the deepest pit in the sample from the Knowledge Center.

TABLE 18.2
Bacterial Analysis of the Water from the FPS and Drinking Water from the Municipal Network in Trondheim

Parameter	Result for Water from the FPS (Cfu/ml)	Result for Water Supply (Cfu/ml)	Method
HPC-22	550	1200	M103
Coliform bacteria	0	0	NS-EN ISO 9308-2
E. coli	0	0	NS-EN ISO 9308-2
Intestinal enterococci	0	0	NS-EN ISO 7899-2

TABLE 18.3
Mineralogical Analysis of the Water from the FPS and Drinking Water from the Municipal Network in Trondheim

Parameter	Unit	Result for Water from the FPS	Result for Drinking Water	Method
Dry Solutes	mg/l	120	138	NS 4764
pH	–	8.0	7.9	NS-EN ISO 10523
Turbidity	NTU	31	65	NS-EN ISO 7027
Conductivity	mS/m	12.9	13.3	NS ISO 7888
Alkalinity	mmol/l	1.0	1.0	Internal method
Color Number	–	14	15	ISO 7887:2011
Chlorides	mg/l	7.26	7.26	NS-EN ISO 10304-1
Sulfates	mg/l	2.39	2.62	NS-EN ISO 10304-1
Nitrates	μg/l	250	360	NS-EN ISO 13395
Calcium	mg/l	20.0	21.8	NS-EN ISO 17294-2
Magnesium	mg/l	0.90	1.09	NS-EN ISO 17294-2
Iron	μg/l	11000	7060	Internal method
Hardness	dH°	3.0	3.3	German hardness degrees

fraction of the microbial environment in the corrosion product and indicates that SRB are not dominant in the biofilm.

A microbial community analysis was performed to identify the different types of bacteria (diversity) present in the corrosion product from the pipe in the Knowledge Center. The analysis is based on amplicon sequencing of the 16S rRNA gene present in all bacteria. Table 18.4 shows a list of observed bacteria in decreasing order by prevalence, and emphasis is placed on bacteria that have a known link to MIC. The table shows that there are multiple MIC-related bacteria present, which

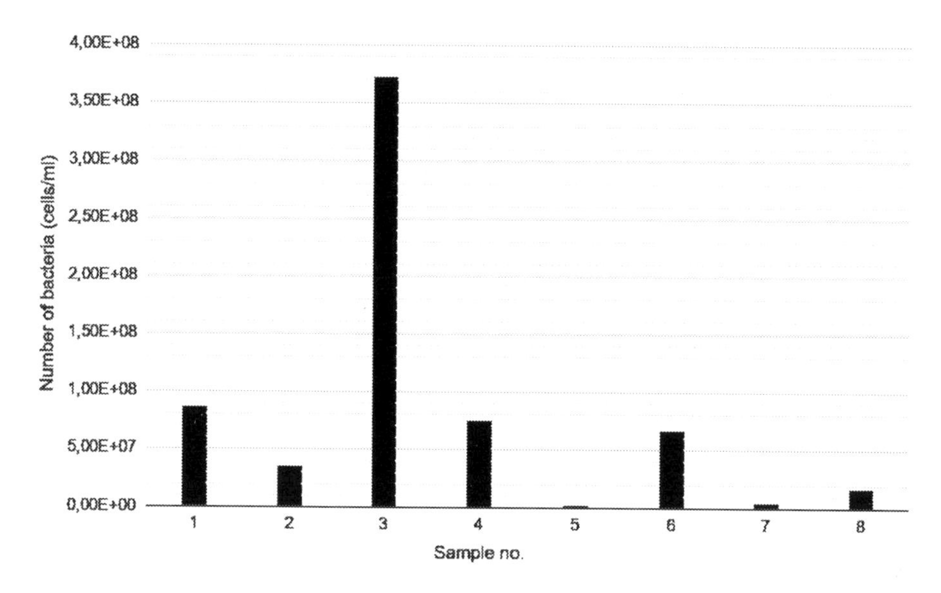

FIGURE 18.17 Total number of bacteria in the corrosion product for samples 1–8 measured by qPCR for total bacteria.

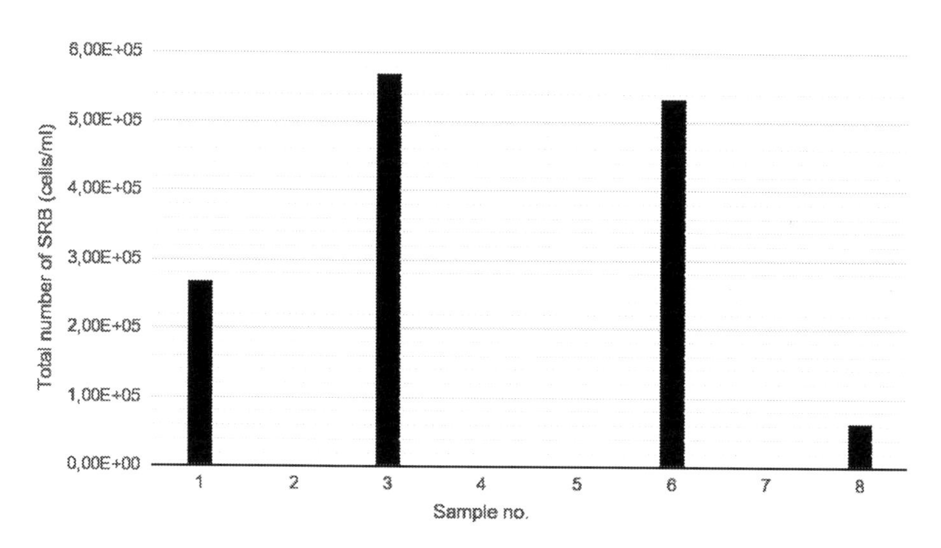

FIGURE 18.18 Total number of SRB in the corrosion product for samples 1, 3, 6, and 8 measured by DSR-qPCR. Samples 2, 4, 5, and 7 were not analyzed.

produce various products through their metabolic processes. Both iron-reducing bacteria (IRB) and iron-oxidizing bacteria (IOB) are present, which produce different oxidized forms of iron and can sustain each other. IOB is also known to contribute to tubercle formation (Ray et al., 2010). Both sulfate-reducing bacteria (SRB) and sulfate-oxidizing bacteria (SOB) were observed, which contribute to

TABLE 18.4
Microbial Community Analysis of the Corrosion Product from the Pipe in the Knowledge Center, in Decreasing Order of Prevalence

Bacteria	Grouping	Metabolic Function	Environment	Relevance to MIC	References
Bradyrhizobiacea sp.	Aerobe bacteria	–	Freshwater	Biofilm formation	(Chao et al., 2015)
Desulfovibrio sp.	SRB (anaerobe)	Reduce sulfate to H_2S	Freshwater	Cathodic depolarization	(Enning et al., 2013; Kakooei et al., 2012)
Geothrix sp.	IRB (anaerobe)	Reduce Fe^{3+} to Fe^{2+}	Freshwater	Reduce iron at the metal surface	(Qin et al., 2017)
Pedomicrobium sp.	Aerobe bacteria	Oxidize Mn and Fe	Freshwater	Accumulation of metal oxides in the biofilm	(Lührig et al., 2015; Braun et al., 2009)
Hyphomicrobium sp.	SOB (aerobe)	Oxidize sulfide	Wastewater	Reduce pH locally	(Skovhus et al., 2017a; Huber et al., 2016)
Sulfuricella sp.	SOB (aerobe)	Oxidize sulfide	Wastewater	Reduce pH locally	(Skovhus et al., 2017a; Dong et al., 2017)
Nitrospira sp.	NOB (aerobe)	Oxidize nitrite	Freshwater	Contributes to acidification	(Ehrich et al., 1995; Nardy et al., 2017)
Ferriphaselus sp.	IOB (aerobe)	Oxidize Fe^{2+} to Fe^{3+}	Freshwater	Tubercle formation	(Ray et al., 2010; Barco et al., 2015)

acidification and an aggressive environment at the metal surface (Enning and Garrelfs, 2013; Skovhus et al., 2017a). *Bradyrhizobiaceae* was found in the highest concentration. It is an aerobe bacterium that likely contributed to biofilm formation. Aerobe bacteria like *Bradyrhizobiaceae* and *Pedomicrobium* contribute to consumption of oxygen, which creates a favorable environment for anaerobe bacteria such as *Desulfovibrio* and *Geothrix*.

18.5 DISCUSSION

18.5.1 Initial Inspection

The visual inspection of the pipes showed that voluminous tubercles and deposits were present on the pipe surfaces. The pipe from the Women's and Children's Center contained the largest deposits in terms of measured height. A smell test was performed using diluted hydrochloric acid to stimulate the formation of H_2S, should FeS be present in the corrosion product. The result from the smell test indicated that sulfur was present in the pipe from the Knowledge Center and Women's and Children's Center. The presence of FeS suggests formation of sulfide, S^{2-}, due to the presence of sulfate-reducing bacteria (Kakooei et al., 2012). The pipe from the Women's and Children's Center was a dead-end U-bend, and here biological activity may have flourished (especially anaerobe bacteria) due to the water not being replaced for 10 years. Stagnant conditions and dead ends in pipe systems can give rise to an environment favorable to microbial activity (Su and Fuller, 2014). The results from the IFM indicate local corrosion in the form of pits. The formation of pits can be caused by frequent exposure to oxygen, which cause oxygen depleted areas under the corrosion product (Kochelek, 2009). Due to how infrequently the water in the fire system is replaced, this is not considered a likely cause of pitting. Fresh supply of oxygen should not occur. The formation of pits may also be caused by formation of biofilms and biological activity on the metal surface (Skovhus et al., 2017a).

18.5.2 Comparison of Water Analysis and Bacterial Analysis

The results from the bacterial analysis of the water show that the water from the public drinking water supply has a higher number of bacteria (HPC-22 = 1,200 cfu/ml) than the water from the fire water system (HPC-22 = 550 cfu/ml) and gives an indication that the water source is not sufficiently protected from contamination (Eurofins, 2018). Tronhus (2020) at Trondheim municipality suggests that the high value for heterotrophic plate counts is due to the water supply having little to no circulation over time. This leads to an increase in HPC-22. A reduction in the HPC-22 in the FPS gives an indication that the bacteria and microorganisms have attached to the pipe surface rather than floating freely in the water. This indicates formation of biofilms and potential MIC in the pipes.

The results from the chemical analysis of the water show that the water taken from the hospital does not contain considerable amounts of Cl^- and SO_4^{2-}, so the water can be classed as non-corrosive (Eurofins, 2018). Since the water does not contain oxygen and is non-corrosive, the documented tubercle formations and deposits are most likely caused by MIC (Su and Fuller, 2014). The results also give an indication that the local corrosion (pits) is caused by initiation and acceleration as a result of bacterial activity, as supposed to pitting caused by chlorides (Bardal, 1994). Table 18.3 shows that the value for turbidity and iron is above the acceptable limit (4 NTU and 200 μg/l, respectively) (Helse- og omsorgsdepartementet, 2016). This give an indication that the water contains a greater amount of iron, which can be caused by corroded pipe material. According to Clarke and Aguilera (2007) the pH inside a tubercle can fall to 2–3 as a result of bacterial activity, where acids are excreted as a biproduct from the bacterial metabolism. The water analysis showed a pH of 8. However, the water inside the cavities of tubercles and other deposits was not analyzed, and the water may be more corrosive than the results shows. Neutral pH and temperature ranging from 20 to 40°C often provide optimal growth of bacteria, which would include a pH of 8 and temperature in the pipes ranging from 20 to 30°C (Borenstein, 1994).

The results from the qPCR for the total number of bacteria shows that the number of bacteria is the highest at the bottom of the pipe, which correlates to the larger amounts of deposits and tubercles at the bottom of the pipe as compared to the top. The measured total number of bacteria is considered to be high (10^6–10^8 cells/ml) and gives an indication that bacterial activity has influenced the corrosion and tubercle formation in the pipe (Parow 2018). The results from the qPCR SRB show that the proportion of SRB only constitutes a small fraction of the total number of bacteria. However, the amount of SRB is still high (> 10^5 cells/ml), and gives an indication that although SRB is not dominant in the biofilm, it still contributes to MIC.

Bacterial community analysis conducted on the corrosion products taken from the pipe at the Knowledge Center show that there are different types of MIC-related bacteria present in the fire protection system. According to Pope and Pope (2000), MIC occurs most aggressively when the biofilm contains different types of bacteria, which gives an indication that the local corrosion in the pipes is caused by bacterial activity. From Table 18.4, bacteria living in symbiosis with each other can be observed; for example IRB and IOB, which derives their energy from various oxidized forms of iron and help each other to grow (Barco et al., 2015; Qin et al., 2017). The activity of bacteria such as *Geothrix, Ferriphaselus,* and *Pedomicrobium* might have contributed to the high levels off turbidity and iron observed (Table 18.3). The results from the community analysis supports earlier claims that the tubercle formations in the pipes is due to bacterial activity, where iron-oxidizing bacteria utilize Fe^{2+} from the metal surface and excrete Fe^{3+} which forms a hard shell that protects other bacteria (Emerson, 2019; Stoecker, 2001). It can also be seen from Table 18.4 that sulfate-reducing bacteria (SRB) occur at the second-highest frequency in the corrosion product. SRB such as *Desulfovibrio* are often connected to MIC due to their ability to reduce sulfate to sulfide, which can lead to cathodic depolarization and an aggressive environment at the metal surface (Kakooei et al., 2012). Sulfate-oxidizing bacteria (SOB) are also present, *Sulfuricella*, which oxidize sulfide to sulfate and can lead to local lowering of pH at the metal surface (Skovhus et al., 2017a). Based on previous observations and the bacterial analysis, MIC is considered the most likely cause for the severe corrosion attacks detected in the fire protection system at St. Olavs Hospital.

18.6 CONCLUSION

The conclusion is summarized in Table 18.5. MIC is considered the most likely cause for the corrosion attacks discovered in the pipes at St. Olavs hospital, and the table shows all different factors that contributed to this conclusion.

TABLE 18.5
Factors Contributing to the Determination of MIC

Chemical Conditions	Microbiological Conditions
• Untreated water • No corrosive compounds in the water (Cl^-, SO_4^{2-}) • No oxygen present in the FPS • High iron content and turbidity • pH = 8	• High number of bacteria present in the corrosion product (10^7–10^8 cells/ml) • Several taxa of MIC-related bacteria (IRB, IOB, SRB, SOB) • High value for Heterotrophic Plate Count (HPC-22)
Corrosion and Metallurgical Information	**Design and Operating Information**
• Localized corrosion and pits • Voluminous tubercles and deposits • Presence of sulfur in corrosion product	• Stagnant conditions • Dead-end piping • Temperatures ranging between 20–30°C

REFERENCES

Barco, R. A., D. Emerson, J. B. Sylvan, et al. 2015. New insights into microbial iron oxidation as revealed by the proteomic profile of an obligate iron-oxidizing chemolithoautotroph. *Applied and Environmental Microbiology* 81: 5927–5937.

Bardal, E. 1994. *Korrosjon og korrosjonsvern*. 2nd. Trondheim: Tapir Akademiske Forlag.

Borenstein, S. W. 1994. *Microbiologically influenced corrosion handbook*. Cambridge. Woodhead Publishing Limited.

Braun, B., I. Richert and U. Szewzyk. 2009. Detection of iron-depositing *Pedomicrobium* species in native biofilms from the Odertal National Park by a new, specific FISH probe. *Journal of Microbiological Methods* 79: 37–43.

Cangelosi, G. A. and J. S. Meschke. 2014. Dead or alive: Molecular assessment of microbial viability. *Applied Environmental Biology* 80: 5884–5891.

Chao, Y., Y. Mao, Z. Wang, et al. 2015. Diversity and functions of bacterial community in drinking water biofilms revealed by high-through sequencing. *Scientific Reports* 5: 10044.

Clarke, B. and A. Aguilera. 2007. *Microbiologically influenced corrosion*. 1st. Hoboken, New Jersey: John Wiley and Sons.

Cote, A. E. (editor). 2008. *Fire protection handbook – Volume 2*. 20th. Massachusetts: NFPA.

Dong, Q., H. Shi and Y. Liu. 2017. Microbial character related sulfur cycle under dynamic environmental factors based on the microbial population analysis in sewage systems. *Frontiers in Microbiology* 8: 64.

Ehrich, S., D. Behrens, E. Lebedeva, et al. 1995. A new obligately chemolithoautotrophic, nitrite-oxidizing bacterium, *Nitrospira moscoviensis* sp. nov. and its phylogenetic relationship. *Archives of Microbiology* 164: 16–23.

Emerson, D. 2019. The role of iron-oxidizing bacteria in biocorrosion: a review. *Biofouling* 34: 989–1000.

Enning, D. and J. Garrelfs. 2013. Corrosion of iron by sulfate-reducing bacteria: new views of an old problem. *Applied and Environmental Microbiology* 80: 1226–1236.

Eurofins. 2018. Forklaring til drikkevannsanalyser. Norway: Eurofins Environment Testing Norway AS. https://gulen.custompublish.com/getfile.php/4592845.1286.kujt7maa7qptwm/Analyseforklaring+drikkevatn.pdf

Folkehelseinstituttet (FHI). 2018. Veileder for stoffer i drikkevann. Web publication: Folkehelseinstituttet. https://www.fhi.no/nettpub/stoffer-i-drikkevann/ (accessed May15, 2020)

Helse- og omsorgsdepartementet. 2016. Forskrift om vannforsyning og drikkevann (drikkevannsforskriften). *Norway: Helse- og omsorgsdepartementet.* https://lovdata.no/dokument/SF/forskrift/2016-12-22-1868, https://cdn2.hubspot.net/hubfs/4773030/White%20papers/White-Paper-MIC-is-NOT-the-Primary-Cause-of-Corrosion-in-Fire-Sprinkler-Systems_2016-10.pdf

Huber, B., B. Herzog, J. E. Drewes, et al. 2016. Characterization of sulfur oxidizing bacteria related to biogenic sulfuric acid corrosion in sludge digesters. *BMC Microbiology* 16: 153.

Kakooei, S., M. Ismail, B. Ariwahjoedi. 2012. Mechanisms of microbiologically influenced corrosion: a review. *World Applied Sciences* 17: 524–531.

Kochelek, J. T. 2009. Microbiologically influenced corrosion (MIC) is not the primary cause of corrosion in the fire sprinkler system. ECS, White paper (October): 1–7.

Little, B., R. Ray, and J. Lee. 2011. Tubercles and localized corrosion on carbon steel. Naval Research Laboratory, Stennis Space Center, MS 39529-5004, USA. https://apps.dtic.mil/dtic/tr/fulltext/u2/a537293.pdf

Lührig, K., B. Canbäck, C. J. Paul, et al. 2015. Bacterial community analysis of drinking water biofilms in southern Sweden. *Microbes Environment* 30: 99–107.

Nardy, K., S. Jansen, M. F. A. Leite, et al. 2017. Methanogens predominate in natural corrosion protective layers on metal sheet piles. *Scientific Reports* 7: 11899.

Parow, H. 2018. Korrosjon i inntaksrørene for brannvann på St. Olavs Hospital – undersøkelse, omfang og årsak. MSc diss., NTNU Trondheim. https://ntnuopen.ntnu.no/ntnu-xmlui/handle/11250/2561055

Pope, D. H. and R. M. Pope. 2000. Microbiologically influenced corrosion in fire protection sprinkler systems. *NACE International* 00401.

Qin, K., I. Struewing, J. S. Domingo, et al. 2017. Opportunities pathogens, and microbial communities and their association with sediment physical parameters in drinking water storage tank sediments. *Pathogens* 6: 54.

Ray, R., B. J. Little and J. S. Lee. 2010. Iron-oxidizing bacteria: a review of corrosion mechanisms in fresh water and marine environments. *NACE International* 10218.

Skovhus, T. L., R. B. Eckert and E. Rodrigues. 2017a. Management and control of microbiologically influenced corrosion (MIC) in the oil and gas industry – Overview and a North Sea case study. *Journal of Biotechnology* 256: 31–45.

Skovhus, T. L., D. Enning and J. S. Lee. 2017b. *Microbiologically influenced corrosion in the upstream oil and gas industry*. 1st. Boca Raton, FL: Taylor and Francis Group.

Stoecker, J. G. 2001. *A practical manual on microbiologically influenced corrosion – Volume 2*. 2nd. Texas: NACE International.

Su, P. and D. B. Fuller. 2014. Corrosion and corrosion mitigation in fire protection systems. Norwood, MA: FM Global. https://sprinkler.nl/wp-content/uploads/2018/04/P14180.pdf

Tronhus, A. 2020. Personal communication. Trondheim Bydrift: Vann og avløp.

19 Microbiologically Influenced Corrosion in Fire Protection Systems

A Material Problem or a Problem of Microbial Activity?

N. Noël-Hermes and J.W. Klijnstra
Endures B.V.

CONTENTS

DOI: 10.1201/9780429355479-22

19.1 INTRODUCTION

19.1.1 MIC and Its Diagnosis in General

MIC is an extremely rapid form of localized corrosion which is difficult to predict and account for. MIC comes from the interplay between the environment, microorganisms and the material exposed to it. Not all environments are the same and the so-called biogeochemistry of the site plays an important role on the potential for corrosion processes. MIC fundamentally occurs as a result from close interactions between microorganisms and the metallic surface, specifically, biofilm formation is known to play a key role (Lewandowski and Beyenal, 2009).

Although the exact mechanisms involved in accelerated corrosion by microorganisms are complex, certain processes can be identified as fingerprints of corrosive microbial activity. In the presence of a metallic structure the following two actions are required in order to proof MIC:

- Demonstrate the presence of MIC-relevant microorganisms and a close relationship between them and the metal being corroded.
- Collect sufficient evidence to build a reasonable hypothesis on the specific MIC process that plays a major role in a particular case.

MIC diagnosis and prediction in different environments remains a complex issue. Defining numbers of microorganisms in water and/or sediment is not enough for MIC prediction and its determination, particularly because microorganisms are generally present in soil and water. It is important to demonstrate that they are interacting with the metal and accelerate the corrosion process.

Proper MIC diagnosis is difficult and based on a case-by-case approach. A combination of different fields of knowledge e.g., failure analysis (metallurgy), microbial analysis, chemistry, and operational conditions, is required to make a reliable diagnosis possible. For precise diagnosis there are several methods which can be used to define if MIC is the actual corrosion mechanism or not. It should not only be proven that microorganisms are present but also that they are active at the affected structure. Furthermore, other corrosion mechanisms and abiotic factors need to be excluded as the reason for the damage case under investigation.

It is important to do a complete case documentation and to put all available data into the right context. Often little details regarding operational conditions can be a key point later on in the diagnosis phase. After a good case documentation, one of the first steps of MIC diagnosis is to identify if MIC-relevant microorganisms are present in the sample taken. The sample should include corrosion products or a part of the affected surface and should be compared with a sample from the environment (Lee and Little, 2017).

Microbial analysis can be done using several methods which can be culture- or non-culture-based techniques. Traditional culture based techniques have been used for decades but the disadvantage is that many environmental microorganisms are not cultivable in the laboratory. New techniques such as quantitative polymerase chain reaction (qPCR) and next generation sequencing (NGS) are more and more common to complement traditional cultivation techniques. However, DNA-based techniques have also their limitations such as the correct choice of the primer and its specificity as well as detection limits and reproducibility. Also the choice of the correct detection method is a case-by-case approach and it is reported by several researchers that often more than one method needs to be applied to understand the impact of microorganisms on the affected structure (Sharma and Voordouw, 2017).

Microbial-related failure analysis focuses on processes that are taking place on the material surface. To that end an investigation of the corrosion products and the attacked surface is useful to characterize the corrosion process and morphology. 3D microscopy can be done to identify pit depths/morphology or other corrosion forms that can hint on different corrosion mechanisms. Next to that, scanning electron microscopy (SEM) coupled with energy dispersive X-ray spectroscopy (EDS), photometric or chromatographic methods can help to identify microbial metabolites in corrosion product samples (Sharma and Voordouw, 2017).

Next to microbial-related failure mechanisms it is also necessary to investigate the environment (water, sediment analysis) to exclude possible abiotic factors as reason for the damage (Su and Fuller, 2014).

Proper MIC diagnosis is just like a puzzle; piece by piece needs to be collected to finalize the whole picture and to answer the question if MIC is really the corrosion mechanism responsible for the damage pattern or not.

19.1.2 MIC in Fire Protection Systems

Corrosion in fire protection systems (FPS) is one of the major issues for maintenance and operation and the presence of corrosion products and damages will have impact on the operation in case of a fire. It doesn't matter which type of system (dry or wet pipe systems) is considered, corrosion can be present due to many different reasons or a combination of factors. Corrosion in FPS can be induced amongst others by a corrosive water quality, incorrect choice of material, supply of oxygenated water, stagnant water or also be due to microbiologically influenced corrosion (Werkgroep Corrosie, 2015; Su and Fuller, 2014).

MIC is mostly a localized form of corrosion accelerated or influenced by microbial activity. Microorganisms can influence corrosion by creation of localized concentration cells which may lead to the dissolution of metals and pit formation. Unlike other problems in FPS, MIC is still difficult to predict, the presence of tubercles does not necessarily mean that MIC is present and active and finding the best approach for treatment of MIC problems is not easy (Su and Fuller, 2014).

Factors which influence MIC in FPS are the presence of MIC relevant microorganisms, the water quality, temperature, and also the system design. Most often water in FPS stands still for a long time without refreshment of water and often there are also dead-ends present in the systems. Especially pH and salinity of the water will have an influence on corrosion in the system but also concentrations of others salts, metals, and amount of organics are important factors. Presence of oxygen does not only have an influence on bacterial growth but is also important for other forms of corrosion such as oxygen corrosion in combination with the use of specific materials.

Next to system design and material choice, also flushing and testing regimes are important factors influencing corrosion in FPS (Su and Fuller, 2014).

19.2 CASE STUDY: MIC OR THE WRONG MATERIAL FOR THE USED SYSTEM?

19.2.1 Introduction

An incident during the testing of a fire extinguishing systems at an oil and gas storage terminal in the Netherlands revealed that there were many small holes in the pipes in the tank wells. Subsequently, all the pipes were tested hydrostatically and marked if leakages occurred. Based on these data, locations were chosen for sampling of water and pipe material from the system.

In total five water samples and two pipe sections were taken from the fire protection system, see Table 19.1.

TABLE 19.1
Overview of Samples Taken from the System

Sample ID	Sample Type	Comments
#1	Water	Clear water
#2	Water	Slightly turbid
#3	Water	Slightly turbid
#4	Water	Slightly turbid
#5	Water	Rusty-brown precipitations
#6	Pipe 1	Outside coated, interior galvanized
#7	Pipe 2	Outside coated, interior galvanized

19.2.2 Material and Methods

19.2.2.1 Water Analysis (Physico-Chemical)

Water samples were analyzed on parameters listed in Table 19.2.

19.2.2.2 Quantification of Corrosion-Related Microorganisms

The fractions of total microorganisms with potential corrosive activity were determined by the Most Probable Number (MPN) method in selective growth media and by quantitative polymerase chain reaction (qPCR).

The MPN technique delivers semi-quantitative data on concentrations of cells in analysed samples. It involves performing ten-fold serial dilutions of a microbial culture or an environmental sample into media until the sample is diluted to extinction (Sutton, 2010). Several growth media were used for specific enrichment of MIC relevant microorganisms including:

TABLE 19.2
Overview of the Parameters Analyzed in the Water Samples

Parameter	Method
Cl^-	Spectrometry NEN-ISO 15923-1
Conductivity (20°C)	Conductometry
Fe	ICP-MS NEN-EN-ISO 17294-2
Cu	ICP-MS NEN-EN-ISO 17294-2
Zn	ICP-MS NEN-EN-ISO 17294-2
SO_4	Spectrometry NEN-ISO 15923-1

- acid-producing bacteria (APB),
- iron-reducing bacteria (IRB),
- sulphur-oxidizing bacteria (SOB),
- iron-oxidizing bacteria (IOB), and
- sulphate-reducing bacteria (SRB).

Microbial growth in different dilutions gives a rough estimate of the number and activity of the microorganisms capable of growth in the selected media and conditions (typically 1% of the total microbial population). The MPN technique was combined with qPCR. qPCR is a genetic-level-based technique which can provide information on microbial populations without the need of cultivation. The method is based on amplification of target DNA fragments which are unique of certain microbial species or microbial processes (e.g. sulphate reduction). The main difference with classical PCR is that in the latter technique it is only possible to assess the end product of the reaction which does not distinguish well the quantity of the target sequence amplified. In contrast, in qPCR the amplification product can be quantified. This provides quantitative information of the presence of the target organism or process (Brunk et al., 2002).

DNA was extracted using commercially available biofilm or power soil extraction kits.

The same samples as for MPNs were analyzed by qPCR for the following target organisms/activities:

- sulphate-reducing activity (dsr gene),
- iron-reducing bacteria (*Geobacter*),
- acid-producing bacteria (agn gene),
- sulfur-oxidizing bacteria (*Thiobacillus, Thiothrix, Beggiatoa, Achromatium, Thiomicrospira*).

These analyses provide the fraction of microorganisms that could be accelerating or facilitating corrosion. DNA from standard microorganisms (known cell numbers) hosting the target gene of interest were used to obtain standard curves so that cell numbers in unknown samples can be calculated based on the specific standard curves. Hereby, it is assumed that one DNA copy is equal to one cell.

Not only the water samples were analyzed for presence of MIC relevant microorganisms but similar analyses were done on corrosion products at the pipe samples. After visual inspection of the pipes, corrosion products were scraped from the surface for microbial analysis.

19.2.2.3 SEM-EDS Analysis

Scanning electron microscopy (SEM) of corrosion products was done in combination with energy dispersive spectroscopy (EDS). For this, a Jeol JSM 5800LV instrument equipped with a Noran instruments EDS system was used. EDS provides semi-quantitative information on the elementary composition of the samples, it does not give information on composition of chemical compounds.

19.2.2.4 3D Microscopy

3D microscopy (Keyence VHX-5000) was used for the determination of micro-size damages in the material. Corrosion morphology was determined and pit dimensions were measured. Pits were quantified and measured after corrosion products had been removed chemically according to ISO8407:2009.

19.2.3 Results and Discussion

19.2.3.1 Water Analysis (Physico-Chemical)

The results of the analyses of the five water samples can be found in Table 19.3, values higher than expected for freshwater quality are shown in bold.

In all five samples the values for conductivity were high with values between 140 and 1,000 mS/m. Furthermore, the chloride values of all five samples were remarkably high with 320–3,110 mg/L. The high chloride content and the conductivity are obviously connected to each other, these are typical for brackish water (Gorrell, 1958) and probably originate from the canal close by or the connected

TABLE 19.3
Results of Water Analyses

Sample ID & Type	Conductivity (20°C) (mS/m)	Cl^- (mg/L)	Fe (mg/L)	Zn (mg/L)	SO_4^{2-} ($mgSO_4/L$)
#1, water	**1,000**	**3,110**	**0.79**	1.6	**460**
#2, water	**940**	**2,940**	0.070	**38.0**	**370**
#3, water	**340**	**920**	**6.1**	0.54	**150**
#4, water	**140**	**320**	0.080	**15.0**	71
#5, water	**950**	**2,930**	**6.4**	**17.0**	**370**

harbours close to the location. The samples definitely do not have freshwater quality. Chloride is harmful to most metals and can cause corrosion. For standard stainless steel (304), the limit value of chloride is 200 mg/L and this goes up until 1,000 mg/L for other stainless steel types such as 316 or 2101 (Mackey and Seacord, 2017) .

Moreover, iron values in samples 1, 3, and 5 were higher as expected. Next to iron, zinc showed elevated concentrations in samples 2, 4 and 5 with levels between 17 and 38 mg/L when compared with normal fresh water (Bodar, 2007). These high values indicate dissolution of metal components. Sulphate levels in samples 1, 2, 3, and 5 are high as well and for sulphate it is known that it can facilitate microbial growth (Hao, 2003) but also cause corrosion (Yang et al., 2016).

19.2.3.2 Microbial Analysis: Quantification of Corrosion-Related Microorganisms

An overview of the results of the MPN and qPCR analysis of corrosion-relevant bacteria is given in Table 19.4.

The quantification of MIC-relevant microorganisms shows different results when MPN and qPCR technique are compared. This demonstrates once again that it is necessary to obtain data from more than one technique to come to reliable conclusions regarding the quantification of microorganisms (Kilbane, 2017). As explained earlier for growth-based techniques only 1% of all environmental samples are cultivable in the laboratory, whereas the use of qPCR has the disadvantage that not for all microorganisms a functional gene is known or available that allows detection of a specific group of microorganisms such as for instance the dsr gene in sulphate-reducing bacteria. If no functional gene of a specific group is available, different strains that are likely to be present in the sample are chosen as a target gene for the qPCR analysis.

TABLE 19.4
Overview of Results of MPN and qPCR Techniques: Quantification of MIC-Relevant Microorganisms

Sample [cells/mL] or [cells/g]	MPN Growth-Based Technique					qPCR DNA-Based Technique			
	IRB	SRB	APB	SOB	IOB	IRB	SRB	APB	SOB
#1, water	1.0E+03	(-)	5.0E+04	(-)	(-)	(-)	(*)	(-)	1.0E+04
#2, water	(-)	(-)	(-)	(-)	(-)	(-)	(*)	(-)	3.6E+03
#3, water	1.9E+03	(-)	5.0E+04	1.0E+03	(-)	(-)	(*)	(-)	1.2E+04
#4, water	1.9E+03	(-)	5.0E+04	(-)	(-)	(-)	(*)	(+)	1.3E+04
#5, water	4.0E+02	1.0E+03	1.9E+03	(-)	(-)	(-)	7.1E+04	(-)	7.1E+01
#6, pipe 1	(-)	(-)	1.0E+03	(-)	(-)	(-)	(*)	(-)	3.6E+03
#7, pipe 2	(-)	(-)	1.0E+03	(-)	(-)	(-)	(*)	(-)	7.5E+04

(-) = not detected, (*) = detected but not quantifiable

The results in Table 19.4 show that IRB were positive with MPN in samples 1, 3, 4, and 5 but could not be detected using qPCR. The latter technique gave a positive response for SRB in all samples but only sample 5 was quantifiable and also positive using MPN. With MPN, acid-producing bacteria (APB) were positive in all samples except for sample 2 but qPCR did not reveal their presence. The only exception is sample 4 that was also positive using qPCR. For SOB, the MPN technique gave a positive response only in sample 3, whereas with qPCR all samples gave a positive signal for SOB. Iron-oxidizing bacteria (IOB) were not detected at all.

All MIC-relevant groups detected contribute directly to corrosion. The end products of their metabolism such as H_2S or acids lead to a fast attack of the metal. SRB, IRB, and APB are growing under anaerobic conditions, so without oxygen. When covered with a biofilm and corrosion products, they find optimal conditions for survival and growth. SOB can contribute to corrosion by production of sulphate which forms sulphuric acid or which can be used by SRB as nutrient (Watkins Borenstein, 1994). Summarizing the results, MIC-relevant microorganisms were found in low to moderate numbers both in water samples and in corrosion products.

19.2.3.3 Failure Analysis of Pipes

The pipe samples were cut in longitudinal direction and visually inspected. Pipes showed external as well as internal corrosion as can be seen in Figures 19.1 and 19.2, respectively. In Figure 19.1, several leakage points, perforations, can be seen. Both pipes were heavily corroded at the 6 o'clock position, which corresponds to the location which most likely was covered with water. In Figure 19.2, the internal condition of both pipe segments is shown. There is a clear difference in surface condition between the 12 o'clock and 6 o'clock position in both pipe segments. The 6 o'clock part showed strong brown and black precipitations whereas the top part of the pipe was covered with white precipitates.

Composition of the corrosion products found on various spots of both pipes were analyzed with SEM-EDS and the results are shown in Table 19.5.

The corrosion products are mainly iron oxides (brown haematite in aerated environment and black magnetite in anaerobe conditions), zinc oxide (white), aluminum oxide (white), and additionally some corrosion products containing chlorine and sulphur. In the presence of high amounts of chlorine and sulphur, pitting corrosion of steel can be accelerated, when the zinc layer is fully dissolved. The dissolution of zinc can also be accelerated by the presence of relatively high amounts of chlorine and sulphur.

The presence of aluminum in the corrosion products and copper in water samples as well as in the corrosion products indicate that these materials are corroding in the system.

The white products mainly consist of aluminum, oxygen, and zinc, which indicates that the zinc layer had dissolved due to the presence of water and oxygen in the pipes. Zinc provides only optimal corrosion protection when the environmental conditions show regular variation between wet and dry. When zinc is permanently exposed to wet or humid conditions, it will dissolve until no zinc is present anymore; thereafter, the steel will start to corrode (Su and Fuller, 2014).

FIGURE 19.1 Details of corrosion damage on pipe 1 (left) and pipe 2 (right), 6 o'clock position.

FIGURE 19.2 Pipe 1 (up) and pipe 2 (down). For both pipes, the 12 o'clock position with white precipitation is placed on top and the 6 o'clock position with rusty-brown precipitations is shown below.

Higher zinc concentrations were found at the 12 o'clock position compared to the 6 o'clock position which is another indication that the pipes were partially filled with water. These types of systems should be either fully filled with water (wet condition) or completely dry. From literature it is well known that dry conditions are rarely achieved in these systems (Su and Fuller, 2014). Also in this case the

TABLE 19.5
Elemental Analysis of the Different Corrosion Products Found in Pipe 1 and Pipe 2

Element (wt%)	Pipe 1		Pipe 2	
	Rusty-Brown Spots	Black Spots	White Spots	Rusty-Brown Spots
O	25.84	41.48	12.67	28.54
Na	0.54	-	-	-
Mg	0.23	0.43	0.54	0.25
Al	22.40	-	66.28	28.92
C	-	3.4	-	-
Si	0.06	-	0.21	-
Cr	-	-	-	0.13
S	0.07	0.19	-	0.52
Cl	0.32	-	0.15	0.15
Mn	0.21	-	0.39	0.23
Fe	43.02	51.14	3.33	35.81
Cu	1.23	-	2.69	1.54
Zn	6.08	3.36	13.74	3.91
Total:	100	100	100	100

system has been partially filled with water, resulting in severe corrosion in the lower part. In a partially filled pipe system the water will never get anaerobic because the upper part contains humid air. This humidity can give some slight corrosion also in the upper part of the pipe.

For further investigation of the morphology of corrosion damage at the surface of the pipe walls, the samples were cleaned chemically to remove all corrosion products and subsequently observed under a 3D microscope. In Figure 19.3, a clear line can be seen where pits seem to develop preferentially; this clear line is the longitudinal weld of the pipe. Preferential corrosion of the longitudinal weld is a well-known and often-occurring phenomenon (Lee et al., 2005).

Figure 19.4 shows a detailed picture of a severe pit in pipe 2. Pit depth measured here reached 1.5 mm. With an original wall thickness of 3 mm, this means that the residual wall thickness is approximately 1.5 mm.

Figure 19.5 (left image) shows a picture with multiple pits observed on pipe sample 2. The maximum pit depth found here was 2.2 mm, which means that this part of the pipe had a residual wall thickness of only 0.8 mm. Actually both pipes suffered from severe pitting corrosion, so both had a low residual wall thickness. Moreover, again at the longitudinal weld a perforation was found (see Figure 19.5, right picture). The cross-cut picture clearly shows how the pit surrounds the perforation of the pipe.

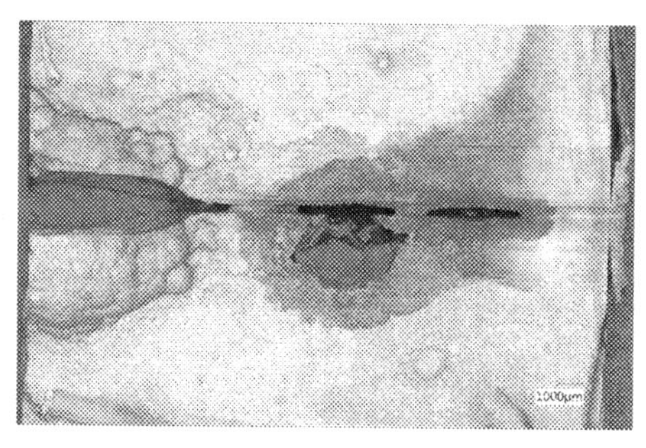

FIGURE 19.3 Example of damage on longitudinal weld of the pipe.

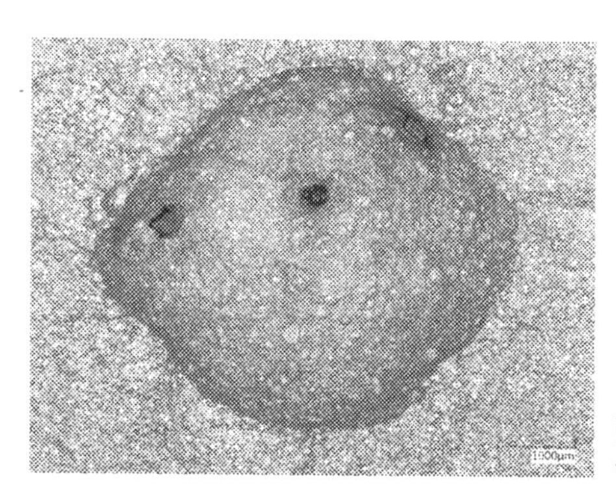

FIGURE 19.4 Severe corrosion pit in the wall of pipe 1. Pit depth is 1.5 mm.

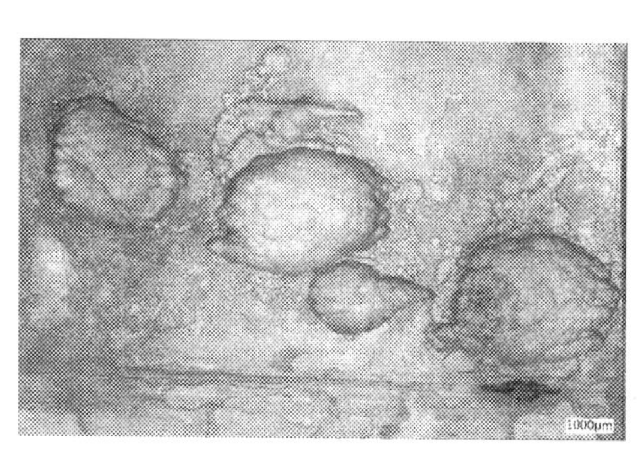

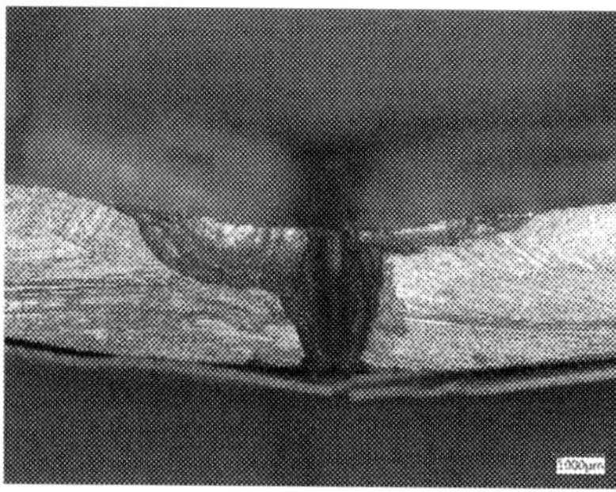

FIGURE 19.5 Example of pitting damage on pipe 2. Left: Multiple pits with max. pit depth of 2.2 mm (bottom right in this picture). Right: Detail of cross cut at perforation at the junction point.

Looking at pit locations and morphology, it is clear that pitting corrosion may occur throughout the pipes but there is a clear tendency for pits to develop around the longitudinal welds.

The welds created are sensitive for pitting corrosion.

Summarizing all results, it can be stated that the present corrosion problem is caused by oxygen corrosion of the steel. This corrosion process occurs after complete dissolution of the zinc layer that originally was present on the inside of this pipe. This happens in a pipe system that is partially filled with aerated water for a long period of time or perhaps continuously.

Whenever pit-shaped attack of the steel takes place this will create separate micro-environments under a layer of iron oxide and hydroxide.

First there is the anodic reaction or dissolution of iron on the one hand and oxygen reduction as a cathodic reaction, on the other hand, to be jointly written as:

$$2Fe + 2H_2O + O_2 \rightarrow 2Fe(OH)_2 \tag{19.1}$$

With increasing thickness of the layer of corrosion products, oxygen supply from the surrounding water to the corroding surface is reduced and the initial cathodic reaction (reduction of oxygen), is replaced by a different reaction. The following hydrolysis reaction then takes place:

$$Fe_2 + (Cl^-)_2 + 2H_2O \rightarrow Fe(OH)_2 + 2H^+ + 2Cl^- \tag{19.2}$$

With increasing availability of iron ions, chloride ions are attracted to the bottom of the pit. This results in a micro-environment with hydrogen and chloride ions, in fact diluted hydrochloric acid, which lowers the pH in the pit and will increase the corrosion rate. This is a process that maintains itself, a so-called autocatalytic process. Due to the higher ion content under the layer of corrosion products, an osmosis process occurs in which the space underneath the corrosion products swells and after drying a pit is present, as found in this failure case (Reddering, 2004).

The leakages were found at the weakest spots, namely the localized imperfections at the weld of the tubes. The quality of the incoming water is accelerating the localized corrosion too. In theory, system performance should be checked once a year and this is usually done by filling the system with brackish water from the harbour. Afterwards the system should be flushed with fresh water and then stored dry. However, it is very difficult to create a really dry system and therefore dry conditions are rarely achieved. Very often remaining water that cannot be tapped from the system will stay locally in the system or in dead ends that cannot be flushed at all with fresh water.

In this case, it was possible to sample residual water from the system and the water that was obtained was clearly brackish (high Cl^- and conductivity values). Especially in water samples 1, 2, and 5 chlorine values and conductivity were very high and at these locations also most of the leakages were found. That shows that these spots were not flushed (completely) with fresh water and that brackish water remained in the system causing accelerated corrosion.

Chlorine and sulphur were not only found in the water, both were also present in corrosion products obtained from the surface. When accumulating these elements will even stronger accelerate localized attacks.

With the presence of MIC-relevant microorganisms localized corrosion can be accelerated too. These microorganisms enter the system with untreated water. Given the fact that this water remained present in the system for a long time, the microorganisms are very well able to adapt and establish a biofilm on the metal

TABLE 19.6
Summary of Findings in this Failure Case

Chemical conditions	Microbiological conditions
Presence of high conductivity, chloride, iron, zinc, and sulfate concentrations found in residual water of the system.	MIC relevant microorganisms were found in low to moderate numbers in water (sampled from residual water in the system) and corrosion products (scratched from the surface of two damaged pipes)
Corrosion and metallurgical information	**Design and operation information**
Presence of chloride and sulfate in corrosion products of damaged pipes.	Once a year due to the system check, the system is filled with brackish water from the harbour. Afterwards it is flushed with fresh water and stored dry. Galvanized pipes are used in the system. However, permanent presence of aerated brackish water in the partially filled galvanized piping system was detected.
Pitting and full perforations detected.	
Residual wall thickness of ~1.5 mm detected	
Preferential corrosion of longitudinal weld detected.	
Use of galvanized steel pipes in continuously wet system	

surface. But in this case the bacteria were not the initial cause for the corrosion damage in the system. They may have given an additional contribution to the final corrosion damage that had started with an incorrect choice of material and inappropriate flush and storage procedures for this fire protection system.

19.3 Final Conclusion

Multiple lines of evidence such as chemistry, microbiology, operating/design conditions, and materials/corrosion are required to establish a proper diagnosis if MIC is the actual root cause of corrosion in a damaged system. Here MIC was not the root cause of the corrosion problem. A summary of the findings in this failure case is given in Table 19.6.

REFERENCES

Bodar, C.W.M. (2007). Environmental risk limits for zinc. RIVM letter report 11235/2007. National Institute for Public Health and the Environment, The Netherlands.

Brunk, Clifford F., Li, Jinliang, Avaniss-Aghajani, Erik. (2002). Analysis of specific bacteria from environmental samples using a quantitative polymerase chain reaction. *Current Issues Molecular Biology* 4: 13–18.

Gorrell, H.A. (1958). Classification of formation waters based on sodium chloride content-geological notes. *AAPG Bulletin*, 2(10): 2513.

Hao, O.J. (2003). Sulphate reducing bacteria. *Chapter 25, Handbook of Water and Wastewater Microbiology*, 1st edition, Academic press, edited by D. Mara and N. Horan.

Ilhan-Sunghur, E., Cansever, N., Cotuk, A. (2006). Microbial corrosion of galvanized steel by a freshwater strain of sulphate reducing bacteria (*Desulfovibrio* sp.). *Corrosion Science* 49 (2007): 1097–1109.

Kilbane II, J.J. (2017). Determining the source of H_2S on an offshore oil production platform. *Chapter 17, Microbiologically influenced corrosion in the upstream oil and gas industry*, CRC Press, edited by T.L. Skovhus, D. Enning and J.S. Lee.

Lee, C.M., Bond, S., Woolin, P. (2005). Preferential weld corrosion: Effects of weldment microstructure and composition. *NACE, Houston, Texas.*

Lee, J.S., Little, B.J. (2017). Diagnosing microbiologically influenced corrosion. *Chapter 8, Microbiologically influenced corrosion in the upstream oil and gas industry*, CRC Press, edited by T.L. Skovhus, D. Enning and J.S. Lee.

Lewandowski, Z., Beyenal, H. (2009). Mechanisms of microbially influenced corrosion. *Marine and industrial biofouling*, 35–64, Springer Berlin Heidelberg, edited by Hans-Curt Flemming, P. Sriyutha Murthy, R. Venkatesan and Keith Cooksey.

Mackey, E.D., Seacord, T.F. (2017). Guidelines for using stainless steel in the water and desalination industries. *Journal AWWA* 109 (5): E158–E169.

Reddering, E. (2004). Onderzoek oorzaak lekkage sprinklerleidingen dealingroom of a bank in Amsterdam. *Endures internal report.*

Sharma, M., Voordouw, G. 2017. MIC detection and assessment- a holistic approach. *Chapter 9, Microbiologically influenced corrosion in the upstream oil and gas industry*, CRC Press, edited by T.L. Skovhus, D. Enning and J.S. Lee.

Su, P., Fuller, D.B. (2014). Corrosion and corrosion mitigation in fire protection systems. *FM Global*, Johnston, RI.

Sutton, S. (2010). The most probable number method and its uses in enumeration, qualification and validation. *Journal of Validation Technology* 16 (3): 35–38.

Watkins Borenstein, S. (1994). Microbiologically influenced corrosion handbook. Woodhead Publishing, *Industrial Press*, New York.

Werkgroep Corrosie. (2015). Whitepaper- Corrosiebeheersing in Sprinklerinstallaties. https://sprinkler.nl/wp-content/uploads/2018/04/Whitepaper-Corrosiebeheersing-in-Sprinklerinstallaties-03-12-2015-1.pdf.

Yang, L., Xu, Y., Zhu, Y., Liu, L., Wang, X., Huang, Y. (2016). Evaluation of interaction effect of sulfate and chloride ions on reinforcements in simulated marine environment using electrochemical methods. *International Journal of Electrochemical Science.*11 (2016): 6943–6958.

Part IV

MIC Failure Analysis Processes and Protocols

20 Determining the Root Cause for Corrosion Failures

Susmitha Purnima Kotu, Katherine M. Buckingham, and Richard B. Eckert
DNV GL USA

CONTENTS

20.1 INTRODUCTION

Corrosion of pipelines, facilities, and storage tanks when left unmitigated often results in leaks or ruptures (Eckert 2015). The NACE IMPACT study published in 2016 determined that the annual global costs associated with corrosion are $2.5 trillion USD (Koch et al. 2016). Reducing these enormous costs associated with corrosion and preventing failures of equipment and industrial systems can be achieved by implementing effective corrosion management procedures and programs.

Broadly, corrosion management involves three steps; threat assessment, barrier identification, and effectiveness monitoring (Skovhus and Eckert 2014). In the first step, all possible corrosion threats and mechanisms in a system are assessed. In the second step, different barriers implemented in a system to prevent and mitigate corrosion are identified. In the last step, the effectiveness of barriers in mitigating corrosion is monitored. In addition to developing and implementing corrosion management procedures, conducting root cause analysis (RCA) on corrosion failures is a crucial aspect for effective corrosion management to assess the corrosion threat(s) that led to the failure, identify the barriers implemented to mitigate the threat(s), identify barriers that may have been missing to mitigate the threat(s), and to measure the effectiveness of those barriers implemented.

DOI: 10.1201/9780429355479-24

Microbiologically influenced corrosion (MIC) is a commonly encountered mechanism for corrosion failures. MIC occurs by the complex interplay of various environmental factors such as microbiological communities, materials and corrosion products, and chemical and physical environments (Kotu et al. 2019, Little and Lee 2014). Hence, the diagnosis of MIC as a corrosion threat and conducting failure analysis for MIC is also an elaborate and multistep process (Starosvetsky et al. 2007). Several MIC failure analysis frameworks (Kotu and Eckert 2019), corrosion management methodologies (Papavinasam 2017), and systems-based approaches for modeling MIC (Taleb-Berrouane et al. 2018, Kannan et al. 2020) have been proposed previously to address this complexity. However, a detailed framework for conducting RCA on MIC failures has not been previously published.

This chapter highlights the methodology for applying RCA to corrosion failures with an emphasis on MIC as the corrosion threat. This framework can be also be applied when investigating the root cause of any other corrosion-related failure with MIC as a probable corrosion threat. A case study of an external corrosion leak of an injection well tubing in an oilfield is presented in this chapter to demonstrate the application of the RCA methodology.

20.2 RCA METHODOLOGY

Conducting an RCA of a failure requires a thorough and comprehensive evaluation of the corrosion threat, followed by evaluation of the mitigation or prevention barriers used to prevent or mitigate corrosion, and finally a review of management systems in the context of the failure (as shown in Figure 20.1). This three-step methodology for RCA is based on the previously published loss causation model and barrier-based systematic causal analysis technique (Pitblado et al. 2011).

In the first step of the RCA, the corrosion threat(s) that resulted in the failure is identified. If MIC is one of the probable corrosion threats, then a systematic framework as described in earlier studies (Kotu and Eckert 2019) should be followed to reliably diagnose MIC as the corrosion threat. Samples of corrosion deposits from the corroded regions should be carefully collected and preserved appropriately to avoid contamination and to prevent sample integrity losses. Microbiological abundance, activity, and composition of biofilms should be determined using corrosion deposit samples. Because MIC is mediated through biofilms (Beech and Sunner 2004, Wrangham and Summer 2013), obtaining microbiological information from biofilms and fluid samples rather than fluid samples alone is essential to infer about the corrosion mechanism. Chemical composition analysis of the liquid and gas is required to interpret if the necessary conditions such as pH, carbon

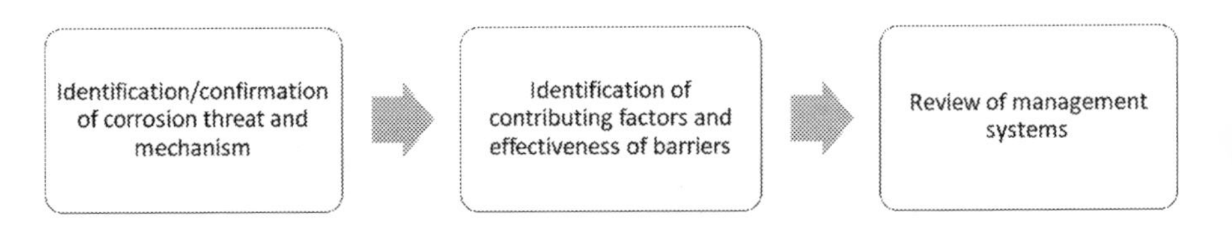

FIGURE 20.1 Three-step methodology for root cause analysis of a failure.

sources, and electron donors and acceptors are available to support MIC. Additionally, the composition of the corrosion products and metallurgical information of the corroding material is important to accurately interpret the corrosion mechanism. Operational conditions such as temperature, flow rates, and duration of no flow also impact the corrosion. Hence, diagnosis of MIC as the corrosion mechanism requires integration of microbiological information along with chemical composition of the fluids, corrosion products, and physical/operating conditions. Comparison of the above-mentioned data from corroded and non-corroded locations provides additional insights on the corrosion mechanism. Lessons learned from the asse's past corrosion threats or failures can also be valuable in determining the corrosion mechanism.

Once MIC is identified as the corrosion threat that resulted in the failure, the different factors contributing to the failure should be evaluated. These contributing factors can be broadly grouped into preventive and assurance factors. The existing corrosion mitigation and monitoring program and its effectiveness should be evaluated. Lessons learned from past corrosion failures can also provide valuable insights on the corrosion threats and the effectiveness of existing mitigation strategies. Design guidelines and maintenance activities specific to corrosion control should be assessed. The timeline of corrosion failures should be compared to changes in the operational history and mitigation to infer any possible links. Lastly, the efficiency of the corrosion management program, corrosion control procedures, and management of change procedures should be assessed.

In the last step of the RCA, the information on the identified corrosion mechanism and the contributing factors to the failure should be integrated to determine the root cause of the failure (Figure 20.2). The effectiveness of the corrosion management program for diagnosing the various corrosion threats and ensuring the effectiveness of monitoring and mitigation programs should be determined and appropriate recommendations should be proposed.

Bowtie analysis is a visual analysis tool used to identify threats, consequences, barriers to prevent corrosion failures, and barriers to mitigate the consequences of

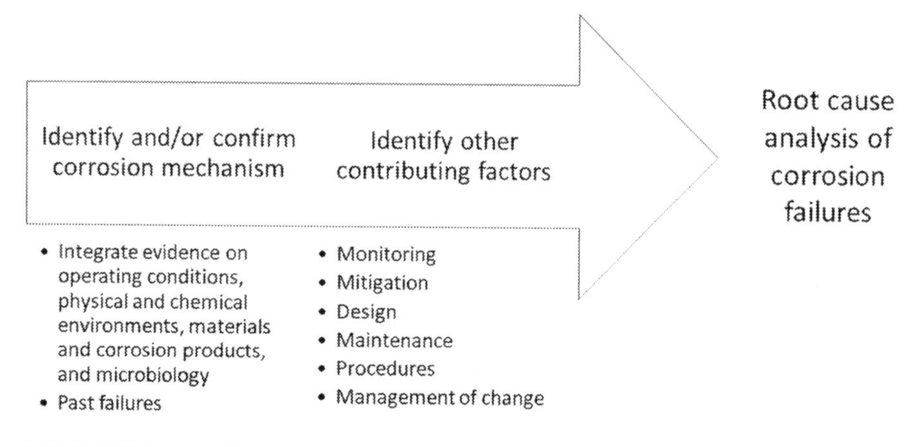

FIGURE 20.2 Detailed methodology for RCA of corrosion failures.

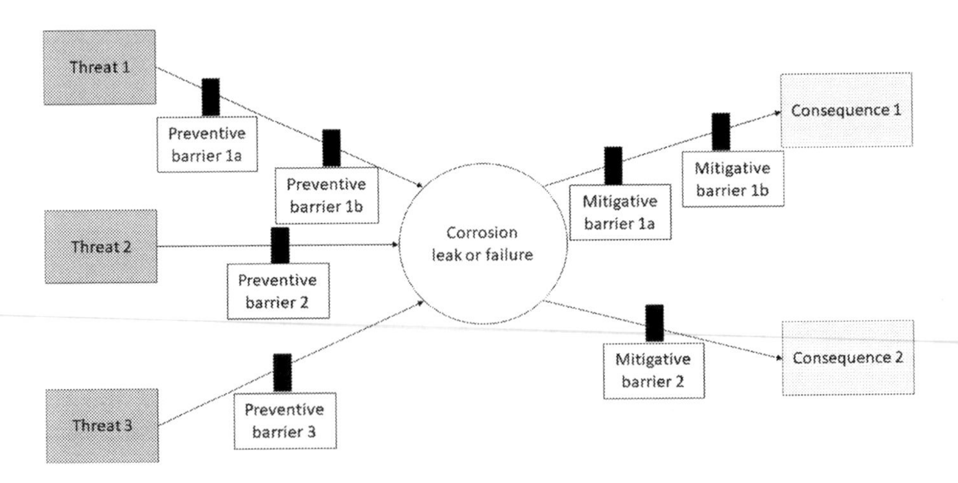

FIGURE 20.3 Schematic of bowtie analysis for a corrosion failure.

failures. The corrosion failure is displayed at the center of the bowtie with the threat (s) on the left side and consequences on the right side (Figure 20.3). In short, a bowtie diagram combines the elements of a fault tree on the left side and elements of an event tree on the right side with the top event or the failure at the center (Yang et al. 2017).

20.3 CASE STUDY: ROOT CAUSE ANALYSIS OF A MIC FAILURE

20.3.1 Overview of the Failure

This section of the chapter describes the RCA methodology as applied to an external corrosion failure of an injection well tubing in an oilfield. The oilfield associated with the corrosion has been in production for more than 50 years. Water produced from the oilfield was used for injection into the well for enhanced oil recovery for over a decade. This led to persistent corrosion issues of the well tubing and subsequently resulted in a leak. Photographs, in the field and the lab, of the tubing that leaked are shown in Figure 20.4.

20.3.2 Corrosion Mechanism Analysis

A detailed corrosion mechanism analysis was performed on the tubing by collecting information about the microbiological conditions, chemical conditions, corrosion products and metallurgy, and design and operating conditions. A summary of the information collected to assess the corrosion mechanism is provided in Table 20.1.

Abundance and activity levels of microorganisms identified through microbiological testing were categorized as low, moderate, and high for 10^2–10^3 cells/cm^2, 10^4–10^5 cells/cm^2, and 10^6–10^8 cells/cm^2, respectively. Culture based tests of the corrosion deposits indicated the presence of aerobic and anaerobic bacteria, acid producing bacteria (APB), sulfate reducing bacteria (SRB), iron reducing bacteria

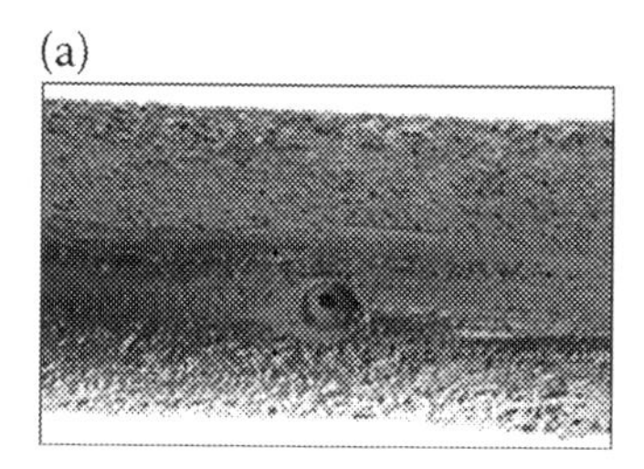

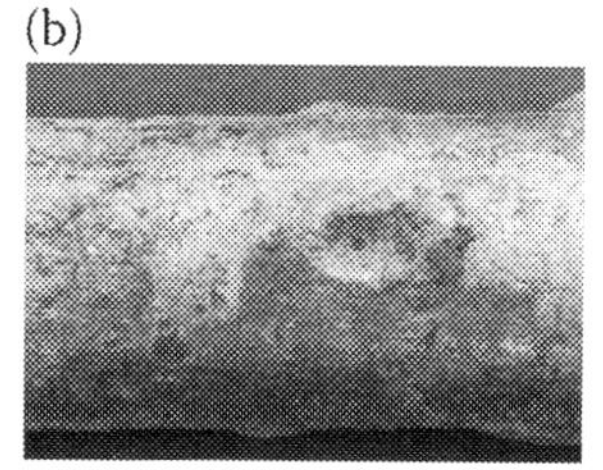

FIGURE 20.4 Photographs of the leaked tubing: (a) In the field after collecting the black corrosion deposits for analysis; (b) in the lab after cleaning.

TABLE 20.1
Summary of the Corrosion Mechanism Analysis for the Injection Well Tubing Leak

Microbiological Conditions • Culture testing of deposits (bacteria/cm^2): Aerobic bacteria (10^2), anaerobic bacteria (10^2), APB (10^3), SRB (10^3), iron related (10^4), NRB (10^2) • qPCR of deposits (cells/cm^2): Total bacteria 10^7, total archaea 10^6, SRB 10^7, MGN 10^6, FER 10^7, IRB 10^6, IO 10^4, SOB 10^5, NRB 10^5 • ATP of produced water: 10^5 ME/mL	**Chemical conditions** • No dissolved oxygen in produced water • Produced gas content: H_2S_g is 500 ppm, no produced CO_2 • Produced water: pH 7, sulfates 30 ppm, bicarbonates 3,500 ppm, chlorides 8,500 ppm, total dissolved solids 20,000 mg/L, treated with glutaraldehyde • Issues with iron sulfide solids
Corrosion products and metallurgy • Black, porous deposits of sulfides and carbonates at location of pits, pH of 7 • EDS: O (38%), Fe (45%), S (10%) • XRD: Presence of siderite, calcite, mixed carbonates, mackinawite, lepidocrocite, non-crystalline phase of sulfur • No metallurgical defects or grain deformation but evidence of significant corrosion	**Design and operating conditions** • Carbon steel used for tubing. Meets specifications for tubing steel • Flow rate in the annulus is 40 m^3/day temperatures in the oilfield 10Â°C–45Â°C

(IRB), and nitrate reducing bacteria (NRB) at low to moderate abundance levels of 100 to 10,000 bacteria/cm^2. Quantitative polymerase chain reaction (qPCR) assay performed on the DNA obtained from the corrosion deposits indicated total bacteria in the order of 10^7 cells/cm^2 and total archaea in the order of 10^6 cells/cm^2. The microbial community composition obtained from qPCR of the targeted functional groups revealed high abundance levels of several MIC related functional groups of microorganisms such as SRB, methanogens (MGN), acid producers or fermenters (FER), and IRB; and moderate abundance levels of a few other MIC related functional groups such as iron oxidizers (IO), sulfate oxidizing bacteria (SOB), and

NRB. Microbial activity of the produced water measured by adenosine triphosphate assays (ATP) showed a moderate microbial activity concentration of 10^5 microbial equivalents (ME)/mL. The results of these microbiological analyses suggest an active role of microorganisms in the observed corrosion.

Chemical composition analysis revealed that the external surface of the tubing was exposed to produced water that indicated a pH of 7, sulfates at 30 ppm, chlorides at 8,500 ppm, bicarbonates at 3,500 ppm, total dissolved solids at 20,000 mg/L, and no dissolved oxygen. Glutaraldehyde was added to the produced water before injection to reduce microbial abundance and prevent MIC. Carbon dioxide was not identified in the gas produced in the oilfield and hydrogen sulfide was produced at 500 ppm. Iron sulfide solids were also observed in the produced water. The results of these chemical analyses indicate that oxygen or carbon dioxide did not result in the corrosive conditions and the leak.

Black, porous corrosion deposits of iron sulfides and carbonates were identified at the locations of the pits using qualitative spot testing in the field with 2N HCl and lead acetate paper. These deposits showed a low pH of 3.5. Energy dispersive spectroscopy (EDS) analyses of the solids removed from the external surface of the tubing indicated high concentrations of iron (Fe) and oxygen (O), moderate concentrations of sulfur (S), and very low concentrations of calcium (Ca), sodium (Na), magnesium (Mg), aluminium (Al), silicon (Si), phosphorus (P), chlorine (Cl), and manganese (Mn). X-ray diffraction (XRD) of the solids showed the presence of siderite ($FeCO_3$), calcite ($CaCO_3$), mixed carbonates ((Ca, Mn, Fe) CO_3), mackinawite (Fe_9S_8), lepidocrocite (γ-FeO(OH)), and non-crystalline phase of sulfur. Mackinawite could have been formed either from SRB activity or the presence of iron sulfide in the produced water, while lepidocrocite is a common corrosion product that forms in the presence of oxygen and is a decomposition product of mackinawite, and siderite and calcite form in the presence of carbon dioxide (Burns 2016). All of these corrosion products are by-products of possible microbial metabolism and support MIC as the likely corrosion mechanism. Metallurgical examinations of cross sections removed from the tubing around the leak did not show metallurgical defects or grain deformation but instead evidence of significant external corrosion at and around the leak.

The leaked tubing was made of carbon steel and meets the industry specifications for tubing steel. The flow rate of produced water in the tubing was approximately 40 m^3/day. The average temperatures in the oilfield range from 10°C to 45°C. These design and operating conditions are highly conducive to the occurrence of MIC.

In the last step of the corrosion failure analysis, integration of microbiological data, corrosion and metallurgical data, and chemical, design, and operating conditions determined that MIC was the likely corrosion mechanism on the external surface of the tubing that eventually resulted in the leak.

20.3.3 RCA of the Corrosion Failure

MIC was identified as the likely corrosion mechanism for the injection well tubing leak in the first step of the RCA. Next, the different contributing factors that resulted in this leak were evaluated. These contributing factors can be grouped into two broad categories: preventive factors including design, operation, water quality, and chemical treatment; and assurance factors including corrosion and water quality monitoring and corrosion program management.

Carbon steel, the material of construction for the tubing, is well known for its susceptibility to MIC. Despite this, alternative materials for the tubing were not identified. Operating controls were not employed for corrosion prevention despite corrosion being a known threat in the injection well system.

The injection system did not have controls for monitoring and managing water quality (dissolved solids, microorganisms, sulfates, bicarbonates, production chemicals) and lacked procedures for documenting water quality. Procedures used for chemical screening and selection were unavailable and compatibility or effectiveness of the production chemicals on corrosion and other downstream processes were not evaluated. The cause of the high microbial abundance, despite biocide use, was not investigated.

Periodic corrosion monitoring using coupons or probes was not performed on the injection well tubing to assess the threat and severity of corrosion. Protocols for documenting visual corrosion observations during well logging were also absent. Procedures and key performance indicators were not used for monitoring water quality.

An established corrosion management program for the injection system was missing. Closer analysis also revealed the absence of corrosion control procedures, data collection and management procedures, and strategic planning for corrosion control.

As a consequence of this injection well tubing leak, both the oilfield operations and environment were impacted. These consequences could be accounted as business impact and health, safety, and environment impact, respectively.

Bowtie analysis was employed to visualize the threat, preventive barriers for the tubing leak, and consequences because of the tubing leak (Figure 20.5). MIC was the primary corrosion threat for the injection well tubing with the tubing leak as the top event that impacted both the business and health, safety and environment. Various preventive barriers such as design and operation, water quality and chemical treatment, corrosion and water quality monitoring, and corrosion program management, if implemented appropriately could prevent tubing leaks in the future.

20.3.4 Conclusions of the RCA of the Corrosion Failure

The RCA methodology was applied to the injection well tubing corrosion leak and revealed an obvious lack of a corrosion management system for threat assessment and evaluation of mitigation and monitoring programs. The injection well system displayed a history of corrosion issues associated with MIC. This incident highlights the need for implementation of a corrosion management system to periodically monitor the microbiology, physical and operating conditions, chemical environment, and materials and corrosion products for effective corrosion assessment and corrosion mitigation.

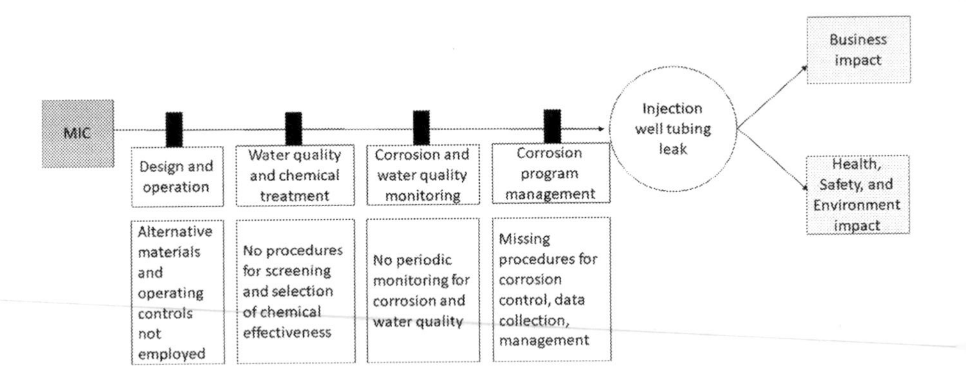

FIGURE 20.5 Bowtie analysis of the injection well tubing leak.

20.4 SUMMARY AND CONCLUSIONS

RCA of corrosion failures can provide insights for implementing effective corrosion management of assets and prevention of similar corrosion failures. A detailed framework for conducting RCA of corrosion failures was described in this chapter. This framework can be utilized when investigating corrosion failures where MIC is a suspected corrosion threat. RCA of corrosion failures should be performed in three steps: identification of the corrosion threat(s) that caused the failure; identification of the contributing factors that led to the failure, highlighting the barriers to prevent the failure and mitigate the consequence of failure; and finally, a review of the management systems in place to identify systematic issues so as to prevent reoccurrence. Identifying the corrosion threat(s) that caused the failure is a critical step in the RCA process as it requires careful examination and integration of the multiple lines of evidence such as microbiological conditions, chemical conditions, corrosion products and metallurgy, and design and operating conditions to validate MIC as the corrosion threat or highlight other abiotic threats. Unless the right corrosion threat is identified, subsequent steps of RCA cannot be applied effectively.

REFERENCES

Beech, I. B., and J. Sunner. 2004. Biocorrosion: Towards understanding interactions between biofilms and metals. *Current Opinion in Biotechnology* 15(3): 181–186.

Burns, M. G. 2016. The tales that rust can tell: The use of corrosion product analyses in corrosion failure investigations. SPE International Oilfield Corrosion Conference and Exhibition. Society of Petroleum Engineers, Aberdeen, UK.

Eckert, R. B. 2015. Corrosion: An integrity threat to the entire O&G asset value chain. *Inspectioneering* 21(6).

Kannan, P., S. P. Kotu, H. Pasman, et al. 2020. A systems-based approach for modeling of microbiologically influenced corrosion implemented using static and dynamic Bayesian networks. *Journal of Loss Prevention in the Process Industries* 65: 104108.

Koch G., J. Varney, N. Thompson, et al. 2016. NACE International IMPACT (International Measures of Prevention, Application, and Economics of Corrosion Technologies) study. NACE International, Houston, TX. http://impact.nace.org/economic–impact.aspx

Kotu, S. P., and R. B. Eckert. 2019. A framework for conducting analysis of microbiologically influenced corrosion failures. *Inspectioneering* 25(4).

Kotu, S.P., M. S. Mannan, and A. Jayaraman. 2019. Emerging molecular techniques for studying microbial community composition and function in microbiologically influenced corrosion. *International Biodeterioration & Biodegradation* 144:104722.

Little B. J., and J. S. Lee. 2014. Microbiologically influenced corrosion: An update. *International Materials Reviews* 59(7): 384–3935.

Papavinasam, S. 2017. Corrosion management. In *Trends in oil and gas corrosion research and technologies*, ed. A. M. El-Sherik, 53–76. Woodhead Publishing, UK.

Pitblado, R., M. Fisher, and A. Benavides. 2011. Linking incident investigation to risk assessment. Mary Kay O'Connor Process Safety Conference, College Station, TX.

Skovhus, T. L., and R. B. Eckert. 2014. Practical aspects of MIC detection, monitoring and management in the oil and gas industry. *CORROSION 2014*. NACE International, San Antonio, TX.

Starosvetsky, J., D. Starosvetsky, and R. Armon. 2007. Identification of microbiologically influenced corrosion (MIC) in industrial equipment failures. *Engineering Failure Analysis* 14(8): 1500–1511.

Taleb-Berrouane, M., F. Khan, K. Hawboldt, et al. 2018. Model for microbiologically influenced corrosion potential assessment for the oil and gas industry. *Corrosion Engineering, Science and Technology* 53(5): 378–392.

Wrangham, J. B., and E. J. Summer. 2013. Planktonic microbial population profiles do not accurately represent same location sessile population profiles. *CORROSION 2013*. NACE International, Orlando, Florida.

Yang, Y., F. Khan, P. Thodi, et al. 2017. Corrosion induced failure analysis of subsea pipelines. *Reliability Engineering & System Safety* 159: 214–222.

21 MIC Sampling Strategies

T. Zintel
TC Energy Pipelines

CONTENTS

21.1 INTRODUCTION

Establishing a logical plan to obtain information through a comprehensive sampling and analysis program is critical in understanding whether MIC is adversely affecting an industrial system, especially at a particular failure location that is suspected to be microbiological in nature. This initial step must be performed in order to be successful in determining the proper actions to take going forward that will lead to an effective solution. Although there are many elements to consider in developing such a plan, there are six basic questions that should be asked: how, what, when, who, where, and why? The first four questions have largely been addressed elsewhere throughout this book; therefore, the last two – *where* and *why* – will be the primary focus of this chapter.

21.2 WHERE TO SAMPLE?

So where should MIC-related sampling be performed? This depends on the system in question or whether a failure has occurred and there is any information available

DOI: 10.1201/9780429355479-25

regarding the microbiology or chemistry at the failure site or upstream of it. Let's begin with sampling an industrial system – they can vary considerably, but a production or gathering system for either natural gas and/or liquid hydrocarbons is a common one that can be used to demonstrate the tenets of a comprehensive sampling program that is useful to the corrosion investigator.

Typically, a gathering system will have several inputs into the main pipeline from smaller lateral lines where production wells or inputs from other operators' products are received to take to a terminal, a plant for processing or other facility to then be transported to the next phase of the operation. Some operators may limit their monitoring to such locations where the gas or liquids are co-mingled from multiple sources in order to minimize the number of samples, as well as the manpower, material and logistical costs associated with the sampling work. There is some information to be gained from such sampling from a "watchdog" perspective, but, due to the co-mingling of gas and/or liquids, the microbiological and chemical constituent values obtained from product sampling will likely be influenced by the largest contributor volume-wise. Thus, potential MIC or other internal corrosion indicators can be affected to a level that may not be indicative of the actual situation. Because data from co-mingled samples can be misleading, it is therefore strongly recommended that, if feasible, all upstream contributors of product into the system be surveyed initially and then monitored on a spot basis at a routine frequency going forward, which is established following the trending of analytical data that has been gathered for evaluation.

As mentioned previously, initially the sampling program for the system should be a survey of all locations providing product inputs. This typically involves a period of more aggressive sample collection to gather enough data, so the results are able to be trended over time to provide a baseline that will allow future monitoring recommendations to be made. In addition, if results indicating a corrosion threat are consistently obtained from the baseline testing, mitigation measures can also be implemented from these data.

21.2.1 Where in a Pipeline System Are Samples Typically Derived?

21.2.1.1 Cleaning Pigs/Inline Inspection (ILI) Tools

Collection of the liquid, sludge, or dry solid materials that are mechanically removed from the interior pipe wall and received with the pigging tool during pipe cleaning or inspection operations of a line segment can provide useful information as to whether microbes are present (or not) within these pigging materials, as well as provide evidence of water and/or corrosion products if chemical analysis is also performed. Unfortunately, if pigging is not performed on a frequent basis (e.g., weekly, monthly, quarterly, etc.) and tested each time, it is often difficult to determine the time frame that any microbes that are detected actually appeared in the pipeline. Also, because it is a composite sample of the materials brought in from the entire length of the line segment being pigged, it will be not be possible to pinpoint the location(s) in the line that have microbes present and whether corrosion has occurred at those locations.

21.2.1.2 Vessels and Equipment

There are various vessels (drips, tanks, etc.) or other equipment that are used to remove liquids and solids from gas systems (e.g., separators, scrubbers, slug catchers, filters), which are able to accumulate these materials allowing samples to be collected for chemical and/or microbiological analyses. Such vessels and equipment are typically located at stations or other processing facilities and so again will only offer co-mingled (composite) samples for testing; however, sometimes they are present at individual production or storage well locations and allow more site-specific sample analysis information to be obtained and utilized for making decisions on future actions to be taken.

21.2.1.3 Routine Maintenance Locations

There are pipeline components that require routine maintenance (e.g., meters, regulators, siphons, valves, etc.) that should be checked for the presence of liquids or solids at prescribed frequencies, and, if found to be present, can then be sampled and analyzed. Depending on their locations in the system, these will be either composite or site-specific sample locations.

21.2.1.4 Corrosion Coupons

Since corrosion coupons (NACE, 2018) are specimens of metal that are able to be inserted into the pipeline or the other aforementioned pipeline components, microbiological samples can be obtained even if accumulated liquids or solids are not present to collect because the coupon surfaces can be swabbed for testing or, in some cases, examined directly on the coupon itself if properly handled and preserved. Coupons have been used at a variety of locations, historically at sites that can be isolated from system pressure, so they can be serviced at atmospheric pressure. Additionally, coupons may be installed and removed under system pressures without interruption of service using high-pressure extraction tools and retractable coupon holder devices, which has expanded their usage and provided more representative internal monitoring conditions.

21.2.2 Sample Types

21.2.2.1 Gas Samples

Gas sample collection is typically accomplished using three methods; spot samples, composite or grab samples, and continuous sampling with online analysis. Special sampling methods may be required for some gas components; for example, biomethane or samples for hydrogen sulfide or organic sulfur analysis should be performed using non-reactive containers (Siltek- and Sulfinert-lined stainless-steel cylinders).

21.2.2.1.1 Spot Samples

Spot samples may be collected in appropriate types of gas cylinders for laboratory analysis or tested in the field using stain tubes or portable analyzers. Spot samples provide compositional information at a single point in time, and frequent resampling is needed to identify trends in gas composition. Spot samples are not useful for detecting short-term gas quality events or occasional upsets.

21.2.2.1.2 Composite or "Grab" Samples

Composite or "grab" samples represent the "average" gas composition over a period of time. A composite gas sampler is a device used to collect and store (over a period of time or volume of flow) a representative gas sample at line conditions. A composite gas sampling system consists of a probe, a sample collection pump, an instrumentation supply system, a timing system (timed or proportional-to-flow) and a collection cylinder for sample transportation. After a period of time, the composite sample is transported to a laboratory for analysis. Composite samples may be able to detect occasional upsets in quality, but the severity of the gas quality excursion won't be accurately represented.

21.2.2.1.3 Continuous Online Sampling

Continuous online sampling and analysis provides near real-time gas quality data, depending on the parameters being measured. Some operators currently use online analyzers for water vapor, hydrogen sulfide, total sulfur, carbon dioxide, and oxygen. The online analyzers are usually located in secure locations, such as a compressor station or meter station where power is also available.

21.2.2.2 Bacteria Samples

Planktonic bacteria samples, which are the "free swimming" microorganisms that are typically present in bulk liquids or sometimes in upper layers of loose solids that have accumulated, are the most commonly collected samples for microbiological analysis (NACE, 2014). Although planktonic bacteria levels can be trended over time and may give a basic idea of bacteria types and numbers that are present in these materials, samples of *sessile* microorganisms obtained directly from metal surfaces are most diagnostic (Wrangham and Summer, 2013). However, the collection of sessile samples is usually much more of a challenge to obtain on a routine or frequent basis since the piping or components need to be opened to get such a sample, unless coupons are in service or a side stream device has been installed.

21.2.3 Conducting a System Survey for Baseline Data

Initially, a survey of all upstream input locations as well as downstream terminus locations with mixed product streams should be performed for a specified period of time (e.g., several weeks or months) to get a sufficient number of samples collected and tested for microbiological and chemical parameters. This large sample set is needed to be able to trend the analytical data for use in making decisions on actions to take for future monitoring and mitigation. A greater number of testing methods are often employed in this phase of monitoring so that sufficient microbiological analysis data can be gathered for evaluation. This additional information can then allow the investigator(s) to better diagnose the presence or severity of MIC or other corrosion issues that are present, which could then indicate the types of tests that should be continued (or not). The same approach can also be taken for the chemical

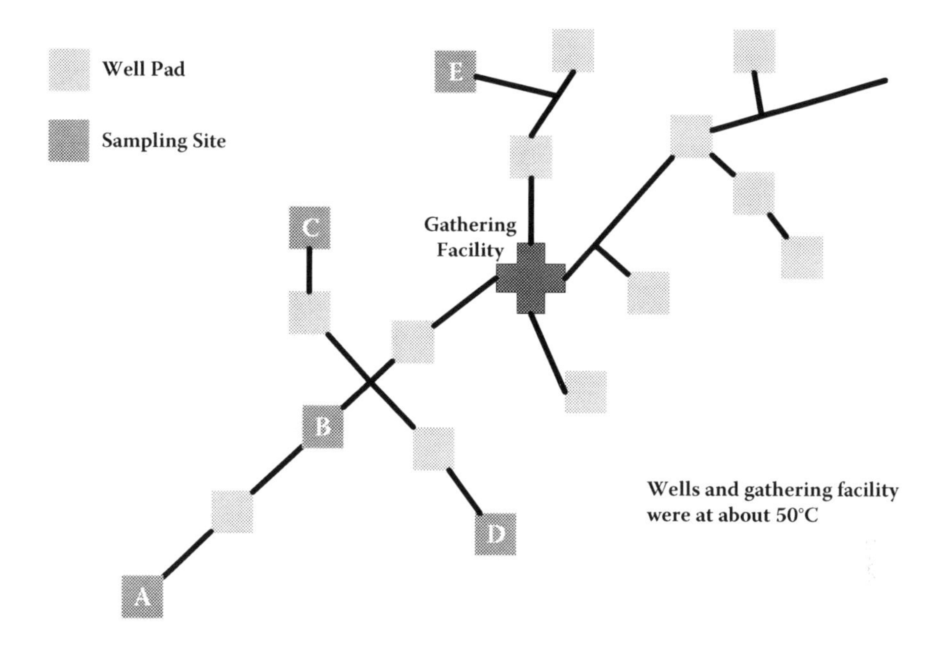

FIGURE 21.1 Schematic of a gathering system showing well pads and sampling sites.

testing performed; however, this may be dependent on the type of sample that is able to be collected, which can then dictate the standard compositional analyses that are able to be performed. An example of the sample collection locations identified for a gathering system is provided in Figure 21.1. In this particular natural gas production system, the individual wells A through E are all sites upstream of the gathering facility, where liquids are produced with the gas. Samples of these produced liquids were collected for chemical and microbiological analysis, from which the presence of microbes and other corrosive constituents were able to be determined vs. sampling only the co-mingled liquids at the downstream gathering site. As a result, mitigation measures were then able to be implemented at the source and this "targeted treatment" allowed for better efficacy as well as reduced costs as compared to treating the entire gathering system.

21.2.4 Routine Sampling

Following the initial survey of the system to determine the baseline microbiological and chemical trend information at each sample location, development of a routine monitoring program to implement going forward should be formulated so any changes in the analytical data can be captured and observed.

So, at what frequency should routine monitoring be conducted? If the industry is regulated by the government, the minimum monitoring frequency obviously must be met. For example, in the U.S. oil and gas industry if the process stream is considered corrosive, monitoring is prescribed at intervals not to exceed 7-1/2 months, twice per calendar year. In many cases, if operations are consistent and survey data trends are consistent, a twice per year frequency is sufficient. In other cases where data trends are more variable, additional monitoring is recommended – where quarterly sampling often will provide the necessary data points to understand what is happening at a location with regard to microbiology and chemical parameters. If an operation is seasonal and rather different conditions occur during the year (e.g., such as in gas storage where dry, pipeline-quality gas is injected into the formation or cavern during much of the year and wet gas and liquids are taken out of the storage field during the winter months), then an adjustment to the monitoring frequency/timing may be in order. This adjustment is needed because internal corrosion, including MIC, is usually more prevalent during the withdrawal cycle when water is present, and increased monitoring frequencies (e.g., monthly, bi-monthly) during that period of time should be considered. More frequent sampling will allow the operator to identify when any changes might occur so that mitigation measures can be applied, if warranted.

21.2.5 Optimizing a Sampling Program

As mentioned previously, once a system sample survey has been conducted and a routine sampling frequency is identified that meets or exceeds regulatory requirements or internal company policies, a sampling program that is cost-efficient and fits within budget constraints can typically be set up at strategic locations with monitoring frequencies that can be stretched out over time, given acceptable test results. If necessary, when monitoring results indicate a potential corrosion concern at co-mingled locations, additional investigation and monitoring may be implemented at individual contributor sites that are located upstream in order to try and determine where a "bad actor" may have developed. If microbial levels are determined to be unacceptable, and especially if MIC has been identified as a threat of concern at individual upstream locations, mitigation measures should be targeted at these upstream sites to maximize treatment effectiveness and potentially control treatment costs. As a part of program optimization where mitigation measures are in place, it is important that downstream monitoring be conducted to determine the effectiveness of the treatment(s); whether it is based on measurements of sufficient chemical residuals, reduction in microbial levels, and particularly – whether corrosion (i.e., MIC in this case) has been mitigated.

21.2.6 Failure Analysis Sampling

In the case of a failure investigation, some of the same principles as described previously are typically applied; however, additional sample locations related to the failure site will also need to be targeted (Kotu and Eckert, 2019). If there is no monitoring data available from upstream inputs, then those samples should be taken right away to try and determine if a particular location, or multiple locations, are the contributor(s) to the failure. Assuming the failed pipe is in an accessible location and the affected section can be cut out and removed from service vs. being clamped, as is sometimes done for offshore leaks, samples of the interior of the pipe need to be sampled as soon as possible upon removal. The leak site is the first location to collect a sample, either by swabbing the affected pit or, if possible, removing scrapings of corrosion deposits that can be tested for evidence of microorganisms. Many times, however, the perforated area will have had most or all of the corrosion products that were present at the failed area be removed from the pit during the release of pressure and product from the interior of the line. As a result of this, the cut-out pipe section should be checked for evidence of similar looking attack, and, if present, sampling of the corrosion deposits (at one or more discrete pits) can be done there in lieu of the failure site that has been altered. In addition to the pitted areas, it is recommended that a sample of any accumulated residue (solids or sludge) on the pipe or component's internal surface be taken from a nearby uncorroded area of the pipe to act as a sort of "control" sample that will provide information as to the local chemistry and microbiology there that can then be compared with those parameters at the failure site and/or the other adjacent corrosion pits.

21.3 WHY SAMPLE?

The remaining question is why should sampling be performed? The answer to this question is fairly straightforward, and that is: to be able to understand an issue and find a solution, one needs to know the root cause of the problem. Subsequently, it is important for the corrosion investigator to determine through sampling and analysis the environmental conditions present at a particular site or at various sites within the system being investigated. This goal is accomplished by identifying the types and numbers of microorganisms that are present, monitoring for localized corrosion and determining whether other abiotic corrosive constituents may be the causative agent. With this information, one can then rule out one or more mechanisms, so the most likely mechanism of corrosive attack can be determined when a failure has occurred, or significant internal corrosion is detected by inspections or corrosion monitoring devices.

In order to determine the cause of corrosion and whether it is due to biotic or abiotic means, it is critical that liquid or solid samples be collected when available at wells, liquid/solid removal equipment, drips, tanks, or other system vessels and components. If the industrial system being investigated is a pipeline and it is designed to be able to be cleaned using pigs, samples of any liquids, solids, or sludge materials received with the pigs should be collected in order for compositional analyses to be performed.

The compositions as well as the volumes of the materials received during the pigging of a line segment can be used to determine frequencies for future pig cleaning operations within that segment. Certainly, if the volume of liquid, sludge, or solid materials received is relatively large, pigging frequencies will need to be increased as necessary for operational reasons, such as: improved product flow efficiencies and maintenance of cleaner pipe surfaces internally, which will often result in minimizing corrosion, including MIC. On the other hand, when small volumes of pigging materials are received relative to the diameter and length of the pipeline being cleaned, time frames between cleaning pig runs can be lengthened since there is likely little in the way of liquid or solid accumulations in the line. However, in this case the analytical results of pigging samples can be important in how pig cleaning frequency recommendations are affected. For example, recommended pigging frequencies will be increased when analytical data shows significant levels of water and/or corrosion products being present, and further concern for recommending shorter time frames between runs will occur when microorganisms are detected. Testing of liquids from storage or production wells for the presence and levels of certain microbes can be diagnostic of potential downhole issues such as souring, plugging, or fouling in addition to MIC problems.

If biocide treatments are in place at pipeline locations to mitigate MIC attack or to reduce microbial populations for other reasons as mentioned above, the testing of any liquid samples that can be obtained upstream of the chemical treatment to compare with similar type liquid samples that are collected downstream of the treatment can identify whether a chemical residual exists in the downstream sample and/or whether the planktonic microbial populations have been reduced following exposure to biocide. However, to better determine the effectiveness of the chemical in use, it is typically more important to test whether sessile microbial levels downstream of treatment have been reduced on metal surfaces as compared to the sessile bacteria levels detected upstream of chemical treatment. Finally, it is also worth mentioning that to better identify whether a MIC attack is being mitigated, the use of extended analysis (NACE, 2018, 2020) corrosion coupons located upstream and downstream of treatment can be a more diagnostic method to employ, since these devices allow biotic pit initiation to be viewed microscopically on the coupons' metal surfaces and the results compared between treated and untreated specimens.

REFERENCES

Jodi B. Wrangham and Elizabeth J. Summer. 2013. Planktonic Microbial Population Profiles Do Not Accurately Represent Same Location Sessile Population Profiles. CORROSION 2013, Paper 2780. NACE International, Houston, TX.

NACE, Test Method TM0194. 2014. Field Monitoring of Bacterial Growth in Oilfield Systems. NACE International, Houston, TX.

NACE, SP0775. 2018. Preparation Installation Analysis and Interpretation of Corrosion Coupons in Oilfield Operations. NACE International, Houston, TX.

NACE, Test Method TM0212. 2018. *Detection Testing and Evaluation of Microbiologically Influenced Corrosion on Internal Surfaces of Pipelines*. NACE International, Houston, TX.

NACE, Technical Report TR3T199. 2020. Techniques for Monitoring Corrosion and Related Parameters in Field Applications. NACE International, Houston, TX.

S.P. Kotu and R.B. Eckert. 2019. A Framework For Conducting Analysis of Microbiologically Influenced Corrosion Failures. *Inspectioneering Journal* 25(4): 2–7.

22 Microbiological Sampling and Preservation for Evaluating Microbial Communities in Oilfield and Other Biological Samples Using Molecular Microbiological Methods

Natalie M. Rachel and Lisa M. Gieg
University of Calgary

CONTENTS

DOI: 10.1201/9780429355479-26

22.1 INTRODUCTION

The adoption and integration of molecular microbiological methods (MMM), including high-throughput sequencing (HTS) and quantitative PCR (qPCR), is enabling an improved understanding of many complex biological ecosystems, including those associated with oilfields and detrimental processes such as microbiologically influenced corrosion (MIC) within oilfield infrastructure (Hirsch et al., 2010; Caporaso et al., 2011; Suflita et al., 2012). When evaluating a sample for the presence of potentially corrosive microorganisms, it is important to analyze the chemical and physical properties as well, since biological, chemical, and physical factors all play a role in MIC (Skovhus et al., 2017). Microorganisms are living catalysts that will alter their environment, and detecting these changes may corroborate the nature and severity of the threat. Highly acidic or basic metabolic products produced by some microbes may disrupt the pH of a system, resulting in the formation of cracks, indicating that pH should be monitored. Monitoring metal surfaces is also important, as cracking can also occur independently of pH changes owing to microbes facilitating the entry of cathodic hydrogen into the metal, which can accumulate and rupture. Accumulation of corrosive products or degradation of protective coatings and chemicals may also be observed; physical pitting features are characteristic of MIC (Jack, 2002). However, MMM are crucial to understanding the root cause of MIC, by intimately the profiling the characteristics of microbes themselves.

The use of MMM, which do not rely on cultivation, has enabled the identification and enumeration of previously unknown microbial types present in oilfield systems, as many microorganisms not amenable to culture-based techniques would otherwise go undetected (Orphan et al., 2000; Grabowski et al., 2005; Larsen et al., 2010; Zhu and Al-Moniee, 2017). MMM typically begin with extracting nucleic acids (most commonly DNA) from a given sample, followed by conducting qPCR for specific microbial groups and/or identifying the entire microbial community composition based on an analysis of the 16S rRNA gene. This genetic-based data contributes to creating an assessment of the microbial community present within oilfield operations and infrastructure, and critically, can provide information as to whether an operational intervention is necessary to manage certain types of microorganisms known to be problematic. Thus, ensuring an accurate diagnosis of the microbiological "health" of a system is crucial, and starts with robust sample collection and processing for microbial community member identification.

In order to process an oilfield sample for MMM, sample collection would ideally be followed immediately by nucleic acid extraction. While there are some options emerging for nucleic acid extraction and processing to be conducted on-site (Sharma and Huang, 2019a), practical limitations often make immediate sample processing for MMM difficult or impossible. The result is a lag between sample collection and processing, wherein the sample will sit for sometimes an extended period (ranging from days to weeks) while transportation to a laboratory occurs, especially if the collection site is remote (Kilbane, 2014). If the integrity of the sample is altered during this time, either from nucleic acid degradation or by a live microbial community adapting to their newfound storage conditions (such as thermophiles being overwhelmed by mesophiles, De Paula et al., 2018), the resulting analysis may produce a microbial community profile that is not truly reflective of the state of the system from which it came, potentially negatively influencing operational decisions. Numerous strategies to prevent DNA degradation or microbial community changes during periods of stasis have been employed for decades, each with its own limitations. It is important to be aware of these limitations, but it ultimately complicates sample collection as one method may not be suitable for all. For example, cold or freezing storage is considered one of the most robust preservation methods (Shikama, 1965), but is often impractical when collecting field samples, especially if large quantities of material are required, and can result in nucleic acid degradation (Oldham et al., 2019). The type of sample collected may also vary; with respect to oilfield systems, produced waters (PW) are one of the most common sampling sources, which frequently contain a low concentration of biomass, requiring large volumes to be collected. Solid samples, such as scrapings from the inside of pipelines or coupons, are also common. These different sample properties mean that a preservative that is optimal for a liquid sample may not be as effective for a solid sample, or vice versa. The selection of a sample preservation agent or protocol should be regarded as a critical piece of experimental design, but is often overlooked or assumed. This indicates that standardized protocols within laboratories for their biological sample collection and handling are paramount to ensuring consistency and accuracy as much as possible. In this chapter, we overview various sample preservation strategies used for MMM, initially focusing on the few studies that have compared various preservatives or preservation protocols using oilfield samples. We extend this review to other biological systems that have more frequently been the subject of preservation studies, focusing on the agent used, sample type, and analysis metric used to evaluate the effectiveness of the preservative. We sought to identify any patterns or consensus among various preservative agents, with the aim of gaining a better understanding to guide the formulation of laboratory protocols so that environmental microbial analysis may be accurate and comparable, contributing to the confidence of genetic data quality that MMM provides.

22.2 SAMPLING AND PRESERVATION STUDIES FOR MMM WITHIN OILFIELD SYSTEMS: WHAT DO WE KNOW?

High temperatures, pressures, complex (and potentially toxic and/or volatile) chemical compositions, low biomass; these distinctive characteristics of oilfield systems can make microbiological sample collection a complicated affair. Out of the reviewed nucleic acid preservation studies for MMM analysis, only a handful have been done using samples related to oilfield systems.

A notable contribution for establishing sample handling directives for microbiological and molecular study related to MIC was published in the ASM handbook nearly two decades ago (Jack, 2002). Practical considerations such as utilizing sterile sampling vessels and sample collection tools, collecting physical samples from multiple locations in addition to a leak site, collecting the liquid phase as well as surface-attached samples, limiting air exposure, maintaining sterility, and chilling the samples for transportation prior to analysis were all recommended. These recommendations, aimed at obtaining as much information from multiple sources as possible while preventing microbial contamination, are also highlighted in more recent industry standard documents related to assessing microorganisms in oilfield samples (NACE, 2014) and evaluating internal pipeline samples for MIC (NACE, 2018). In these latter standards, the importance of minimizing transportation time and/or time between sample collection and microbial analysis is also emphasized, in order to minimize changes in microbial numbers or microbial community composition can otherwise occur over time. As an illustration of such changes, a study was performed wherein an oilfield sample was subject to different storage conditions following collection prior to microbiological analysis (Kilbane, 2014). Using a serial dilution approach that enumerated different groups of microorganisms in the sample immediately after collection or following storage for up to 7 days at 4°C, 25°C, or 35°C, the study showed that microbial numbers increased by approximately two orders of magnitude when samples were stored at 25°C or 35°C for only two days after sample collection, while numbers decreased by up to 1.5 orders of magnitude with 4°C storage. Though MMM were not used in this study, results such as these underline the importance of using sample preservation protocols that minimize microbial changes between the time that samples are collected and then analyzed.

Preservation protocols used for oilfield samples have consisted of either keeping samples cold (De Paula et al., 2018; Salgar-Chapporo and Machuca, 2018; Rachel and Gieg, 2020), freezing samples (Ridley and Voordouw, 2018; Rachel and Gieg, 2020), adding chemical preservatives such as alcohols or buffer solutions (Duncan et al., 2017; Sharma and Huang, 2019a; Rachel and Gieg, 2020) or proprietary nucleic acid preservatives (Duncan et al., 2017; De Paula et al., 2018; Rachel and Gieg, 2020) prior to conducting MMM analyses. These approaches have advantages and limitations (Sharma and Huang, 2019b), which will be communicated throughout this chapter.

A few studies have recently emerged comparing the effects of various handling and preservation methodologies for oilfield samples processed by MMM. The limited number of these comparative studies have primarily focused on oilfield produced water samples. In a study that examined various offshore platform locations for their biocorrosion potential, produced water samples were filtered from the same location. The effects of preserving the filters using 60% ethanol versus a proprietary chemical preservative DNAzol® (Thermo Fisher), which uses a guanidine-detergent lysing solution for its preservative effect, on microbial community composition were compared as determined using 16S rRNA gene sequencing (Duncan et al., 2017). No significant differences were observed in microbial composition at the order level, showing that both chemical preservatives performed similarly to preserve filtered biomass. In another study, produced water samples were either treated DNAzol® and kept on ice during transport, or were filtered and transported at ambient temperature, and submerged within DNAzol® prior to freezing storage (De Paula et al., 2018). The two sets of samples were also shipped to two different laboratories for processing, introducing additional variables with respect to nucleic acid extraction and amplification procedures. Dramatic differences in taxonomic community profiles were observed as a result of these different approaches to sample handling and processing, despite the two sets of samples having been collected from the same wells. Considering the number of methodological variables at play, these results are not surprising, but the study does clearly illustrate that differences in sample handling and processing lead to vastly different microbiological results. The authors of this study clearly articulated the need for standardized sample handling protocols; without them, it becomes incredibly difficult or impossible to accurately compare results (De Paula et al., 2018).

Filtration and DNAzol® were further investigated, in addition to RNA lysis buffer containing β-mercaptoethanol (RLA+, Promega), as preservation strategies in a comparative study using seawater, which is commonly injected into offshore oilfield systems (Oldham et al., 2019). Seawater was immediately filtered and treated with one of the two chemical preservatives, placed on dry ice for transport, and frozen immediately upon arriving at the laboratory 3–6 days later. These samples were compared with unfiltered seawater that had been shipped as is to the laboratory, where filtering then occurred (3–6 days later). Microbial community data, generated following nucleic acid extraction, amplification, and 16S rRNA gene sequencing, was extensively characterized, providing numerous diversity metrics. Differences between immediate filtering and the shipped seawater were observed; a greater number of operational taxonomic units (OTUs) were detected in the filters treated with DNAzol®, and consistently for those that had been immediately filtered. The relative abundance of specific taxa was found to be more impactful than their presence or absence, suggesting that a greater number of taxa may only represent a minute portion of the population, in which case their existence may not be consequential. Importantly, the study demonstrated that a few days of lag time between sample collection and processing will alter DNA quantity as well as composition. There were also differences between the chemical preservatives that were used; in the end, DNAzol® was recommended for DNA extraction and RLA+ for RNA (Oldham et al., 2019).

The effects of immediate filtration and chemical preservation of the filter (using RNAprotect®, active ingredient, tetradecyl-trimethyl-ammonium bromide, or TTAB; Qiagen) versus cold storage (4°C for one week) of unfiltered produced water samples collected from an oil production facility were compared in another study (Salgar-Chapporo and Machuca, 2018). Unfiltered samples were also stored at room temperature for 24 hours prior to cold storage for one week. Following 16S rRNA gene sequencing, the samples initially stored for 24 hours at room temperature followed by cold storage shifted dramatically from their filtered and chemically preserved counterparts. Samples kept cold for the entire week-long storage period also shifted in taxonomic composition, but to a lesser extent. The authors also compared similar preservation approaches on a sludge sample collected from a corroded pipe at same facility. For this sample type, cold or room temperature storage resulted in some shifts in relative abundances of taxa compared to the chemically preserved sample, rather than the more dramatic changes in taxa observed with the produced water (Salgar-Chapporo and Machuca, 2018). These results highlight that different types of samples can behave differently in response to similar preservation and storage approaches.

Finally, another study recently compared nine different storage conditions on produced water collected from the same location within a heavy-oil producing field (Rachel and Gieg, 2020). Over the course of a month, changes to microbial community profiles were monitored by comparing to produced water that had not been treated with a preservative. In the study, 1 L of produced water was initially collected in a sterile bottle, mixed, and half poured into a separate bottle already containing a measured quantity of preservative. Water stored at a final concentration of 20% ethanol, isopropanol, Everclear®, or with DMSO-EDTA-salt (DESS, a homemade preservation buffer) displayed the most consistent and conserved microbial community when compared to aliquots that had been processed on the same day of collection, despite some deviation in terms of relative abundance of some specific taxa. The authors took a rigorous methodological approach to sample handling, demonstrating effective community preservation for produced water. Even so, the results may not extend to other possible combinations of conditions, including the introduction of filtration or other sample types.

Given these few studies, we can see how the handling of oilfield samples is not a trivial matter; standardization with respect to preservation in addition to molecular techniques is key. However, the choice of preservative may vary depending on the nature of the sample, so this variable may remain open to multiple possibilities. Collecting an unpreserved sample is a valuable reference to quantify community divergence and should always be a part of a study, as well as replicate samples whenever possible to aid in accounting for variation.

22.2.1 Comparing Sample Preservation Studies for Biological Systems

The limited number of preservation studies conducted with oilfield samples prompted us to investigate to what extent the topic had been explored within the general context of biological and environmental sampling. After reviewing the literature, it quickly became obvious that the studies are highly diverse in that numerous combinations of variables change between each one. This means that even if a study is exceptionally rigorous in its technical methodology, it remains in its own microcosm, ultimately complicating comparisons and knowledge transfer. For example, if multiple studies investigate the efficacy of ethanol as a preservative, even using the same type of samples, differences in molecular techniques from DNA extraction protocols, amplicon synthesis, and the selected analysis metric (DNA quantity, quality, gene copy numbers, sequencing) will make it difficult to interpret big-picture results; all these steps have been demonstrated to have an impact on DNA community analysis (Kennedy et al., 2014; Fouhy et al., 2016; Teng et al., 2018). By extension, judgment as to which preservative may be most effective becomes muddled. In an attempt to draw and clearly communicate shared conclusions from these divergent studies, they have been divided into smaller clusters, focusing on one variable at a time: each section has been grouped according to the type of preservative. Within each section, sample type, analysis metric, and length of storage are also examined. These have also been summarized in tables (Tables 22.1–22.3). It should be noted that many studies assessed effectiveness by quantifying either the genomic DNA or subsequent PCR products that were produced; relatively few analyzed and compared community composition. Therefore, in these studies, amplicon quality with respect to its contents of individual microbial community members remains largely unknown.

22.2.2 Alcohols as Chemical Preservatives

Alcohol, most frequently high-percentage ethanol (≥95%), has been commonly used to preserve biological specimens for decades owing to its antiseptic properties, availability, and affordability. Outside of the context of oilfield systems, numerous investigations into the effectiveness of alcohol as a nucleic acid preservative have been conducted (Table 22.1).

Approximately half of these studies determined its efficacy by quantifying DNA before and after a given storage period; in some cases, genomic DNA was quantified, while in others, the resulting amplicons were quantified. This is worth keeping in mind when evaluating efficacy, because while DNA quantification reflects degradation, it does not reflect how taxonomic community profiles may be altered; some nucleic acid sequences may be more prone to degradation than others, resulting in biased and inaccurate taxonomic community profiles. Despite this limitation, DNA quantification remains a relevant metric.

In the studies wherein DNA was quantified as an analysis metric for the efficacy of alcohol preservation, the general consensus is that while the presence of ethanol is beneficial over its absence, the results were on par or less successful than buffers or low temperatures. One group conducted a straightforward study in which they created and monitored an *in vitro* environmental sediment sample: two types of sterile soils were inoculated with a culture of *E. coli* (Harry et al., 2000). Subsamples, along with a control lacking *E. coli*, were divided up for storage at 4°C, 28°C, as dried DNA, or in absolute ethanol, and kept for room temperature for up to one year. The type of soil into which *E. coli* was inoculated proved to be the most critical factor affecting DNA recovery; in both cases, ethanol consistently yielded the highest values, with 4°C giving comparable values. Conversely, in another study that also examined sediment samples, it was determined that ethanol use was the least favorable strategy investigated (Rissanen et al., 2010). Three different types of environmental sediments of varying carbon contents were collected in either ethanol, ammonium sulfate-based RNAlater™ (Invitrogen), or phenol-chloroform-isoamyl alcohol (PCIAA) and stored for up to a month. Unlike the previous work utilizing *E. coli* (Harry et al., 2000), these true environmental samples would have vastly increased microbial diversity, and by extension a highly diverse nucleic acid composition rather than of a single sequence. The authors went considerably further in assessing the DNA quality by preparing and sequencing clone libraries of their 16S rRNA gene amplicons, analyzing the microbial community profile. Both ethanol-stored and RNAlater™-stored sediments diverged heavily from controls, which had been processed immediately, although coupling ethanol to storage at -80°C helped. Freezing the sediments without a storage solution, or with PCIAA, was most effective at preserving community diversity with

TABLE 22.1
Summary of Studies That Tested Alcohol as a Preservation Agent for Biological Samples

Reference	Sample Type	Incubation Time	Temperature (°C)	Analysis Method	Efficacy Compared to Other Techniques (Excellent, Average, Poor)
Rachel and Gieg (2020)	Oilfield produced waters	28 days	Ambient, 4, –20	16S rRNA amplicon sequencing	Excellent; various alcohols effectively maintained community integrity
Duncan et al. (2017)	Oilfield produced waters	1–2 weeks	Ambient	qPCR DNA copy number, 16S rRNA amplicon sequencing	Average; comparable results to DNAzol®
Harry et al. (2000)	Soil inoculated with *E. coli*	1 year	28, 4	Genomic DNA quantification	Excellent; on par with cold storage
Rissanen et al. (2010)	Sediments	1 month	4, –20	Genomic DNA quantification; 16S rRNA amplicon sequencing	Poor
Hinlo et al. (2017)	Filtered freshwater	28 days	Ambient, -20	qPCR DNA copy number	Excellent; on par with freezing
Majaneva et al. (2018)	Filtered freshwater	1 week	Ambient, on ice, –20	F230 and BE amplicon sequencing	Average; beneficial, but produced greater community composition variation and lower number of taxa
Minamoto et al. (2016)	Filtered freshwater	6 days	25, –25	qPCR DNA copy number	Excellent
Ladell et al. (2019)	Freshwater	6 days	Ambient, 4	qPCR DNA copy number	Excellent; on par with cold storage
Gaither et al. (2011)	Coral	28 months	Ambient	Genomic DNA visualization; amplicon quantification; qPCR	Average; higher copy numbers, but poorer quality
Kilpatrick (2002)	Mouse liver	2 years	Ambient	Genomic DNA and PCR	Average; degradation during

(*Continued*)

TABLE 22.1 (Continued)
Summary of Studies That Tested Alcohol as a Preservation Agent for Biological Samples

Reference	Sample Type	Incubation Time	Temperature (°C)	Analysis Method	Efficacy Compared to Other Techniques (Excellent, Average, Poor)
				amplicon quantification	DNA extraction observed
Camacho-Sanchez et al. (2013)	Rat tissues	10 months	Ambient	Genomic DNA quantification; Bioanalyzer	Average; comparable to buffers and cryopreservation
Vink et al. (2005)	Whole arachnid specimens	6 weeks	40, ambient, 4, −20, −80	Genomic DNA and PCR amplicon quantification	Poor; only effective when also kept at freezing temperatures

respect to having a similar composition to the controls. These two were the only studies involving ethanol as a preservative utilizing bacteria as the source of DNA, making their findings the most relevant when considering preservation of microbial communities from oilfield sources. It is also evident that while ethanol may improve DNA stability, or allow for successful amplicon synthesis, it does not necessarily provide an accurate representation of its composition.

Other studies utilized a diverse array of specimens as a source of DNA such as filtered freshwater (Minamoto et al., 2016; Hinlo et al., 2017; Ladell et al., 2019; Majaneva et al., 2018), mouse liver (Kilpatrick, 2002), whole arachnids (Vink et al., 2005), coral (Gaither et al., 2011), and rat tissues (Camacho-Sanchez et al., 2013). Monitoring the storage of environmental DNA (eDNA) from freshwater is of particular interest owing to its comparability to samples from oilfield systems; similarly, they commonly come in the form of dilute water requiring filtration to concentrate biomass. The effects of numerous variables on filtered eDNA such as filter type and DNA extraction methods were investigated, and included experimenting with ethanol preservation (Minamoto et al., 2016). After up to six days of storage at room temperature, the number of eDNA copies amplified from the filters dramatically decreased without ethanol, indicating that including ethanol is vastly preferable over no preservative when comes to protecting genomic DNA from degradation. When comparing the freezing of filters containing eDNA to ethanol, it was determined that both strategies are equally effective (Hinlo et al., 2017), allowing for storage for up to four days without significantly affecting DNA copy numbers, although the DNA extraction strategy affected copy numbers recovered. Similarly, after six days, DNA copy numbers of ethanol or ice-preserved water samples remained consistent (Ladell et al., 2019). But unlike the work of Hinlo et al. (2017), these samples were not filtered prior to preservation, suggesting that ethanol or cold temperatures can contribute greatly to protecting genomic DNA present in water samples from degradation when filtration is not possible. Recently, amplicon libraries of filtered eDNA were subjected to sequencing using MiSeq; after one week, filters stored in ethanol produced libraries with greater variation in their community composition, as well as a lower number of taxa than other preservation strategies, which included the use of lysis buffer, freezing, and drying on silica gel (Majaneva et al., 2018). Filters dried using silica gel at room temperature produced the most consistent community composition profiles, but may not be viable for longer storage periods; in this case, the authors recommended storage in lysis buffer if the filter will not be processed for several weeks (Majaneva et al., 2018).

Studies using whole specimens as a source of DNA share the general conclusion that other preservation strategies outperform ethanol with respect to the quantity of genomic DNA recovered. DNA extracted from mouse liver and stored for up to two years at room temperature yielded relatively large amounts of low molecular weight DNA, indicative of degradation (Kilpatrick, 2002). However, degradation appeared to occur during the extraction procedure, rather than due to storage. By comparison, either DMSO-salt solution or Longmire buffer, the latter being a lysis buffer that has been used extensively for nucleic acid storage (Longmire et al., 1997), appear to protect DNA during extraction. The presence of EDTA in buffers, which indirectly inhibits DNase activity (Yasuda et al., 1990; Chen et al., 1999; Barra et al., 2015), are likely responsible for this added DNA protection. Another study also hypothesized that the presence of inhibitory compounds, such as EDTA, is beneficial when they preserved rat tissues in either a homemade nucleic acid preservation (NAP) buffer, Longmire buffer, 95% ethanol, or at –80°C (Camacho-Sanchez et al., 2013). The three former conditions were stored at room temperature, and all samples were stored for up to 10 months. Buffers slightly outperformed ethanol and cryopreservation; however, all samples contained detectable amounts of DNA, and cryopreservation was also found superior for RNA storage. When paired with cold or freezing temperatures, ethanol was determined to be effective for storing DNA from whole arachnid specimens, but RNAlater™ and propylene glycol were best for room temperature storage (Vink et al., 2005). Ethanol yielded higher numbers of 16S rRNA gene copies from coral specimens as quantified by qPCR, but were of poorer quality (Gaither et al., 2011). Amplification from specimens stored in saturated DMSO-salt solution was more effective, owing to the fact that these samples yielded significantly higher quantities of higher molecular weight DNA; once again, this indicates protection from genomic DNA degradation.

In most cases, the presence of alcohol as a sample preservation agent is beneficial compared to storage without it, frequently performing comparably to freezing samples, particularly for genomic DNA coming from filtered water. Ethanol protects against genomic DNA degradation, but may affect community composition more drastically than temperature-controlled storage or chemical agents. In addition, ethanol does not protect DNA against degradation as effectively as buffers. DNases may not be completely inactivated in the presence of ethanol, which may be inconsequential for short-term storage (a few days or less), but potentially of concern for samples stored for more than a few days. It is unfortunate that only two studies employed amplicon sequencing, especially since these studies reported bias and poor reproducibility with ethanol-treated samples. An explanation for ethanol's poor efficacy with sediment samples could be owing to the precipitative nature of the solvent, fixing proteins or organic compounds such as humic substances onto the DNA. As humic substances are common in sediment and soil samples and are known PCR inhibitors (Tsai and Olson, 1992; Tebbe and Vahjen, 1993; Sidstedt et al., 2015), they could be a source of inconsistent amplification and bias.

22.2.3 Other Non-Alcohol Chemical Preservation Agents

Chemical preservatives are diverse. They may be proprietary, commercially available products, or homemade non-proprietary solutions of varying compositions. The resulting performance of each chemical preservative varies in each study, with the nature of the sample often being a key variable. As with the previous section, we will commence by discussing studies investigating chemical preservative effects on samples utilizing microorganisms as the source of DNA (Table 22.2).

An in-depth, controlled study was conducted in which the authors grew liquid cultures of two bacterial mock communities they had constructed, and preserved them using five different methods: RNAlater™ (ammonium sulfate-based), DMSO–EDTA–salt (DESS), DNAgard® (guanidine hydrochloride, Biomatrica), FTA® and FTA elute® cards (Whatman) (Gray et al., 2013). Over a period of six months, the presence or absence of the community members was determined. Liquid methods performed more consistently than cards, producing less variation. The liquid methods were all comparable, with DNAgard® edging slightly out front, and DESS being the most affordable, as it is a non-proprietary solution. The recovery of DNA specific to the mock community organisms was the key evaluation metric to monitor degradation, rather than analyzing how community composition may shift by taking relative abundance into consideration. While ethanol did not perform well with environmental sediments (Rissanen et al., 2010), the authors of a separate study also investigated RNAlater™ and phenol-chloroform isoamyl alcohol (PCIAA). Like ethanol, RNAlater™ also performed poorly, its community composition shifting dramatically from the same-day controls. This could be owing to the fact that RNAlater™ contains a high concentration of ammonium sulfate, commonly used for protein precipitation. Similar to ethanol, this property could adversely affect the quality of the DNA and introduce bias during PCR amplification. In the study conducted by Rissanen et al. (2010), PCIAA was the most successful preservative investigated, exhibiting high similarity to the same-day controls. When we compare these results to those using liquid culture mock communities (Gray et al., 2013), it becomes apparent that the formulation differences of chemical preservatives are not trivial; what functions for one type of samples may not apply to others. It also demonstrates a limitation of studies using synthetic biological mock communities as they are not truly representative of a more complex environmental sample.

TABLE 22.2
Summary of Studies That Tested Non-Alcohol Chemical Preservatives for a Variety of Biological Samples

Reference	Sample Type	Incubation Time	Temperature (°C)	Analysis Method	Efficacy Compared to Other Techniques (Excellent, Average, Poor)
De Paula et al. (2018)	Oilfield produced waters	3–28 days	Ambient, on ice	16s rRNA amplicon sequencing	Average; large taxonomic differences between samples, but numerous variables also involved
Rachel and Gieg (2020)	Oilfield produced waters	28 days	Ambient, 4, −20	16S rRNA amplicon sequencing	Average; DESS effective, other solutions were not
Duncan et al. (2017)	Oilfield produced waters	1–2 weeks	Ambient	qPCR DNA copy number, 16S rRNA amplicon sequencing	Average; comparable results to ethanol
Gray et al. (2013)	Bacterial cultures	6 months	Ambient, −20	16S rRNA fingerprinting (ARISA)	Average; less variation with liquids, all comparable.
Rissanen et al. (2010)	Sediments	1 month	4, −20	Genomic DNA quantification; 16S rRNA amplicon sequencing	Average; RNAlater™ performed poorly, but PCIAA gave best results
Wegleitner et al. (2015)	Filtered freshwater	150 days	Ambient	Genomic DNA quantification; qPCR DNA copy number	Excellent; no difference in recovered DNA after 150 days compared to Day 0
McCarthy et al. (2015)	Filtered freshwater	10 months	On dry ice, −80	16S rRNA amplicon sequencing	Average; all comparable
Williams et al. (2016)	Freshwater	56 days	Ambient, −80	qPCR DNA copy number	Excellent; Longmire requires highest concentration,

(*Continued*)

TABLE 22.2 (Continued)
Summary of Studies That Tested Non-Alcohol Chemical Preservatives for a Variety of Biological Samples

Reference	Sample Type	Incubation Time	Temperature (°C)	Analysis Method	Efficacy Compared to Other Techniques (Excellent, Average, Poor)
					comparable to freezing
Renshaw et al. (2015)	Filtered freshwater	2 weeks	45, 20, −20	qPCR DNA copy number	Longmire gives most consistent results
Yamanaka et al. (2017)	Freshwater	10 days	Ambient	qPCR DNA copy number	Average; degradation still observed
Hunter et al. (2019)	Filtered freshwater	8 days	4	qPCR DNA copy number	Average; Longmire buffer compatible with PCR inhibitor removal
Majaneva et al. (2018)	Filtered freshwater	1 week	Ambient, on ice, −20	F230 and BE amplicon sequencing	Excellent; buffer gave more coherent results for detection and community composition
Oldham et al. (2019)	Filtered seawater	6 days	−20	16S rRNA amplicon sequencing	Excellent; diversity of preserved samples maintained
Sales et al. (2019)	Freshwater	8 hours	Ambient, on ice	12S rRNA amplicon sequencing	Average; cold storage outperformed BAC for number of OTUs
Gaither et al. (2011)	Coral	28 months	Ambient	Genomic DNA visualization; amplicon quantification; qPCR	Excellent; higher quality DNA
Kilpatrick (2002)	Mouse liver	2 years	Ambient	Genomic DNA and PCR amplicon quantification	Excellent; protection from degradation

(*Continued*)

TABLE 22.2 (Continued)
Summary of Studies That Tested Non-Alcohol Chemical Preservatives for a Variety of Biological Samples

Reference	Sample Type	Incubation Time	Temperature (°C)	Analysis Method	Efficacy Compared to Other Techniques (Excellent, Average, Poor)
Camacho-Sanchez et al. (2013)	Rat tissues	10 months	Ambient	Genomic DNA quantification; Bioanalyzer	Average; comparable to ethanol and cryopreservation
Vink et al. (2005)	Whole arachnid specimens	6 weeks	40, ambient, 4, −20, −80	Genomic DNA and PCR amplicon quantification	Excellent; RNAlater™, propylene glycol most effective
Florell et al. (2001)	Human skin	6 weeks	4	Denaturing gels, Northern blots, microarray	Excellent
Mäki et al., (2017)	Phytoplankton	2 weeks	Ambient, −80	18S rRNA amplicon sequencing	Average; DNA yields equal, community composition affected
Menke et al. (2017)	Sheep feces	10 days	Ambient, −20	16S rRNA amplicon sequencing	Average; freezing ideal, NAP buffer next best
Nechvatal et al. (2008)	Human feces	5 days	Ambient	DNA quantification, qPCR	Average; RNAlater™ most effective, but all methods worked

Chemical preservatives have been commonly used on filtered freshwater as well. One that was investigated within multiple studies is Longmire buffer, whose key components are EDTA, SDS, and sodium azide, has a long usage history. A long-term study indicated that Longmire buffer is highly successful at protecting against DNA degradation; after incubating filters used to processes freshwater for 150 days in the buffer, all samples successfully amplified, and there was no difference in the quantity of genomic DNA (Wegleitner et al., 2015). In another study focusing primarily on removing inhibitory compounds that may be retained during DNA extraction, Longmire buffer was determined to be an important component of the strategy that ultimately yielded the highest amount of genomic DNA, and successful removal of inhibitors (Hunter et al., 2019). Longmire buffer was also demonstrated to be effective at preserving unfiltered freshwater samples at various ratios (Williams et al., 2016), with the highest ratio tested of 1:3 being the most effective, nearly producing the same number of DNA copies as freezing. When compared to another non-proprietary surfactant solution, cetyltrimethyl ammonium bromide (CTAB), Longmire buffer produced significantly higher DNA copy numbers (Renshaw et al., 2015), although both solutions successfully protected against DNA degradation, even at 45°C for up to two weeks. When taken together, these results indicate that within the context of preventing genomic DNA degradation within dilute aqueous samples, Longmire buffer is effective. Other chemical preservatives have also seen success. RNAlater™, RNAprotect®, and RLT+ lysis buffer (Qiagen; guanidinium thiocyanate based) were compared by storing filters used for processing freshwater into each of these solutions for 10 months; DNA yield was better with the presence of a preservative compared to a sample stored without (McCarthy et al., 2015). All three solutions were comparable, even with respect to community composition, demonstrating their viability as chemical preservation agents. Owing to their antiseptic properties, three cationic surfactants were investigated by adding them directly to freshwater after sampling (Yamanaka et al., 2017). They were either immediately filtered after being transported to the laboratory six hours later, or stored for 10 days at ambient temperature. Multiple versions of these stored samples were prepared, each containing varying concentrations of surfactant. All of the surfactants were effective in halting or slowing DNA degradation, with benzalkonium chloride (BAC) giving the best results; after 10 days, roughly 50% of the genomic DNA was retained. While the surfactants were helpful, this was a substantial DNA loss compared to other protection agents. Another study built upon this finding: BAC was used to preserve freshwater samples for an eight-hour incubation period before they were filtered, and were compared to samples subjected to cold storage (Sales et al., 2019). With respect to the number of OTUs detected, cold storage outperformed BAC in this study, providing a greater number of OTUs. However, no unpreserved controls were analyzed, making it difficult to know how the community composition compared between the two preservation methods.

Studies examining chemical preservation and storage of whole specimens also included phytoplankton (Mäki et al., 2017), human skin (Florell et al., 2001), sheep feces (Menke et al., 2017), and human feces (Nechvatal et al., 2008) in addition to ones already described in the previous section. Phytoplankton are microscopic aquatic organisms with tough cell walls, making nucleic acid extraction difficult; a combination of processing techniques, including comparing chemical preservation to freezing, were evaluated (Mäki et al., 2017). Mock phytoplankton communities were prepared and either stored at –80°C or in Lugol's solution, a non-proprietary disinfectant solution composed of elemental iodine and potassium iodine. Lugol's solution gave equal DNA yields and PCR performance to freezing, but affected community composition; no other preservation reagents were included in the study that may have provided better protection. Two additional studies reported RNAlater™ as being effective at preserving solid mammalian biological samples (Florell et al., 2001; Nechvatal et al., 2008). Human tissue samples were indistinguishable from tissues processed in routine fashion, and produced high-quality RNA. Feces samples stored in RNAlater™ gave the highest DNA recovery compared to other methods, including cryopreservation; however, these studies did not include community profiling. Conversely, a study that not only conducted community profiling, but also thoroughly analyzed the diversity of swabbed fecal samples, investigated the efficacy of homemade nucleic acid preservation buffer (NAP), DNA/RNAshield™ (Zymo Research, active ingredient not disclosed), and RNAlater™ in comparison to immediate freezing as well as unpreserved controls (Menke et al., 2017). DNA extraction yield was lower for all treatments that were not frozen, with the exception of DNA/RNAshield™; this buffer maintained DNA integrity for room temperature storage. In addition, alpha diversity among all non-frozen samples were lower than those that were frozen; commercial preservatives greatly affected OTU proportions. Ultimately, freezing is recommended, and failing that, homemade preservation buffer outperformed commercial buffers with respect to maintaining community diversity metrics.

Based on these reports, non-alcohol-based chemical preservation is an effective method, however the efficacy of the chemical agent of choice will vary heavily depending on the properties of the sample. RNAlater™ is an example: it did not effectively preserve environmental sediments, possibly owing to the presence of humic substances, but filtered freshwater and liquid bacterial cultures were effectively preserved. Longmire buffer has been investigated the most frequently in numerous independent studies, and performed consistently well with freshwater samples, whether they had been filtered or not. Considering this, it is crucial to consider both the composition of the samples and preservative in an attempt to prevent any unfavorable chemical interactions. Conducting a pilot study comparing chemical agents with samples is advised; establishing standard protocols through first-hand data collection for the specific system to be studied is imperative to ensuring accuracy.

22.2.4 Temperature

Cold or frozen storage is an established, longstanding preservation strategy, and has the advantage that it does not require the addition of an agent to samples, retaining it in the same form as it was when collected or processed. Owing to this, it cold storage has long been the preferred means to store samples, making it a great standard to compare the efficacy of other preservation agents. While ice or ice packs are often available for keeping samples cold, specialized equipment for freezing samples is rarely feasible for field sample collection.

Several studies have examined the use of cold and/or frozen storage for preserving biological samples (Table 22.3). Two studies investigating preservative effects on sediments, already discussed, compared the effects of alcohol and chemical protection agents to cold or frozen storage (Harry et al., 2000; Rissanen et al., 2010) and determined that freezing was the preferred storage method as determined by comparing community composition from sequencing 16S rRNA amplicon clone libraries, with some chemical agents performing equally as well, depending on the nature of the sediments. Freshwater samples, whether they were filtered or not, all yielded their best results by being kept cold or frozen (Renshaw et al., 2015; Eichmiller et al., 2016; Williams et al., 2016; Hinlo et al., 2017; Ladell et al., 2019; Sales et al., 2019), with the exception of one study in which the addition of lysis buffer proved to be further beneficial (Majaneva et al., 2018). In numerous cases, a protective agent performed equally as well as temperature control. This efficacy becomes more muddled when dealing with other specimens); one study reported that cold storage did not make a difference in transporting human tissue, fecal, or sediments after 14 days (Lauber et al., 2010). This is contradictory to the results presented by all other studies discussed. Even so, it stands to reason that cold storage did not dramatically alter community composition, so it was ultimately not disadvantageous.

Three studies investigated the combined effect of temperature and a chemical preservative. There was no significant advantage observed by having the presence of both preservation methods with respect to the percent amount DNA copy numbers quantified from filtered freshwater; the presence of ethanol and sodium acetate solution, or ice storage, was sufficient on their own (Ladell et al., 2019). The combination of CTAB and freezing at –20°C on filtered freshwater actually produced approximately half the number of DNA copy numbers compared to storage with CTAB at 20°C or 45°C, but even so, the number of copies obtained remained viable (Renshaw et al., 2015). Sediments treated with either ethanol or RNAlater™ displayed different community compositions depending on what temperature at which they were stored. Freezing, the addition of PCIAA, or both, yielded similar results (Rissanen et al., 2010). Finally, studying fecal swabs, frozen samples consistently provided larger DNA yields than unfrozen ones, even with a chemical preservative added (Menke et al., 2017). Each chemical preservative also altered community composition differently when compared to each other and to those that were frozen only. As only a few studies examined these combinatorial effects, there remains many details to be elucidated, particularly within the context of oilfield microbiological samples.

Overall, the consistency observed from these studies reinforce that cold or freezing is a trustworthy form of storage for maintaining the integrity of nucleic acids in many types of biological samples. Some practical hurdles (such as freezing in the field) prevent this preservation approach from being the standard method for field sample collection; however, the fact that numerous studies determined that it is possible to have comparable results from the addition of a chemical preservative is encouraging.

22.2.5 Gaps in Knowledge and Considerations for Developing Best Practices for Oilfield Sample Preservation

Following collection, environmental samples can shift dramatically in terms of their microbial community composition or microbial numbers if not processed immediately or stored without adequate preservation for any length of time. Here, the efficacy of various preservation methodologies on a variety of biological samples, including oilfield samples, were overviewed. Overall, ethanol

TABLE 22.3
Summary of Studies That Tested Cold or Freezing Storage for Preserving a Variety of Biological Samples

Reference	Sample Type	Incubation Time	Temperature (°C)	Analysis Method	Efficacy Compared to Other Techniques (Excellent, Average, Poor)
Rachel and Gieg (2020)	Oilfield produced waters	28 days	Ambient, 4, −20	16S rRNA amplicon sequencing	Average; comparable to alcohols and DESS
Salgar-Chapporo and Machuca (2018)	Oilfield produced waters and sludge	1 week	25, 4	16S rRNA amplicon sequencing	Excellent; freezing ideal over cold storage, both superior to room temperature
Kilbane (2014)	Mixed culture from oilfield sample	1 week	4, 25, 35	Serial dilution (MPN)	Average; cold storage superior to other temperatures but numbers decreased
Harry et al. (2000)	Sterile soil inoculated with *E. coli*	1 year	28, 4	Genomic DNA quantification	Excellent; cold storage on par with ethanol
Rissanen et al. (2010)	Sediments	1 month	4, −20	Genomic DNA quantification; 16S rRNA amplicon sequencing	Excellent; freezing or PCIAA gave best results
Williams et al. (2016)	Freshwater	56 days	Ambient, −80	qPCR DNA copy number	Excellent; freezing on par with high concentration of Longmire
Hinlo et al. (2017)	Filtered freshwater	28 days	Ambient, −20	qPCR DNA copy number	Excellent; freezing on par with chemical agents
Eichmiller et al. (2016)	Freshwater	28 days	35, 25, 15, 5	qPCR DNA copy number	Excellent; DNA degradation inhibited by cold

(*Continued*)

TABLE 22.3 (Continued)
Summary of Studies That Tested Cold or Freezing Storage for Preserving a Variety of Biological Samples

Reference	Sample Type	Incubation Time	Temperature (°C)	Analysis Method	Efficacy Compared to Other Techniques (Excellent, Average, Poor)
Renshaw et al. (2015)	Filtered freshwater	2 weeks	45, 20, −20	qPCR DNA copy number	Excellent; freezing and Longmire gave most consistent results
Majaneva et al. (2018)	Filtered freshwater	1 week	Ambient, on ice, −20	F230 and BE amplicon sequencing	Average; ice storage comparable to buffer, which performed best
Ladell et al., (2019)	Freshwater	6 days	Ambient, 4	qPCR DNA copy number	Excellent; cold storage on par with ethanol
Sales et al. (2019)	Freshwater	8 hours	Ambient, on ice	12S rRNA amplicon sequencing	Excellent; cold storage outperformed BAC for number of OTUs
Vink et al. (2005)	Whole arachnid specimens	6 weeks	40, ambient, 4, −20, −80	Genomic DNA and PCR amplicon quantification	Average; freezing temperatures effective and enhanced by protection agents
Mäki et al. (2017)	Phytoplankton	2 weeks	Ambient, −80	18S rRNA amplicon sequencing	Excellent; freezing storage ideal, most coherent community compositions
Lauber et al. (2010)	Soil, human tissue	14 days	20, 4, −20, −80	16S rRNA amplicon sequencing	Average; relative abundance of taxa may change, but room temperature storage

(*Continued*)

TABLE 22.3 (Continued)
Summary of Studies That Tested Cold or Freezing Storage for Preserving a Variety of Biological Samples

Reference	Sample Type	Incubation Time	Temperature (°C)	Analysis Method	Efficacy Compared to Other Techniques (Excellent, Average, Poor)
					comparable to freezing
Sekar et al. (2009)	Bacterial films from coral	1 hour	Ambient, −20	16S rRNA amplicon sequencing	Average; species detection varied for frozen vs. unfrozen samples

prevents DNA degradation, but research probing its effect on microbial community composition is lacking. Chemical preservatives can be excellent solutions, but require testing before conducting a formal study to ensure compatibility with the samples to which it will be applied. Finally, cold and especially freezing storage are highly effective, but may have limited capacity during field sample collection.

While several preservation studies have been conducted for various types of biological and environmental samples, comparatively few have examined the effects of various preservation agents and techniques on oilfield samples. To date, the few studies that have been conducted have primarily focused on water (such as seawater) or oilfield produced water samples. Other sample types such as pigging sludge and other pipeline solids more applicable to MIC have not undergone similar rigorous examination for effective preservation strategies suitable for MMM. Thus, there is a clear need for additional studies into sample preservation efficacies on a broader range of sample types from oilfield systems.

The importance of conducting a preliminary study investigating the effectiveness of a preservation method and establishing it as a standardized laboratory protocol, before routinely applying it, cannot be understated; doing so is imperative to ensuring accuracy of microbiological analysis. Different kinds of samples, including those from oilfield systems, can be complex and interact differently with a given additive. In addition, we highly recommend replicates of samples and unpreserved controls to make the interpretation of results as confident and conclusive as possible. Finally, there is a clear need for additional studies that examine the effects of preservatives such as ethanol on microbial taxonomic composition (e.g., using HTS) rather than solely on DNA quantities/recovery when preserving oilfield samples for MMM.

REFERENCES

Barra, G. B., T. H. Santa Rita, J. de Almeida Vasques, C. F. Chianca, L. F. A. Nery, and S. S. S. Costa. 2015. EDTA-mediated inhibition of DNases protects circulating cell-free DNA from ex vivo degradation in blood samples. *Clinical Biochemistry* 48: 976–981.

Camacho-Sanchez, M., P. Burraco, I. Gomez-Mestre, and J. A. Leonard. 2013. Preservation of RNA and DNA from mammal samples under field conditions. *Molecular Ecology Resources* 13: 663–673.

Caporaso, J. G., C. L. Lauber, W. A. Walters, D. Berg-Lyons, C. A. Lozupone, P. J. Turnbaugh, N. Fierer, and R. Knight. 2011. Global patterns of 16S rRNA diversity at a depth of millions of sequences per sample. *Proceedings of the National Academy of Sciences USA* 108: 4516–4522.

Chen, B., H. R. Costantino, J. Liu, C. C. Hsu, and S. J. Shir. 1999. Influence of calcium ions on the structure and stability of recombinant human deoxyribonuclease in the aqueous and lyophilized states. *Journal of Pharmaceutical Sciences* 88: 477–482.

De Paula, R., C. St Peter, A. Richardson, J. Bracey, E. Heaver, K. Duncan, M. Eid, and R. Tanner. 2018. DNA sequencing of oilfield samples: impact of protocol choices on the microbiological conclusions. In: *Corrosion 2018*, Paper # 11662, NACE Corrosion 2018 Conference, Phoenix AZ, USA.

Duncan, K. E., I. A. Davidova, H. S. Nunn, B. W. Stamps, B. S. Stevenson, P. J. Souquet, and J. M. Suflita. 2017. Design features of offshore oil production platforms influence their susceptibility to biocorrosion. *Applied Microbiology and Biotechnology* 101: 6517–6529.

Eichmiller, J., S. E., Best, and P. W. Sorensen. 2006. Effects of temperature and trophic state on degradation of environmental DNA in lake water. *Environmental Science and Technology* 50: 1859–1867.

Florell, S. R., C. M. Coffin, J. A. Holden, J. W. Zimmermann, J. W. Gerwels, B. K. Summers, D. A. Jones, and S. A., Leachman. 2001. Preservation of RNA for functional genomic studies: a multidisciplinary tumor bank protocol. *Modern Pathology* 14: 116–128.

Fouhy, F., A. G. Clooney, C. Stanton, M. J. Claesson, and P. D. Cotter. 2016. 16S rRNA gene sequencing of mock microbial populations - Impact of DNA extraction method, primer choice and sequencing platform. *BMC Microbiology* 16: 123.

Gaither, M. R., Z. Szabó, M. W. Crepeau, C. E. Bird, and R. J. Toonen. 2011. Preservation of corals in salt-saturated DMSO buffer is superior to ethanol for PCR experiments. *Coral Reefs* 30: 329–333.

Grabowski, A., O. Nercessian, F. Fayolle, D. Blanchet, and C. Jeanthon. 2005. Microbial diversity in production waters of a low-temperature biodegraded oil reservoir. *FEMS Microbiology Ecology* 54: 427–443.

Gray, M.A., Z. A. Pratte, and C. A. Kellogg. 2013. Comparison of DNA preservation methods for environmental bacterial community samples. *FEMS Microbiology Ecology* 83: 468–477.

Harry, M., B. Gambier, and E. Garnier-Sillam. 2000. Soil conservation for DNA preservation for bacterial molecular studies. *European Journal of Soil Biology 36*: 51–55.

Hinlo R., D. Gleeson, M. Lintermans, and E. Furlan. 2017. Methods to maximise recovery of environmental DNA from water samples. *PLOS One* 12: e0179251.

Hirsch, P. R., T. H. Mauchline, and I. M. Clark. 2010. Culture-independent molecular techniques for soil microbial ecology. *Soil Biology and Biochemistry* 42: 878–887.

Hunter, M. E., J. A. Ferrante, G. Meigs-Friend, and A. Ulmer. 2019. Improving eDNA yield and inhibitor reduction through increased water volumes and multi-filter isolation techniques. *Scientific Reports* 9: 5259.

Jack, T. R. (2002). Biological corrosion failures. In: *ASM handbook: failure analysis and prevention*, vol. 11, 881–898. ASM International, Metal Park, OH, USA.

Kennedy, N. A., A. W. Walker, S. H. Berry, S. H. Duncan, F. M. Farquarson, P. Louis, J. M. Thomson, J. Satsangi, H. J. Flint, J. Parkhill, C. W. Lees, and G. L. Hold. 2014. The impact of different DNA extraction kits and laboratories upon the assessment of human gut microbiota composition by 16S rRNA gene sequencing. *PLOS One*, 9: e88982.

Kilbane, J. 2014. Effect of sample storage conditions on oilfield microbiological samples. In: *Corrosion 2014*, Paper # 3788, NACE Corrosion 2014 Conference, San Antonio, TX, USA.

Kilpatrick, C. W. 2002. Noncryogenic preservation of mammalian tissues for DNA extraction: an assessment of storage methods. *Biochemical Genetics* 40: 53–62.

Ladell, B. A., L. R. Walleser, S. G. McCalla, R. A. Erickson, and J. J. Amberg. 2019. Ethanol and sodium acetate as a preservation method to delay degradation of environmental DNA. *Conservation Genetics Resources* 11: 83–88.

Larsen, J., K. Rasmussen, H. Petersen, K. Sorensen, T. Lundgaard, and T. L. Skovus. 2010. Consortia of MIC bacteria and archaea causing pitting corrosion in top side oil production facilities. In: *Corrosion 2010*, Paper #10252, NACE Corrosion 2010 Conference, San Antonio, TX, USA.

Lauber, C. L., N. Zhou, J. I. Gordon, R. Knight, and N. Fierer. 2010. Effect of storage conditions on the assessment of bacterial community structure in soil and human-associated samples. *FEMS Microbiology Letters* 307: 80–86.

Longmire, J. L., M. Maltbie, and R. J. Baker. 1997. Use of "lysis buffer" in DNA isolation and its implications for museum collections. Occasional Papers. *The Museum of Texas Tech University*, 163: 1–3.

Majaneva, M., O. H. Diserud, S.H.C. Eagle, E. Boström, M. Hajibabaei, and T. Ekrem, T. 2018. Environmental DNA filtration techniques affect recovered biodiversity. *Scientific Reports* 8: 4682.

Mäki, A., P. Salmi, A. Mikkonen, A. Kremp, A., and M. Tiirola. 2017. Sample preservation, DNA or RNA extraction and data analysis for high-throughput phytoplankton community sequencing. *Frontiers in Microbiology* 8: 1848.

McCarthy, A., E. Chiang, M. L. Schmidt, and V. J. Denef. 2015. RNA preservation agents and nucleic acid extraction method bias perceived bacterial community composition. *PLOS One* 10: e0121659.

Menke, S., M. A. F. Gillingham, K. Wilhelm, and S. Sommer. 2017. Home-made cost effective preservation buffer is a better alternative to commercial preservation methods for microbiome research. *Frontiers in Microbiology* 8: 102.

Minamoto, T., T. Naka, K. Moji, and A. Maruyama. 2016. Techniques for the practical collection of environmental DNA: filter selection, preservation, and extraction. *Limnology* 17: 23–32.

NACE. (2014). *NACE TM0194-2014. Field monitoring of bacterial growth in oil and gas systems*. NACE International, Houston, TX.

NACE. (2018). *NACE TM0212-2018. Detection, testing, and evaluation of microbiologically influenced corrosion on internal surfaces of pipelines*. NACE International, Houston, TX.

Nechvatal, J.M., J. L. Ram, M. D. Basson, P. Namprachan, S. R. Niec, K. Z. Badsha, L. H. Matherly, A. P. N. Majumdar, and I. Kato. 2008. Fecal collection, ambient preservation, and DNA extraction for PCR amplification of bacterial and human markers from human feces. *Journal of Microbiological Methods* 72: 124–132.

Oldham, A. L., V. Sandifer, and K. E. Duncan. 2019. Effects of sample preservation on marine microbial diversity analysis. *Journal of Microbiological Methods* 158: 6–13.

Orphan, V. J., L. T. Taylor, D. Hafenbradl, and F. F. Delong. 2000. Culture-dependent and culture-independent characterization of microbial assemblages associated with high-temperature petroleum reservoirs. *Applied and Environmental Microbiology* 66: 700–711.

Rachel, N. M., and L. M. Gieg. 2020. Preserving microbial community integrity in oilfield produced water. *Frontiers in Microbiology* 11: 581387.

Renshaw, M. A., B. P. Olds, C. L. Jerde, M. M. McVeigh, and D. M. Lodge. 2015. The room temperature preservation of filtered environmental DNA samples and assimilation into a phenol-chloroform-isoamyl alcohol DNA extraction. *Molecular Ecology Resources* 15: 168–176.

Ridley, C. M., and G. Voordouw. 2018. Aerobic microbial taxa dominate deep subsurface cores from the Alberta oil sands. *FEMS Microbiology Ecology* 94: fiy073.

Rissanen, A. J., E. Kurhela, T. Aho, T. Oittinen, and M. Tiirola. 2010. Storage of environmental samples for guaranteeing nucleic acid yields for molecular microbiological studies. *Applied Microbiology and Biotechnology* 88: 977–984.

Sales, N. G., O. S. Wangensteen, D. C. Carvalho, and S. Mariani. 2019. Influence of preservation methods, sample medium and sampling time on eDNA recovery in a neotropical river. *Environmental DNA* 1: 119–130.

Salgar-Chapporo, S., and L. L. Machuca. 2018. Effect of sample storage conditions on the molecular assessment of MIC. In: *Corrosion & Prevention 2018*, Paper # 46, Adelaide, Australia.

Sekar, R., L. T., Kaczmarsky, and L. L. Richardson. 2009. Effect of freezing on PCR amplification of 16S rRNA genes from microbes associated with black band disease of corals. *Applied and Environmental Microbiology* 78: 2581–2584.

Sharma, N., and W. Huang. 2019a. Expanding industry access to molecular microbiological methods: development of an off-the-shelf laboratory workflow for qPCR and NGS analysis. In: *Corrosion 2019*, Paper # 13033, NACE Corrosion 2019 Conference, Nashville, TN, USA.

Sharma, N., and W. Huang. 2019b. Rapid in-field collection and ambient temperature preservation of corrosion-related microbial samples for downstream molecular analysis. In: *Oilfield microbiology*, eds.T. L. Skovhus and C. Whitby. CRC Press, Boca Raton, FL, USA.

Shikama, K. (1965). Effect of freezing and thawing on the stability of double helix of DNA. *Nature* 207: 529–530.

Sidstedt, M., L. Jansson, E. Nilsson, L. Noppa, M. Forsman, P. Rådström, and J. Hedman. 2015. Humic substances cause fluorescence inhibition in real-time polymerase chain reaction. *Analytical Biochemistry* 487: 30–37.

Skovhus, T. L., R. B. Eckert, and E. Rodrigues. 2017. Management and control of microbiologically influenced corrosion (MIC) in the oil and gas industry – Overview and a North Sea case study. *Journal of Biotechnology* 256: 31–45.

Suflita, J. M., D. F. Aktas, A. L. Oldham, B. M. Perez-Ibarra, and K. E. Duncan. 2012. Molecular tools to track bacteria responsible for fuel deterioration and microbiologically influenced corrosion. *Biofouling* 28: 1003–1010.

Tebbe, C. C., and W. Vahjen. 1993. Interference of humic acids and DNA extracted directly from soil in detection and transformation of recombinant-DNA from bacteria and a yeast. *Applied and Environmental Microbiology* 59: 2657–2665.

Teng, F., S. S. D. Nair, P. Zhu, S. Li, S. Huang, X. Li, J. Xu, and F. Yang. 2018. Impact of DNA extraction method and targeted 16S-rRNA hypervariable region on oral microbiota profiling. *Scientific Reports* 8: 16321.

Tsai, Y. L., and B. H. Olson. 1992. Rapid method for separation of bacterial-DNA from humic substances in sediments for polymerase chain-reaction. *Applied and Environmental Microbiology* 58: 2292–2295.

Vink, C. J., S. M. Thomas, P. Paquin, C. Hayashi, and M. Hedin. 2005. The effects of preservative and temperature on arachnid DNA. *Invertebrate Systematics* 19: 99–104.

Wegleitner, B. J., C. L. Jerde, A. Tucker, W. L. Chadderton, and A. R. Mahon. 2015. Long duration, room temperature preservation of filtered eDNA samples. *Conservation Genetics Resources* 7: 789–791.

Williams, K. E., K. P. Huyvaert, and A. J. Piaggio. 2016. No filters, no fridges: a method for preservation of water samples for eDNA analysis. *BMC Research Notes* 9: 298.

Yamanaka, H., T. Minamoto, J. Matsuura, S. Sakurai, S. Tsuji, H. Motozawa, M. Hongo, Y. Sogo, N. Kakimi, I. Teramura, M. Sugita, M. Baba, and A. Kondo. 2017. A simple method for preserving environmental DNA in water samples at ambient temperature by addition of cationic surfactant. *Limnology*, 18: 233–241.

Yasuda, T., S. Awazu, W. Sato, R. Iida, Y Tanaka, and K. Kishi. 1990. Human genetically polymorphic deoxyribonuclease: purification, characterization, and multiplicity of urine deoxyribonuclease. *Journal of Biochemistry* 108: 393–398.

Zhu, X., and M. Al-Moniee. 2017. Molecular microbiology techniques. In: *Trends in oil and gas corrosion research and technologies*, ed.A.M. El-Sherik, 513–536. Woodhead Publishing, Duxford, UK.

23 Implications of Sampling and Chemistry

Kelly A. Hawboldt, Christina S. Bottaro, Abdulhaqq Ibrahim, Mahsan Basafa, and Angham Saeed
Memorial University of Newfoundland

CONTENTS

23.1 INTRODUCTION

MIC is impacted by many factors including the chemistry of the environment (e.g.,, produced water vs crude oil), temperature, pressure, pH, material degradation, in addition to the composition of the microbial community. These factors are well understood with respect to promoting or inhibiting MIC, but there is less knowledge on the chemical interaction between the fluid, microbes, and corrosion products and subsequent impact on rate and extent of MIC. This is particularly true for the dynamic nature of the chemistry of the surrounding environment (due to changes in temperature, pressure, pH, etc.) on microbial growth or corrosion products and the subsequent impact on MIC. The first steps in attempting to capture these interactions are the accurate detection and measurement of key compounds involved in MIC in the process stream, as well as the dominant chemical reactions/interactions as a function of temperature, pressure, and pH. This chapter is focused on discussing proper sampling protocols to maintain sample integrity in the field and for transport so key compounds can be

DOI: 10.1201/9780429355479-27

identified. We provide a summary of analytical methods (field and lab) to detect and measure these compounds and a summary of key chemical reactions occurring in oil and gas production platforms/sites that could influence MIC.

23.2 SAMPLING AND ANALYSIS

Sampling protocols for chemical analyses are a function of the type of fluids (produced water, crude oil, injection water, etc.), type of analyses (metals, nutrients, sulfur compounds, etc.), stability of sample, and distance to analysis point (e.g., on-site or transported off-site). Other factors that must be considered are conditions at the point of sampling. For instance, if samples are taken at atmospheric conditions from a high-pressure separator, then this must be corrected for at some point in the analysis of the sample, either adjusting analytical procedures or calculating compositions by correcting for temperature, pressure and pH. While there are published sampling protocols, these do not typically address the challenges associated with sampling in remote locations such as off-site analyses, as well as lack of on-line or in-situ analysis that complies with existing regulations. These challenges are common on off-shore oil and gas operations in remote regions and particularly important for unmanned facilities, where all sampling must be done by automated systems. As such, any sampling or monitoring program must be cognizant of these challenges.

As indicated previously, proper sampling, storage, and transport are key steps in maintaining the integrity of the samples and can be responsible for a significant proportion of the overall analysis error (Stanley and Smee, 2007; Yang and McEwan, 2005). Chapter 26 contains a sample protocol for collecting samples for chemical and corrosion analyses. The suite of analyses to be performed is a function of the type of fluid (e.g., produced water, soured crude, etc.), production conditions (e.g., seawater injection and other enhanced oil recovery methods, etc.), on-site laboratory capacity, and regulations.

23.2.1 Sampling and Preservation

There are a number of compounds that can be measured as indicators of MIC, its onset, and/or could be integrated into a risk assessment model for MIC. However, during the sampling or transport of samples, if the samples are not properly handled there can be speciation/reaction of the compounds with time, temperature, pH, and pressure. Any changes in chemical composition due to reaction could result in lack of action or inappropriate actions with respect to corrosion management, which is why sampling and preservation procedures are critical. There are well established protocols and methods for sampling of crude oil, produced water, and related reservoir fluids (e.g., ASTM D4057: Standard Practice for Manual Sampling of petroleum and petroleum products). In general, depending on the compounds to be analyzed, different sampling bottles are used. Metal analysis typically requires polyethylene bottles, except where mercury and arsenic are to be analyzed. In these cases, colorless glass bottles are used with an additional cleaning step with nitric acid and rinse with deionized water prior to sampling (OSPAR, 2005; ISO, 2003; OLF (Norwegian Oil Industry Association), n.d.). Dark glass bottles with Teflon or Teflon-lined caps are used for where fluids are to be analyzed for organics, with gas tight sample containers or tubes for volatile organic compounds (Utvik, 1999). Sampling for analysis of radioactive compounds requires glass bottles with Teflon caps that have been previously rinsed with deionized water. During sampling the bottle is flushed three times with the batches from the sample point. To preserve the integrity of the sample, samples are ideally stored in a dark cold environment, at about 4°C and transported as fast as possible to the laboratory for analyses (OSPAR, 2005; ISO, 2003).

Various preservation methods can be used to prevent metals from precipitation and adsorption (by sample container or microbial activity) and aid in analysis (sample digestion), the most common are acidification with HNO_3, HCl, or H_2SO_4, and low temperatures (e.g., 1–4°C) (ISO, 2003). ISO 5667–3 outlines preservation methods in detail. However, addition of these preservatives can result in speciation or reaction in solution. For instance, sulfur species including sulfur oxyanions (such as thiosulfate) and sulfides are sensitive to changes in temperature, pH, and redox environment; leading to rapid changes in speciation conditions change or if the sample is exposed to oxidizing environments or during sample preparation procedures such as degassing (Miranda-Trevino et al., 2013).

23.2.1.1 Chemicals of Interest in MIC and Related Corrosion

The chemical environment (i.e., produced water, crude oil, gas/oil mixtures) impacts the type of microbes that are present, the acceleration or inhibition of corrosion (electrochemical activity) and in turn, metabolites released or nutrient absorbed through microbial activity and corrosion by-products impact the chemistry of the environment (Figure 23.1).

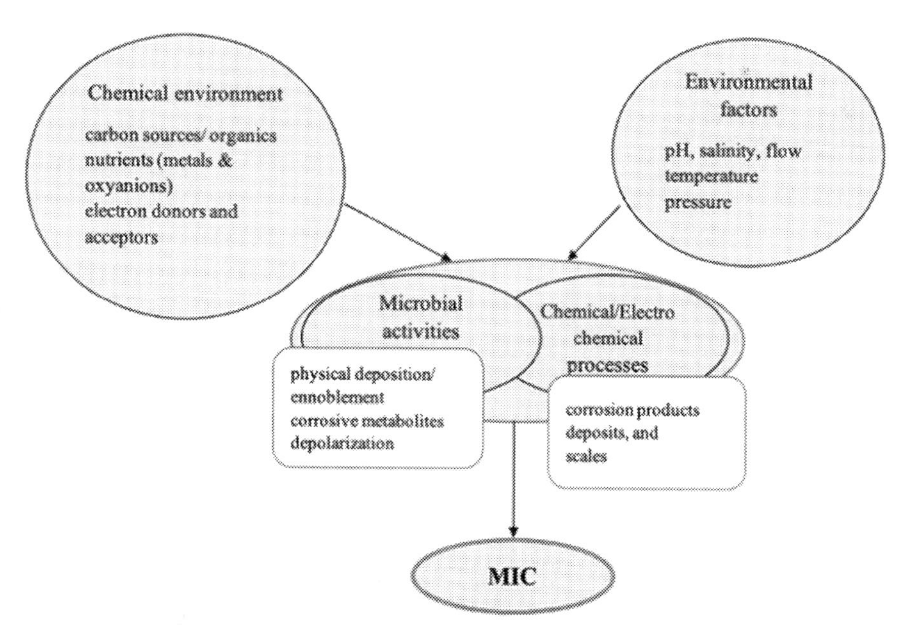

FIGURE 23.1 Interactions between chemical environment, environmental factors, electrochemical, and microbial processes.

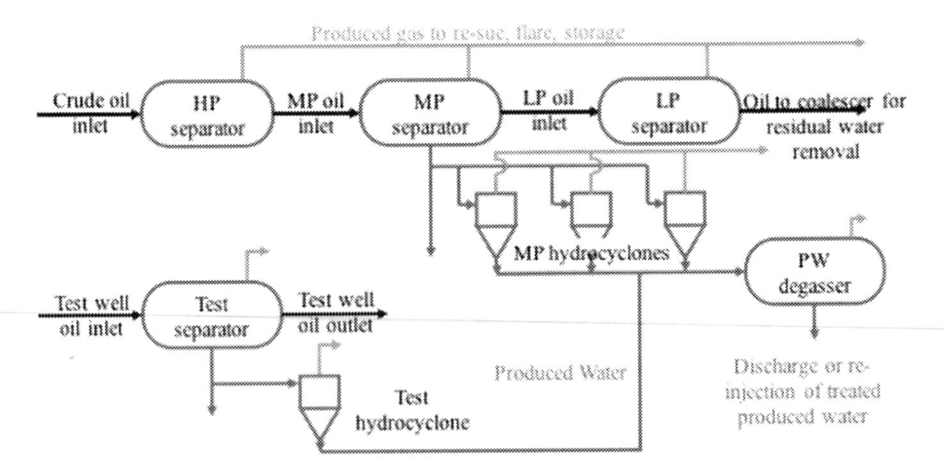

FIGURE 23.2 Typical separator train on oil producing offshore platform (HP – High Pressure, MP – Medium Pressure, LP – Low Pressure, PW – Produced Water).

In MIC, in addition to corrosion being a function of temperature, pressure, pH, composition, etc., in MIC we have the added effects of microbes which could enhance or inhibit the rate of corrosion. The purpose of this section is to attempt to review the impact of the chemistry on MIC. The information presented is modified from published and unpublished work in our lab as well as relevant literature from other studies.

23.2.1.2 Chemistry of Produced Fluids

Reservoir fluids are pumped to the surface and depending on composition go through a series of separators to remove water, gas, and/or other compounds that impact oil/gas quality. The process flow for a typical offshore oil producing platform is outlined in Figure 23.2.

The crude oil inlet is largely oil, water, dissolved gases, and, to a lesser extent, production fluids, biocides, corrosion inhibitors, oxygen scavengers, naturally occurring radioisotopes, inorganic salts, etc. As the fluid moves through the separator train, the temperature and pressure change results in changes in pH and composition of oil, aqueous, and gas phases. By the low-pressure separator (typically operates between 1 and 1.5 bar), most of the gases and water have been removed from the crude oil. The produced water goes through a separate treatment system to degas and remove oil to discharge or reuse levels (as low as 20 ppm in some regions). Produced gas can be used on the platform in gas turbines, re-injected for enhanced oil recovery (EOR), and/or flared. If the platform is close to shore or connected to a pipeline, the gas can be transported to land for use. The gas can also be liquefied to LNG for transport. However, both of these options are limited when the platform is remotely located or in harsh environments where space on the platform for processing is limited. The amount of processing of the reservoir fluids and any EOR methods directly impacts the type of fluids present in the fluid and the partitioning behaviour (phase transitions) through the separator train. For instance, flowlines upstream of the HP separator will be more acidic if CO_2 is present in the reservoir fluids, as this will be in a dissolved state in the aqueous fraction of the fluids. Downstream of the MP separator, the pH will rise due to the degassing of the CO_2. This shift in pH may be small (e.g., from 6.3 to 7) but this small shift can have significant impacts on the partitioning and speciation of the compounds and hence the risk of corrosion.

23.2.1.3 Petroleum Fraction

Organic compounds including hydrocarbons such as alkanes and alkenes (C_1–C_{30}), and aromatics dominate oil and gas. They can serve as substrates in the form of carbon sources for aerobic or anaerobic microbial species including but not limited to denitrifying, sulfate-reducing, and ferric iron-reducing bacteria, which are important bacteria in MIC (Heider et al., 1998). The presence of organic substrates coupled with electron donors and acceptors present in the reservoir fluids can create favorable conditions for microbial growth and therefore offshore infrastructure are highly prone to MIC.

As indicated previously, the corrosivity of petroleum is a function of the chemistry, water:oil:gas ratios, oil emulsion type, and wettability. Specifically in the case of the corrosivity of crude oil compounds such as inorganic salts, sulfur, organic acids, dissolved gasses, solids, and paraffins drive the corrosion potential (Kui et al., 2008).

Crude may contain 100–2,000 mg/L of inorganic salts (NACE, 1958), with sulfur content typically less than 1 wt% but higher for heavier crude oils (e.g., 2.0–3.5 wt%) (Easton, 1960). The form of sulfur is a key factor in corrosion, not necessarily the amount of total sulfur (Piehl, 1960). Sulfur compounds such as sulfate are associated with corrosion; however, if a stable sulfide layer is formed on the metal surface, it can inhibit corrosion. In crude oil, naphthenic acid has been identified as the key contributor to corrosivity at higher temperatures (e.g., 220–370°C) (Kui et al., 2008). Other organic acids may play a role in the corrosivity at low temperatures including formic, acetic, and propionic acids (Greco and Griffin, 1946). Solids can also contribute to corrosion. Crude oil can contain fine solid particles which in turn settle at the bottom of the pipe at low flow velocities and facilitate under-deposit corrosion (Lotz et al., 1991).

23.2.1.4 Gas

The main corrosive compounds in the gaseous phase or dissolved in the fluid phase include O_2, H_2S, and CO_2, which depends on solubility or partitioning. For instance, H_2S is more soluble in oil than water with a partition coefficient of 3 at 30°C (Eden et al., 1993). O_2 and CO_2 are also less soluble in water than in hydrocarbons (Lotz et al., 1991). The concentration of the gases, pH of solution, presence of dissolved water in the crude, and temperature and pressure of the fluid will determine the degree of corrosion due to these gases.

23.2.1.5 Water (Aqueous) Phase

Produced water is the aqueous phase produced with the oil and/or gas phases and may be formation water (water present in the pores of hydrocarbon-producing rock

TABLE 23.1
Inorganic Constituents in Produced Water

Compound/Element/Ion	References
Na, Ca, Mg, K, Sr, Ba, B, $LiNH_4^+$, NH_3,Cl^-, Br^-, $I^-SO_4^{2-}$, HSO_4^- SO_3^{2-}, HSO_3^-, S^0, H_2S, HS^-, S^{2-}, $S_2O_3^{2-}$, polysulfide (S_2^{2-}, S_3^{2-}, S_4^{2-}, S_5^{2-}) CO_3^{2-}, HCO_3^-, dissolved O_2, OH^-, NO_3^-, NO_2^-, PO_4^{3-}	(Laredo et al., 2004; Lotz et al., 1991; Davies and Scott, 2006; Schmitt and Feinen, 2000; Revie and Uhlig, 2008)

layers) or a combination of seawater and formation water, depending on the production scheme. Produced water is a complex mixture of dissolved gases, solids, and organic and inorganic compounds. Produced water is predominantly made up of dissolved organics (including hydrocarbons), minerals, and gases (O_2, CO2, H_2S; suspended oil (nonpolar); suspended solids (corrosion products, scale, sand, silt, etc. (; traces of heavy metals; production chemicals (treating chemicals, kill fluids, acids, etc.); and bacteriological matter (Juniel, 2003) The inorganic constituents that have been measured in produced water are summarized in Table 23.1.

Produced water also contains organic acids (mono- and di-carboxylic acids of saturated and aromatic hydrocarbons). Organic acids of low molecular weight carboxylic acids (formic, acetic, propanoic, butanoic, pentanoic, and hexanoic acids) (Utvik, 1999), make up the bulk of organic carbon (TOC) in produced water (Lee and Neff, 2011). Microorganisms such as bacteria and fungi can utilize these low molecular weight organics.

The salinity varies widely but has a much higher concentration of dissolved inorganic salts compared to seawater (Neff, 1997). The corrosivity is a function of the concentration and types of ions as they influence conductivity, electrochemical reactions, and alter metal surface layers (Kui et al., 2008). The most common compound, NaCl, disassociates to Na^+ and Cl^- which in turn increases conductivity and subsequent transport of ions to and from the metal. Cl^- can penetrate and destroy the metal's protective oxide film or other surface layers and as such increases the susceptibility to localized pitting corrosion. The rate of pit development increases as Cl^- concentrations increase 10,000 to 120,000 ppm (Papavinasam et al., 2010). Other halides may also impact pitting corrosion, however Cl has the largest effect on localized pitting corrosion followed by Br, I, and F (Schimdt and Fenien, 2000). A study by Revie and Uhlig (2008), showed the impact of chlorine on corrosion. The corrosion rate increases with Cl^- concentration until 3% NaCl. As NaCl levels exceed 3% the dissolved oxygen in solution decreases (due to salinity) and therefore the corrosion rate drops.

Phosphate ions can decrease the susceptibility of metal to localized pitting corrosion. Phosphoric acid ($H_2PO_4^-$) has the most impact on decreasing corrosion followed by the more reduced forms (HPO_4^{-2} and PO_4^{-3}) of the ion.

Univalent cations, including Na^+, K^+, Rb^+, and Cs^+ increase the susceptibility of iron and steel to pitting corrosion. Li^+ has not shown does not show the same pitting behavior. In general, this effect decreases as valence of the cation increases; in fact, of the bivalent cations, only Zn^{2+} showed increases in the susceptibility of metals to pitting corrosion. There was no pitting corrosion observed in the presence of trivalent cation such as Al^{+3} (Revie and Uhlig, 2008).

The combined presence of cations and anions in aqueous phase can form salt precipitate or scale. Scale formation may cause underdeposit corrosion which could in turn plug flowlines and equipment. Common scale deposits include salts such as calcium carbonate ($CaCO_3$), calcium sulfate ($CaSO_4$), strontium sulfate ($SrSO_4$), and barium sulfate ($BaSO_4$). The amount and persistence of scale depends on the solubility of the salt which in turn is a function of composition, temperature, pressure, and pH (Papavinasam, 2014).

23.2.1.6 Sampling

As outlined previously, given the complex nature of petroleum fluids, multiple phases, stability of compounds, and impact that shifting conditions can have on partitioning and species of compounds, careful sampling and preservation is key. The preservation methods outlined in the first part of this chapter, while effective for ensuring the stability of the targeted compounds, can shift the partitioning and/or species of other key MIC-related chemical compounds of interest. This would result in an inaccurate picture of the process sample point and consequently ineffective or incorrect action taken to manage the corrosion. The protocols outlined in the appendix of this chapter were developed with this in mind. Preservatives are kept to a minimum or eliminated in order to give an accurate picture of the compounds present and their relative amounts in each phase while still ensuring the samples are stable. The protocols were developed based on sets of tests where samples were taken in duplicate, one with required preservatives added for analyses of key compounds (for example addition of acids to control microbial growth) and the second where the sample was maintained at approximately 4°C. Samples from multiple points in the separator train were taken to ensure the breadth of the process was represented. The samples were shipped from offshore platforms where storage on the platform and/or transport time varied from a few hours to a several days. The two sets of samples were then compared with respect to compounds that need preservatives (e.g., organics) for stability and compounds that are impacted by shifts in composition and pH (e.g., sulfur compounds). These samples were also compared with analyses completed on the platform. Overall, the samples stored at 4°C not only showed stability with respect to compounds that degrade without preservation but also showed good agreement in composition with target compounds measured on the platform. The protocols were then developed based on the breadth of target compounds that are associated with MIC. While sampling is key, the type of analyses (i.e., how it is measured) are also important and often overlooked.

23.2.1.7 Chemical Analysis of Key MIC Related Chemical Compounds

Analysis of chemical species implicated in MIC can be complex with a wide range of concentrations, interferences, and phases to consider. With respect to nutrients, oil and gas operators routinely analyze the composition of crude oil and process fluids for process control and regulatory compliance. They are often guided by regulatory agencies who mandate the features of analytical methods for analysis of discharge from oil and gas extraction and processing facilities. The array of chemical species measured along with the variety of analytical tools involved are too vast to cover here, and to a large extent the methods are well reported in many comprehensive reviews and monographs (Nadkarni and Nadkarni, 2007; Tibbetts et al., 1992). However, these methods should be drawn upon when completing compositional analysis as part of the assessment for MIC potential. Comprehensive chemical analysis is important for diagnosis and treatment of MIC, but the resulting data are also crucial in building appropriate experimental models that reflect the real-world chemical environment. In this section, we highlight some of the common and emerging technologies for analysis of the labile species implicated in MIC and other corrosion processes with an eye to help chose the effective methods for timely and accurate data for applications from modelling to mitigation.

23.2.1.8 Inorganic Anions and Organic Acids

Ion chromatography (IC) has been used extensively to analyze all types of cations and anions. For the reactive sulfur species, IC was the first viable analytical method for simultaneous determination of trace inorganic anions with good reliability (Chen and Naidu, 2003; Haddad et al., 1999). Despite its a good sensitivity, IC exhibits only moderate separation efficiency and speed, and low tolerance for complex matrices which can quickly erode column performance. Capillary electrophoresis (CE) has emerged as a versatile alternative to IC (Padarauskas et al., 2000; Motellier and Descostes, 2001; Safizadeh and Larachi, 2014; Hissner et al., 1999). CE is an operationally simple instrument with nearly unparalleled separation efficiency and speed. These features combined with the low cost of consumables and tolerance for complex sample matrices make it well-suited for the analysis of nearly any charged species, even in highly saline complex samples (Kaniansky et al., 1999; Tůma et al., 2008; Soga and Ross, 1999; Donkor et al., 2015). Figure 23.3 shows analysis of produced water with direct and indirect detection of anions, with a focus on the products of sulfate reduction, which are abundant. The indirect method allows for detection of species invisible by UV-vis; the direct method is suitable for most thiosalts and allows for correction of chloride signal for the co-migrating thiosulfate. Perhaps more importantly, simple sample preparation combined with the speed of separation can dramatically improve the precision of analysis for unstable sulfur-containing species like thiosulfate and polythionates (De Carvalho and Schwedt, 2005; Miranda-Trevino et al., 2013). Since the flow that drives CE separations is generated through application of an electric field within the capillary (or microfluidic channel) rather than by an external pump, it is also the preferred separation mode for lab-on-a-chip applications, which are seen as the best path to development of field portable systems for complex analyses. The main caveat is that CE methods rarely meet the sensitivity and linearity of ion-selective electrodes, which are the more common field portable detection systems for electroactive species (Crespo, 2017; Pandey et al., 2012). Nevertheless, CE can be highly specific and as well provides a platform with numerous strategies to improve

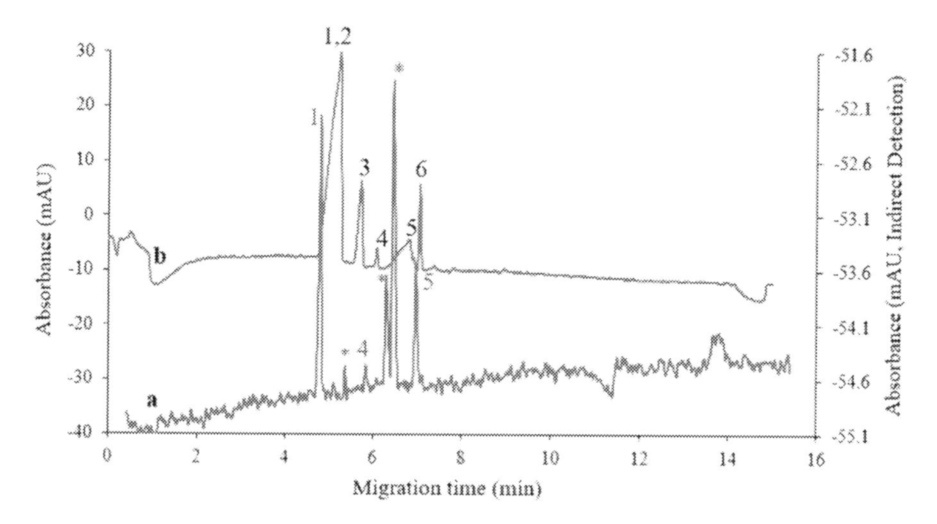

FIGURE 23.3 Produced water with: 1) $S_2O_3^{2-}$ 0.81 mM, 2) Cl^- 3.42 mM; 3) SO_4^{2-} 1.41 mM, 4) HS^- 2.56 mM, 5) $S_4O_6^{2-}$ 0.66 mM, 6) SO_3^{2-} 1.59 mM. BGE: a) direct method: 1.0 mM HMOH, 16.0 mM NH_4OH/CH_2O_2, pH 9.0, b) indirect method: 1.0 mM HMOH, 16.0 mM NH_4OH, 6.0 mM PMA, pH 9.0. Spectra acquired at 214 nm with reference at 360 nm (Saeed, 2017). *Unidentified.

sensitivity and portability (Lara et al., 2016; Liu et al., 2016; Swinney and Bornhop, 2000; Wang, 2005). Moreover, performance of electrochemical sensors can be dramatically impaired by fouling, which is not a concern for robust CE strategies (Hanssen et al., 2016; Manica et al., 2003; Salimi et al., 2006; Yang et al., 2013).

23.2.1.9 Organic Biomarkers

Organic biomarkers of MIC are usually confined to extracellular metabolites produced in systems with known corrosion. Some of the compounds that have already been identified as MIC indicators are simple metabolic products of bacterial activity in general; however, some are specific to nutrient sources intrinsic to oil and gas pipelines. For example, in a study of MIC in copper water transmission lines, three-dimensional fluorescence spectroscopy revealed protein-like and humic-like fluorophores as specific indicators of biological activity (Beale et al., 2012). The fluorometric methods are unlikely to be successful for MIC in petrogenic systems given the complex mixture of fluorescent compounds, such as polycyclic aromatic hydrocarbons phenols and naphthenic acids. In the same study, derivatization gas chromatography with mass spectrometry (GC-MS) could be used to identify numerous small organic acids (e.g., acetic, propionic, maleic, and oxalic acids), amino acids, and fatty acids, some increasing and others decreasing with microbial activity. Suflita and co-workers (Lenhart et al., 2014) showed that samples from oil pipelines with confirmed MIC presented an assortment of aromatic hydrocarbons (i.e., toluene, xylene, light polycyclic hydrocarbons and their alkylated analogues). Given that most of the metabolites feature ionizable proton on the acidic functionalities in contrast to typical oil hydrocarbons, CE is a good option for biomarker separations. Nevertheless, although GC requires derivatization of the acid moiety, it provides exceptional peak capacities and proven sensitivity and specificity when combined with MS especially when non-targeted analysis is desired. For targeted analysis, high-performance chromatography with tandem MS (HPLC-MS^n) is a good choice and analysis can proceed without the safety and analytical pitfalls associated with derivatization.

The chemical composition and resulting reactivity of fluids is a major determinant of the initiation, extent, and impact of MIC. However, the chemical reactions are often ignored in favour of microbial reactions and/or "lumped in" as part of the microbial activity, overlooking or discounting the key role that chemical reactions play in the production of compounds that could serve as microbial growth inhibitors versus growth enhancers and products that accelerate versus impede corrosion. By identifying the chemical compounds via robust and fast analytical methods and identifying the key reactions that can impact MIC we can "decouple" the chemical reactivity from the microbial activity. Although MIC is frequently detected after a failure, increasingly oil and gas producers and pipeline operators are equipped with a range of approaches to monitor and detect MIC early, such as the use of well-placed coupons and a range of electrochemical techniques (Little and Lee, 2007). These techniques are most effective once microbial communities have been established and biofilms form, which makes treatment and prevention of corrosion more difficult. Approaches for detecting MIC ahead of significant corrosion will require methods that detect trace quantities of critical nutrients and/or metabolites of MIC-specific organisms. Both genomics and metabolomics research will be central to establishing a set of indicators that can be broadly monitored in systems at risk for MIC (Li et al., 2017; Gieg et al., 2011; Beale et al., 2012; Procópio, 2020). Armed with a good indicator list, analysts can develop the suite of methods for sensitive, fast targeted screening analytical protocols to allow for timely interventions. Currently in-line, on-line, and remote monitoring systems mainly use spectrophotometric or electrochemical methods (Mukhopadhyay and Mason, 2013; Zulkifli et al., 2018; Bertone et al., 2018). None of these will meet the detection limits needed in these complex systems. Separations of some fashion will be required. Online mass spectrometry that use membrane introduction may be viewed as an option, but it is well-recognized that typical permeable membrane introduction systems work well with light hydrophobic compound, though not for polar or ionized metabolites, particularly acids already associated with MIC (Bell et al., 2015; Johnson et al., 2000). Research into the use of condensed phase membrane introduction interfaces with MS that better allow for the transfer of these compounds is progressing (Vandergrift et al., 2017; Krogh and Gill, 2014). Also emerging is the use of molecular imprinted polymer (MIP) adsorbents for targeted monitoring applications. MIPs have been engineered for a wide range of targets, including some of the metabolic products associated with MIC or closely related compounds (Pichon and Chapuis-Hugon, 2008; Ramström et al., 2001; Michailof et al., 2008; Sun et al., 2001; Abu-Alsoud and Bottaro, 2021; Shahhoseini et al., 2020). These can be integrated into sensors for selective sequestration with detection by any method sensitive for the target; electrochemical detection, quartz microsensors, fluorescence, and mass spectrometry are routinely used (Yáñez-Sedeño et al., 2017;

Dickert et al., 1999; Dickert et al., 2001; Huang et al., 2015). Since MIPs can be prepared as films, they can be deployed for in situ sampling with analysis in a traditional laboratory or interrogated on-site using field portable detectors, e.g., mass spectrometers, fluorimeters, spectrophotometers (Van Biesen et al., 2010; Santos et al., 2020; Henry et al., 2005). As multidisciplinary research teams continue cooperative efforts to both identify good early-warning indicators and the methods to detect them, it is anticipated that better tools will result for MIC mitigation.

REFERENCES

Abu-Alsoud, G. F., & Bottaro, C. S. (2021). Porous thin-film molecularly imprinted polymer device for simultaneous determination of phenol, alkylphenol and chlorophenol compounds in water. *Talanta, 223*. 10.1016/j.talanta.2020.121727

Beale, D. J., Dunn, M. S., Morrison, P. D., Porter, N. A., & Marlow, D. R. (2012). Characterisation of bulk water samples from copper pipes undergoing microbially influenced corrosion by diagnostic metabolomic profiling. *Corrosion Science, 55*, 272–279. 10.1016/j.corsci.2011.10.026

Bell, R. J., Davey, N. G., Martinsen, M., Collin-Hansen, C., Krogh, E. T., & Gill, C. G. (2015). A field-portable membrane introduction mass spectrometer for real-time quantitation and spatial mapping of atmospheric and aqueous contaminants. *Journal of the American Society for Mass Spectrometry, 26*(2), 212–223. 10.1021/jasms.8b04961

Bertone, E., Burford, M. A., & Hamilton, D. P. (2018). Fluorescence probes for real-time remote cyanobacteria monitoring: A review of challenges and opportunities. *Water Research, 141*, 152–162. 10.1016/j.watres.2018.05.001

Chen, Z., & Naidu, R. (2003). Separation of sulfur species in water by co-electroosmotic capillary electrophoresis with direct and indirect UV detection. *International Journal of Environmental Analytical Chemistry, 83*(9), 749–759. 10.1080/0306731031000114938

Collins, A. G. (1975). *Geochemistry of oilfield waters*. Elsevier Scientific Pub. Co.

Crespo, G. A. (2017). Recent advances in ion-selective membrane electrodes for in situ environmental water analysis. *Electrochimica Acta, 245*, 1023–1034. 10.1016/j.electacta.2017.05.159

Davies, M., & Scott, P. J. B. (2006). *Oilfield water technology: Types of waters* (1st ed.). NACE International.

De Carvalho, L. M., & Schwedt, G. (2005). Sulfur speciation by capillary zone electrophoresis: Determination of dithionite and its decomposition products sulfite, sulfate and thiosulfate in commercial bleaching agents. *Journal of Chromatography A, 1099*(1–2), 185–190. 10.1016/j.chroma.2005.08.084

Dickert, F. L., Halikias, K., Hayden, O., Piu, L., & Sikorski, R. (2001). Sensors based on fingerprints of neutral and ionic analytes in polymeric materials. *Sensors and Actuators B: Chemical, 76*(1), 295–298. 10.1016/S0925-4005(01)00588-3

Dickert, Franz L., Tortschanoff, M., Bulst, W. E., & Fischerauer, G. (1999). Molecularly imprinted sensor layers for the detection of polycyclic aromatic hydrocarbons in water. *Analytical Chemistry, 71*(20), 4559–4563. 10.1021/ac990513s

Donkor, K. K., Guo, Z. C., Soliman, L. C., Law, Y. T., Risley, J. M., Schmidt, K. J., Crabtree, H. J., & Warrender, N. A. (2015). Determination of sulfate and chloride ions in highly saline oilfield water by capillary electrophoresis using bilayer-coated capillaries and indirect absorption detection. *International Journal of Environmental Analytical Chemistry*, *95*(2), 175–186. 10.1080/03067319.2014.1002489

Easton, C. L. (1960). Corrosion control in petroleum refineries processing Western Canadian crude oils. *Corrosion*, *16*(6), 275t–280t. 10.5006/0010-9312-16.6.109

Eden, B., Laycock, P. J., & Fielder, M. (1993). Oilfield reservoir souring. Health and Safety Executive - Offshore Technology Report, OTH 92 385.

Gieg, L. M., Jack, T. R., & Foght, J. M. (2011). Biological souring and mitigation in oil reservoirs. *Applied Microbiology and Biotechnology*, *92*(2), 263–282. 10.1007/s00253-011-3542-6

Greco, E. C., & Griffin, H. T. (1946). Laboratory studies for determination of organic acids as related to internal corrosion of high pressure condensate wells. *Corrosion*, *2*(3), 138–152.

Haddad, P. R., Doble, P., & Macka, M. (1999). Developments in sample preparation and separation techniques for the determination of inorganic ions by ion chromatography and capillary electrophoresis. *Journal of Chromatography A*, *856*, 145–177.

Hanssen, B. L., Siraj, S., & Wong, D. K. Y. (2016). Recent strategies to minimise fouling in electrochemical detection systems. *Reviews in Analytical Chemistry*, *35*(1), 1–28. 10.1515/revac-2015-0008

Heider, J., Spormann, A. M., Beller, H. R., & Widdel, F. (1998). Anaerobic bacterial metabolism of hydrocarbons. *FEMS Microbiology Reviews*, *22*(5), 459–473. 10.1016/S0168-6445(98)00025-4

Henry, O. Y. F., Cullen, D. C., & Piletsky, S. A. (2005). Optical interrogation of molecularly imprinted polymers and development of MIP sensors: A review. *Analytical and Bioanalytical Chemistry*, *382*(4), 947–956. 10.1007/s00216-005-3255-8

Higashi, R. M., & Jones, A. D. (1997). Identification of bioreactive compounds from produced water discharge/Characterization of organic constituent patterns at a produced water discharge site. *OCS Study MMS 97-0023. Coastal Research Center, Marine Science Institute, University of California, Santa Barbara, California., MMS Cooper*, 43.

Hissner, F., Mattusch, J., & Heinig, K. (1999). Quantitative determination of sulfur-containing anions in complex matrices with capillary electrophoresis and conductivity detection. *Journal of Chromatography A*, *848*(1–2), 503–513. 10.1016/S0021-9673(99)00458-6

Huang, D.-L., Wang, R.-Z., Liu, Y.-G., Zeng, G.-M., Lai, C., Xu, P., Lu, B.-A., Xu, J.-J., Wang, C., & Huang, C. (2015). Application of molecularly imprinted polymers in wastewater treatment: A review. *Environmental Science and Pollution Research*, *22*(2), 963–977. 10.1007/s11356-014-3599-8

Ibrahim, A., Hawboldt, K., Bottaro, C., & Khan, F. (2018). Review and analysis of microbiologically influenced corrosion: The chemical environment in oil and gas facilities. *Corrosion Engineering, Science and Technology*, *53*(8), 549–563. 10.1080/1478422X.2018.1511326

ISO. (2003). Water quality–Sampling, Part 3: Guidance on the preservation and handling of water samples (3rd ed.). ISO.

Johnson, R. C., Cooks, R. G., Allen, T. M., Cisper, M. E., & Hemberger, P. H. (2000). Membrane introduction mass spectrometry: Trends and applications. *Mass Spectrometry Reviews*, *19*(1), 1–37. 10.1002/(SICI)1098-2787(2000)19:1<1::AID-MAS1>3.0.CO;2-Y

Juniel, K. A. (2003). *Practical application of produced water treating technology for land-based injection operations | corrosion | filtration.* Houston, TX: NATCO Group.

Kaniansky, D., Masár, M., Marák, J., & Bodor, R. (1999). Capillary electrophoresis of inorganic anions. *Journal of Chromatography A*, *834*(1), 133–178. 10.1016/S0021-9673(98)00789-4

Krogh, E. T., & Gill, C. G. (2014). Membrane introduction mass spectrometry (MIMS): A versatile tool for direct, real-time chemical measurements. *Journal of Mass Spectrometry*, *49*(12), 1205–1213.
Kui, X., Chao-fang, D., Xiao-gang, L., & Fu-ming, W. (2008). Corrosion products and formation mechanism during initial stage of atmospheric corrosion of carbon steel. *Journal of Iron and Steel Research International*, *15*(5), 42–48.
Lara, F. J., Airado-Rodríguez, D., Moreno-González, D., Huertas-Pérez, J. F., & García-Campaña, A. M. (2016). Applications of capillary electrophoresis with chemiluminescence detection in clinical, environmental and food analysis. A review. *Analytica Chimica Acta*, *913*, 22–40. 10.1016/j.aca.2016.01.046
Laredo, G. C., López, C. R., Álvarez, R. E., Castillo, J. J., & Cano, J. L. (2004). Identification of naphthenic acids and other corrosivity-related characteristics in crude oil and vacuum gas oils from a Mexican refinery. *Energy and Fuels*, *18*(6), 1687–1694.
Lee, K., & Neff, J. (2011). *Produced water: Environmental risks and advances in mitigation technologies*. Springer Science.
Lenhart, T. R., Duncan, K. E., Beech, I. B., Sunner, J. A., Smith, W., Bonifay, V., Biri, B., & Suflita, J. M. (2014). Identification and characterization of microbial biofilm communities associated with corroded oil pipeline surfaces. *Biofouling*, *30*(7), 823–835. 10.1080/08927014.2014.931379
Li, X.-X., Liu, J.-F., Zhou, L., Mbadinga, S. M., Yang, S.-Z., Gu, J.-D., & Mu, B.-Z. (2017). Diversity and composition of sulfate-reducing microbial communities based on genomic DNA and RNA transcription in production water of high temperature and corrosive oil reservoir. *Frontiers in Microbiology*, *8*, 1011. https://www.frontiersin.org/article/10.3389/fmicb.2017.01011
Little, B. J., & Lee, J. S. (2007). *Microbiologically influenced corrosion*. Wiley. https://books.google.ca/books?id=1i8AJW6ziKMC
Liu, Y., Huang, X., & Ren, J. (2016). Recent advances in chemiluminescence detection coupled with capillary electrophoresis and microchip capillary electrophoresis. *Electrophoresis*, *37*(1), 2–18. 10.1002/elps.201500314
Lotz, U., Van Bodegom, L., & Ouwehand, C. (1991). The effect of type of oil or gas condensate on carbonic acid corrosion. *Corrosion*, *47*(8), 635–644. 10.5006/1.3585301
Manica, D., Mitsumori, Y., & Ewing, A. (2003). Characterization of electrode fouling and surface regeneration for a platinum electrode on an electrophoresis microchip. *Analytical Chemistry*, *75*(17), 4572–4577. 10.1021/ac034235f
Michailof, C., Manesiotis, P., & Panayiotou, C. (2008). Synthesis of caffeic acid and p-hydroxybenzoic acid molecularly imprinted polymers and their application for the selective extraction of polyphenols from olive mill waste waters. *Journal of Chromatography A*, *1182*(1), 25–33. 10.1016/j.chroma.2008.01.001
Miranda-Trevino, J. C., Pappoe, M., Hawboldt, K., & Bottaro, C. (2013). The importance of thiosalts speciation: Review of analytical methods, kinetics, and treatment. *Critical Reviews in Environmental Science and Technology*, *43*(19), 2013–2070.
Motellier, S., & Descostes, M. (2001). Sulfur speciation and tetrathionate sulfitolysis monitoring by capillary electrophoresis. *Journal of Chromatography A*, *907*(1–2), 329–335. 10.1016/S0021-9673(00)01046-3
Mukhopadhyay, S. C., & Mason, A. (2013). *Smart sensors for real-time water quality monitoring*. Springer.
NACE. (1958). Theoretical aspects of corrosion in low water producing sweet oil wells. *Corrosion*, *14*(1), 51.
Nadkarni, R. A., & Nadkarni, R. A. (2007). *Guide to ASTM test methods for the analysis of petroleum products and lubricants* (Vol. 44). ASTM International.
Neff, J. (1997). *Metals and organic chemicals associated with oil and gas produced water:*

Bioaccumulation, fates, and effects in the marine environment. Continental Shelf Associates.

OLF (Norwegian Oil Industry association). (n.d.). Recommended guidelines for the sampling and analysis of produced water. OLF.

OSPAR. (2005). *Oil in produced water analysis–guideline on criteria for alternative method acceptance and general guidelines on sample taking and handling*. OSPAR.

Padarauskas, A., Paliulionyte, V., Ragauskas, R., & Dikcius, A. (2000). Capillary electrophoretic determination of thiosulfate and its oxidation products. *Journal of Chromatography A*, *879*(2), 235–243. 10.1016/S0021-9673(00)00298-3

Pandey, S. K., Kim, K.-H., & Tang, K.-T. (2012). A review of sensor-based methods for monitoring hydrogen sulfide. *TrAC Trends in Analytical Chemistry*, *32*, 87–99. 10.1016/j.trac.2011.08.008

Papavinasam, Sankara. (2014). *Corrosion control in the oil and gas industry* (1st ed.). Gulf Professional Publishing.

Papavinasam, S, Doiron, A., & Revie, R. W. (2010). Model to predict internal pitting corrosion of oil and gas pipelines. *Corrosion Science*, *9312*(March), 1–11. 10.5006/1.3360912

Pichon, V., & Chapuis-Hugon, F. (2008). Role of molecularly imprinted polymers for selective determination of environmental pollutants—A review. *Analytica Chimica Acta*, *622*(1), 48–61. 10.1016/j.aca.2008.05.057

Piehl, R. L. (1960). Correlation of corrosion in a crude distillation unit with chemistry of the crudes. *Corrosion*, *16*(6), 305. 10.5006/0010-9312-16.6.139

Procópio, L. (2020). The era of 'omics' technologies in the study of microbiologically influenced corrosion. *Biotechnology Letters*, *42*(3), 341–356. 10.1007/s10529-019-02789-w

Ramström, O., Skudar, K., Haines, J., Patel, P., & Brüggemann, O. (2001). Food analyses using molecularly imprinted polymers. *Journal of Agricultural and Food Chemistry*, *49*(5), 2105–2114. 10.1021/jf001444h

Revie, R. W., & Uhlig, H. H. (2008). *Corrosion and corrosion control: an introduction to corrosion science and engineering* (4th ed.). John Wiley & Sons, Inc. 978–0-471–73279–2

Saeed, A. H. (2017). *Analysis of sulfide and sulfur oxyanions in water and wastewater using capillary zone electrophoresis with detection by indirect and direct UV-Vis spectrophotometry* (Issue July). Memorial University of Newfoundland. http://research.library.mun.ca/id/eprint/12931

Safizadeh, F., & Larachi, F. (2014). Speciation of sulfides and cyanicides in the cyanidation of precious metal bearing ores using capillary electrophoresis—A review. *Instrumentation Science & Technology*, *42*(3), 215–229. 10.1080/10739149.2013.869756

Salimi, A., Roushani, M., & Hallaj, R. (2006). Micromolar determination of sulfur oxoanions and sulfide at a renewable sol–gel carbon ceramic electrode modified with nickel powder. *Electrochimica Acta*, *51*(10), 1952–1959. 10.1016/j.electacta.2005.07.002

Santos, H., Martins, R. O., Soares, D. A., & Chaves, A. R. (2020). Molecularly imprinted polymers for miniaturized sample preparation techniques: Strategies for chromatographic and mass spectrometry methods. *Analytical Methods*, *12*(7), 894–911.

Schmitt, G., & Feinen, S. (2000). Effect of anions and cations on the pit initiation in CO2 corrosion of iron and steel. *Corrosion*, *00149*, Paper No.00031.

Shahhoseini, F., Azizi, A., Egli, S. N., & Bottaro, C. S. (2020). Single-use porous thin film extraction with gas chromatography atmospheric pressure chemical ionization tandem mass spectrometry for high-throughput analysis of 16 PAHs. *Talanta*, *207*, 120320. 10.1016/j.talanta.2019.120320

Soga, T., & Ross, G. A. (1999). Simultaneous determination of inorganic anions, organic acids, amino acids and carbohydrates by capillary electrophoresis. *Journal of Chromatography A*, *837*(1), 231–239. 10.1016/S0021-9673(99)00092-8

Stanley, C. R., & Smee, B. W. (2007). *Geochemistry: Exploration, environment, analysis*. (Vol. 7), 329–340.The Geological Society of London. https://doi.org/10.1144/1467-7873/07-128

Sun, B. W., Li, Y. Z., & Chang, W. B. (2001). Molecularly imprinted polymer usingp-hydroxybenzoic acid, p-hydroxyphenylacetic acid and p-hydroxyphenylpropionic acid as templates. *Journal of Molecular Recognition*, *14*(6), 388–392. 10.1002/jmr.550

Swinney, K., & Bornhop, D. J. (2000). Detection in capillary electrophoresis. *Electrophoresis*, *21*(7), 1239–1250. 10.1002/(SICI)1522-2683(20000401)21:7<1239::AID-ELPS1239>3.0.CO;2-6

Tibbetts, P. J. C., Buchanan, I. T., Gawel, L. J., & Large, R. (1992). A comprehensive determination of produced water composition. In *Produced water* (pp. 97–112). Springer.

Tůma, P., Samcová, E., & Duška, F. (2008). Determination of ammonia, creatinine and inorganic cations in urine using CE with contactless conductivity detection. *Journal of Separation Science*, *31*(12), 2260–2264. 10.1002/jssc.200700655

Utvik, T. I. R. (1999). Chemical characterisation of produced water from four offshore oil production platforms in the North Sea. *Chemosphere*, *39*(15), 2593–2606. 10.1016/S0045-6535(99)00171-X

Van Biesen, G., Wiseman, J. M., Li, J., & Bottaro, C. S. (2010). Desorption electrospray ionization-mass spectrometry for the detection of analytes extracted by thin-film molecularly imprinted polymers. *Analyst*, *135*(9), 2237–2240.

Vandergrift, G. W., Krogh, E. T., & Gill, C. G. (2017). Polymer inclusion membranes with condensed phase membrane introduction mass spectrometry (CP-MIMS): Improved analytical response time and sensitivity. *Analytical Chemistry*, *89*(10), 5629–5636. 10.1021/acs.analchem.7b00908

Wang, J. (2005). Electrochemical detection for capillary electrophoresis microchips: A review. *Electroanalysis*, *17*(13), 1133–1140. 10.1002/elan.200403229

Witter, A. E., & Jones, A. D. (1998). Comparison of methods for inorganic sulfur speciation in a petroleum production effluent. *Environmental Toxicology and Chemistry*, *17*(11), 2176. 10.1897/1551-5028(1998)017<2176:COMFIS>2.3.CO;2

Yáñez-Sedeño, P., Campuzano, S., & Pingarrón, J. M. (2017). Electrochemical sensors based on magnetic molecularly imprinted polymers: A review. *Analytica Chimica Acta*, *960*, 1–17. 10.1016/j.aca.2017.01.003

Yang, M., & McEwan, D. (2005). Oil-in-water analysis method (OIWAM) JIP, A JIP report for 10 organisations including 8 operators and 2 government bodies. TUV NEL Report No: 2005/96, July 2005.

Yang, X., Kirsch, J., Fergus, J., & Simonian, A. (2013). Modeling analysis of electrode fouling during electrolysis of phenolic compounds. *Electrochimica Acta*, *94*, 259–268. 10.1016/j.electacta.2013.01.019

Zulkifli, S. N., Rahim, H. A., & Lau, W.-J. (2018). Detection of contaminants in water supply: A review on state-of-the-art monitoring technologies and their applications. *Sensors and Actuators B: Chemical*, *255*, 2657–2689. 10.1016/j.snb.2017.09.078

24 Workflow of Transportation, Sampling, and Documentation of Topsides Pipework with a Leak from an Offshore Oil Platform

Torben Lund Skovhus
VIA University College

CONTENTS

24.1 INTRODUCTION

When a section of an offshore pipe installation is identified with a leak that requires failure investigation, a process of events starts to secure the pipe sample to be removed. First, the part of the production containing the pipe leak under investigation will need to be by-passed, isolated from pressure and drained, so the section with the leak can be removed and replaced with a new pipe section. Secondly, the removed pipe section needs to be preserved to prevent damage and contamination and transported to shore in a safe and controlled manner. Finally, the removed pipe section will need to be examined at the shore base where samples for microbiology and other analyses will be secured. The most optimal procedure would be to take samples for microbiological and chemical analyses right after the section was removed from the process system offshore to avoid contamination and changes to the microbial community during transportation; however, for a number of reasons, this may not be possible.

DOI: 10.1201/9780429355479-28

In this particular case, sampling for microbiological analyses was not possible offshore, and therefore samples were taken when the pipe sections arrived at the shore base 10 days after it was shipped from the oil platform. After these microbiological samples were secured, the pipe section was shipped to the metallurgical laboratory as the final destination. Figure 24.1 shows a schematic illustration of the water injection system from the topside oil production platform.

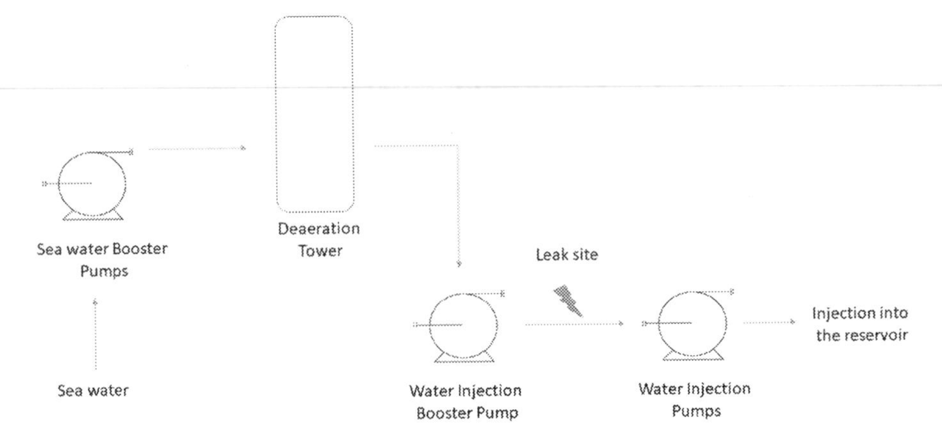

FIGURE 24.1 Location of the leak in the offshore process system.

This chapter serves as a *general guide* on how to transport, document, and sample from such offshore pipe sections, so both metallurgical and microbiological evidence is obtained at the best possible quality. Figure 24.2 shows an example of a removed pipe section for both metallurgical and microbiological analysis. Further details on the actual failure investigation can be found in Chapter 7.

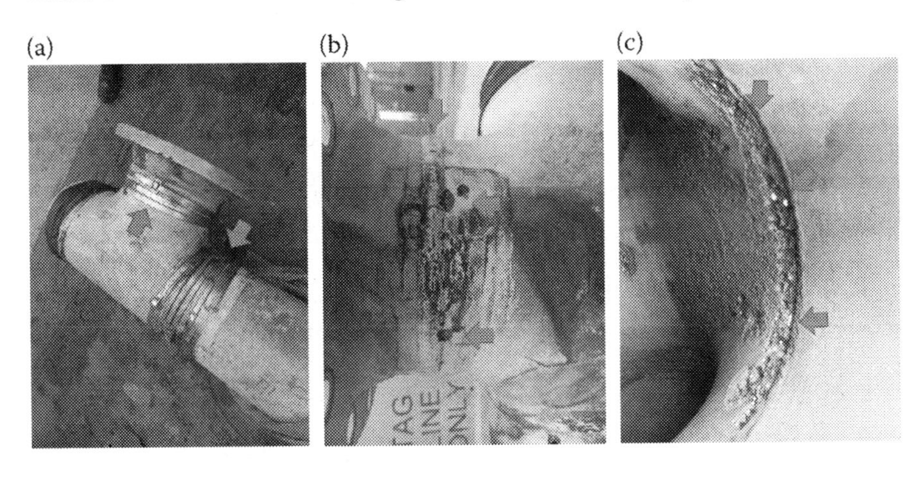

FIGURE 24.2 Two clamps were covering the leak sites (A). Two deep penetrating pits were visible from the outside close to the weld (B). Internal view of the pipe showing the two deep penetrating pits and additional corrosion close to the weld (C). The nominal pipe dimension is 8" (219.1 mm), and the wall thickness is 8.18 mm.

24.2 PROCEDURE

The procedure falls in three stages, as seen in Figure 24.3. Offshore Sampling (stage 1) and Shore Base Sampling (stage 2) are described below. Stage 3 on how the individual samples were collected and analyzed in the two specialized laboratories is described in in detail in Chapter 7.

FIGURE 24.3 Workflow of samples from offshore to shore base and further on to the two specialized laboratories.

24.2.1 Offshore Sampling

The operator followed these instructions to remove the pipe section offshore:

1. Locate and mark up the leak site on the failed pipe section (Sample X1).
2. Provide external photo documentation of pipe sections (use cm ruler to show size range).
3. Markup orientation of pipe (12, 3, 6, and 9 o'clock positions; use white waterproof pen).
4. Perform cold cutting 40 cm on each site of the leak site (total length of cutout approx. 1 m, see Figure 24.4).
5. Select a *control* pipe section downstream of the leak in the same system about 1 m in length. Mark, document, and cold cut (as above) the control section of pipe (in this case including the last flange to be disassembled) (Sample X2). See Figure 24.5.
6. Provide internal photo documentation of both sections (use cm ruler to show size range).

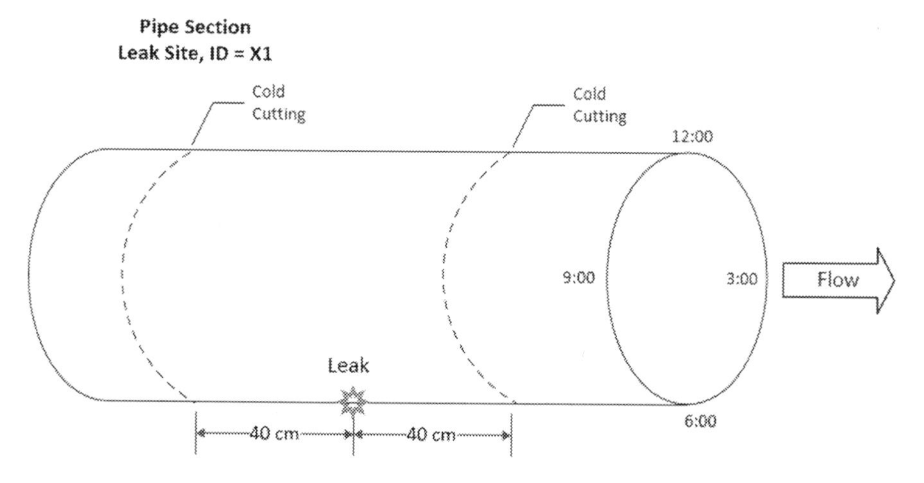

FIGURE 24.4 Pipework with the identified leak is called X1. The figure shows where the pipe should be cut with cold cutting.

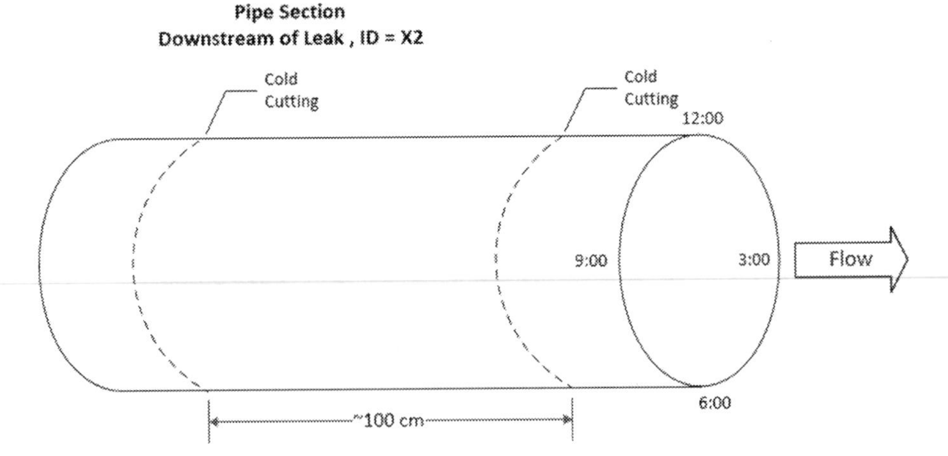

FIGURE 24.5 Pipework without any leak downstream of the leak is called X2. The figure shows where the pipe should be cut with cold cutting.

Before the two pipe sections were shipped to shore, they were preserved and packaged according to the following instructions:

1. Seal the pipe sections (individually) with clean plastic sheeting at both ends and seal with duct tape (see Figure 24.6).
2. Place the pipe sections on pallets and strap them firmly (see Figure 24.6).
3. Make sure the pipe sections are marked correctly on the outside.
4. Notify the onshore sampling specialist as soon as the samples leave the platform and provide the anticipated arrival time at the shore base.

24.2.2 Shore Base Sampling

When the pipe sections arrived on shore, the following instructions were followed:

1. Transfer staff will hand over the pipe sections to the sampling specialist at the pipe yard or shore base.
2. Photo documentation taken offshore should be provided to sampling specialist for further analysis.
3. The sampling specialist will take out samples for microbiological analyses (from both X1 and X2). Sterile sampling containers and equipment should be used.
   - Samples from corroded and non-corroded locations were sampled. Top and bottom of pipe were sampled.

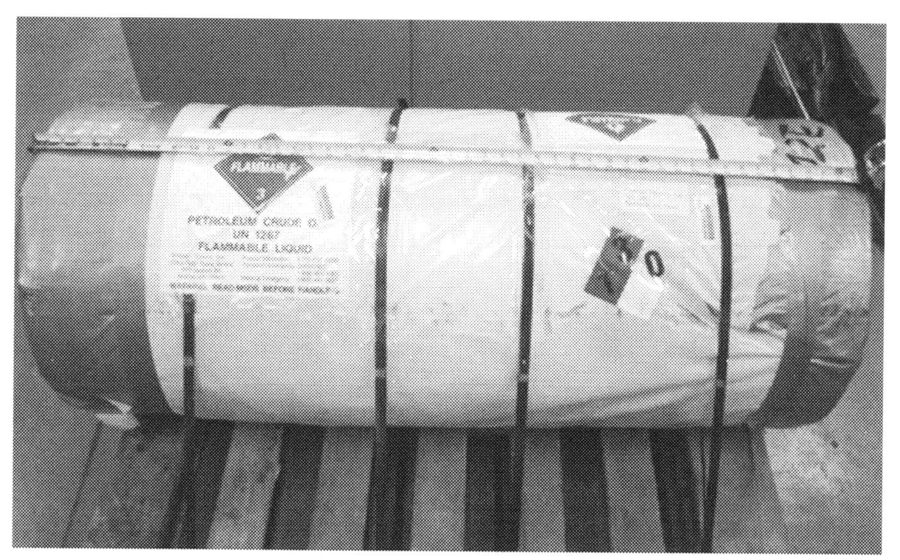

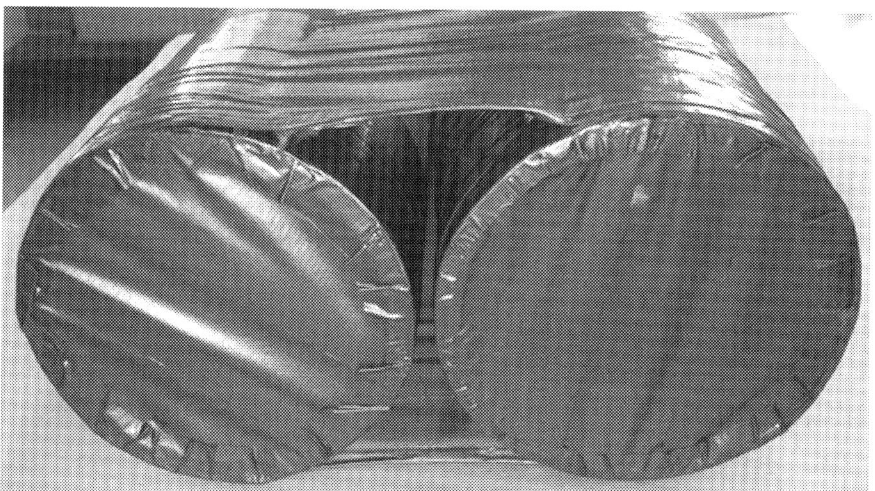

FIGURE 24.6 Examples of how the pipe sections are sealed (individually) with clean plastic and duct tape at the ends. Pipe sections should be placed on pallets and strapped firmly before shipment.

- Samples were transferred to the laboratory right after they were sampled for analysis.

4. For each pipe section, collect a number of internal and external photos to document the sampling performed for microbiological and chemical analysis.
5. Reseal the ends of the pipe with the plastic to avoid contamination.

6. The pipe sections are then shipped to the metallurgical laboratory for corrosion analysis with the same markings as provided offshore (X1 and X2).

When possible, the most optimal way of sampling for microbiological and chemical evidence is when the pipe sections are first removed from the original process system. However, when this is not possible, these simple and straightforward procedures may be used for sections of pipe from both offshore and onshore. This process includes removal of pipe samples from the original system, protection from contamination, transportation, documentation and sub-sampling for metallurgical and microbiological evidence. The key to successful results is rigorous documentation, short transportation time, and keeping the pipe sections clean during transportation over a long distance.

The references below are for the reader seeking additional theoretical and practical input on this topic.

ADDITIONAL RESOURCES

ASTM E1459-13(2018), Standard Guide for Physical Evidence Labeling and Related Documentation, ASTM International, West Conshohocken, PA, 2018, www.astm.org

ASTM E1492-11(2017), Standard Practice for Receiving, Documenting, Storing, and Retrieving Evidence in a Forensic Science Laboratory, ASTM International, West Conshohocken, PA, 2017, www.astm.org

ASTM G161-00(2018), Standard Guide for Corrosion-Related Failure Analysis, ASTM International, West Conshohocken, PA, 2018, www.astm.org

C Whitby and TL Skovhus (2011). *Applied Microbiology and Molecular Biology in Oil Field Systems.* Switzerland: Springer Publisher. www.springer.com/chemistry/biotechnology/book/978-90-481-9251-9

I Comanescu, C Taxen, and RE Melchers (2012, January 1) Assessment of MIC in Carbon Steel Water Injection Pipelines. Richardson, TX:Society of Petroleum Engineers. doi: 10.2118/155199-MS

J Larsen, TL Skovhus, AM Saunders, B Højris, and M Agerbæk (2008) Molecular Identification of MIC Bacteria from Scale and Produced Water: Similarities and Differences. Corrosion 2008, paper 08652. Houston, TX: NACE International.

J Kilbane (2014, May 13) *Forensic Analysis of Failed Pipe: Microbiological Investigations.* Houston, TX: NACE International. NACE-2014-3789

J Larsen, K Rasmussen, H Pedersen, K Sørensen, T Lundgaard,, and TL Skovhus (2010) *Consortia of MIC Bacteria and Archaea causing Pitting Corrosion in Top Side Oil Production Facilities.* Corrosion 2010, paper 10252. Houston, TX: NACE International.

M Al Muaisub (2020, August 3) *Is It MIC or Pitting Corrosion? An Insight in a Common Overlapping.* Houston, TX: NACE International. NACE-2020-14719

MA Al-Saleh, PF Sanders, T Lundgaard, KB Sørensen, and S. Juhler (2012, January 1). *General Characterization of Microbiologically Influenced Corrosion (MIC) Related Microorganisms in Crude Oil Samples.* Houston, TX: NACE International. NACE-2012-1267

RB Eckert, "Field Guide for Internal Corrosion Monitoring and Mitigation of Pipelines", NACE Press, 2016, NACE International, Houston, TX. ISBN 978-1-57590-328-6

RB Eckert and TL Skovhus (2019) Pipeline Failure Investigation: Is it MIC? *Materials Performance*. Houston, TX: *NACE International*.
RB Eckert, TL Skovhus, and B. Graver (2015, May 12) *Corrosion Management of MIC Contributes to Pipeline Integrity*. Houston, TX: NACE International. NACE-2015-5729
TL Skovhus, L Holmkvist, K Andersen, H Pedersen, and J Larsen (2012) MIC Risk Assessment of the Halfdan Oil Export Spool. SPE-155080, SPE International Conference and Exhibition on Oilfield Corrosion, Aberdeen, UK, 28–29 May 2012.

25 Standards for MIC Management in Engineered Systems

Torben Lund Skovhus
VIA University College

Richard B. Eckert
DNV GL USA

CONTENTS

25.1 INTRODUCTION

Effective management of microbiologically influenced corrosion (MIC) in engineered systems begins in the design stage and continues through construction, commissioning, and operations until the point of abandonment and decommissioning. Throughout each of these stages of life cycle, asset operators, and owners seek guidance from industry consensus standards that often distill decades of learning from many individuals and organizations into a process that is used to improve the success of future activities.

Corrosion management is a structured process that is applied concurrently to abiotic and biotic corrosion threats throughout all stages of asset life. The essence of the corrosion management process consists of the assessment of corrosion threats; the identification of mitigation and preventive measures to reduce the severity and likelihood of those threats; and monitoring the performance and then the effectiveness of the control measures. Ideally, assets owners can leverage past experience and industry knowledge by employing established standards and procedures to optimize their corrosion management activities. The development of management systems is described in ISO 55001 (2014) and corrosion management systems are detailed in NACE SP-21430 (2019).

In addition to MIC, engineered systems, e.g., oil and gas producing assets, are often exposed to abiotic internal corrosion threats, resulting from CO_2, H_2S,

DOI: 10.1201/9780429355479-29

oxygen, and other corrosive species, as well as flow-assisted corrosion, erosion, under deposit corrosion, and environmentally assisted cracking. These corrosion threats may occur concurrently or in synergy with one another, and they may change over time in the different stages of asset life. To respond to changes in the types or severity of corrosion threats the corrosion management process must be a continuing cycle of assessment, mitigation, and monitoring, and; it must also incorporate learning from failures.

While microorganisms exist in essentially all engineered assets, determining their role in corrosion and developing effective corrosion management strategies requires a consistent and methodical approach. For example, in the oil and gas industry, MIC mitigation efforts have historically focused on enumeration of planktonic rather than sessile microorganisms in biofilms, thereby discounting surface phenomena that directly influence both localized abiotic corrosion and MIC. Preventive and mitigative measures that are based on an understanding of biofilms, surface microbiological consortia and activity, and their relationship to corrosion, are significantly more effective in managing MIC than measures based only on planktonic culture data (Eckert 2015).

A number of recent case histories illustrate the benefits of improved MIC diagnostics based on integrating data from molecular microbiological methods (MMM) with corrosion monitoring and operational information. Although many industries are witnessing growing use of MMMs for managing MIC, a standard procedure for clearly linking microbiological conditions with electrochemical corrosion initiation and propagation mechanisms is still emerging.

This chapter briefly discusses the vital role of standards in MIC management and incorporates the latest efforts by industry, academia, and international standards organizations to improve the application of MIC diagnostic methods and data integration within the corrosion management process. Improving asset owners' and operators' ability to manage MIC threats through well-developed, state-of-the-art consensus standards will ultimately drive the reduction of operating risk, help extend asset life, and move the engineered environment in the direction of sustainability.

25.2 CURRENT STANDARDS

Several standards and guidance documents have been published over the past decades on the topics of corrosion management, risk-based inspection, and integrity management in the oil, gas, and petroleum industry as outlined and discussed in Skovhus and Eckert (2017). Most recently a significant development has taken place in this field, initiated by the introduction of MMM as a new diagnostic tool to the oil and gas industry (Skovhus et al. 2007, 2010). Table 25.1 summarizes the most recent key industry standards and guidance documents in this field.

In particular, two key documents from NACE International (TM0106-2016 2016 and TM0212-2018 2018) have now both been updated with the latest MMM and biotechnology methods and are widely used in the energy sector since they were published. Mistakenly, TM0194-2014 (2014) has often been seen as a document dealing with MIC assessment, which is not the case. The document has not been updated since 2014 and mainly relates to bacterial growth of some selected

TABLE 25.1
Most Recent Published Standards on Oilfield Corrosion Management, Inspection, Oilfield Microbiology, and MIC

Standard/Guidance Document	Source	Year of Release
TM0106-2016, Detection, Testing, and Evaluation of Microbiologically Influenced Corrosion (MIC) on External Surfaces of Buried Pipelines	NACE International	2016
TM0212-2018, Standard Test Method Detection, Testing, and Evaluation of Microbiologically Influenced Corrosion on Internal Surfaces of Pipelines	NACE International	2018
DNVGL-RP-G101, Risk-based inspection of offshore topsides static mechanical equipment	DNV GL	2021
DNVGL-RP-F116, Integrity management of submarine pipeline systems	DNV GL	2019

microorganisms that thrive on culture media. The document is therefore not recommended for MIC assessments and will not be discussed further here. However, two recently updated documents from DNV GL (DNV GL-RP-F116 and DNV GL-RP-G101) now incorporate and apply MMM in integrity management of submarine pipeline systems and in risk-based inspection of offshore topsides static equipment. This is a major step forward for operators of these assets to increase safety and control corrosion.

A limited number of publications have been produced on how to obtain biofilm samples of high value for microbiological analyses; however, three publications stand out in particular. Keasler and Chatterjee (2014) pointed out the importance of combining measures (weight loss and pitting formation) from corrosion coupons from the field with quantitative polymerase chain reaction (qPCR) and denaturing gradient gel electrophoresis (DGGE) of biofilms. Galvão and Lutterbach (2014) showed the importance of obtaining biofilm samples from the inside of a fuel storage tank suffering from MIC. In a recent paper by Eroini et al. (2015), sampling and shipment of biofilm samples was explained in detail. The biofilm/corrosion products were transferred to the laboratory in a glass bottle closed by screw cap with two hose connections. The atmosphere was purged with inert gas that limited ingress of oxygen to avoid additional oxidation and reduced further growth of microorganisms. The obtained samples were analyzed with qPCR and X-ray diffraction and fluorescence (XRD/XRF) when they arrived in the laboratory. The transportation of the preserved samples could take from weeks to months to arrive at the laboratory.

Further publications including theories and case stories have been a driver to help scope and draft existing standards and formulate new standards. These include the newly published book on MIC in the upstream oil and gas industry (Skovhus, Enning, and Lee 2017) and a couple of journal publications on MIC management in the North Sea (Skovhus, Eckert, and Rodrigues 2017) and how to integrate MMM in MIC management in the oil and gas industry worldwide (Eckert and Skovhus 2018). It is highly recommended to study these key documents for a deeper understanding of how the existing standards has been updated and expanded.

25.3 FUTURE STANDARDS

At the time of this writing, several standards-developing organizations are working on new standards that involve MIC and the practical application of MMM. Some examples of these forthcoming standards and guidelines are provided below and are also listed in Table 25.2.

TABLE 25.2
Standards under Development Related to Management of MIC (as of February 2021)

Standard/Guidance Document under Development	Source	Expected Year of Release
TM21465, Molecular Microbiological Methods – Sample Handling and Laboratory Processing	NACE International	2021
TM21495, Laboratory Evaluation of the Effect of Biocides on Biofilms	NACE International	2022
Quantification of Microbial Contamination in Liquid Fuels and Fuel Associated Water by Quantitative Polymerase Chain Reaction (qPCR)	ASTM	2021
Guidance on the Use of Biocides in the Oil Industry	Energy Institute	2021
Guidelines on Managing Microbiologically Influenced Corrosion (MIC) in Water Injection Systems	Energy Institute	2021
Guidance on selection, applicability, and use of molecular microbiological methods (MMM) in the upstream and downstream oil industry	Energy Institute	2022

NACE International presently has two standards under development that will support improvements in MIC management, including Molecular Microbiological Methods – Sample Handling and Laboratory Processing (TM21465) and Laboratory Evaluation of the Effect of Biocides on Biofilms (TM21495). NACE standard TM21465 will include information about collection and preservation of samples for microbiological analysis using MMM. Guidelines provided in the standard intend to provide common procedures and best practices for microbiological analysis of environmental samples by different laboratories in order to generate comparable results. NACE standard TM21495 will include information about how to establish biofilms in the lab, analytical methods to apply and data to collect, and data interpretation.

As of February 2021, ASTM is developing a procedure titled "Quantification of Microbial Contamination in Liquid Fuels and Fuel Associated Water by Quantitative Polymerase Chain Reaction (qPCR)".

Two new documents were under development in 2020 by the Energy Institute, the first being Guidelines on Managing Microbiologically Influenced Corrosion (MIC) in Water Injection Systems and the second being Guidance on the Use of Biocides in the Oil Industry. These are anticipated to be released in 2021. Another new Energy Institute standard that will begin development in 2021 is "Guidance on selection, applicability and use of molecular microbiological methods (MMM) in the upstream and downstream oil industry". This standard is anticipated to be published in 2022.

25.4 WHAT IS NEEDED TO FILL THE GAP?

According to a recent bibliometric study by Hashemi et al. (2018), that analyzed several thousand publications on MIC research, they concluded that "MIC research is multidisciplinary in nature; however, it is siloed between two main subject areas: material/corrosion sciences and microbiology/environmental sciences." From our experience working in this field for several decades, MIC research and knowledge is also highly fragmented and siloed between organizations, inside organizations, and among technical disciplines. This makes true collaboration a great challenge in the sphere of MIC research, development, and management.

One way to overcome this problem is to initiate true collaborative MIC research with a team consisting of highly diverse members. It should consist of all relevant stakeholders for the success of the project, such as manufacturers, academics, operators/end-users, and consultants, all with different business perspectives and technical backgrounds. There should also be a balance between technical disciplines, such as engineers, microbiologists, chemists, and material scientists onboard the team.

Two recently established and very different initiatives are trying to bridge these identified gaps and stimulate a more holistic research tradition for MIC research and development. The first initiative is the European MIC Network, which is an open network for everybody interested in MIC research of any kind. The network hosts free online webinars, workshops, and surveys among its members. It was initiated during the spring of 2020 – while the first wave of coronavirus caused worldwide lockdown of e.g., international conferences and symposia. The network discusses technology, sampling, and funding opportunities; all to bridge the various gaps in the MIC research community. The network was founded in 2020 and is managed by Dr. Torben Lund Skovhus (Denmark) and Dr. Andrea Koerdt (Germany).

The second initiative is "NACE SC-22 Biodeterioration", which is a new standards committee at NACE International (now AMPP). SC-22 is tasked to develop and maintain standards and reports for measuring, monitoring, and mitigating biodeterioration in engineered systems and assets.

The objectives of SC-22 are to:

- Produce standards and technical reports on the diagnosis, monitoring, and mitigation of biodeterioration at the request of industry (e.g., manufacturers, end users, asset owners), regulators, and others.

- Create, prepare, and maintain documents that contain technical information, specifications, and standard practices for the management of biodeterioration through assessment, monitoring, and mitigation activities.
- Identify areas where knowledge is lacking and recommend research or standards development to fill in knowledge gaps.
- Provide consistent technical support to other standard committees and standards development efforts in the area of biodeterioration of engineered systems and assets.
- Provide a forum for the exchange of technical information and best practices relating to the practice of managing biodeterioration.

The end products of this committee will be standards and technical reports, which will be published by AMPP as Standard Practices, Test Methods, Material Requirements, and Technical Reports. SC-22 is managed by Dr. Torben Lund Skovhus (Denmark) and Dr. Jason Lee (USA).

25.5 CONCLUSIONS

While the broader implementation of MMM in various industries is underway, end users of need sound guidance for practically employing these methods to diagnose and control MIC, and for realizing the benefits of the data produced. The process of developing industry standards and best practices around MMM can provide a vital platform for collaboration by stakeholders, including manufacturers, academics, operators/end-users, service providers, and consultants with different business perspectives and technical backgrounds. Since MIC is a multidisciplinary subject, standards should be collaboratively developed by teams with varied technical disciplines, including engineers, microbiologists, chemists, and material scientists, to improve the application of MIC diagnostic methods and data integration. As discussed in this chapter, that process has begun on several new standards focusing on MMM. It is projected that the next generation working to develop clean energy will need to manage MIC threats even more efficiently in oil and gas production until at least mid-century. To achieve this goal, asset managers will require young smart minds and well-developed, state-of-the-art consensus standards to drive reduction of operating risk, help extend asset life, and move the engineered environment toward greater sustainability.

REFERENCES

DNVGL-RP-F116. 2019. *Integrity management of submarine pipeline systems*. Høvik: DNV GL.

DNVGL-RP-G101. 2021. *Risk based inspection of offshore topsides static mechanical equipment*. Høvik: DNV GL.

Eckert, R. B. 2015. Emphasis on biofilms can improve mitigation of microbiologically influenced corrosion in the oil and gas industry. *Corrosion Engineering Science and Technology* 50(3):163–168.

Eckert, R. B., and T. L. Skovhus. 2018. Advances in the application of molecular microbiological methods in the oil and gas industry and links to microbiologically influenced corrosion. *International Biodeterioration and Biodegradation* 126:169–176. 10.1016/j.ibiod.2016.11.019

Eroini, V., H. Anfindsen, and A. F. Mitchell. 2015. Investigation, classification and remediation of amorphous deposits in oilfield systems (SPE-173719). *SPE International Symposium on Oilfield Chemistry 2015*, The Woodlands, Texas, 13–15 April 2015.

Galvão, M., and M. Lutterbach. 2014. Application of the qPCR technique for SRB quantification in samples from the oil and gas industries. In *Applications of molecular microbiological methods*, ed.T. L. Skovhus, S. Caffrey, and C. Hubert, 69–76. Norfolk, UK: Caister Academic Press.

Hashemi, S. J., N. Bak, F. Khan, K. Hawboldt, L. Lefsrud, and J. Wolodko. 2018. Bibliometric analysis of microbiologically influenced corrosion (MIC) of oil and gas engineering systems. *Corrosion* 74(4):468–486. 10.5006/2620

International Standards Organization (ISO) 55001. 2014. Asset management-Management Systems-requirements. Geneva: International Standards Organization.

Keasler, V., and I. Chatterjee. 2014. Using the power of molecular microbiological methods in oilfield corrosion management to diagnose MIC. In *Applications of molecular microbiological methods*, ed.T. L. Skovhus, S. Caffrey, and C. Hubert, 23–32. Norfolk, UK: Caister Academic Press.

Skovhus, T. L. and R. B. Eckert. 2017. Management of MIC in the oil and gas industry. In *Microbiologically influenced corrosion in the upstream oil & gas industry*, ed.T. L. Skovhus, D. Enning, and J. Lee, 141–155. Boca Raton, FL: CRC Press. ISBN 978-1-49872-656-6

Skovhus, T. L., R. B. Eckert, and E. Rodrigues. 2017. Management and control of microbiologically influenced corrosion (MIC) in the oil and gas industry - overview and a North Sea case study. *Journal of Biotechnology* 256:31–45. 10.1016/j.jbiotec.2017.07.003

Skovhus, T. L., D. Enning, and J. Lee. 2017. *Microbiologically influenced corrosion in the upstream oil & gas industry*. Boca Raton, FL: CRC Press. ISBN 978-1-49872-656-6.

Skovhus, T. L., B. Højris, A. M. Saunders, T. R. Thomsen, M. Agerbæk, and J. Larsen. 2007. Practical use of new microbiology tools in oil production (SPE 109104). *SPE Offshore Europe Conference 2007*, Aberdeen, UK, 4–7 September 2007.

Skovhus, T. L., K. B. Sørensen, J. Larsen, K. Rasmussen, and M. Jensen. 2010. Rapid determination of MIC in oil production facilities with a DNA-based Diagnostic Kit (SPE 130744). *SPE International Conference on Oilfield Corrosion 2010,* Aberdeen, UK, 24–25 May 2010.

SP-21430. 2019. *Standard framework for establishing corrosion management systems.* Houston, TX: NACE International.

TM0106-2016. 2016. *Detection, testing, and evaluation of microbiologically influenced corrosion (MIC) on external surfaces of buried pipelines.* Houston, TX: NACE International.

TM0194-2014. 2014. *Standard test method field monitoring of bacterial growth in oil and gas systems.* Houston, TX: NACE International.

TM0212-2018. 2018. *Standard test method detection, testing, and evaluation of microbiologically influenced corrosion on internal surfaces of pipelines.* Houston, TX: NACE International.

26 Standard Operating Procedures for Sampling Onshore and Offshore Assets for Genomic, Microbial, and Chemical Analyses and/or Experiments

Lisa Gieg and Mohita Sharma
University of Calgary

Tesfa Haile
InnoTech Alberta

Kelly Anne Hawboldt, Christina Bottaro, and Ali Modir
Memorial University of Newfoundland

Kathleen Duncan
University of Oklahoma

Trevor Place
Enbridge

Richard B. Eckert
DNV GL USA

Torben Lund Skovhus
VIA University College

Jennifer Sargent
Suez – Water Technologies and Solutions

DOI: 10.1201/9780429355479-30

Thomas R. Jack and Nuno Fragoso
University of Calgary

CONTENTS

These SOPs were developed as part of geno-MIC (Managing Microbial Corrosion in Canadian Offshore and Onshore Oil Production Operations), a large-scale applied research project funded by Genome Canada, Genome Alberta, Alberta Innovates, InnoTech Alberta, Government of Newfoundland and Labrador, and Mitacs, with in-kind support from Natural Resources Canada (CanmetMATERIALS), Genome Atlantic, Baker Hughes, BP, Brenntag, CRC, DNV GL, Dupont, Enbridge, Husky Energy, LuminUltra, Marathon, Microbial Analysis, Nalco Champion, OSP, Promega, Schlumberger, Shell, Suez, TransMountain, and United Initiators.

Version 1.2.3 – January 2021

26.1 IMPORTANT GENERAL NOTES

1. Avoid contamination of samples at all times.
   a. Wear lab/latex gloves when collecting samples.
   b. Do not touch the inside of containers and lids, or the sampling end of sampling tools (such as swabs, spatulas, spoons).
   c. Do not reuse needles, syringes, and sampling tools (such as spoons, swabs).
2. Some sample containers may contain a chemical preservative that is used to preserve nucleic acids (DNA and/or RNA) or acidify samples (for chemical analysis of crude oil). Do not discard this preservative prior to sampling.

3. Wherever possible, containers should be filled to the brim to minimize headspace/air pockets.
4. Samples should be placed in a cooler with ice/ice packs immediately after collection.
   a. Do not freeze samples.
   b. Do not expose samples to extreme heat or sunlight.
   c. Ship samples as soon as possible and with enough ice/ice packs to last at least four days in transit.
5. The procedures outlined here provide a general overview of how to collect samples for microbial and genomic analysis, as well as for chemical analysis of crude oil (liquids sampling section only). They are applicable to both onshore and offshore systems. These procedures are not intended to replace industry standards or recommended guidelines. Some industry standards related to this document that are recommended for the reader to consult include:
   a. NACE TM0194-2014. Field monitoring of bacterial growth in oil and gas systems. (NACE International, Houston, TX, 2014)
   b. NACE TM0212-2018. Detection, testing, and evaluation of microbiologically influenced corrosion on internal surfaces of pipelines. (NACE International, Houston, TX, 2018)
6. As each site operator has company-specific rules and regulations around safety and site management, the coordination of procuring sampling supplies, collecting samples, organizing sample shipments, and other logistics must be done in collaboration with the company providing samples and the lab analyzing the samples.

Table 26.1 summarizes the types of samples that can be collected and analyses that can be conducted from the samples. Table 26.2 overviews the type of supplies needed for collecting and preserving samples. The protocols are written with the assumption that the appropriate types of supplies have been added to a sampling kit prepared by a laboratory, and that the collected samples will be shipped back to that laboratory for appropriate analyses.

26.2 LIQUIDS SAMPLING

This sampling protocol is to be used for collecting liquids (e.g., produced waters or crude oils either from flowing or static systems) from onshore and offshore oil and gas operations.

These samples can be used for the following:

1. Genomic analysis (sampling vessels will contain a nucleic acid preservative)
2. Live microbiological analysis (sampling vessels will not contain a preservative)
3. On-site initiation of corrosion tests
4. Bench-top MIC testing (if enough field water sample is provided)
5. Chemical analysis (specifically for crude oil, sampling vessels will contain an acid preservative)

TABLE 26.1
Summary of Samples and Analyses

Sample Type	Source	Sample Container	Sample Amount	Number of Samples	Preservative in Container	Analysis to Be Performed
Liquids	source or produced waters	1 L bottles	1 L	1	yes	genomic – DNA or RNA
		1 L bottles	1 L	3	no	microbial assays
		100 mL serum bottles	50 mL	2	no (beads/ coupon)	on-site corrosion tests
Sludge, Solids	pigging material, internal swabs	50 mL tube	5–20 g	2	yes	genomic – DNA or RNA
		250 mL jar/ 50 mL tube	50–200 g	1	No	microbial assays & corrosion tests
	surface of coupon	15 mL tube	1 swab	1	yes	genomic – DNA or RNA
	coupon itself	50 mL tube/ plastic bag	1 coupon	1	no	weight loss

Samples may be collected from either flowing stream (e.g., pipeline) or static (e.g., storage tank) systems.

1. From flowing process liquid – samples shall be collected by slowly opening the sample point and adjusting the flow to a steady rate.
2. From dead space fluid – sample shall be collected with precaution so as not to introduce other variables, such as air (oxygen), and to avoid turbulent flow while sampling/opening valves.
3. Unless dead-space fluids represent the desired sample, the fluid should be allowed to flow to thoroughly flush out dead-space fluids and solids debris before the samples are collected. It is recommended to purge the water line with at least a couple of volumes of liquid (e.g., minimum of 2 L) before collecting a sample.
4. During sampling of systems containing both oil and water, phase separation should be permitted to occur before the water is used. However, it is satisfactory to directly use an emulsion for microbial analysis.

26.2.1 Liquids Sampling Using 1 L Bottles

Note: Sampling procedure is similar for collecting samples for live microbiological analysis, or for molecular microbiological analysis.

TABLE 26.2
Summary of Sampling Supplies

Item	Description	*Non-Scientific Alternative
15 or 50 mL centrifuge tube	sterile, plug cap	new 125 mL mason jar
**100 mL or 1 L bottle	sterile, plastic, wide mouth or amber glass bottle	new Nalgene™ water bottle
250 mL Jar	sterile, plastic/glass, cap with PTFE liner	new 250 mL mason jar
swab	sterile, cotton swab with wood or plastic shaft	new cotton swab (Q-Tip®)
blade spatula/ lab scoop	stainless steel	new cake spatula/ putty knife/spoon
chemical preservative	***commercially available product, homemade buffer, alcohol, acid (HCl)	rubbing alcohol (70% isopropanol)/ Everclear™
125 mL serum vial	sterile, glass vial with rubber stopper	na
50 mL syringe & 21 G needle	sterile, plastic syringe & stainless-steel needle	na

Non-lab specific supplies

permanent marker	lab gloves	field notebook/paper
plastic zip-sealed bags	ice packs	sharps disposal
bubble wrap pouches	cooler	rubber bands

*can be purchased from local store (e.g., grocery store, pharmacy, general store, etc.)
**bottles for crude oil collection should be glass; bottles for water collection should be high-grade plastic
***several commercially available nucleic acid preservatives are available from different vendors, and different laboratory-prepared buffer solutions can also be used; depending on the sample type, chemical preservatives may vary in their efficacy (for example, see De Paula et al. 2018, Oldham et al. 2019, Rachel & Gieg 2020).
na = not available

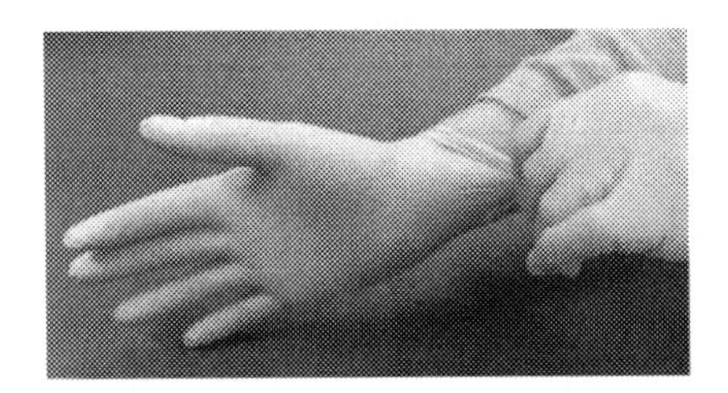

1. Put on latex/nitrile gloves.
   - Use a clean pair for each sampling location.
   - Do not touch the insides of bottles or lids.

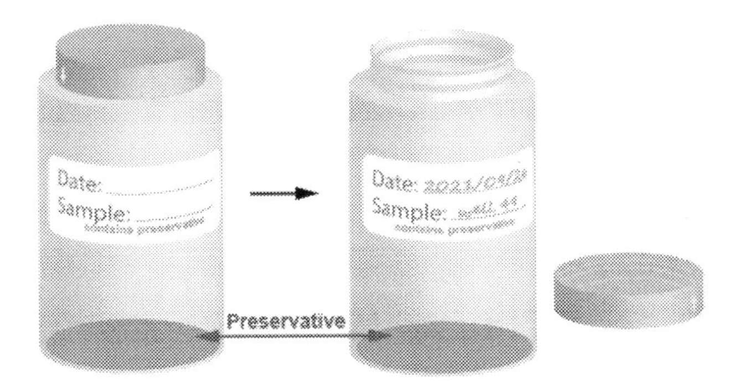

2. Label each bottle with the sampling location, date (and other relevant information if known, such as temperature, pH, etc.).
3. Unscrew the sample bottle, taking care not to touch the brim of the sample bottle or the inside of the lid; set lid down in an inverted manner on a flat surface.
4. Fill each bottle to the brim (so that it contains no air/headspace), and close the lid, taking care not to touch the inside of the lid. Seal the cap using the tape provided.
   Note: For bottles containing the chemical preservative – do not overflow the bottle as the preservative will be lost.

5. For bottle containing preservative, invert the sampling bottle to gently mix.
   Note: Oil can cause tape to come off and permanent marker writing to blur, so be sure to write ID information in two different locations, such as on the side of the bottle and on the lid.
   Note: Keep a more detailed record of sample information in a notebook and/or Excel spreadsheet or data collection sheet.

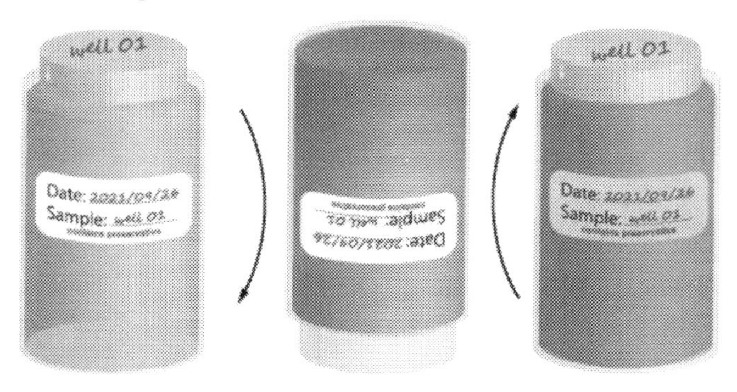

6. If glass bottles are used to collect samples (such as for crude oil collection), wrap them in bubble wrap (ideally, a bubble wrap sleeve) and put a rubber band around it.

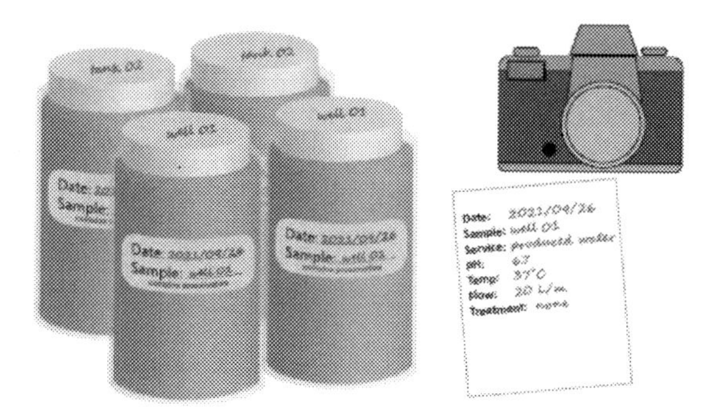

7. Take a digital photograph of the filled sample containers and your field notes.

8. Place samples in a cooler with ice packs for shipping, tape up the cooler, and ship to the appropriate lab as soon as possible in order to maintain sample integrity.
9. If requested, proceed to initiate on-site experiments as described below.

26.2.2 On-site Corrosion Initiation Experiments

1. Put on latex/nitrile gloves.
   - Use a clean pair for each sampling location.
   - Do not touch the insides of bottles or lids.

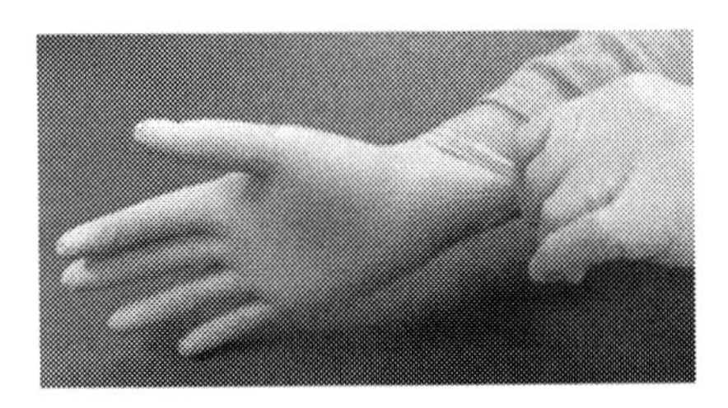

2. Collect liquid samples as outlined in Liquids sampling using 1 L bottles. **Note:** You will be using samples without preservative.

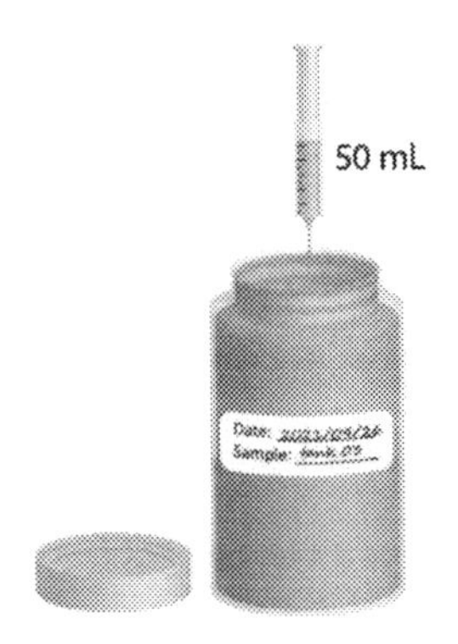

3. Remove syringe and needles from sterile packaging and connect them together.
4. Open the 1 L sample bottle kept out of the cooler (again, placing lid down in an inverted manner) and remove 50 mL using the syringe and needle.
5. Dispense this sample into the empty, butyl rubber stopper-sealed glass serum bottles containing test metal inside (beads/coupon). Insert the needle through the stopper, then push on the syringe piston to add liquid into the bottle. This will become difficult near the end as there will be backpressure. Once the liquid is added, let go of the syringe piston in order to release the back-pressure (will be about 50 mL) and remove the syringe from the stopper. Dispose of syringe and needle properly.

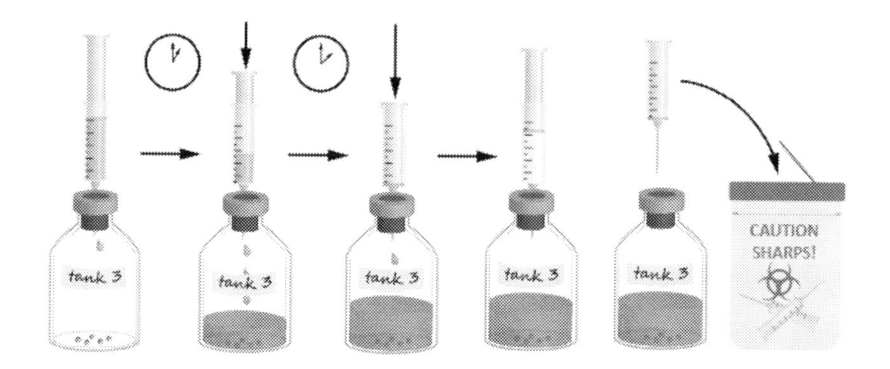

6. Repeat for additional bottles and samples as required.
7. Confirm bottles are labeled with the sample ID and date. Then place in bubblewrap bags for shipping. Depending on the temperature of these experiments, they may need to be shipped in a separate cooler without ice packs.
8. Close lid of 1 L bottle and add to the original cooler with the other 1 L sample bottles.
9. Tape up cooler for shipping, and ship to the appropriate lab as soon as possible in order to maintain sample integrity.

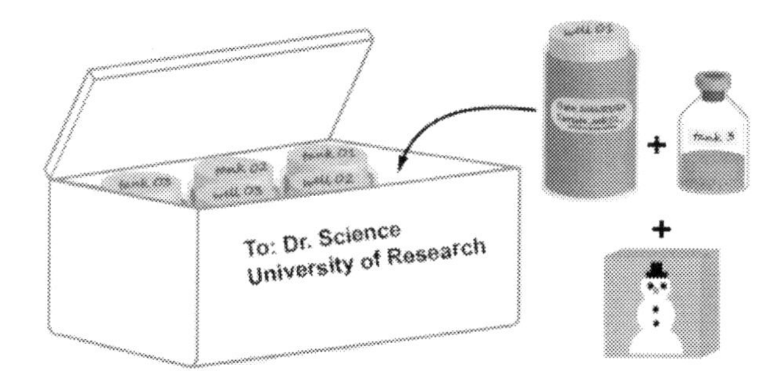

26.3 SOLIDS SAMPLING FROM PIGGING OPERATIONS

This sampling protocol is to be used for collecting solids (sludge, debris) from pig receivers (pig traps) following mechanical cleaning (e.g., a pigging operation).

These samples will be used for:

A. Live microbiological analysis (sampling vessels will not contain a preservative)
B. Genomic analysis (sampling vessels will contain a nucleic acid preservative)

26.3.1 Solids Sampling for Microbiological Analysis

1. Put on latex/nitrile gloves.
   - Use a clean pair for each sampling location.
   - Do not touch the insides of bottles or lids.

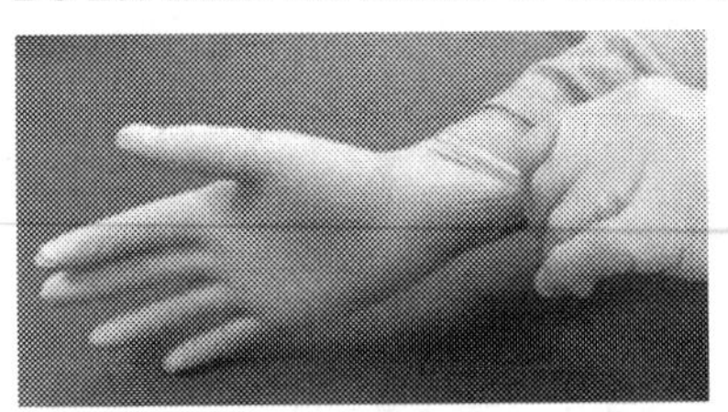

2. Loosen the lid of one of the jars (or 50 mL conical tubes; not shown). Set the lid down, thread side-up. Do not to touch the inside of the tube or lid.

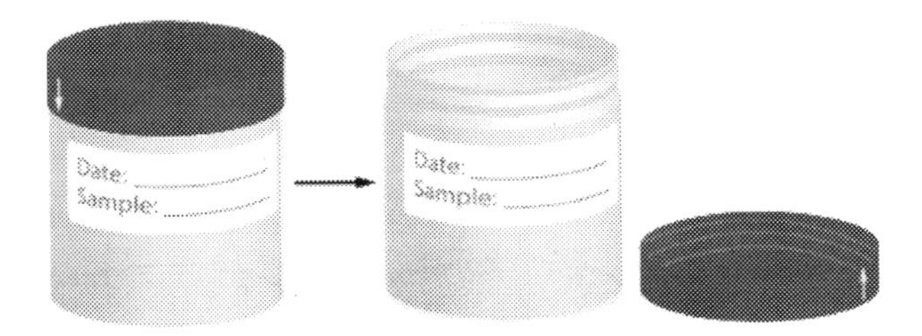

3. Tear open the large sterile sampling tool package, near handle end.

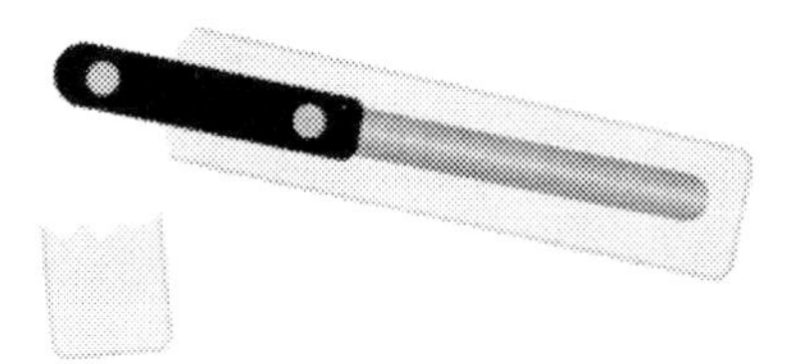

4. Remove the sterile sampling tool from the package and use it to take a sample from the pig receiver (the sample shall be representative of the fresh solids collected).

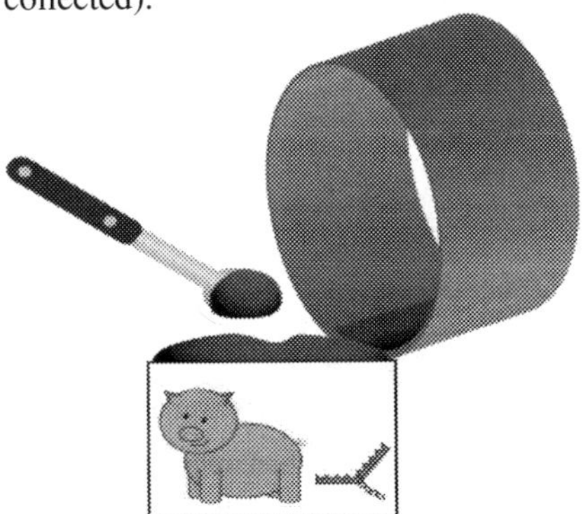

5. If possible, fill the jar with solids/sludge and leave no head space. Close the cap tightly.

6. Clearly label the bottle on the side and lid with sampling location, pig number (1, 2, etc.), and date.
7. Take a digital photograph of the sample container (after filling) along with your field notes.
8. Place used spoons in individual plastic zip-sealed bags (labeled) and return to the lab with samples.
9. Place samples in a cooler with ice packs for shipping, tape up the cooler, and ship to the appropriate lab as soon as possible in order to maintain sample integrity.

26.3.2 Solids Sampling for Genomic Analysis

1. Put on a clean pair of latex/nitrile gloves.
   - Use a clean pair for each sampling location.
   - Do not touch the insides of bottles or lids.

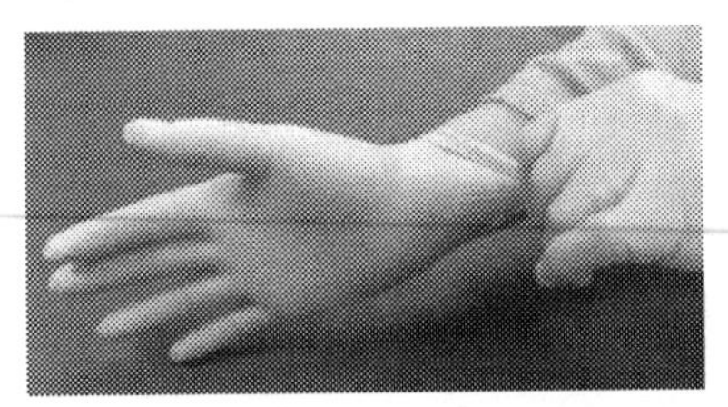

2. Loosen the lid of one of the conical tubes labeled "**Preservative**". Tubes contain a small volume of a chemical preservative – do not discard. Set the lid down, thread side-up. Do not touch the inside of the tube or lid.

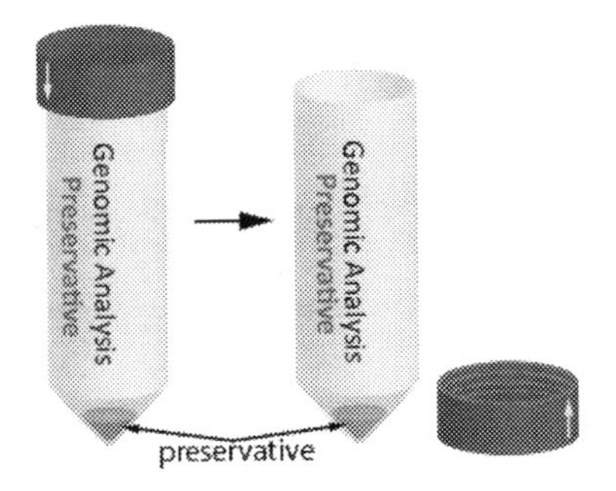

3. Tear open the sterile sampling tool package, near handle end.

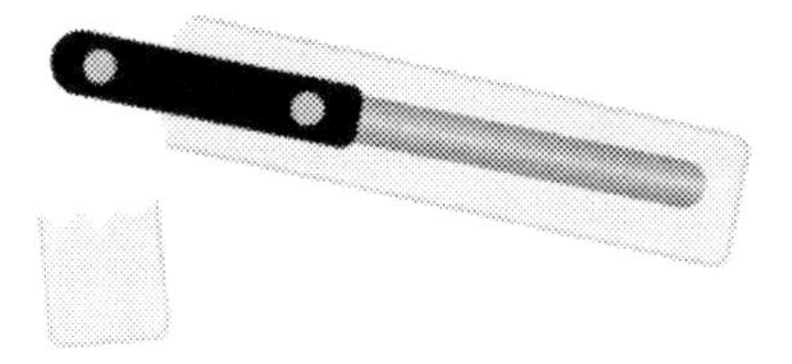

4. Remove one sterile sampling spoon and use it to take a sample (e.g., ~1 to 5 g, if possible) from the pig receiver (the sample shall be representative of the fresh solids collected).
5. Insert the sample into the conical tube, and swish around to make sure the preservative contacts the sample material. Close the tube tightly with the lid.

6. Place used sampling tools into a zip-sealed bag and return with samples.
7. Clearly label the tubes with sampling location and date.
8. Take a digital photograph of the sample containers (after filling) along with your field notes.
9. Place samples in a cooler with ice packs for shipping, tape up the cooler, and ship to the appropriate lab as soon as possible in order to maintain sample integrity.

26.4 SAMPLING FROM INTERNAL PIPELINE SURFACES OR CORROSION COUPONS

This sampling protocol is to be used for collecting solids from pipeline surfaces or corrosion coupon surfaces.

These samples will be used for the following:

A. Genomic analysis (sampling vessels will contain a nucleic acid preservative)
B. Live microbiological analysis (samples will not contain a preservative)

Given that pipeline solids may not always be plentiful, it is acknowledged there may not be enough sample for this analysis.

26.4.1 Ideal Locations for Solid Samples Collected from Internal Pipeline Surfaces

1. *Leak Site:* The location of the leak (where MIC may have occurred). Please indicate whether the samples were obtained from inside, outside, or at a distance from the pits. It is preferred that samples are collected from the pits as the location where the leak occurred may have washed away the solids samples or has caused cross contamination from the surrounding area.
2. *Adjacent Site:* A location directly adjacent to the leak. This sample should be taken because the leak itself or the leak testing may have washed away or introduced microbes at original leak site. A site adjacent to the leak will still have a chance of containing MIC-associated microorganisms.
3. *Control Site:* A "control" location away from the pitting area or where there is no pitting (uncorroded control).

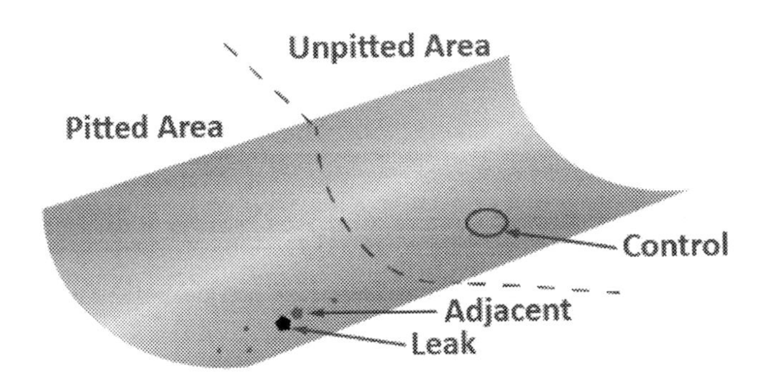

Sampling locations around a leak location on the inside of pipe.

Note: Make sure you take pictures before sampling and indicate the location on the pictures where the samples were taken and indicate the orientation of the pipe (e.g., pit at the 6 o'clock position) and the direction of flow.

26.4.2 Protocol for Sampling Internal Pipeline Solids for Genomics Analysis

Perform the following steps for EACH of the three pipeline sites indicated above:

1. Put on a clean pair of latex/nitrile gloves.
   - Use a clean pair for each sampling location.
   - Do not touch the insides of bottles or lids.

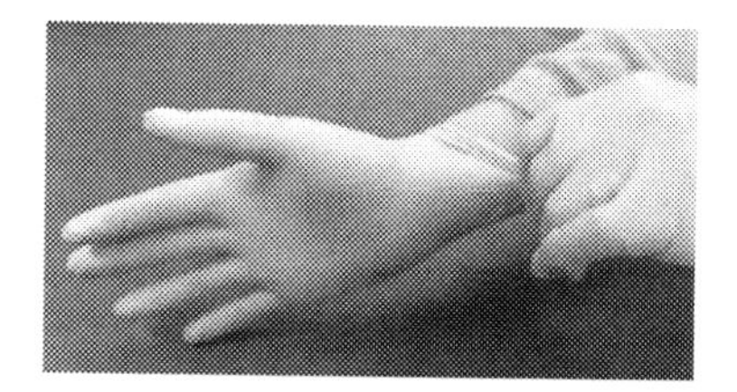

2. Two conical tubes will be in a zip-sealed bag. Loosen the lid of one of the conical tubes labeled **"SWAB with preservative."** Tubes contain a small volume of a chemical preservative – do not discard. Set the lid down, thread side-up. Do not touch the inside of the tube or lid.

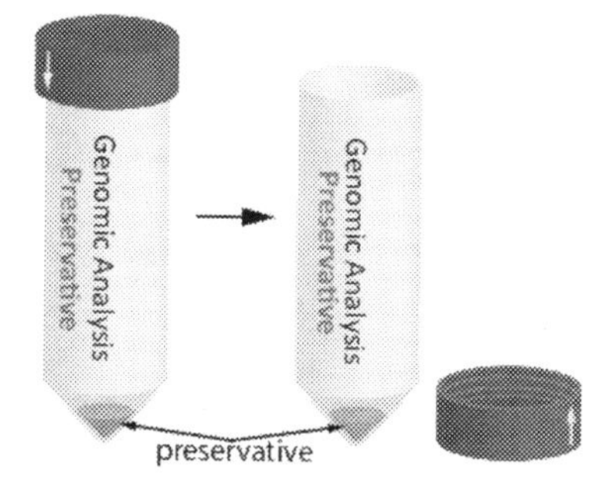

3. Tear open the swab package near the stick end. Avoid touching the swab end.

4. Remove one sterile swab, ensuring to hold the swab near the top 2.5–3 cm. Use to swab on the inside of the pipe, going back and forth and upwards and downwards in an area of about 6 cm^2 (~ 2.5 cm x 2.5 cm).
Note: If the pipe appears dry, dip the swab into the conical tube containing the liquid preservative first to moisten, then use to swab.

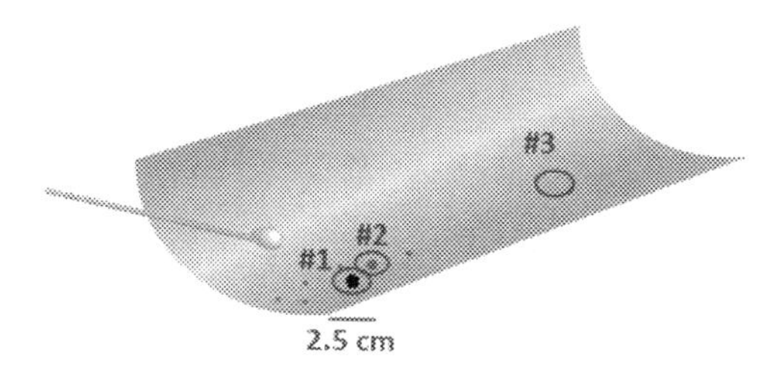

5. Insert the swab into the conical tube, and swish around to make sure the preservative contacts the sample material.

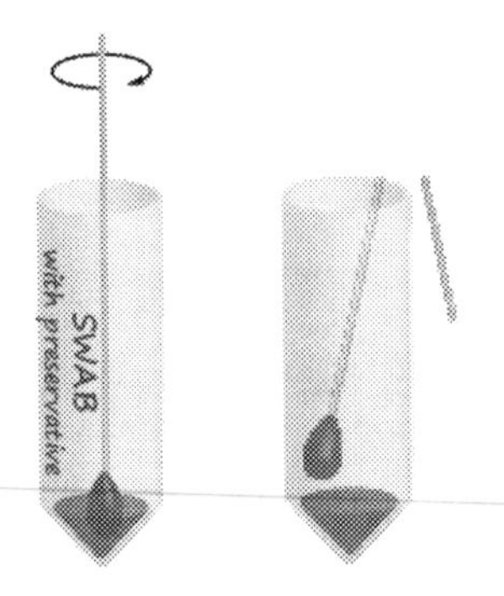

6. Break off excess wooden portion of the cotton swab so that it fits into the conical tube when closed – try to break at a point lower than your fingers have touched. Close the lid tightly.
7. Repeat the same protocol (steps 2–6) with the second 50 mL tube and the other swab from the package.

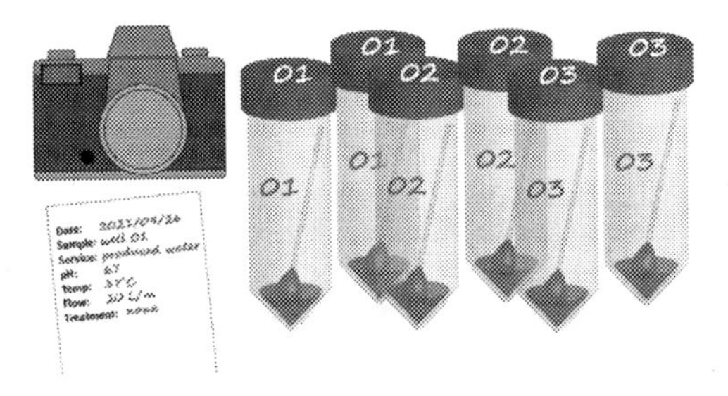

8. Clearly label the tubes with sampling location and date.
9. Take a digital photograph of the sample containers (after filling) along with your field notes.
10. For each site, it is ideal to first take two swab samples for genomic analysis, and also to collect solids for live microbiological analysis if enough sample is present. Take samples for genomic analysis first, then collect any additional solids for live microbiological tests.

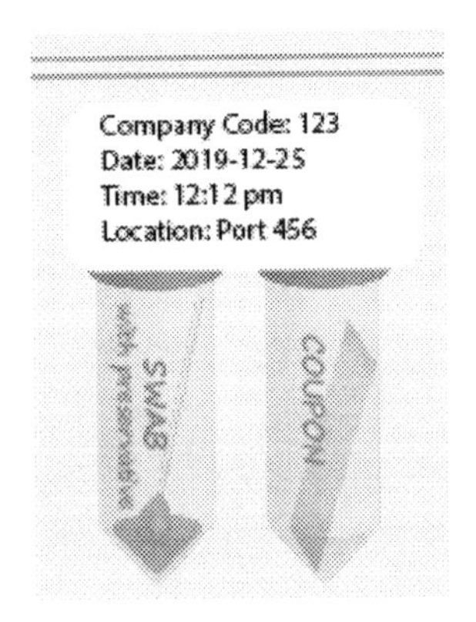

11. Place duplicate conical tubes back in the zip-sealed bag, then clearly label the bag with sampling location, data, and time of collection.
12. Place samples in a cooler with ice packs for shipping, tape up the cooler, and ship to the appropriate lab as soon as possible in order to maintain sample integrity.

26.4.3 Protocol for Sampling Coupons for Genomics Analysis

1. Put on a clean pair of latex/nitrile gloves.
   - Use a clean pair for each sampling location.
   - Do not touch the insides of bottles or lids.

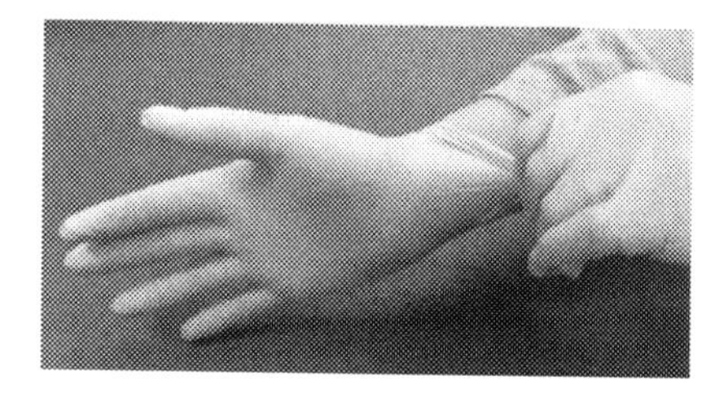

2. Two conical tubes will be in a zip-sealed bag. Loosen the lid of one of the conical tubes labeled **"SWAB with preservative."** Tubes contain a small volume of a chemical preservative – do not discard. Set the lid down, thread side-up. Do not touch the inside of the tube or lid.

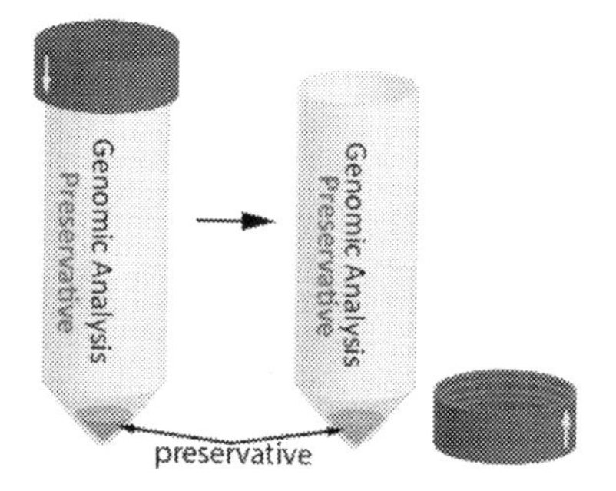

3. Tear open the swab package near the stick end. Avoid touching the swab end.

4. Remove the sterile swab, ensuring to hold the swab near the top 1–1.5 inches and ensuring not to touch the swab to anything. Holding the coupon on the sides only, swab the exposed surface of the coupon. Start at one end and swab from side to side (in a back and forth motion as indicated in the diagram and/or in an upward and downward motion) several times until you reach the other end of the end of the coupon.

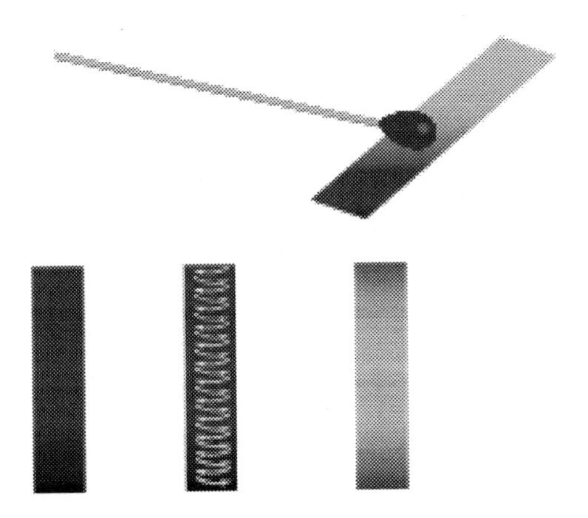

NOTE: If the coupon appears dry, dip the swab into the conical tube containing the liquid preservative first to moisten, then use to swab.

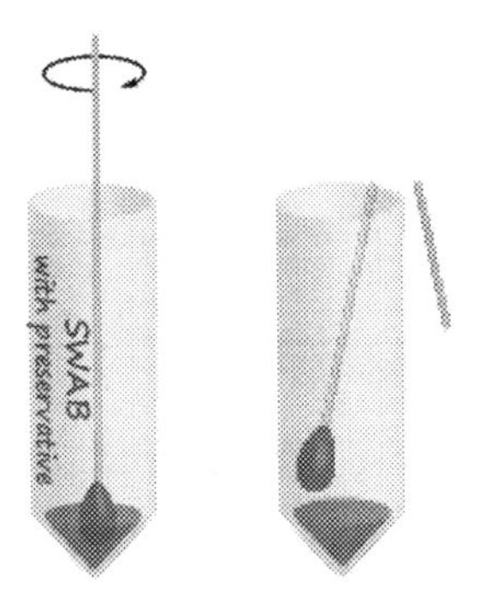

5. Insert the swab into the conical tube, and swish around to make sure the preservative contacts the sample material.
6. Break off excess wooden portion of the cotton swab so that it fits into the conical tube when closed – try to break at a point lower than your fingers have touched. Close the lid tightly.
7. Repeat the same protocol (steps 2–6) with the second 50 mL tube and the other swab from the package.

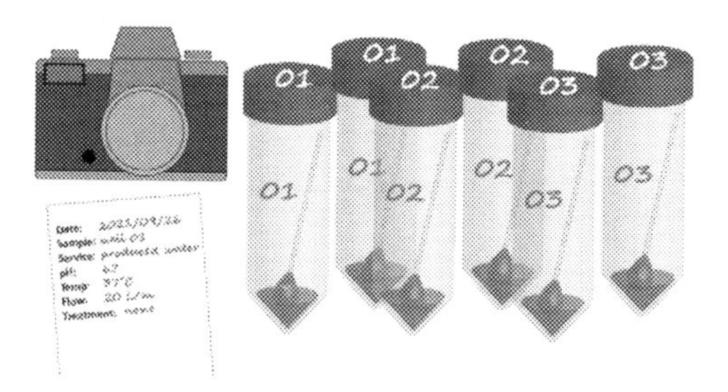

8. Clearly label the tubes with sampling location and date.
9. Take a digital photograph of the sample containers (after filling) along with your field notes.
10. For each site, it is ideal to first take two swab samples for genomic analysis, and also to collect solids for live microbiological analysis if enough sample is present. Take samples for genomic analysis first, then collect any additional solids for live microbiological tests.

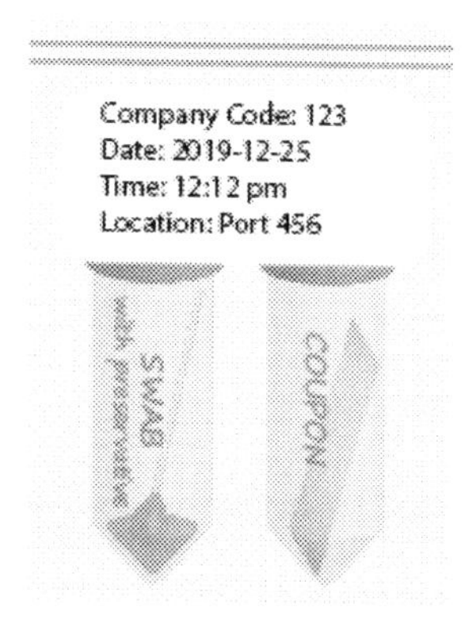

11. Place duplicate conical tubes back in the zip-sealed bag, then clearly label the bag with sampling location, data, and time of collection.
12. Place samples in a cooler with ice packs for shipping, tape up the cooler, and ship to the appropriate lab as soon as possible in order to maintain sample integrity.

26.5 SAMPLE SHIPMENT

1. Samples in glass bottles should be placed in bubblewrap sleeves ahead of shipping.
2. After collection, samples should be stored in the dark.
3. Place ice packs around sample containers to lower temperatures to within the range of 1 to 4°C.
4. Prolonged storage of longer than a few days should be avoided.
   Note: Store all samples on ice once taken – **keep cold but do not freeze** (this can ruin the integrity of DNA).
5. Most samples should be shipped expedited/overnight to the lab in a cooler with ice packs (or ice in sealed bags). The exception is incubation experiments, which should be shipped at room (ambient) temperature.

REFERENCES

De Paula, R., St. Peter, C., Richardson, A., Bracey, J., Heaver, E., Duncan, K., et al. (2018). DNA sequencing of oilfield samples: impact of protocol choices on the microbiological conclusions. *In Corrosion 2018, Paper # 11662, NACE Corrosion 2018 Conference*; April 15–19, 2018; Phoenix AZ, USA.

Oldham, A. L., Sandifer, V., and Duncan, K. E. (2019). Effects of sample preservation on marine microbial diversity analysis. *Journal of Microbiological Methods* 158, 6–13. doi: 10.1016/j.mimet.2019.01.006

Rachel, N. M., and Gieg, L. M. (2020). Preserving microbial community integrity in oilfield produced water. *Frontiers in Microbiology* 11, 2536. doi: 10.3389/fmicb.2020.581387

Index

C

N

O

P

S

W